*The Ultimate Challenge*:
# The 3$x$+1 Problem

*The Ultimate Challenge:*

# The 3x+1 Problem

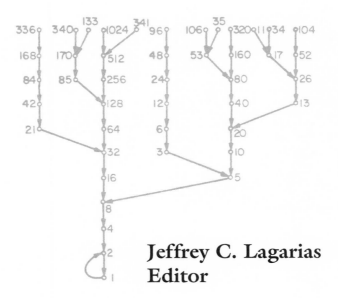

Jeffrey C. Lagarias
Editor

AMERICAN MATHEMATICAL SOCIETY
Providence, Rhode Island

2000 *Mathematics Subject Classification.* Primary 11B83, 37A45; Secondary 11B37, 68Q99.

For additional information and updates on this book, visit
**www.ams.org/bookpages/mbk-78**

The AMS is grateful to the many publishers that have permitted the reprint of their papers in this volume. For specific acknowledgements, please see the editor's Preface.

**Library of Congress Cataloging-in-Publication Data**

The ultimate challenge : the $3x + 1$ problem / Jeffrey C. Lagarias, editor.
    p. cm.
    Includes bibliographical references.
    ISBN 978–8218-4940-8 (alk. paper)
    1. Sequences (Mathematics) 2. Polynomials. 3. Harmonic analysis. I. Lagarias, Jeffrey C., 1949– II. Title: Ultimate challenge : the $3x + 1$ problem. III. Title: Ultimate challenge : the triple plus one problem.
  QA292.U48  2010
  515′.24—dc22

2010035270

Fourier believed that the main goals of mathematics were to serve the common good and to explain natural phenomena. However a philosopher like him should have known that science is solely for the honor of the human spirit. Therefore a question about numbers is as important as a question about the world.

*Carl Gustav Jacob Jacobi*

It is already remarkable and philosophically significant that the first and simplest questions about the numbers $1, 2, 3, \ldots$ present such profound difficulties. These difficulties must be overcome.

*David Hilbert*

"Hopeless. Absolutely hopeless."

*Paul Erdős*

# Contents

# Preface

The $3x + 1$ problem, also known as the Collatz problem, is a notorious unsolved problem in arithmetic. Consider the operation on positive integers $x$ given by: if $x$ is odd, multiply it by 3 and add 1; while if $x$ is even, divide it by 2. The $3x + 1$ problem asks whether, starting from any positive integer $x$, repeating this operation over and over will eventually reach the number 1. The answer appears to be "yes" for all such $x$ but this has never been proved. Despite its simple appearance, this problem is believed to be extraordinarily difficult.

A goal of this book is to report on what is known about the problem. It is divided into five parts. This book contains two introductory papers on the problem, three survey papers on the problem's connections to various fields, two papers devoted to stochastic models and computational results for the problem, six reprinted papers of historical interest, and a paper which giving an annotated bibliography of work on this problem up to 2000. We now describe in more detail the papers appearing in this volume.

**PART I. Overview and Introduction.** Part I of this volume contains two introductory papers on the $3x + 1$ problem.

(1) Jeffrey C. Lagarias, *The $3x + 1$ problem: an overview* ([**15**]).

This paper gives a brief review of the $3x + 1$ problem, its history, its connection to various fields of mathematics, and its current status. It discusses generalizations of the problem, and describes areas for mathematical research that it impacts. It summarizes current "world record" on various aspects of the problem. It formulates some directions for future research. Finally it discusses whether or not this problem is a good mathematical problem.

(2) Jeffrey C. Lagarias, *The $3x + 1$ problem and its generalizations*, American Math. Monthly **92** (1985), 3–23 ([**14**]).

This paper describes the $3x + 1$ problem and related problems. It presents, with proofs, basic results on the behavior of iterates modulo powers of 2, and shows that most integers $n$ iterate to some value less than $n$. It formulates basic conjectures on trajectories and cycles, all of them still unsolved. It focuses on number theoretic aspects of the problem. Although this paper only covers work through 1984, it is still up to date as an introduction to the basic features of the problem. This reprinted version includes minor corrections and updates the reference citations.

**PART II. Three Survey Papers.** Part II of this volume presents three current survey papers on aspects of work on the $3x + 1$ problem.

(3) Marc Chamberland, *A $3x+1$ survey: number theory and dynamical systems* ([**3**]).

This paper reports on research progress in various directions on the $3x+1$ problem covering the period 1985 through 2007. It includes number theory results and dynamical systems results, as well as other connections. It emphasizes the problem viewed as a dynamical system, and discusses generalizations to dynamical systems on larger spaces. This paper is a revised and extended version of the author's earlier survey paper [**2**], written in the Catalan language.

(4) Keith R. Matthews, *Generalized $3x + 1$ mappings: Markov chains and ergodic theory* ([**19**]).

This paper summarizes work on Markov chain models for iterating generalized $3x + 1$ maps, much of it due to the author. These models supply heuristics for the behavior of iteration of a general class of functions that include the $3x+1$ function. It contains many interesting examples whose limiting behavior under iteration is not understood, even on a conjectural level. This is a potentially fruitful area for further research.

(5) Maurice Margenstern and Pascal Michel, *Generalized $3x + 1$ functions and the theory of computation* ([**18**]).

This paper surveys the appearance of $(3x + 1)$−like functions in mathematical logic and in the theory of computation. In 1972 John Conway [**5**] exhibited a $3x + 1$-like function that gives an undecidable computational problem, and he later ([**6**]) expanded on this construction to show how to formulate any computer program using such functions, presented as a programming language FRACTRAN. Other $3x + 1$-like functions have been used to show that various "small" Turing machines, those with few states and small alphabets, which are not known to be universal computers, nevertheless can encode some apparently difficult problems.

**PART III. Stochastic Modelling and Computation Papers.** Part III of this volume presents two papers on mathematical modelling and empirical results from large computations.

(6) Alex Kontorovich and Jeffrey C. Lagarias, *Stochastic models for the $3x + 1$ problem and related problems* ([**13**]).

This paper reports on various stochastic models for the behavior of $3x + 1$ iterates and for comparison, results on iterates for the $5x + 1$ function. A remarkable feature of the $3x + 1$ map is that, although the iteration is deterministic, the best models for the behavior of the iteration are probabilistic. Such stochastic models make predictions about the behavior for iterating a "generic" integer, and also make predictions about the extreme behavior of some integers that may be observed. These models include random walk models for forward iteration, branching processes and branching random walk models for backwards iteration. A third set of branching process models can model the growth of $3x+1$ trees, viewed 3-adically.

(7) Tomas Olivera e Silva, *Empirical verification of the $3x + 1$ conjecture and related conjectures* ([**20**]).

This paper reports on the latest computations of the $3x + 1$ problem, and some related functions. In particular the $3x + 1$ conjecture is verified for all $n \leq 5.764 \times 10^{18}$. Results of the computations include tests of some of the predictions of the stochastic models above. It is an interesting challenge to write efficient progams to verify the $3x + 1$ Conjecture to some bound and also collect statistics on the conjecture.

**PART IV. Reprinted Early Papers.** Part IV of this volume presents six early papers of historical interest, in chronological order. These papers are short, easy to read, and most are hard to obtain in their original source. We provide an editorial commentary after each paper, including additional references and biographical information.

(8) H. S. M. Coxeter, *Cyclic sequences and frieze patterns: The Fourth Felix Behrend Memorial Lecture*, Vinculum **8** (1971), 4–7 ([**7**]).

This paper is the written version of a 1970 lecture, which was published in 1971 in Vinculum, the journal of the Mathematical Association of Victoria (Australia). To my knowledge it is the earliest published paper that explicitly states the $3x + 1$ problem. It presents the problem at the end of the lecture as "mathematical gossip." Coxeter offers a $50.00 prize for its solution. The main subject of the paper, cyclic sequences and frieze patterns, is of interest in its own right.

(9) John H. Conway, *Unpredictable iterations*, in: *Proc. 1972 Number Theory Conference (Univ. Colorado, Boulder, Colo., 1972)*, Univ. Colorado, Boulder, Colo. 1972, pp. 49–52 ([**5**]).

This 1972 paper, from the proceedings of a number theory conference held at the University of Colorado, shows that a generalization of the $3x + 1$ problem is undecidable. Conway later used this encoding to design a computer language FRACTRAN for universal computation using multiplication of fractions, see paper (13) below.

(10) C. J. Everett, *Iteration of the number theoretic function $f(2n) = n$, $f(2n + 1) = 3n + 2$*, Advances in Math. **25** (1977), 42–45. ([**9**]).

This 1977 paper gives an elegant proof of a basic result showing that almost positive integers $n$ iterate to a smaller value under action iteration of the $3x+1$ function. A similar result was independently obtained in 1976 by Riho Terras [**21**], [**22**].

(11) Richard K. Guy, *Don't try to solve these problems!*, American Math. Monthly **90** (1983), 35–41 ([**10**]).

This 1983 paper, written for the Unsolved Problems column of the American Mathematical Monthly, presents a potpourri of $3x + 1$-like problems, including the original problem. True to its name, so far none of the four problems formulated there have been solved.

(12) Lothar Collatz, *On the motivation and origin of the* $(3n + 1)$ *problem* (Chinese), J. Qufu Normal University, Natural Science Edition [Qufu shi fan da xue xue bao] **12** (1986), No. 3, 9–11 ([**4**]).

This 1986 paper, written in Chinese, is the only paper of Lothar Collatz that discusses his work on the $3x + 1$ problem. It is based on a talk that Collatz gave at Qufu Normal University, Qufu, Shandong, China. Here we present an English translation of this paper, using Collatz's original illustrations.

(13) John H. Conway, *FRACTRAN: A simple universal programming language for arithmetic*, In: *Open Problems in Communication and Computation* (T. M. Cover and B. Gopinath, Eds.), Springer-Verlag: New York 1987, pp. 3-27 ([**6**]).

This 1987 paper of Conway expands on his 1972 paper to show how to encode any computational problem in terms of iteration of a suitable $3x + 1$-like function. The programming language name FRACTRAN is a pun on FORTRAN (The IBM Mathematical Formula Translating System). This is not the only pun in this paper.

**PART V. Annotated Bibliography.** Part V of this volume an an annotated bibliography of work on the $3x + 1$ problem and related iteration problems, from 1963-1999.

(14) Jeffrey C. Lagarias, *The* $3x + 1$ *problem: An annotated bibliography (1963-1999)* ([**16**]).

This bibliography attempts to be relatively complete over the period cited. It includes a number of papers from the "prehistory" of the problem, in the 1960's. It also covers many papers appearing in unusual places, not covered by Mathematical Reviews or Zentralblatt für Mathematik. It groups papers on the problem into ten year subintervals. The growth of the number of papers in these time intervals, which total 8, 34, 52 and 103 papers, respectively, show increasing effort devoted to the $3x + 1$ problem and generalizations. A follow-up bibliography, currently covering the period 2000–2009 ([**17**]) is posted on the math arXiv.

**Book Title: The Ultimate Challenge.** The results known about the $3x + 1$ problem strongly suggest that it does not fit in the scope of classical "structural" mathematics. Instead it seems to lie in a wilderness between the well-organized part of mathematical knowledge, typified by the subjects covered in the volumes of Bourbaki, and the boundary of undecidable problems, those problems that can encode the action of a universal computer. The title does not assert the problem is "ultimate" in its importance. Rather, "ultimate challenge" refers to the contrast between the simplicity of the statement of the problem and the apparent difficulty (perhaps impossibility) of resolving the problem. The papers in this volume give ample warning that the problem shows no sign of being solvable at present. Remember, not all challenges need to be accepted!

**Epigraphs: References.** The statement of G. C. J. Jacobi is taken from a letter to Legendre written in 1830, published in 1875 in Borchardt [**1**, p. 272]

Il est vrai que M. Fourier avait l'opinion que le but principal de mathématiques était l'utitlité publique et l'explications des phénomès naturels; mais un philosophe comme lui aurait dû savoir que le but unique de la science, c'est l'honneur de l'esprit humain, et que sous ce titre, une question des nombres vaut autant qu'une question du système du monde.

The statement of D. Hilbert is taken from a 1931 paper on foundations on mathematics (Hilbert [**12**, p. 486]).

Es ist schon an sich merkwürdig und philosophich bedeutsam, dass die ersten und einfachsten Fragen über die Zahlen $1, 2, 3, \ldots$ so tieflegende Schwierigkeiten bieten. Diese Schwierigkeiten müssen überwunden werden.

The statement of P. Erdős was made in conversation, sometime after publication of the 1985 survey paper ([**8**]).

**Acknowledgments.** I think Gerasimos Ladas for encouragement to prepare a volume on this topic. I thank Andreas Blass, Mark Conger, Alex Kontorovich, Stephen R. Miller, Chris Xiu and Mike Zieve for various forms of assistance in this project. The three survey papers in this volume, and that of T. Oliveira e Silva, have been peer-reviewed. I thank the anonymous reviewers for their efforts. I thank Sergei Gelfand for his positive suggestions about the title and production of this volume. During the period of this work I received support from NSF Grants DMS-0500555 and DMS-0801029.

*Jeffrey C. Lagarias*
Ann Arbor, Michigan
August 2010

# References

[1] C. W. Borchardt, Editor, Correspondence mathématique entre Legendre et Jacobi, J. reine Angew. Math. **80** (1875), 205–279.

[2] M. Chamberland, Una actualizachio del problema $3x + 1$ [An update on the $3x + 1$ problem] (Catalan), Butlletí Societat Catalana de Matemàtiques **18** (2003), No.1, 19–45.

[3] M. Chamberland, A $3x + 1$ survey: number theory and dynamical systems, paper in this volume.

[4] L. Collatz, On the motivation and origin of the $(3n + 1)$- problem (Chinese), J. Qufu Normal University, Natural Science Edition [Qufu shi fan da xue xue bao] **12** (1986), No. 3, 9–11. [Translation included in this volume]

[5] J. H. Conway, Unpredictable Iterations, Proc. 1972 Number Theory Conference (Univ. Colorado, Boulder, Colo., 1972 ), pp. 49–52. Univ. Colorado, Boulder, Colo. 1972.

[6] J. H. Conway, FRACTRAN: A Simple Universal Computing Language for Arithmetic, In: *Open Problems in Communication and Computation* (T. M. Cover and B. Gopinath, Eds.), Springer-Verlag: New York 1987, pp. 3-27 [Reprinted in this volume]

[7] H. S. M. Coxeter, Cyclic sequences and frieze patterns: The Fourth Felix Behrend Memorial Lecture), Vinculum **8** (1971), 4–7. [Reprinted in this volume]

[8] P. Erdős, Private communication with J. C. Lagarias.

[9] C. J. Everett, Iteration of the number theoretic function $f(2n) = n, f(2n + 1) = 3n + 2$, Advances in Math. **25** (1977), 42–45. [Reprinted in this volume]

[10] R. K. Guy, Don't try to solve these problems!, American Math. Monthly **90** (1983), 35–41. [Reprinted in this volume]

[11] R. K. Guy, Conway's prime-producing machine, Math. Magazine **56** (1983), no. 1, 26–33.

[12] D. Hilbert, *Die Grundlagen der elementaren Zahlentheorie,* Math. Annalen **104** (1931), 484–494. [English Translation: Chap. 17 of P. Mancosu, *From Brouwer to Hilbert. The debate on the foundations of mathematics in the 1920's,* Oxford Univ. Press: New York 1998]

[13] A.V. Kontorovich and J. C. Lagarias, Stochastic models for the $3x + 1$ problem and $5x + 1$ problems, paper in this volume.

[14] J. C. Lagarias, The $3x + 1$ problem and its generalizations, Amer. Math. Monthly **92** (1985), 3–23. [Reprinted with corrections in this volume].

[15] J. C. Lagarias, The $3x + 1$ problem: an overview, paper in this volume.

[16] J. C. Lagarias, The $3x + 1$ Problem: An Annotated Bibliography (1963-1999), paper in this volume.

[17] J. C. Lagarias, The $3x + 1$ Problem: An Annotated Bibliography, II (2000-2009), `arXiv:math/0608208`.

[18] P. Michel and M. Margenstern, Generalized $3x + 1$ functions and the theory of computation, paper in this volume.

[19] K. R. Matthews, Generalized $3x + 1$ mappings: Markov chains and ergodic theory, paper in this volume.

[20] T. Oliveira e Silva, Empirical verification of the $3x + 1$ and related conjectures, paper in this volume.

[21] R. Terras, A stopping time problem on the positive integers, Acta Arithmetica **30** (1976), 241–252.

[22] R. Terras, On the existence of a density, Acta Arithmetica **35** (1979), 101–102.

# PART I.

## Overview and Introduction

# The $3x + 1$ Problem: An Overview

## Jeffrey C. Lagarias

## 1. Introduction

The $3x + 1$ problem concerns the following innocent seeming arithmetic procedure applied to integers: If an integer $x$ is odd then "multiply by three and add one", while if it is even then "divide by two". This operation is described by the *Collatz function*

$$C(x) = \begin{cases} 3x + 1 & \text{if } x \equiv 1 \pmod{2}, \\ \dfrac{x}{2} & \text{if } x \equiv 0 \pmod{2}. \end{cases}$$

The $3x+1$ problem, which is often called the *Collatz problem*, concerns the behavior of this function under iteration, starting with a given positive integer $n$.

$3x + 1$ **Conjecture.** *Starting from any positive integer $n$, iterations of the function $C(x)$ will eventually reach the number 1. Thereafter iterations will cycle, taking successive values $1, 4, 2, 1, \dots$.*

This problem goes under many other names, including the *Syracuse problem*, *Hasse's algorithm*, *Kakutani's problem* and *Ulam's problem*.

A commonly used reformulation of the $3x + 1$ problem iterates a different function, the $3x + 1$ *function*, given by

$$T(x) = \begin{cases} \dfrac{3x + 1}{2} & \text{if } x \equiv 1 \pmod{2}, \\ \dfrac{x}{2} & \text{if } x \equiv 0 \pmod{2}. \end{cases}$$

From the viewpoint of iteration the two functions are simply related; iteration of $T(x)$ simply omits some steps in the iteration of the Collatz function $C(x)$. The relation of the $3x + 1$ function $T(x)$ to the Collatz function $C(x)$ is that:

$$T(x) = \begin{cases} C(C(x)) & \text{if } x \equiv 1 \pmod{2}, \\ C(x) & \text{if } x \equiv 0 \pmod{2}. \end{cases}$$

As it turns out, the function $T(x)$ proves more convenient for analysis of the problem in a number of significant ways, as first observed independently by Riho Terras ([**88**], [**89**]) and by C. J. Everett [**27**].

The $3x + 1$ problem has fascinated mathematicians and non-mathematicians alike. It has been studied by mathematicians, physicists, and computer scientists. It remains an unsolved problem, which appears to be extremely difficult.

This paper aims to address two questions:

(1) *What can mathematics currently say about this problem?*

(2) *How can this problem be hard, when it is so easy to state?*

To address the first question, this overview discusses the history of work on the problem. Then it describes generalizations of the problem, and lists the different fields of mathematics on which the problem impinges. It gives a brief summary of the current strongest results on the problem.

Besides the results summarized here, this volume contains more detailed surveys of mathematicians' understanding of the $3x + 1$ problem and its generalizations. These cover both rigorously proved results and heuristic predictions made using probabilistic models. The book includes several survey articles, it reprints several early papers on the problem, with commentary, and it presents an annotated bibliography of work on the problem and its generalizations.

To address the second question, let us remark first that the true level of difficulty of any problem can only be determined when (and if) it is solved. Thus there can be no definitive answer regarding its difficulty. The track record on the $3x + 1$ problem so far suggests that this is an extraordinarily difficult problem, completely out of reach of present day mathematics. Here we will only say that part of the difficulty appears to reside in an inability to analyze the pseudorandom nature of successive iterates of $T(x)$, which could conceivably encode very difficult computational problems. We elaborate on this answer in §7.

Is the $3x + 1$ problem an important problem? Perhaps not for its individual sake, where it merely stands as a challenge. It seems to be a prototypical example of an extremely simple to state, extremely hard to solve, problem. A middle of the road viewpoint is that this problem is representative of a large class of problems, concerning the behavior under iteration of maps that are expanding on part of their domain and contracting on another part of their domain. This general class of problems is of definite importance, and is currently of great interest as an area of mathematical (and physical) research; for some perspective, see Hasselblatt and Katok [**45**]. Progress on general methods of solution for functions in this class would be extremely significant.

This overview describes where things currently stand on the $3x + 1$ problem and how it relates to various fields of mathematics. For a detailed introduction to the problem, see the following paper of Lagarias [**58**] (in this volume). In §2 we give some history of the problem; this presents some new information beyond that given in [**58**]. Then in §3 we give a flavor of the behavior of the $3x + 1$ iteration. In §4 we discuss various frameworks for generalizing the problem; typically these concern iterations of functions having a similar appearance to the $3x + 1$ function. In §5 we review areas of research: these comprise different fields of mathematics and computer science on which this problem impinges. In §6 we summarize the current best results on the problem in various directions. In §7 we discuss the hardness of the $3x + 1$ problem. In §8 we describe some research directions for future progress.

In §9 we address the question: "Is the $3x + 1$ problem a good problem?" In the concluding section §10 we offer some advice on working on $3x + 1$-related problems.

## 2. History and Background

The $3x + 1$ problem circulated by word of mouth for many years. It is generally attributed to Lothar Collatz. He has stated ([14]) that he took lecture courses in 1929 with Edmund Landau and Fritz von Lettenmeyer in Göttingen, and courses in 1930 with Oskar Perron in Munich and with Issai Schur in Berlin, the latter course including some graph theory. He was interested in graphical representations of iteration of functions. In his notebooks in the 1930's he formulated questions on iteration of arithmetic functions of a similar kind (cf. [58, p. 3]). Collatz is said by others to have circulated the problem orally at the International Congress of Mathematicians in Cambridge, Mass. in 1950. Several people whose names were subsequently associated with the problem gave invited talks at this International Congress, including H. S. M. Coxeter, S. Kakutani, and S. Ulam. Collatz [15] (in this volume) states that he described the $3x + 1$ problem to Helmut Hasse in 1952 when they were colleagues at the University of Hamburg. Hasse was interested in the problem, and wrote about it in lecture notes in 1975 ([44]). Another claimant to having originated the $3x + 1$ problem is Bryan Thwaites [90], who asserts that he came up with the problem in 1952. Whatever is its true origin, the $3x + 1$ problem was already circulating at the University of Cambridge in the late 1950's, according to John H. Conway and to Richard Guy [43].

There was no published mathematical literature about the $3x + 1$ problem until the early 1970's. This may have been, in part, because the 1960's was a period dominated by Bourbaki-style mathematics. The Bourbaki viewpoint emphasized complete presentations of theories with rich internal structure, which interconnect with other areas of core mathematics (see Mashaal [65]). In contrast, the $3x + 1$ problem initially appears to be an isolated problem unrelated to the rest of mathematics. Another obstacle was the difficulty in proving interesting results about the $3x + 1$ iteration. The results that could be proved appeared pathetically weak, so that it could seem damaging to one's professional reputation to publish them. In some mathematical circles it might have seemed in bad taste even to show interest in such a problem, which appears déclassé.

During the 1960's, various problems related to the $3x + 1$ problem appeared in print, typically as unsolved problems. This included one of the original problems of Collatz from the 1930's, which concerned the behavior under iteration of the function

$$U(2n) = 3n, \ U(4n + 1) = 3n + 1, \ U(4n + 3) = 3n + 2.$$

The function $U(n)$ defines a permutation of the integers, and the question concerns whether the iterates of the value $n = 8$ form an infinite set. This problem was raised by Murray Klamkin [52] in 1963 (see Lagarias [58, p. 3]), and remains unsolved. Another such problem was posed by Ramond Queneau, a founder of the French mathematical-literary group Oulipo (Ouvroir de littérature potentielle), which concerns allowable rhyming patterns generalizing those used in poems by the 12-th century troubadour, Arnaut Daniel. This problem turns out to be related to a $(3x + 1)$-like function whose behavior under iteration is exactly analyzable, see

Roubaud [**80**]. Concerning the $3x + 1$ problem itself, during the 1960's large computations were done testing the truth of the conjecture. These reportedly verified
the conjecture for all $n \leq 10^9$.

To my knowledge, the $3x + 1$ problem first appeared in print in 1971, in the
written version of a 1970 lecture by H. S. M. Coxeter [**22**] (in this volume). It
was presented there "as a piece of mathematical gossip." In 1972 it appeared in six
different publications, including a Scientific American column by Martin Gardner
[**32**] that gave it wide publicity. Since then there has been a steady stream of work
on it, now amounting to several hundred publications.

Stanislaw Ulam was one of many who circulated the problem; the name "Ulam's
problem" has been attached to it in some circles. He was a pioneer in ergodic
theory and very interested in iteration of functions and their study by computer;
he formulated many problem lists (e.g. [**92**], [**21**]). A collaborator, Paul Stein [**87**,
p. 104], wrote about Ulam:

> Stan was not a number theorist, but he knew many number-theoretical
> facts. As all who knew him well will remember, it was Stan's par-
> ticular pleasure to pose difficult, though simply stated, questions
> in many branches of mathematics. Number theory is a field par-
> ticularly vulnerable to the "Ulam treatment," and Stan proposed
> more than his share of hard questions; not being a professional in
> the field, he was under no obligation to answer them.

Ulam's long term collaborator C. J. Everett [**27**] wrote one of the early papers
about the $3x + 1$ problem in 1977.

The $3x + 1$ problem can also be formulated in the backwards direction, as that
of determining the smallest set $S_0$ of integers containing 1 which is closed under
the affine maps $x \mapsto 2x$ and $3x + 2 \mapsto 2x + 1$, where the latter map may only
be applied to inputs $3x + 2$ whose output $2x + 1$ will be an integer. The $3x + 1$
conjecture then asserts that $S_0$ will be the set of all positive integers. This connects
the $3x + 1$ problem with problems on sets of integers which are closed under the
action of affine maps. Problems of this sort were raised by Isard and Zwicky [**51**]
in 1970. In 1970-1971 David Klarner began studying sets of integers closed under
iteration of affine maps, leading to joint work with Richard Rado [**54**], published in
1974. Interaction of Klarner and Paul Erdős at the University of Reading in 1971
led to the formulation of a (solved) Erdős prize problem: Does the smallest set $S_1$ of
integers containing 1 and closed under the affine maps $x \mapsto 2x + 1$, $x \mapsto 3x + 1$ and
$x \mapsto 6x + 1$ have a positive (lower asymptotic) density? This set $S_1$ was proved to
have zero density by D. J. Crampin and A. J. W. Hilton (unpublished), according
to Klarner [**53**]. The solvers collected £10 from Erdős ([**50**]). Later Klarner [**53**,
p. 47] formulated a revised problem:

**Klarner's Integer Sequence Problem.** *Does the smallest set of integers $S_2$
containing 1 and closed under the affine maps $x \mapsto 2x$, $x \mapsto 3x + 2$ and $x \mapsto 6x + 3$
have a positive (lower asymptotic) density?*

This problem remains unsolved; see the paper of Guy [**40**] (in this volume) and
accompanying editorial commentary.

Much early work on the problem appeared in unusual places, some of it in
technical reports, some in problem journals. The annotated bibliography given in
this book [**60**] covers some of this literature, see also its sequel [**61**]. Although

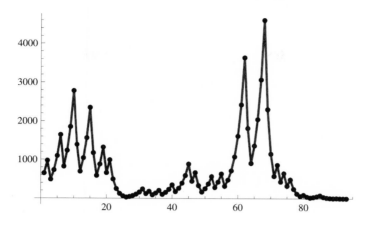

FIGURE 1. Trajectory of $n = 649$ plotted on standard vertical scale

the problem began life as a curiosity, its general connection with various other areas of mathematics, including number theory, dynamical systems and theory of computation, have made it a respectable topic for mathematical research. A number of very well known mathematicians have contributed results on it, including John H. Conway [16] and Yakov G. Sinai [84], [85].

## 3. $3x + 1$ Sampler

The fascination of the $3x + 1$ problem involves its simple definition and the apparent complexity of its behavior under iteration: there seems to be no simple relation between the input value $n$ and the iterates of $n$. Exploration of its structure has led to the formulation of a web of subsidiary conjectures about the behavior of iterates of the $3x + 1$ function and generalizations; these include conjectures (C1)–(C5) listed in §8. Many of these conjectures seem to be extremely difficult problems as well, and their exploration has led to much further research. Since other papers in this volume give much more information on this complexity, here we give only a brief sampler of $3x + 1$ function behavior.

**3.1. Plots of Trajectories.** By the *trajectory* of $x$ under a function $T$, we mean the forward orbit of $x$, that is, the sequence of its forward iterates $(x, T(x), T^{(2)}(x), T^{(3)}(x), ...)$. Figure 1 displays the $3x + 1$-function iterates of $n = 649$ plotted on a standard scale. We see an irregular series of increases and decreases, leading to the name "hailstone numbers" proposed by Hayes [46], as hailstones form by repeated upward and downward movements in a thunderhead.

To gain insight into a problem it helps to choose an appropriate scale for picturing it. Here it is useful to view long trajectories on a logarithmic scale, i.e., to plot $\log T^{(k)}(n)$ versus $k$. Figure 2 displays the iterates of $n_0 = 100\lfloor \pi 10^{35} \rfloor$ on such a scale. Using this scale we see a decrease at a certain geometric rate to the value of 1, indicated by the trajectory having roughly a constant slope. This is characteristic of most long trajectories. As explained in §3.3 a probabilistic model predicts that most trajectories plotted on a logarithmic scale will stay close to a line of constant slope $-\frac{1}{2} \log \frac{3}{4} \sim -0.14384$, thus taking about $6.95212 \log n$ steps

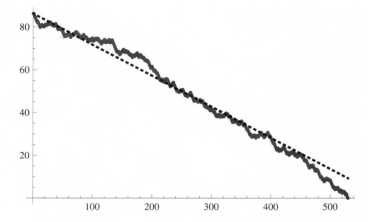

FIGURE 2. Trajectory of $n_0 = 100\lfloor \pi \cdot 10^{35} \rfloor$ plotted on a logarithmic vertical scale. The dotted line is a probability model prediction for a "random" trajectory for this size $N$.

to reach 1. This line is pictured as the dotted line in Figure 2. This trajectory takes 529 steps to reach $n = 1$, while the probabilistic model predicts about 600 steps will be taken.

On the other hand, plots of trajectories suggest that iterations of the $3x + 1$ function also seem to exhibit pseudo-random features, i.e. the successive iterates of a random starting value seem to increase or decrease in an unpredictable manner. From this perspective there are some regularities of the iteration that appear (only) describable as statistical in nature: they are assertions about the majority of trajectories in ensembles of trajectories rather than about individual trajectories.

**3.2. Patterns.** Close examination of the iterates of the $3x + 1$ function $T(x)$ for different starting values reveals a myriad of internal patterns. A simple pattern is that the initial iterates of $n = 2^m - 1$ are

$$T^{(k)}(2^m - 1) = 3^k \cdot 2^{m-k} - 1, \text{ for } 1 \leq k \leq m.$$

In particular, $T^{(m)}(2^m - 1) = 3^m - 1$; this example shows that the iteration can sometimes reach values arbitrarily larger than the initial value, either on an absolute or a relative scale, even if, as conjectured, the iterates eventually reach 1. Other patterns include the appearance of occasional large clusters of consecutive numbers which all take exactly the same number of iterations to reach the value 1. Some of these patterns are easy to analyze, others are more elusive.

Table 1 presents data on iterates of the $3x + 1$ function $T(x)$ for $n = N_0 + m$, $0 \leq m = 10j + k \leq 99$, with

$$n_0 = 100\lfloor \pi \cdot 10^{35} \rfloor = 31,415,926,535,897,932,384,626,433,832,795,028,800.$$

Here $\sigma_\infty(n)$ denotes the *total stopping time* for $n$, which counts the number of iterates of the $3x + 1$-function $T(x)$ needed to reach 1 starting from $n$, counting $n$ as the 0-th iterate. This number is the same as the number of even numbers appearing in the trajectory of the Collatz function before first reaching 1.

TABLE 1. Values of total stopping time $\sigma_\infty(n)$ for $n = n_0 + 10j + k$, with $n_0 := 100\lfloor \pi \cdot 10^{35} \rfloor = 31,415,926,535,897,932,384,626,433,832,795,028,800$.

|       | $j=0$ | $j=1$ | $j=2$ | $j=3$ | $j=4$ | $j=5$ | $j=6$ | $j=7$ | $j=8$ | $j=9$ |
|-------|-------|-------|-------|-------|-------|-------|-------|-------|-------|-------|
| $k=0$ | 529 | 529 | 529 | 678 | 529 | 529 | 846 | 529 | 846 | 846 |
| $k=1$ | 659 | 659 | 529 | 678 | 659 | 529 | 846 | 529 | 529 | 529 |
| $k=2$ | 846 | 529 | 659 | 529 | 529 | 529 | 659 | 846 | 529 | 659 |
| $k=3$ | 846 | 529 | 659 | 846 | 659 | 529 | 659 | 846 | 529 | 659 |
| $k=4$ | 659 | 659 | 659 | 846 | 678 | 529 | 846 | 846 | 846 | 659 |
| $k=5$ | 659 | 659 | 846 | 846 | 678 | 529 | 529 | 529 | 846 | 659 |
| $k=6$ | 659 | 529 | 659 | 846 | 678 | 846 | 529 | 846 | 659 | 846 |
| $k=7$ | 529 | 529 | 659 | 846 | 659 | 659 | 529 | 846 | 659 | 529 |
| $k=8$ | 529 | 678 | 659 | 846 | 529 | 846 | 529 | 529 | 846 | 846 |
| $k=9$ | 529 | 678 | 659 | 659 | 529 | 529 | 529 | 529 | 659 | 846 |

We observe that the total stopping time function takes only a few different values, namely: 529, 654, 678 and 846, and these four values occur intermixed in a somewhat random-appearing way, but with some regularities. Note that around $n_0 \sim 3.14 \times 10^{37}$ the predicted "average size" of a trajectory is $6.95212 \log n_0 \approx 600$. In the data here we also observe "jumps" of size between the occurring values on the order of 100.

This is not a property of just this starting value. In Table 2 we give similar data for blocks of 100 near $n = 10^{35}$ and $10^{36}$, respectively. Again we observe that there are also four or five values occurring, but now they are different values. In this table we present data on two other statistics: the *frequency* statistic gives the count of these number of occurrences of each value, and the 1-*ratio* statistic denotes the fraction of odd iterates occurring in the given trajectory up to and including when 1 is reached. It is an experimental fact that all sequences in the table having the same total stopping time also have the same 1-ratio. In the first two blocks the value $\sigma_\infty(n) = 481$ (resp. 351) that occurs with frequency 1 is that for the initial value $n = 10^{35}$ (resp. $n = 10^{36}$) in the given interval; these initial values are unusual in being divisible by a high power of 2. Probabilistic models for the $3x+1$-function iteration predict that even and odd iterates will initially occur with equal frequency, so we may anticipate the 1-ratio values to be relatively close to 0.5.

TABLE 2 Values of total stopping time, their frequencies, and 1-ratio for (a) $10^{35} \le n \le 10^{35} + 99$, (b) $10^{36} \le n \le 10^{36} + 99$, (c) $n_0 \le n \le n_0 + 99$.

| | (a) $10^{35}$ | | | (b) $10^{36}$ | | | (c) $n_0$ | |
|---|---|---|---|---|---|---|---|---|
| $\sigma_\infty(n)$ | freq. | 1-ratio | $\sigma_\infty(n)$ | freq. | 1-ratio | $\sigma_\infty(n)$ | freq. | 1-ratio |
| 481 | 1 | 0.47817 | 351 | 1 | 0.41594 | 529 | 38 | 0.48204 |
| 508 | 19 | 0.48622 | 467 | 72 | 0.46895 | 654 | 28 | 0.51138 |
| 573 | 49 | 0.50261 | 508 | 21 | 0.48228 | 678 | 7 | 0.51474 |
| 592 | 10 | 0.50675 | 519 | 6 | 0.48554 | 846 | 27 | 0.53782 |
| 836 | 21 | 0.54306 | | | | | | |

The data in Table 2 suggests the following heuristic: as $n$ increases only a few values of $\sigma_\infty(n)$ locally occur over short intervals; there is then a slow variation in which values of $\sigma_\infty(n)$ occur. However these local values are separated from each other by relatively large "jumps" in size. We stress that this is a purely empirical observation, nothing like this is rigorously proved! Our heuristic did not quantify what is a "short interval" and it did not quantify what "relatively large jumps" should mean. Even the existence of finite values for $\sigma_\infty(n)$ in the tables presumes the $3x + 1$ conjecture is true for all numbers in the table.

### 3.3. Probabilistic Models.
A challenging feature of the $3x+1$ problem is the huge gap between what can be observed about its behavior in computer experiments and what can be rigorously proved. Attempts to understand and predict features of empirical experimentation have led to the following curious outcome: *the use of probabilistic models to describe a deterministic process.* This gives another theme of research on this problem: the construction and analysis of probabilistic and stochastic models for various aspects of the iteration process.

A basic probabilistic model of iterates of the $3x+1$ function $T(x)$ proposes that most trajectories for $3x + 1$ iterates have equal numbers of even and odd iterates, and that the parity of successive iterates behave in some sense like independent coin flips. A key observation of Terras [**88**] and Everett [**27**], leading to this model, is that the initial iterates of the $3x + 1$ function have this property (see Lagarias [**58**, Lemma B]).) This probabilistic model suggests that most trajectories plotted on a logarithmic vertical scale should appear close to a straight line having negative slope equal to $-\frac{1}{2}\log\frac{3}{4} \sim -0.14384$, and should thus take about $6.95212\log n$ steps to reach 1.

The corresponding behavior of iterates of the Collatz function $C(x)$ is more complicated. The allowed patterns of even and odd Collatz function iterates always have an even iterate following each odd iterate. Probabilistic models taking this into account are more complicated to formulate and analyze than that for the $3x + 1$ function; this is a main reason for studying the $3x + 1$ function rather than the Collatz function. Use of the probabilistic model above allows the heuristic inference that Collatz iterates will be even about two-thirds of the time.

A variety of fairly complicated stochastic models, many of which are rigorously analyzable (as probability models), have now been formulated to model various aspects of these iterations, see Kontorovich and Lagarias [**56**] (in this volume). Rigorous results for such models lead to heuristic predictions for the statistical behavior of iterates of the generalized $3x + 1$ map. The model above predicts the behavior of "most" trajectories. A small number of trajectories may exhibit quite different behavior. One may consider those trajectories that that seem to offer maximal value of some iterate of $T^{(k)}(n)$ compared to $n$. Here a probabilistic model (see [**56**, Sec. 4.3] in this volume) predicts that the statistic

$$\rho(n) := \frac{\log(\max_{k\geq 1}\left(T^{(k)}(n)\right)}{\log n}$$

as $n \to \infty$ should have $\rho(n) \leq 2 + o(1)$ for all sufficiently large $n$. Figure 3.3 offers a plot of the trajectory, for the value $n_1 = 19809760576694878447$, which attains the largest value of the statistic $\rho(n)$ over $1 \leq n \leq 10^{18}$; this value was found by Oliveira e Silva [**76**, Table 6] (in this volume). This example has $\rho(n_1) \approx 2.04982$. Probabilistic models suggest that the extremal trajectories of this form

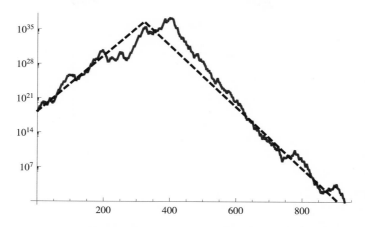

FIGURE 3. Extremal trajectory $n_1 = 1980976057694878447$ given in Oliveira e Silva's Table 6.

will approach a characteristic shape which consists of two line segments, one of length $7.645 \log n$ steps of slope about 0.1308 up to the maximal value of about $2 \log n$, the second of about $13.905 \log n$ steps of slope about $-0.1453$ to 0, taking $21.55 \log n$ steps in all. This shape is indicated by the dotted lines on Figure 3.3 for comparison purposes.

Another prediction of such stochastic models, relevant to the $3x + 1$ conejcture, is that the number of iterations required for a positive integer $n$ to iterate to 1 under the $3x + 1$ function $T(x)$ is at most $41.677647 \log n$ (see [**62**], [**56**, Sect. 4]). In particular such models predict, in a quantitative form, that there will be no divergent trajectories.

These stochastic models can be generalized to model the behavior of many generalized $3x + 1$ functions, and they make qualitatively different predictions depending on the function. For example, such models predict that no orbit of iteration of the $3x + 1$ function "escapes to infinity" (divergent trajectory). However for the $5x + 1$ *function* given by

$$T_5(x) = \begin{cases} \dfrac{5x + 1}{2} & \text{if } x \equiv 1 \ (\text{mod } 2), \\[2ex] \dfrac{x}{2} & \text{if } x \equiv 0 \ (\text{mod } 2), \end{cases}$$

similar stochastic models predict that almost all orbits should "escape to infinity" ([**56**, Sect. 8]). These predictions are supported by experimental computer evidence, but it remains an unsolved problem to prove that there exists even one trajectory for the $5x + 1$ problem that "escapes to infinity".

There remains considerable research to be done on further developing stochastic models. The experiments on the $3x + 1$ iteration reported above in §3.2 exhibit some patterns not yet explained by stochastic models. In particular, the behaviors of total stopping times observed in Tables 1 and 2, and the heuristic presented there, have not yet been justified by suitable stochastic models.

## 4. Generalized $3x + 1$ functions

The original work on the $3x + 1$ problem viewed it as a problem in number theory. Much of the more recent work views it as an example of a special kind of discrete dynamical system, as exemplified by the lecture notes volume of G. J. Wirsching [**95**]. As far as generalizations are concerned, a very useful class of functions has proved to be the set of generalized Collatz functions which are defined below. These possess both number-theoretical and dynamic properties; the number-theoretic properties have to do with the existence of $p$-adic extensions of these maps for various primes $p$.

At present the $3x + 1$ problem is most often viewed as a discrete dynamical system of an arithmetical kind. It can then be treated as a special case, within the framework of a general class of such dynamical systems. But what should be the correct degree of generality in such a class?

There is significant interest in exploring the behavior of dynamical systems of an arithmetic nature, since these may be viewed as "toy models" of more complicated dynamical systems arising in mathematics and physics. There are a wide variety of interesting arithmetic dynamical systems. The book of Silverman [**82**] studies the iteration of algebraic maps on algebraic varieties. The book of Schmidt [**81**] considers dynamical systems of algebraic origin, meaning $\mathbb{Z}^d$-actions on compact metric groups, using ergodic theory and symbolic methods. The book of Fursten-berg [**30**] considers various well structured arithmetical dynamical systems; for a further development see Glasner [**34**]. The generalized $3x + 1$ functions studied in this book provide another distinct type of arithmetic discrete dynamical system.

We present a taxonomy of several classes of functions which represent successive generalizations of the $3x + 1$ function. The simplest generalization of the $3x + 1$ function is the $3x + k$ function, which is defined for $k \equiv 1$ or $5 \pmod 6$, by

$$T_{3,k}(x) = \begin{cases} \dfrac{3x + k}{2} & \text{if } x \equiv 1 \pmod{2}, \\[2mm] \dfrac{x}{2} & \text{if } x \equiv 0 \pmod{2}. \end{cases}$$

The generalization of the $3x + 1$ conjecture to this situation is twofold: first, that under iteration every orbit becomes eventually periodic, and second, that there are only a finite number of cycles (periodic orbits). This class of functions occurs in the study of cycles of the $3x + 1$ function (Lagarias [**59**]). Note that the $3x + 1$ function $T(x)$ can be extended to be well defined on the set of all rational numbers having odd denominator, and a rescaling of any $T$-orbit of such a rational number $r = \frac{n}{k}$ to clear its denominator $k$ will give an orbit of the map $T_{3,k}$. Thus, integer cycles of the $3x + k$ function correspond to rational cycles of the $3x + 1$ function having denominator $k$.

To further generalize, let $d \geq 2$ be a fixed integer and consider the function defined for integer inputs $x$ by

(4.1) $$f(x) = \frac{a_i x + b_i}{d} \quad \text{if } x \equiv i \pmod{d}, \ 0 \leq i \leq d - 1,$$

where $\{(a_i, b_i) : 0 \leq i \leq d - 1\}$ is a collection of integer pairs. Such a function is called *admissible* if the integer pairs $(a_i, b_i)$ satisfy the condition

$$(4.2) \qquad ia_i + b_i \equiv 0 \pmod{d} \text{ for } 0 \leq i \leq d - 1.$$

This condition is necessary and sufficient for the map $f(x)$ to take integers to integers. These functions $f(x)$ have been called *generalized Collatz functions,* or *RCWA functions* (Residue-Class-Wise Affine functions). Generalized Collatz functions have the nice feature that they have a unique continuous extension to the space $\mathbb{Z}_d$ of $d$-adic integers in the sense of Mahler [**64**].

An important subclass of generalized Collatz functions are those of *relatively prime type*. These are the subclass of generalized Collatz functions for which

$$(4.3) \qquad \gcd(a_0 a_1 \cdots a_{d-1}, d) = 1.$$

This class includes the $3x + 1$ function $T(x)$ but not the Collatz function $C(x)$ itself. It includes the $5x + 1$ function $T_5(x)$, which as mentioned above appears to have quite different long-term dynamics on the integers $\mathbb{Z}$ than does the $3x + 1$ function. Functions in this class have the additional property that their unique extension to the $d$-adic integers $\mathbb{Z}_d$ has the $d$-adic Haar measure as an invariant measure. This permits ergodic theory methods to be applied to their study, see the survey paper of Matthews [**67**, Thm. 6.2] (in this volume) for many examples.

As a final generalization, one may consider the class of integer-valued functions, which when restricted to residue classes $(\bmod\, d)$ are given by a polynomial $P_i(x)$ for each class $i$ $(\bmod\, d)$. Members of this class of functions have arisen in several places in mathematics. They are now widely called *quasi-polynomial functions* or *quasi-polynomials*. Quasi-polynomials appear in commutative algebra and algebraic geometry, in describing the Hilbert functions of certain semigroups, in a well known theorem of Serre, see Bruns and Herzog [**9**, pp. 174–175] and Bruns and Ichim [**10**]. In another direction, functions that count the number of lattice points inside dilated rational polyhedra have been shown to be quasi-polynomial functions (on the positive integers), starting with work of Ehrhart [**23**], see Beck and Robins [**6**] and Barvinok [**5**, Chap. 18]. They also have recently appeared in differential algebra in connection with $q$-holonomic sequences, see Garoufalidis [**33**]. Such functions were introduced in group theory by G. Higman in 1960 [**48**] under the name PORC functions (polynomial on residue class functions). Higman's motivating problem was the enumeration of $p$-groups, cf. Evseev [**28**]. The class of all quasi-polynomial functions is closed under addition and pointwise multiplication, and forms a commutative ring under these operations.

We arrive at the following taxonomy of function classes of increasing generality:

$\{3x + 1 \text{ function } T(x)\} \subset \{3x + k \text{ functions } T_{3,k}(x)\}$

$\qquad\qquad \subset \{\text{generalized Collatz functions of relatively prime type}\}$

$\qquad\qquad \subset \{\text{generalized Collatz functions}\}$

$\qquad\qquad \subset \{\text{quasi-polynomial functions}\}.$

For applications in mathematical logic, it has proved useful to further widen the definition of generalized Collatz functions to allow *partially defined functions*. Such functions are obtained by dropping the admissibility condition (4.2); they map integers to rational numbers having denominator dividing $d$. If a non-integer value is encountered, then one cannot iterate such a function further. In this circumstance we adopt the convention that if a non-integer iteration value is encountered, the calculation stops in a special "undefined" state. This framework allows the encoding of partially-defined (recursive) functions. One can use this convention to also define composition of partially defined functions.

## 5. Research Areas

Work on the $3x + 1$ problem cuts across many fields of mathematics. Six basic areas of research on the problem are: (1) *number theory*: analysis of periodic orbits of the map; (2) *dynamical systems*: behavior of generalizations of the $3x + 1$ map; (3)*ergodic theory*: invariant measures for generalized maps; (4) *theory of computation*: undecidable iteration problems; (5) *stochastic processes and probability theory*: models yielding heuristic predictions for the behavior of iterates; and (6) *computer science*: algorithms for computing iterates and statistics, and explicit computations. We treat these in turn.

### (1) *Number Theory*

The connection with number theory is immediate: the $3x + 1$ problem is a problem in arithmetic, whence it belongs to elementary number theory. Indeed it is classified as an unsolved problem in number theory by R. K. Guy [**42**, Problem E16]. The study of cycles of the $3x+1$ map leads to problems involving exponential Diophantine equations. The powerful work of Baker and Masser–Wüstholz on linear forms in logarithms gives information on the non-existence of cycles of various lengths having specified patterns of even and odd iterates. A class of generalized $3x + 1$ functions has been defined in a number theory framework, in which arithmetic operations on the domain of integers are replaced with such operations on the ring of integers of an algebraic number field, or by function field analogues such as a polynomial ring with coefficients in a finite field. Number-theoretic results are surveyed in the papers of Lagarias [**58**] and Chamberland [**11**] in this volume.

### (2) *Dynamical Systems*

The theory of iscrete dynamical systems concern the behavior of functions under iteration; that of continuous dynamical systems concern flows or solutions to differential equations. The $3x + 1$ problem can be viewed as iterating a map, therefore it is a discrete dynamical system on the state space $\mathbb{Z}$. This viewpoint was taken in Wirsching [**95**]. The important operation for iteration is *composition of functions*. One can formulate iteration and composition questions in the general context of universal algebra, cf. Lausch and Nobauer [**63**, Chap. 4.5]. In the taxonomy above, the classes of generalized $3x + 1$ functions, and quasi-polynomial functions are each closed under addition and composition of functions. The iteration properties of the first three classes of functions above have been studied, in connection with the $3x + 1$ problem and the theory of computation. However the iteration of general quasi-polynomial functions remains an unexplored research area.

Viewing the problem this way suggests that it would be useful in the study of the $3x + 1$ function to obtain dynamical systems on larger domains, including the real numbers $\mathbb{R}$ and the complex numbers $\mathbb{C}$. Other extensions include defining analogous functions on the ring $\mathbb{Z}_2$ of 2-adic integers, or, for generalized $3x + 1$

maps, on a ring of $d$-adic integers, for a value of $d$ determined by the function. When one considers generalized $3x+1$ functions on larger domains, a wide variety of behaviors can occur. These topics are considered in the papers of Chamberland [11] and Matthews [67] in this volume. For a general framework on topological dynamics see Akin [1].

(3) *Ergodic Theory*

The connection with ergodic theory arises as an outgrowth of the dynamical systems viewpoint, but adds the requirement of the presence of an invariant measure. It was early observed that there are finitely additive measures which are preserved by the $3x+1$ map on the integers. Extensions of generalized $3x+1$ functions to $d$-adic integers lead to maps invariant under standard measures (countably additive measures). For example, the (unique continuous) extension of the $3x+1$ map to the 2-adic integers has 2-adic measure as an invariant measure, and the map is ergodic with respect to this measure. Ergodic theory topics are considered in the surveys of Matthews [67] and Kontorovich and Lagarias [56] in this volume. An interesting open problem is to classify all invariant measures for generalized $3x+1$ functions on the $d$-adic integers.

(4) *Mathematical Logic and the Theory of Computation*

The connection to logic and the theory of computation starts with the result of Conway that there is a generalized $3x+1$ function whose iteration can simulate a universal computer. Conway [16] exhibited an unsolvable iteration problem for a particular generalized $3x+1$ function: starting with a given input which is a positive integer $n$, decide whether or not some iterate of this map with this input is ever a power of 2. In this connection note that the $3x+1$ problem can be reformulated as asserting that, starting from any positive integer $n$, some iterate $C^{(k)}(n)$ of the Collatz function (or of the $3x+1$ function) is a power of 2. It turns out that iteration of $3x+1$-like functions had already been considered in understanding the power of some logical theories even in the late 1960's; these involved partially defined functions taking integers to integers (with undefined output for some integers), cf. Isard and Zwicky [51]. More recently such functions have arisen in studying the computational power of "small" Turing machines, that are too small to encode a universal computer. These topics are surveyed in the paper of Michel and Margenstern [68] in this volume.

(5) *Probability Theory and Stochastic Processes*

A connection to probability theory and stochastic processes arises when one attempts to model the behavior of the $3x+1$ iteration on large sets of integers. This leads to heuristic probabilistic models for the iteration, which allow predictions of its behavior. Some authors have argued that the iteration can be viewed as a kind of pseudo-random number generator, viewing the input as being given by a probability distribution, and then asking how this probability distribution evolves under iteration. In the reverse direction, one can study trees of inverse iterates (the inverse map is many-to-one, giving rise to a unary-binary tree of inverse iterates). Here one can ask for facts about the structure of such trees whose root node is an integer picked from some probability distribution. One can model this by a stochastic model corresponding to random tree growth, e.g. a branching random walk. These topics are surveyed in the paper of Kontorovich and Lagarias [**56**] in this volume.

(6) *Computer Science: Machine Models, Parallel and Distributed Computation*

In 1987 Conway [**17**] (in this volume) formalized the Fractran model of computation as a universal computer model, based on his earlier work related to the $3x+1$ problem. This computational model is related to the register machine (or counter machine) model of Marvin Minsky ([**70**], [**71**, Sect. 11.1]). Both these machine models have recently been seen as relevant for developing models of computation using chemical reaction networks, and to biological computation, see Soloveichik et al [**86**] and Cook et al. [**20**].

The necessity to make computer experiments to test the $3x+1$ conjecture, and to explore various properties and patterns of the $3x + 1$ iteration, leads to other questions in computation. One has the research problem of developing efficient algorithms for computing on a large scale, using either parallel computers or a distributed computer system. The $3x + 1$ conjecture has been tested to a very large value of $n$, see the paper of Oliveira e Silva [**76**] in this volume. The computational method used in [**76**] to obtain record results can be parallelized. Various large scale computations for the $3x + 1$ problem have used distributed computing, cf. Roosendaal [**79**].

## 6. Current Status

We give a brief summary of the current status of the problem, which further elaborates answers to the two questions raised in the introduction.

**6.1. Where does research currently stand on the $3x + 1$ problem?** The $3x + 1$ problem remains unsolved, and a solution remains unapproachable at present. To quote a still valid dictum of Paul Erdős ([**58**, p. 3]) on the problem:

> "Mathematics is not yet ready for such problems."

Research has established various "world records", all of which rely on large computer calculations (together with various theoretical developments).

(W1) The $3x + 1$ conjecture has now been verified for all $n < 20 \times 2^{58} \approx 5.7646 \times 10^{18}$ (Oliveira e Silva [**76**] (in this volume)).

(W2) The trivial cycle $\{1, 2\}$ is the only cycle of the $3x + 1$ function on the positive integers having period length less than $10,439,860,591$. It is also the only cycle containing less than $6,586,818,670$ odd integers (Eliahou [**24**, Theorem 3.2]*).

(W3) Infinitely many positive integers $n$ take at least $6.143 \log n$ steps to reach 1 under iteration of the $3x + 1$ function $T(x)$ (Applegate and Lagarias [**3**]).

---

*This number is the bound $(21, 0)$ given in [**24**, Table 2]. The smaller values in Table 2 are now ruled out by the computations in item (W1) above.

(W4) The positive integer $n$ with the largest currently known value of $C$, such that it takes $C \log n$ iterations of the $3x + 1$ function $T(x)$ to reach 1, is $n = 7,219,136,416,377,236,271,195$ with $C \approx 36.7169$ (Roosendaal [**79**, $3x + 1$ Completeness and Gamma records]).

(W5) The number of integers $1 \le n \le X$ that iterate to 1 is at least $X^{0.84}$, for all sufficiently large $X$ (Krasikov and Lagarias [**57**]).

There has also been considerable progress made on showing the nonexistence of various kinds of periodic points for the $3x + 1$ function, see Brox [**8**] and Simons and de Weger [**83**]. These bounds are based on number-theoretic methods involving Diophantine approximation.

**6.2. Where does research stand on generalizations of the $3x + 1$ problem?** It has proved fruitful to view the $3x + 1$ problem as a special case of wider classes of functions. These function classes appear naturally as the correct level of generality for basic results on iteration; this resulted in the taxonomy of function classes given in §3. There are some general results for these classes and many unsolved problems.

The $3x + k$ problem seems to be the correct level of generality for studying rational cycles of the $3x + 1$ function ([**59**]). There are extensive results on cycles of the $3x + 1$ function, and the methods generally apply to the $3x + k$ function as well, see the survey of Chamberland [**11**] (in this volume).

The class of generalized $3x + 1$ functions of relatively prime type is a very natural class from the ergodic theory viewpoint, since this is the class on which the $d$-adic extension of the function has $d$-adic Haar measure as an invariant measure. The paper of Matthews [**67**] (in this volume) reports general ergodicity results and raises many questions about such functions.

The class of generalized Collatz functions has the property that all functions in it have a unique continuous extension to the domain of $d$-adic integers $\mathbb{Z}_d$. This general class is known to contain undecidable iteration problems, as discussed in the paper of Michel and Margenstern [**68**] (in this volume). The dynamics of general functions in this class is only starting to be explored; many interesting examples are given in the paper of Matthews [**67**] (in this volume). An interesting area worthy of future development is that of determining the existence and structure of invariant Borel measures for such functions on $\mathbb{Z}_d$, and determining whether there is some relation of their structure to undecidability of the associated iteration problem.

**6.3. How can this be a hard problem, when it is so easy to state?** Our answer is that there are two different mechanisms yielding hard problems, either or both of which may apply to the $3x + 1$ problem. The first is "pseudorandomness"; this involves a connection with ergodic theory. The second is "non-computability". Both of these are discussed in detail in this volume.

The "ergodicity" connection has been independently noted by a number of people, see for example Lagarias [**58**] (in this volume) and Akin [**2**]. The unique continuous extension of the $3x+1$ map $T(x)$ to the 2-adic integers $\mathbb{Z}_2$ gives a function which is known to be ergodic in a strong sense, with respect to the 2-adic measure. It is topologically and metrically conjugate to the shift map, which is a maximum entropy map. The iterates of the shift function are completely unpredictable in the ergodic theory sense. Given a random starting point, predicting the parity of the $n$-th iterate for any $n$ is a "coin flip" random variable. The $3x+1$ problem concerns the behavior of iterating this function on the set of integers $\mathbb{Z}$, which is a dense subset of $\mathbb{Z}_2$, having 2-adic measure zero. The difficulty is then in finding and understanding non-random regularities in the iterates when restricted to $\mathbb{Z}$. Various probabilistic models are discussed in the paper of Kontorovich and Lagarias [56] (in this volume). Empirical evidence seems to indicate that the $3x+1$ function on the domain $\mathbb{Z}$ retains the "pseudorandomness" property on its initial iterates until the iterates enter a periodic orbit. This supports the $3x+1$ conjecture and at the same time deprives us of any obvious mechanism to prove it, since mathematical arguments exploit the existence of structure, rather than its absence.

A connection of a generalized Collatz function to "non-computability" was made by Conway [16] (in this volume), as already mentioned. Conway's undecidability result indicates that the $3x+1$ problem could be close to the unsolvability threshold. It is currently unknown whether the $3x+1$ problem is itself undecidable, however no method is currently known to approach this question. The survey of Michel and Margenstern [68] (in this volume) describes many results on generalized $3x+1$ functions that exhibit undecidable or difficult-to-decide iteration problems. The $3x+1$ function might conceivably belong to a smaller class of generalized $3x+1$ functions that evade undecidability results that encode universal computers. Even so, it conceivably might encode an undecidable problem, arising by another (unknown) mechanism. As an example, could the following question be undecidable: "Is there any positive integer $n$ such that $T^{(k)}(n) > 1$ for $1 \leq k \leq 100 \log n$?"

## 7. Hardness of the $3x+1$ problem

Our viewpoint on hard problems has evolved since 1900, starting with Hilbert's program in logic and proof theory and benefiting from developments in the theory of computation. Starting in the 1920's, Emil Post uncovered great complexity in studying some very simple computational problems, now called "Post Tag Systems". A *Tag system* in the class $TS(\mu, \nu)$ consists of a set of rules for transforming words using letters from an alphabet $\mathcal{A} = \{a_1, ..., a_\mu\}$ of $\mu$ symbols, a deletion number (or shift number) $\nu \geq 1$, and a set of $\mu$ production rules

$$a_j \mapsto w_j := a_{j,0} a_{j,1} \cdots a_{j,n_j}, \ 1 \leq j \leq \mu,$$

in which the output $w_j$ is a finite string (or word) of length $n_j$ in the alphabet $\mathcal{A}$. Starting from an initial string $S$ a Tag system looks at the leftmost symbol of $S$, call it $a_j$, then attaches to the right end of the string the word $w_j$, and finally deletes the first $\nu$ symbols of the resulting string $Sw_j$, thus obtaining a new string $S'$. Here the "tag" is the set of symbols $w_j$ attached to the end of the word, and the iteration halts if a word of length less than $\nu$ is encountered. The *halting problem* is the question of deciding whether for an arbitrary initial word $S$, iteration eventually reaches the empty word. The *reachability problem* is that of deciding whether, given words $S$ and $\tilde{S}$, starting from word $S$ will ever produce word $\tilde{S}$ under iteration. The halting problem is a special case of the reachability problem. Post [78] reports that in 1920–1921 he found a complete decision procedure[†] for the case $\mu = 2, \nu = 2$, i.e. the class $T(2,2)$. He then tried to solve the case $\mu = 2, \nu > 2$, without success.

---

[†]Post did not publish his proof. A decision procedure for both problems is outlined in de Mol [73].

He reported [**78**, p. 372] that the special case $\mu = 2, \nu = 3$ with $\mathcal{A} = \{0,1\}$ and the two production rules

(7.4)                       $0 \mapsto w_0 = 00, \ 1 \mapsto w_1 = 1101$

already seemed to be an intractable problem. We shall term this problem

**Post's Original Tag Problem.** *Is there a recursive decision procedure for the halting problem for the Tag system in $T(2,3)$ given by the rules $0 \mapsto 00$ and $1 \mapsto 1101$?*

Leaving this question aside, Post considered the parameter range $\mu > 2, \nu = 2$. He wrote [**78**, p. 373]:

> For a while the case $\nu = 2$, $\mu > 2$ seemed to be more promising, since it seemed to offer a greater chance of a finitely graded series of problems. But when this possibility was explored in the early summer of 1921, it rather led to an overwhelming confusion of classes of cases, with the solution of the corresponding problem depending more and more on problems of ordinary number theory. Since it had been our hope that the known difficulties of number theory would, as it were, be dissolved in the particularities of this more primitive form of mathematics, the solution of the general problem of "tag" appeared hopeless, and with it our entire program of the solution of finiteness problems.

Discouraged by this, Post reversed course and went on to obtain a "Normal Form Theorem" ([**77**]), published in the 1940's, showing that a general logical problem could be reduced to a form slightly more complicated than Tag Systems. In 1961 Marvin Minsky [**70**] proved that Post Tag Systems were undecidable problems in general. In the next few years Hao Wang [**94**], J. Cocke and M. Minsky [**13**] and S. Ju. Maslov [**66**] independently showed undecidability for the subclass of Post Tag Systems consisting of those with $\nu = 2$, thus showing that Post was right to quit trying to solve problems in that class. At present the recursive solvability or unsolvability in the class $T(2, \nu)$ remains open for all $\nu > 2$. Post's original tag problem, which is the halting problem for one special function in $T(2,3)$, is still unsolved, see Lisbeth De Mol [**72**], [**74**, p. 93], and for further work [**73**], [**75**].

Recently de Mol showed that the $3x+1$ problem can be encoded as a reachability problem for a tag system in $T(3,2)$ ([**74**, Theorem 2.1]). This tag system encodes the $3x + 1$ function, and the reachability problem is:

$3x + 1$ **Tag Problem.** *Consider the tag system $T_C$ in $T(3,2)$ with alphabet $\mathcal{A} = \{0,1,2\}$, deletion number $\nu = 2$, and production rules*

$$0 \mapsto 12, \ 1 \mapsto 0, \ 2 \mapsto 000.$$

*For each $n \geq 1$, if one starts from the configuration $S = 0^n$, will the tag system iteration for $T_C$ always reach state $\tilde{S} = 0$?*

In 1931 Kurt Gödel [**35**] showed the existence of undecidable problems: he showed that certain propositions were undecidable in any logical system complicated enough to include elementary number theory. This result showed that Hilbert's proof theory program could not be carried out. Developments in the theory of computation showed that one of Gödel's incompleteness results corresponded to the unsolvability of the halting problem for Turing machines. This was based on the existence of a universal Turing machine, that could simulate any computation, and in his 1937 foundational paper Alan Turing [**91**] already showed one could be constructed of a not very large size.

We now have a deeper appreciation of exactly how simple a problem can be and still simulate a universal computer. Amazingly simple problems of this sort have been found in recent years. Some of these involve cellular automata, a model of computation developed by John von Neumann and Stansilaw M. Ulam in the 1950's. One of these problems concerns the possible behavior of a very simple one-dimensional nearest neighbor cellular automaton, Rule 110, using a nomenclature introduced by Wolfram [**96**], [**97**]. This rule was conjectured by Wolfram to give a universal computer ([**98**, Table 15], [**99**, pp. 575–577]). It was proved to be weakly universal by M. Cook (see Cook [**18**], [**19**]). Here weakly universal means that the initial configuration of the cellular automaton is required to be ultimately periodic, rather than finite. Another is John H. Conway's game of "Life," first announced in 1970 in Martin Gardner's column in Scientific American (Gardner [**31**]), which is a two-dimensional cellular automaton, having nearest neighbor interaction rules of a particularly simple nature. Its universality as a computer was later established, see Berkelamp, Conway and Guy [**7**, Chap. 25]. Further remarks on the size of universal computers are given in the survey of Michel and Margenstern [**68**] (in this volume).

There are, however, reasons to suspect that the $3x + 1$ function is not complicated enough to be universal, i.e. to allow the encoding of a universal computer in its input space. First of all, it is so simple to state that there seems very little room in it to encode the elementary operations needed to create a universal computer. Second, the $3x + 1$ conjecture asserts that the iteration halts on the domain of all positive integer inputs, so for each integer $n$, the value $F(n)$ of the largest integer observed before visiting 1 is recursive. To encode a universal computer, one needs to represent all recursive functions, including functions that grow far faster than any given recursive function $F(n)$. It is hard to image how one can encode it here as a question about the iteration, without enlarging the domain of inputs. Third, the $3x + 1$ function possesses the feature that there is a nice (finitely additive) invariant measure on the integers, with respect to which it is completely mixing under iteration. This is the measure that assigns mass $\frac{1}{2^n}$ to each complete arithmetic progression (mod $2^n$), for each $n \geq 1$. This fundamental observation was made in 1976 by Terras [**88**], and independently by Everett [**27**] in 1977, see Lagarias [**58**, Theorem B] for a precise statement. This "mixing property" seems to fight against the amount of organization needed to encode a universal computer in the inputs. We should caution that this observation by itself does not rule out the possibility that, despite this mixing property, a universal computer could be encoded in a very thin set of input values (of "measure zero"), compatible with an invariant measure. It just makes it seem difficult to do. Indeed, the 1972 encoding of a universal computer in the iteration of a certain generalized $3x+1$ function found by Conway [**16**] (in this volume) has the undecidability encoded in the iteration of a very thin set of integers. However Conway's framework is different from the $3x + 1$ problem in that the halting function he considers is partially defined.

Even if iteration of the $3x+1$ function is not universal, it could still potentially be unsolvable. Abstractly, there may exist in an axiomatic system statements $F(n)$ for a positive integer predicate, such that $F(1), F(2), F(3), \ldots$ are provable in the system for all integer $n$, but the statement $(\forall n)\, F(n)$ is not provable within the system. For example, one can let $F(n)$ encode a statement that there is no contradiction in a system obtainable by a proof of length at most $n$. If the system is consistent, then $F(1), F(2), \ldots$ will all individually be provable. The statement $(\forall n)\, F(n)$ then encodes the consistency of the system. But the consistency of a system sufficiently complicated to include elementary number theory cannot be proved within the system, according to Gödel's second incompleteness theorem.

The pseudo-randomness or "mixing" behavior of the $3x + 1$ function also seems to make it extremely resistant to analysis. If one could rigorously show a sufficient amount of mixing is guaranteed to occur, in a controlled number of iterations in terms of the input size $n$, then one could settle part of the $3x + 1$ conjecture, namely prove the non-existence of divergent trajectories. Here we have the fundamental difficulty of proving in effect that the iterations actually do have an explicit pseudo-random property. Besides this difficulty, there remains a second fundamental difficulty: solving the number-theoretic problem of ruling out the existence of an enormously long non-trivial cycle of the $3x + 1$ function. This problem also seems unapproachable at present by known methods of number theory. However the finite cycles problem does admit proof of partial results, showing the nonexistence of non-trivial cycles having particular patterns of even and odd iterates.

A currently active and important general area of research concerns the construction of pseudo-random number generators: these are deterministic recipes that produce apparently random outputs (see Knuth [**55**, Chap. 3]). More precisely, one is interested in methods that take as input $n$ truly random bits and deterministically produce as output $n + 1$ "random-looking" bits. These bits are to be "random-looking" in the sense that they appear random with respect to a given family of statistical tests, and the output is then said to be pseudo-random with respect to this family of tests. Deciding whether pseudo-random number generators exist for statistical tests in various complexity classes is now seen as a fundamental question in computer science, related to the $P = NP$ probem, see for example Goldreich [**37**], [**38**]. It may be that resolving the issue of the pseudo-random character of iterating the $3x + 1$ problem will require shedding light on the general existence problem for pseudo-random number generators.

All we can say at present is that the $3x + 1$ problem appears very hard indeed. It now seems less surprising than it might have once seemed that a problem as simple-looking as this one could be genuinely difficult, and inaccessible to known methods of attack.

## 8. Future Prospects

We observe first that further improvements are surely possible on the "world records" (W1)–(W5) above. In particular, concerning (W3), it seems scandalous that it is not known whether or not there are infinitely many positive integers $n$ which iterate to 1 under the $3x + 1$ map $T(x)$ and take at least the "average" number $\frac{2}{\log 4/3} \log n \approx 6.95212 \log n$ steps to do so. Here the stochastic models for the $3x + 1$ iteration predict that at least half of all positive integers should have this property! These "world records" are particularly worth improving if they can shed more light on the problem. This could be the case for world record (5), where there is an underlying structure for obtaining lower bounds on the exponent, which involves an infinite family of nonlinear programs of increasing complexity ([**57**]).

Analysis of the $3x + 1$ problem has resulted in the formulation of a large set of "easier" problems. At first glance some of these seem approachable, but they also remain unsolved, and are apparently difficult. As samples, these include:

(C1) (*Finite Cycles Conjecture*) Does the $3x + 1$ function have finitely many cycles (i.e. finitely many purely periodic orbits on the integers)? This is conjectured to be the case.

(C2) (*Divergent Trajectories Conjecture-1*) Does the $3x + 1$ function have a divergent trajectory, i.e., an integer starting value whose iterates are unbounded? This is conjectured *not* to be the case.

(C3) (*Divergent Trajectories Conjecture-2*) Does the $5x + 1$ function have a divergent trajectory? This is conjectured to be the case.

(C4) (*Infinite Permutations-Periodic Orbits Conjecture*) If a generalized Collatz function permutes the integers and is not globally of finite order, is it true that it has only finitely many periodic orbits? The original Collatz function $U(n)$, which is a permutation, was long ago conjectured to have finitely many cycles. A conjecture of this kind, imposing extra conditions on the permutation, was formulated by Venturini [**93**, p. 303 top].

(C5) (*Infinite Permutations-Zero Density Conjecture*) If a generalized Collatz function permutes the integers, is it true that every orbit has a (natural) density? Under some extra hypotheses one may conjecture that all such orbits have density zero; compare Venturini [**93**, Sec. 6].

Besides these conjectures, there also exist open problems which may be more accessible. One of the most intriguing of them concerns establishing lower bounds for the number $\pi_1(x)$ of integers less than $x$ that get to 1 under the $3x + 1$ iteration. As mentioned earlier it is known ([**57**]) that there is a positive constant $c_0$ such that

$$\pi_1(x) > c_0 x^{0.84}.$$

It remains an open problem to show that for each $\epsilon > 0$ there exists a positive constant $c(\epsilon)$ such that

$$\pi_1(x) > c(\epsilon) x^{1-\epsilon}.$$

Many other specific, but difficult, conjectures for study can be found in the papers in this volume, starting with the problems listed in Guy [**40**].

We now raise some further research directions, related to the papers in this volume. A first research direction is to extend the class of functions for which the Markov models of Matthews [**67**] can be analyzed. Matthews shows that the class of generalized $3x + 1$ functions of relatively prime type ([**67**, Sec. 2]) is analyzable. He formulates some conjectures for exploration. It would be interesting to characterize the possible $d$-adic invariant measures for arbitrary generalized Collatz functions. It may be necessary to restrict to subclasses of such functions in order to obtain nice characterizations.

A second research direction concerns the class of generalized $3x + 1$ functions whose iterations extended to the set of $d$-adic integers are ergodic with respect to the $d$-adic measure, cf. Matthews [**67**, Sec. 6]).

**Research Problem.** *Does the class of generalized Collatz functions of relatively prime type contain a function which is ergodic with respect to the standard $d$-adic measure, whose iterations can simulate a universal computer? Specifically, could it have an unsolvable iteration problem of the form: "Given positive integers*

$(n, m)$ *as input, does there exist* $k$ *such that the* $k$-*th iterate* $T^{(k)}(n)$ *equals* $m$*?"*
*Or does ergodicity of the iteration preclude the possibility of simulating universal*
*computation?*

A third research direction concerns the fact that generalized Collatz functions
have now been found in many other mathematical structures, especially if one
generalizes further to integer-valued functions that are piecewise polynomial on
residue classes (mod $d$). These functions are the quasi-polynomial functions noted
above, and they show up in a number of algebraic contexts, particularly in counting
lattice points in various regions. It may prove worthwhile to study the iteration
of various special classes of quasi-polynomial functions arising in these algebraic
contexts.

At this point in time, in view of the intractability of problems (C1)–(C5) it
also seems a sensible task to formulate a new collection of even simpler "toy problems", which may potentially be approachable. These may involve either changing
the problem or importing it into new contexts. For example, there appear to be
accessible open problems concerning variants of the problem acting on finite rings
(Hicks et al. [**47**]). Another promising recent direction is the connection of these
problems with generating sets for multiplicative arithmetical semigroups, noted by
Farkas [**29**]. This has led to a family of more accessible problems, where various results can be rigorously established ([**4**]). Here significant unsolved problems remain
concerning the structure of such arithmetical semigroups. Finally it may prove
profitable to continue the study, initiated by Klarner and Rado [**54**], of sets of integers (or integer vectors) closed under the action of a finitely generated semigroup
of affine maps.

## 9. Is the $3x + 1$ problem a "good" problem?

There has been much discussion of what constitutes a good mathematical problem. We can not do better than to recall the discussion of Hilbert [**49**] in his famous
1900 problem list. On the importance of problems he said ([**49**, p. 437]):

> The deep significance of certain problems for the advance of math-
> ematical science in general, and the important role they play in
> the work of the individual investigator, are not to be denied. As
> long as a branch of science offers an abundance of problems, so
> long is it alive; a lack of problems foreshadows extinction or the
> cessation of independent development. Just as every human un-
> dertaking pursues certain objects, so also mathematical research
> requires its problems. It is also by the solution of problems that
> the investigator tests the temper of his steel; he finds new methods
> and new outlooks, and gains a wider and freer horizon.

Hilbert puts forward three criteria that a good mathematical problem ought to
satisfy:

> It is difficult and often impossible to judge the value of a problem
> correctly in advance; for the final award depends upon the gain
> which science obtains from the problem. Nevertheless we can ask
> whether there are general criteria which mark a good mathematical
> problem. An old French mathematician said: "A mathematical
> theory is not to be considered complete until you have made it so

clear that you can explain it to the first man that you meet on the street." This clearness and ease of comprehension, here insisted on for a mathematical theory, I should still more demand for a mathematical problem if it is to be perfect; for what is clear and easily comprehended attracts, the complicated repels us.

Moreover a mathematical problem should be difficult in order to entice us, but not completely inaccessible, lest it mock at our efforts. It should be to us a guide post on the mazy paths to hidden truths, and ultimately a reminder of our pleasure in its successful solution.

From the viewpoint of the Hilbert criteria for a good problem, we see that:

(1) The $3x + 1$ problem is a clear, simply stated problem;

(2) The $3x + 1$ problem is a difficult problem;

(3) The $3x + 1$ problem initially seems accessible, in that it possesses a fairly intricate internal structure.

But – and it is a big "but" – the evidence so far suggests that obtaining a proof of the $3x+1$ problem is inaccessible! Not only does this goal appear inaccessible, but various simplified conjectures derived from it appear to be completely inaccessible in their turn, leading to a regress to formulation of a series of simpler and simpler inaccessible problems, namely conjectures (C1)–(C5) listed in §8.

We conclude that the $3x + 1$ problem comes close to being a "perfect" problem in the Hilbert sense. However it seems to fail the last of Hilbert's requirements: It mocks our efforts! It is possible to work hard on this problem to no result. It is definitely a dangerous problem! It could well be that the $3x + 1$ problem remains out of human reach. But maybe not. Who knows?

## 10. Working on the $3x + 1$ probem

Whether or not the $3x + 1$ problem is a "good" problem, it is not going away, due to its extreme accessibility. It offers a large and tantalizing variety of patterns in computer experiments. This problem stands as a mathematical challenge for the 21-st century.

In working on this problem, the most cautious advice, following Richard Guy [40] is:

*Don't try to solve these problems!*

But, as Guy said [40, p. 35], some of you may be already scribbling, in spite of the warning!

We also note that Paul Erdős said, in conversation, about its difficulty ([25]):

*"Hopeless. Absolutely hopeless."*

In Erdős-speak, this means that there are no known methods of approach which gave any promise of solving the problem. For other examples of Erdős's use of the term "hopeless" see Erdős and Graham [26, pp. 1, 27, 66, 105].

At this point we may recall further advice of David Hilbert [49, p. 442] about problem solving:

If we do not succeed in solving a mathematical problem, the reason frequently consists in our failure to recognize the more general standpoint from which the problem before us appears only as a single link in a chain of related problems. After finding this standpoint, not only is this problem frequently more accessible to our investigation, but at the same time we come into possession of a method that is applicable to related problems.

The quest for generalization cuts in two directions, for Hilbert also says [**49**, p. 442]:

He who seeks for methods without having a definite problem in mind seeks for the most part in vain.

Taking this advice into account, researchers have treated many generalizations of the $3x+1$ problem, which are reported on in this volume. One can consider searching for general methods that apply to a large variety of related iterations. Such general methods as are known give useful information, and answer some questions about iterates of the $3x+1$ function. Nevertheless it is fair to say that they do not begin to answer the central question:

*What is the ultimate fate under iteration of such maps over all time?*

My personal viewpoint is that the $3x+1$ problem is somewhat dangerous, and that it is prudent not to focus on resolving the $3x+1$ conjecture as an immediate goal. Rather, one might first look for more structure in the problem. Also one might profitably view the problem as a "test case", to which one may from time to time apply new results arising from the ongoing development of mathematics. When new theories and new methods are discovered, the $3x+1$ problem may be used as a testbed to assess their power, whenever circumstances permit.

To conclude, let us remind ourselves, following Hilbert [**49**, p. 438]:

The mathematicians of past centuries were accustomed to devote themselves to the solution of difficult particular problems with passionate zeal. They knew the value of difficult problems.

The $3x+1$ problem stands before us as a beautifully simple question. It is hard to resist exploring its structure. We should not exclude it from the mathematical universe just because we are unhappy with its difficulty. It is a fascinating and addictive problem.

**Acknowledgments.** I am grateful to Michael Zieve and Steven J. Miller each for detailed readings and corrections. Marc Chamberland, Alex Kontorovich, and Keith R. Matthews also made many helpful comments. I thank Andreas Blass for useful comments on incompleteness results and algebraic structures. The author was supported by NSF Grants DMS-0500555 and DMS-0801029.

## References

[1] E. Akin, *The general topology of dynamical systems,* Graduate Studies in Mathematics 1., American Mathetmatical Society,Providence, RI 1993.

[2] E. Akin, Why is the $3x+1$ Problem Hard?, In: *Chapel Hill Ergodic Theory Workshops* (I. Assani, Ed.), Contemp. Math. vol 356, Amer. Math. Soc. 2004, pp. 1–20.

[3] D. Applegate and J. C. Lagarias, Lower bounds for the total stopping time of $3x+1$ iterates, Math. Comp. **72** (2003), 1035–1049.

[4] D. Applegate and J. C. Lagarias (2006), The $3x+1$ semigroup, J. Number Theory **177** (2006), 146–159.

[5]  A. Barvinok, *Integer Points in Polyhedra*, European Math.Soc. Publishing, ETH, Zürich 2008.

[6]  M. Beck and S. Robins, *Computing the continuous discretely. Integer-point enumeration in polyhedra¡* Springer: New York 2007.

[7]  E. R. Berlekamp, J. H. Conway and R. K. Guy, *Winning Ways for Your Mathematical Plays, Volume 4 (Second Revised Edition)* A. K. Peters, Ltd. 2004.

[8]  T. Brox, Collatz cycles with few descents, Acta Arithmetica **92** (2000), 181–188.

[9]  W. Bruns and J. Herzog, *Cohen-Macaulay rings*, Cambridge Univ. Press: Cambridge 1993.

[10] W. Bruns and B. Ichim, On the coefficients of Hilbert quasipolynomials, Proc. Amer. Math. Soc. **135** (2007), No. 5, 1305-1308.

[11] M. Chamberland, A $3x + 1$ Survey: Number theory and dynamical systems, in this volume.

[12] V. Chvatal, D. Klarner and D. E. Knuth, Selected combinatorial research problems, Stanford Computer Science Dept. Technical Report STAN-CS-72-292 June 1972, 31 pages.

[13] J. Cocke and M. Minsky, Universality of tag systems with $P = 2$, Journal of the ACM **11**, (1964), No. 1, 15–20.

[14] L. Collatz, Letter to Michael E. Mays, dated 17 Sept. 1980.

[15] L. Collatz, On the motivation and origin of the $(3n + 1)$- problem (Chinese), J. Qufu Normal University, Natural Science Edition [Qufu shi fan da xue xue bao] **12** (1986), No. 3, 9–11.

[16] J. H. Conway, Unpredictable Iterations, Proc. 1972 Number Theory Conference (Univ. Colorado, Boulder, Colo., 1972 ), pp. 49–52. Univ. Colorado, Boulder, Colo. 1972.

[17] J. H. Conway, FRACTRAN: A Simple Universal Computing Language for Arithmetic, In: *Open Problems in Communication and Computation* (T. M. Cover and B. Gopinath, Eds.), Springer-Verlag: New York 1987, pp. 3-27 [Reprinted in this volume]

[18] M. Cook, Universality in elementary cellular automata, Complex Systems **15** (2004), 1–40.

[19] M. Cook, A concrete view of rule 110 computation, T. Neary, D. Woods, A. K. Seda and N. Murphy (Eds.), *Proceedings International Workshop on The Complexity of Simple Programs (CSP 2008)*, EPCTS (Electronic Proceedings in Theoretical Computer Science) **1** (2009), 31–55.

[20] M. Cook, D. Soloveichik, E. Winfree and J. Bruck, Programmability of Chemical Reaction Networks, to appear in a Festschrift for Grzegorz Rozenberg, Springer-Verlag.

[21] Necia Grant Cooper (Ed), *From Cardinals to Chaos. Reflections on the life and legacy of Stanslaw Ulam*, Cambridge Univ. Press: Cambridge 1989 [Reprint of Los Alamos Science, Vol. 15.]

[22] H. S. M. Coxeter, Cyclic sequences and frieze patterns: The Fourth Felix Behrend Memorial Lecture), Vinculum **8** (1971), 4–7. [Reprinted in this volume]

[23] L. Ehrhart, Sur un problème de géométrie diophantienne linéaire. II. Systèmes diophantiens linéaires, J. Reine Angew. Math. **227** (1967), 25–49.

[24] S. Eliahou, The $3x+1$ problem: new lower bounds on nontrivial cycle lengths, Discrete Math. **118** (1993), 45–56.

[25] P. Erdős, Private communication with J. C. Lagarias.

[26] Paul Erdős and R. L. Graham, *Old and new problems and results in combinatorial number theory* Monographie No. 28 de L'Enseignement Mathématique, Kundig: Geneva 1980. (Chapter 1 appeared in: Enseign. Math. **25** (1979), no. 3-4, 325–344. )

[27] C. J. Everett, Iteration of the number theoretic function $f(2n) = n, f(2n + 1) = 3n + 2$, Advances in Math. **25** (1977), 42–45.

[28] A. Evseev, Higman's PORC conjecture for a family of groups, Bull. London Math. Soc. **40** (2008), 405–414.

[29] H. M. Farkas, Variants of the $3N+1$ conjecture and multiplicative semigroups, in: *Geometry, spectral theory, groups, and dynamics*, pp. 121–127, Contemp. Math. Vol 387, Amer. Math. Soc.: Providence, RI 2005.

[30] H. Furstenberg, *Recurrence in ergodic theory and combinatorial number theory*, Princeton University Press, Princeton, NJ 1981.

[31] M. Gardner, Mathematical Games, Scientific American **223**  (1970) Number 4 (October), 120–123.

[32] M. Gardner, Mathematical Games, Scientific American **226** (1972) Number 6 (June), 114–118.

[33] S. Garoufalidis, The degree of a $q$-holonomic sequence is a quadratic quasi-polynomial, eprint: `arxiv:1005.4580v1`

[34] E. Glasner, *Ergodic Theory via joinings,* Mathematical Surveys and Monographs, Vol. 101, American Mathematical Society, Providence, RI, 2003.

[35] K. Gödel, Über formal unentscheidbare Sätzes der *Principia Mathematica* und verwandter Systeme I, [On formally undecidable propositions of *Principia mathematica* and related systems I] Mönatshefte für Mathematik und Physik **28** (1931), 173–198. (English translation in [**36**, p. 144–195].)

[36] K. Gödel, *Collected Works, Volume I. Publications 1929–1936,* S. Feferman et al. (Eds.) , Oxford University Press: New York 1986.

[37] O. Goldreich, *Foundations of cryptography. Basic tools.* Cambridge University Press: Cambridge 2001.

[38] O. Goldreich, *A Primer on Pseudorandom Generators,* University Lecture Series, No. 55, American Math. Society: Providence, RI 2010.

[39] K. Greenberg, Integer valued functions on the integers, Math. Medley **17** (1989), 1–10.

[40] R. K. Guy, Don't try to solve these problems!, American Math. Monthly **90** (1983), 35–41. [Reprinted in this volume]

[41] R. K. Guy, Conway's prime-producing machine, Math. Magazine *56* (1983), no. 1, 26–33.

[42] R. K. Guy, *Unsolved Problems in Number Theory. Third Edition.* Problem Books in Mathematics, Springer-Verlag: New York 2004.

[43] R. K. Guy, private communication, 2009.

[44] H. Hasse, *Unsolved Problems in Elementary Number Theory,* Lectures at University of Maine (Orono), Spring 1975, Mimeographed notes.

[45] B. Hasselblatt and A. B. Katok, *Introduction to the modern theory of dynamical systems,* Cambridge Univ. Press, Cambridge 1995.

[46] B. Hayes, Computer recreations: The ups and downs of hailstone numbers, Scientific American **250** , No. 1, (1984), 10–16.

[47] K. Hicks, G. L. Mullen, J. L. Yucas and R. Zavislak, A Polynomial Analogue of the $3N + 1$ Problem?, American Math. Monthly **115** (2008), No. 7, 615–622.

[48] G. Higman, Enumerating $p$-groups II: Problems whose solution is PORC, Proc. London Math. Soc. **10** (1960), 566–582.

[49] D. Hilbert, Mathematische Probleme, Göttinger Nachrichten (1900) 253–297. Reprinted in: Archiv der Mathematik und Physik, 3rd Ser. **1** (1901) 44-63 and 213-237. (English translation: Mathematical Problems, Bull. Amer. Math. Soc. **8** (1902) 437–479. Reprinted in: *Mathematical Developments Arising From Hilbert Problems,* Proc. Symp. Pure Math. Volume 28, AMS: Providence 1976, pp. 1-34.)

[50] A. J. W. Hilton, private communication, 2010.

[51] S. D. Isard and H. M. Zwicky, Three open questions in the theory of one-symbol Smullyan systems, SIGACT News, Issue No. 7, 1970, 11–19.

[52] M. Klamkin, Problem $63 - 13^*$, SIAM Review **5** (1963), 275–276.

[53] D. A. Klarner, A sufficient condition for certain semigroups to be free, Journal of Algebra **74** (1982), 40–48.

[54] D. A. Klarner and R. Rado, Arithmetic properties of certain recursively defined sets, Pacific J. Math. **53** (1974), No. 2, 445–463.

[55] D. E. Knuth, *The Art of Computer Programming. Vol 2. Seminumerical Algorithms. Second Edition.* Addison-Wesley: Reading, MA 1981.

[56] A.V. Kontorovich and J. C. Lagarias, Stochastic models for the $3x + 1$ problem and generalizations, paper in this volume.

[57] I. Krasikov and J. C. Lagarias, Bounds for the $3x + 1$ problem using difference inequalities, Acta Arith. **109** (2003), no. 3, 237–258.

[58] J. C. Lagarias, The $3x + 1$ problem and its generalizations, Amer. Math. Monthly **92** (1985), 3–23. [Reprinted with corrections in this volume].

[59] J. C. Lagarias, The set of rational cycles for the $3x + 1$ problem, Acta Arithmetica **56** (1990), 33–53.

[60] J. C. Lagarias, The $3x + 1$ Problem: An Annotated Bibliography (1963-1999), paper in this volume.

[61] J. C. Lagarias, The $3x + 1$ Problem: An Annotated Bibliography, II (2000-2009), arXiv:math/0608208.

[62] J. C. Lagarias and A. Weiss, The $3x + 1$ problem: Two stochastic models, Annals of Applied Probability **2** (1992), 229–261.

[63] H. Lausch and W. Nöbauer, *Algebra of Polynomials*, North-Holland Publ. Co.: Amsterdam 1973.

[64] K. Mahler, *Lectures on diophantine approximations. Part I. g-adic number and Roth's theorem.* Prepared from notes of R. P. Bambah of lectures given at Univ. of Notre Dame in the Fall of 1957, Univ. of Notre Dame Press, Notre Dame, Ind. 1961, xi+188pp.

[65] M. Mashaal, *Bourbaki: A Secret Society of Mathematicians.* (Translated from 2002 French original by A. Pierrehumbert) American Math. Society, Providence RI 2006.

[66] S. Ju. Maslov, On E. L. Post's "tag problem" (Russian), Trudy Mat. Inst. Steklov **72** 91964), 57–68. [English translation: American Math. Soc. Translations, Series 2, Vol. 97 (1970), 1–14.]

[67] K. R. Matthews, Generalized $3x + 1$ mappings: Markov chains and ergodic theory, paper in this volume.

[68] P. Michel and M. Margenstern, Generalized $3x + 1$ functions and the theory of computation, paper in this volume.

[69] S. J. Miller and R. Takloo-Bighash, *An invitation to modern number theory,* Princeton University Press: Princeton NJ 2006.

[70] M. Minsky, Recursive unsolvability of Post's problem of tag and other topics in the theory of Turing machines, Annals of Mathematics **74** (1961), 437–455.

[71] M. Minsky, *Computation: Finite and Infinite Machines*, Prentice-Hall, Inc: Engelwood Cliffs, NJ 1967.

[72] L. De Mol, Closing the circle: An analysis of Emil Post's Early Work, Bull. Symbolic Logic **12** (2006), No. 2, 267–289.

[73] L. De Mol, Study of limits of solvability in tag systems, pp. 170–181 in: J. Durand-Lose, M. Margenstern (Eds.), *Machines, Computations, Universality, MCU 2007*, Lecture Notes in Computer Science, vol. 4664, Springer-Verlag: New York 2007.

[74] L. De Mol, Tag systems and Collatz-like functions, Theoretical Computer Science **290** (2008), 92–101.

[75] L. De Mol, On the boundaries of solvability and unsolvability in tag systems. Theoretical and experimental results., in: T. Neary, D. Woods, A. K. Seda and N. Murphy (Eds.), *Proceedings International Workshop on The Complexity of Simple Programs (CSP 2008)*, EPCTS (Electronic Proceedings in Theoretical Computer Science) **1** (2009), 56–66.

[76] T. Oliveira e Silva, Empirical verification of the $3x + 1$ and related conjectures, paper in this volume.

[77] E. L. Post, Formal reductions of the general combinatorial decision problem, Amer. J. Math. **65** (1943), 197–215.

[78] E. L. Post, Absolutely unsolvable problems and relatively undecidable propositions- account of an anticipation, in: M. Davis, Ed., *The Undecidable. Basic Papers on Undecidable Propositions, Unsolvable Problems and Computable Functions,* Raven Press: New York 1965. (Reprint: Dover Publications, 2004).

[79] Eric Roosendaal, On the $3x + 1$ problem, web document, available at: http://www.ericr.nl/wondrous/index.html

[80] J. Roubaud, Un probléme combinatoire posé par poésie lyrique des troubadours, Matématiques et Sciences Humaines **27** Autumn 1969, 5–12.

[81] Klaus Schmidt, *Dynamical Systems of Algebraic Origin.* Progress in Math., 128. Birkhäuser Verlag: Basel 1995.

[82] J. Silverman, *The Arithmetic of Dynamical Systems.* Springer-Verlag: New York 2007.

[83] J. Simons and B. de Weger, Theoretical and computational bounds for $m$-cycles of the $3n + 1$ problem, Acta Arithmetica **117** (2005), 51–70.

[84] Ya. G. Sinai, Statistical $(3X + 1)$-Problem, Dedicated to the memory of Jürgen K. Moser. Comm. Pure Appl. Math. **56** No. 7 (2003), 1016–1028.

[85] Ya. G. Sinai, Uniform distribution in the $(3x + 1)$ problem, Moscow Math. Journal **3** (2003), No. 4, 1429–1440. (S. P. Novikov 65-th birthday issue).

[86] D. Soloveichik, M. Cook E. Winfree and J. Bruck, Computation with finite stochastic chemical reaction networks, Natural Computing **7** (2008), 615–633.

[87] Paul R. Stein, Iteration of maps, strange attractors, and number theory-an Ulamian potpourri, pp. 91–106 in [**21**].

[88] R. Terras, A stopping time problem on the positive integers, Acta Arithmetica **30** (1976), 241–252.

[89] R. Terras, On the existence of a density, Acta Arithmetica **35** (1979), 101–102.

[90] B. Thwaites, My conjecture, Bull. Inst. Math. Appl. **21** (1985), 35–41.

[91] A. M. Turing, On computable numbers, with an application to the Entschidungsproblem, Proc. London Math. Soc. **42** (1937), 230–265. Corrections, **43** (1937), 544–546.

[92] S. Ulam, *Problems in Modern Mathematics,* John Wiley and Sons: New York 1964.

[93] G. Venturini, On a generalization of the $3x+1$ problem, Adv. Appl. Math. **19** (1997), 295–305.

[94] Hao Wang, Tag systems and lag systems, Math. Annalen **152** (1963), 65–74.

[95] G. J. Wirsching, *The Dynamical System Generated by the $3n + 1$ Function,* Lecture Notes in Math. No. 1681, Springer-Verlag: Berlin 1998.

[96] S. Wolfram, Statistical mechanics of cellular automata, Rev. Mod. Phys. **55** (1983), 601-644.-

[97] S. Wolfram, Universality and complexity and cellular automata, Physica C **10** (1984), 1–35.

[98] S. Wolfram, *Theory and Applications of Cellular Automata,* World Scientific: Singapore 1987.

[99] S. Wolfram, *Cellular Automata and Complexity: Collected Papers,* Westview Press: Perseus Books Group 1994.

DEPARTMENT OF MATHEMATICS, UNIVERSITY OF MICHIGAN, ANN ARBOR, MI 48109-1109
*E-mail address*: `lagarias@umich.edu`

# The $3x + 1$ Problem and its Generalizations

Jeffrey C. Lagarias

## 1. Introduction

The $3x + 1$ problem, also known as the *Collatz problem, the Syracuse problem, Kakutani's problem, Hasse's algorithm*, and *Ulam's problem*, concerns the behavior of the iterates of the function which takes odd integers $n$ to $3n + 1$ and even integers $n$ to $n/2$. The $3x + 1$ Conjecture asserts that, starting from any *positive* integer $n$, repeated iteration of this function eventually produces the value 1.

The $3x+1$ Conjecture is simple to state and *apparently* intractably hard to solve. It shares these properties with other iteration problems, for example that of aliquot sequences (see Guy [**36**], Problem B6) and with celebrated Diophantine equations such as Fermat's last theorem. Paul Erdős commented concerning the intractability of the $3x + 1$ problem: "Mathematics is not yet ready for such problems." Despite this doleful pronouncement, study of the $3x + 1$ problem has not been without reward. It has interesting connections with the Diophantine approximation of $\log_2 3$ and the distribution ( mod 1) of the sequence $\{(3/2)^k : k = 1, 2, \ldots\}$, with questions of ergodic theory on the 2-adic integers $\mathbf{Z}_2$, and with computability theory — a generalization of the $3x + 1$ problem has been shown to be a computationally unsolvable problem. In this paper I describe the history of the $3x + 1$ problem and survey all the literature I am aware of about this problem and its generalizations.

The exact origin of the $3x + 1$ problem is obscure. It has circulated by word of mouth in the mathematical community for many years. The problem is traditionally credited to Lothar Collatz, at the University of Hamburg. In his student days in the 1930's, stimulated by the lectures of Edmund Landau, Oskar Perron, and Issai Schur, he became interested in number-theoretic functions. His interest in graph theory led him to the idea of representing such number-theoretic functions as directed graphs, and questions about the structure of such graphs are tied to the behavior of iterates of such functions [**25**]. In his notebook dated July 1, 1932, he considered the function

$$g(n) = \begin{cases} \dfrac{2}{3}n \,, & \text{if } n \equiv 0(\text{mod } 3) \,, \\[2ex] \dfrac{4}{3}n - \dfrac{1}{3} \,, & \text{if } n \equiv 1(\text{mod } 3) \,, \\[2ex] \dfrac{4}{3}n + \dfrac{1}{3} \,, & \text{if } n \equiv 2(\text{mod } 3) \,, \end{cases}$$

which gives rise to a permutation $P$ of the natural numbers

$$P = \begin{pmatrix} 1 & 2 & 3 & 4 & 5 & 6 & 7 & 8 & 9 \\ 1 & 3 & 2 & 5 & 7 & 4 & 9 & 11 & 6 & \dots \end{pmatrix}.$$

He posed the problem of determining the cycle structure of $P$, and asked in particular whether or not the cycle of this permutation containing 8 is finite or infinite, i.e., whether or not the iterates $g^{(k)}$ (8) remain bounded or are unbounded [24]. I will call the study of the iterates of $g(n)$ the *original Collatz problem*. Although Collatz never published any of his iteration problems, he circulated them at the International Congress of Mathematicians in 1950 in Cambridge, Massachusetts, and eventually the original Collatz problem appeared in print ([9], [47], [62]). His original question concerning $g^{(k)}$ (8) has never been answered; the cycle it belongs to is believed to be infinite. Whatever its exact origins, the $3x + 1$ problem was certainly known to the mathematical community by the early 1950's; it was discovered in 1952 by B. Thwaites [72].

During its travels the $3x + 1$ problem has been christened with a variety of names. Collatz's colleague H. Hasse was interested in the $3x + 1$ problem and discussed generalizations of it with many people, leading to the name *Hasse's algorithm* [40]. The name *Syracuse problem* was proposed by Hasse during a visit to Syracuse University in the 1950's. Around 1960, S. Kakutani heard the problem, became interested in it, and circulated it to a number of people. He said "For about a month everybody at Yale worked on it, with no result. A similar phenomenon happened when I mentioned it at the University of Chicago. A joke was made that this problem was part of a conspiracy to slow down mathematical research in the U.S. [45]." In this process it acquired the name *Kakutani's problem*. S. Ulam also heard the problem and circulated the problem at Los Alamos and elsewhere, and it is called *Ulam's problem* in some circles ([13], [72]).

In the last ten years the $3x + 1$ problem has forsaken its underground existence by appearing in various forms as a problem in books and journals, sometimes without attribution as an unsolved problem. Prizes have been offered for its solution: $50 by H. S. M. Coxeter in 1970, then $500 by Paul Erdös, and more recently £1000 by B. Thwaites [72]. Over twenty research articles have appeared on the $3x + 1$ problem and related problems.

In what follows I first discuss what is known about the $3x + 1$ problem itself, and then discuss generalizations of the problem. I have included or sketched proofs of Theorems B, D, E, F, M and N because these results are either new or have not appeared in as sharp a form previously; the casual reader may skip these proofs.

## 2. The 3x+1 problem.

The known results on the $3x + 1$ problem are most elegantly expressed in terms of iterations of the function

(2.1)
$$T(n) = \begin{cases} \dfrac{3n + 1}{2}, & \text{if } n \equiv 1 (\text{mod } 2), \\[2ex] \dfrac{n}{2}, & \text{if } n \equiv 0 (\text{mod} 2). \end{cases}$$

One way to think of the $3x + 1$ problem involves a directed graph whose vertices are the positive integers and that has directed edges from $n$ to $T(n)$. I call this graph the *Collatz graph* of $T(n)$ in honor of L .Collatz [**25**]. A portion of the Collatz graph of $T(n)$ is pictured in Figure 1.

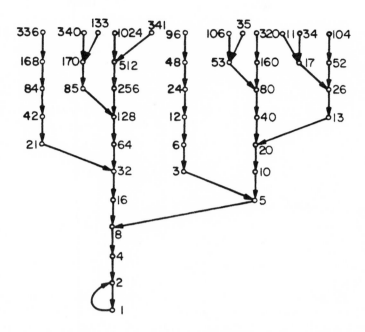

FIGURE 1. A portion of the Collatz graph of iterates of the function $T(n)$.

A directed graph is said to be *weakly connected* if it is connected when viewed as an undirected graph, i.e., for any two vertices there is a path of edges joining them, ignoring the directions on the edges. The $3x + 1$ Conjecture can be formulated in terms of the Collatz graph as follows.

3x+1 CONJECTURE (First form). *The Collatz graph of $T(n)$ on the positive integers is weakly connected.*

We call the sequence of iterates $(n, T(n), T^{(2)}(n), T^{(3)}(n), \ldots)$ the *trajectory* of $n$. There are three possible behaviors for such trajectories when $n > 0$.

(i) *Convergent trajectory.* Some $T^{(k)}(n) = 1$.
(ii) *Non-trivial cyclic trajectory.* The sequence $T^{(k)}(n)$ eventually becomes periodic and $T^{(k)}(n) \neq 1$ for any $k \geq 1$.
(iii) *Divergent trajectory.* $\lim_{k \to \infty} T^{(k)}(n) = \infty$.

The $3x + 1$ Conjecture asserts that all trajectories of positive $n$ are convergent. It is certainly true for $n > 1$ that $T^{(k)}(n) = 1$ cannot occur without some $T^{(k)}(n) < n$ occurring. Call the least positive $k$ for which $T^{(k)}(n) < n$ the *stopping time* $\sigma(n)$ of $n$, and set $\sigma(n) = \infty$ if no $k$ occurs with $T^{(k)}(n) < n$. Also call the least positive $k$ for which $T^{(k)}(n) = 1$ the *total stopping time* $\sigma_\infty(n)$ of $n$, and set $\sigma_\infty(n) = \infty$ if no such $k$ occurs. We may restate the $3x + 1$ Conjecture in terms of the stopping time as follows.

TABLE 1. Behavior of iterates $T^{(k)}(n)$.

| $n$ | $\sigma(n)$ | $\sigma_\infty(n)$ | $s(n)$ |
|---|---|---|---|
| 1 | $\infty$ | 2 | 2 |
| 7 | 7 | 11 | 3.7 |
| 27 | 59 | 70 | 171. |
| $2^{50} - 1$ | 143 | 383 | $6.37 \times 10^8$ |
| $2^{50}$ | 1 | 50 | 1 |
| $2^{50} + 1$ | 2 | 223 | 1.50 |
| $2^{500} - 1$ | 1828 | 4331 | $1.11 \times 10^{88}$ |
| $2^{500} + 1$ | 2 | 2204 | 1.50 |

3x+1 CONJECTURE (Second form). *Every integer $n \geq 2$ has a finite stopping time.*

The appeal of the $3x+1$ problem lies in the irregular behavior of the successive iterates $T^{(k)}(n)$. One can measure this behavior using the stopping time, the total stopping time, and the expansion factor $s(n)$ defined by

$$s(n) = \frac{\sup_{k \geq 0} T^{(k)}(n)}{n} \ ,$$

if $n$ has a bounded trajectory and $s(n) = +\infty$ if $n$ has a divergent trajectory. For example $n = 27$ requires 70 iterations to arrive at the value 1 and

$$s(27) = \frac{\sup_{k \geq 0} T^{(k)}(27)}{27} = \frac{4616}{27} \approx 171 \ .$$

Table 1 illustrates the concepts defined so far by giving data on the iterates $T^{(k)}(n)$ for selected values of $n$.

The $3x + 1$ Conjecture has been numerically checked for a large range of values of $n$. It is an interesting problem to find efficient algorithms to test the conjecture on a computer. The current record for verifying the $3x + 1$ Conjecture seems to be held by Nabuo Yoneda at the University of Tokyo, who has reportedly checked it for all $n < 2^{40} \approx 1.2 \times 10^{12}$ [2]. In several places the statement appears that A. S. Fraenkel has checked that all $n < 2^{50}$ have a finite total stopping time; this statement is erroneous [32].

**2.1. A heuristic argument.** The following heuristic probabilistic argument supports the $3x + 1$ Conjecture (see [28]). Pick an odd integer $n_0$ at random and iterate the function $T$ until another odd integer $n_1$ occurs. Then $\frac{1}{2}$ of the time $n_1 = (3n_0+1)/2$, $\frac{1}{4}$ of the time $n_1 = (3n_0+1)/4$, $\frac{1}{8}$ of the time $n_1 = (3n_0+1)/8$, and so on. If one supposes that the function $T$ is sufficiently "mixing" that successive odd integers in the trajectory of $n$ behave as though they were drawn at random (mod $2^k$) from the set of odd integers (mod $2^k$) for all $k$, then the expected growth in size between two consecutive odd integers in such a trajectory is the multiplicative factor

$$\left(\frac{3}{2}\right)^{1/2} \left(\frac{3}{4}\right)^{1/4} \left(\frac{3}{8}\right)^{1/8} \ldots = \frac{3}{4} < 1 \ .$$

Consequently this heuristic argument suggests that on average the iterates in a trajectory tend to shrink in size, so that divergent trajectories should not exist. Furthermore it suggests that the total stopping time $\sigma_\infty(n)$ is (in some average sense) a constant multiple of $\log n$.

From the viewpoint of this heuristic argument, the central difficulty of the $3x + 1$ problem lies in understanding in detail the "mixing" properties of iterates of the function $T(n) \pmod{2^k}$ for all powers of 2. The function $T(n)$ does indeed have some "mixing" properties given by Theorems B and K below; these are much weaker than what one needs to settle the $3x + 1$ Conjecture.

**2.2. Behavior of the stopping time function.** It is Riho Terras's ingenious observation that although the behavior of the total stopping time function seems hard to analyze, a great deal can be said about the stopping time function. He proved the following fundamental result ([**67**], [**68**]), also found independently by Everett [**31**].

THEOREM A (Terras). *The set of integers $S_k = \{n : n$ has stopping time $\leq k\}$ has limiting asymptotic density $F(k)$, i.e., the limit*

$$F(k) = \lim_{x \to \infty} \frac{1}{x} \#\{n : n \leq x \text{ and } \sigma(n) \leq k\}$$

*exists. In addition, $F(k) \to 1$ as $k \to \infty$, so that almost all integers have a finite stopping time.*

The ideas behind Terras's analysis seem basic to a deeper understanding of the $3x + 1$ problem, so I describe them in detail. In order to do this, I introduce some notation to describe the results of the process of iterating the function $T(n)$. Given an integer $n$, define a sequence of $0 - 1$ valued quantities $x_i(n)$ by

$$(2.2) \qquad T^{(i)}(n) \equiv x_i(n) \pmod{2}, \ 0 \leq i < \infty,$$

where $T^{(0)}(n) = n$. The results of first $k$ iterations of $T$ are completely described by the *parity vector*

$$(2.3) \qquad v_k(n) = (x_0(n), \ldots, x_{k-1}(n)),$$

since the result of $k$ iterations is

$$(2.4) \qquad T^{(k)}(n) = \lambda_k(n)n + \rho_k(n),$$

where

$$(2.5) \qquad \lambda_k(n) = \frac{3^{x_0(n)+\ldots+x_{k-1}(n)}}{2^k}$$

and

$$(2.6) \qquad \rho_k(n) = \sum_{i=0}^{k-1} x_i(n) \frac{3^{x_{i+1}(n)+\ldots+x_{k-1}(n)}}{2^{k-i}}.$$

Note that in (2.5), (2.6) both $\lambda_k$ and $\rho_k$ are completely determined by the parity vector $\mathbf{v} = \mathbf{v}_k(n)$ given by (2.3); I sometimes indicate this by writing $\lambda_k(\mathbf{v})$, $\rho_k(\mathbf{v})$ (instead of $\lambda_k(n)$, $\rho_k(n)$). The formula (2.4) shows that a necessary condition for $T^{(k)}(n) < n$ is that

$$(2.7) \qquad \lambda_k(n) < 1,$$

TABLE 2. Cycle structure and order of permutation $\bar{Q}_k$.

| $k$ | $Q_k$ | order |
|---|---|---|
| 1 | identity | 1 |
| 2 | identity | 1 |
| 3 | (1,5) | 2 |
| 4 | (1,5)(2,10)(9,13) | 2 |
| 5 | (1,21)(2,10)(4,20)(5,17)(7,23)(9,29,25,13)(18,26) | 4 |
| 6 | (1,21)(2,42)(3,35)(4.20)(5,17,37,49) | |
| | (7,23)(8,40)(9,29,25,13)(10,34) | |
| | (18,58,50,26)(19,51)(27,59)(33,53) | |
| | (36,52)(29,55)(41,61,57,45) | 4 |

since $\rho_k(n)$ is nonnegative. Terras [67] defines the *coefficient stopping time* $\omega(n)$ to be the least value of $k$ such that (2.7) holds, and $+\infty$ if no such value of $k$ exists. It is immediate that

$$(2.8) \qquad\qquad\qquad \omega(n) \leq \sigma(n) \ .$$

The function $\omega(n)$ plays an important role in the analysis of the behavior of the stopping time function $\sigma(n)$, see Theorem C.

The formula (2.2) expresses the parity vector $\mathbf{v} = \mathbf{v}_k(n)$ as a function of $n$. Terra's idea is to reverse this process and express $n$ as a function of $\mathbf{v}$.

THEOREM B. *The function* $Q_k : \mathbf{Z} \to \mathbf{Z}/2^k\mathbf{Z}$ *defined by*

$$Q_k(n) = \sum_{i=0}^{k-1} x_i(n)2^i$$

*is periodic with period* $2^k$. *The induced function* $\bar{Q}_k : \mathbf{Z}/2^k\mathbf{Z} \to \mathbf{Z}/2^2\mathbf{Z}$ *is a permutation, and its order is a power of 2.*

*Proof* (sketch). The theorem is established by induction on $k$, using the inductive hypotheses:

(1) $x_i(n)$ is periodic with period $2^{i+1}$ for $0 \leq i \leq k-1$. In fact

$$(2.9) \qquad\qquad\qquad x_i(n+2^i) \equiv x_i(n) + 1 \ (\mathrm{mod}\ 2)$$

for $0 \leq i \leq k-1$.

(2) $Q_k(n)$ is periodic with period $2^k$.

(3) $\lambda_k(n)$ and $\rho_k(n)$ are periodic with period $2^k$.

(4) $\bar{Q}_k$ is a permutation whose order divides $2^k$. Also

$$(2.10) \qquad\qquad \bar{Q}_k(n+2^{k-1}) \equiv \bar{Q}_k(n) + 2^{k-1} \ (\mathrm{mod}\ 2^k) \ .$$

I omit the details. $\square$

The cycle structure and order of the first few permutations $\bar{Q}_k$ are given in Table 2. (One-cycles are omitted.) It is interesting to observe that the order of the permutation $\bar{Q}_k$ seems to be much smaller than the upper bound $2^k$ proved in Theorem B. Is there some explanation of this phenomenon?

Theorem B allows one to associate with each vector $\mathbf{v} = (v_0, \ldots, v_{k-1}) \in (\mathbf{Z}/2\mathbf{Z})^k$ of length $k$ a unique congruence class $S(\mathbf{v})$ (mod $2^k$) given by

$$S(\mathbf{v}) = \{n : \mathbf{v} = (x_0(n), \ldots, x_{k-1}(n))\} .$$

The integer

$$n_0(\mathbf{v}) \equiv (\bar{Q}_k)^{-1} \left( \sum_{i=0}^{k-1} v_i 2^i \right) \pmod{2^k}$$

with $0 \leq n_0(\mathbf{v}) < 2^k$ is the minimal element in $S(\mathbf{v})$ and $S(\mathbf{v})$ is the arithmetic progression:

$$S(\mathbf{v}) = \{n_0(\mathbf{v}) + 2^k i : 0 \leq i < \infty\} .$$

Now I consider the relation between a vector $\mathbf{v}$ and stopping times for integers $n \in S(\mathbf{v})$. Define a vector $\mathbf{v} = (v_0, v_1, \ldots, v_{k-1})$ of length $k$ to be *admissible* if

(1) $(v_0 + \ldots + v_{k-1}) \ln 3 < k \ln 2$,
(2) $(v_0 + \ldots + v_i) \ln 3 > (i+1) \ln 2$, when $0 \leq i \leq k-2$.

Note that all admissible vectors $\mathbf{v}$ of length $k$ have

(2.11) $$v_0 + \ldots + v_{k-1} = [k\theta] ,$$

where $\theta = \ln 2 / \ln 3 = (\log_2 3)^{-1} \approx .63093$ and $[x]$ denotes the largest integer $\leq x$. The following result is due to Terras.

THEOREM C. (Terras). (a) *The set of integers with coefficient stopping time $k$ are exactly the set of integers in those congruence classes $n$ (mod $2^k$) for which there is an admissible vector $\mathbf{v}$ of length $k$ with $n = n_0(\mathbf{v})$.*

(b) *Let $n = n_0(\mathbf{v})$ for some vector $\mathbf{v}$ of length $k$. If $\mathbf{v}$ is admissible, then all sufficiently large integers congruent to $n$ (mod $2^k$) have stopping time $k$. If $\mathbf{v}$ is not admissible, then only finitely many integers congruent to $n$ (mod $2^k$) have stopping time $k$.*

*Proof.* The assertions made in (a) about coefficient stopping times follow from the definition of admissibility, because that definition asserts that

(i) $\lambda_k(\mathbf{v}) < 1$,
(ii) $\lambda_i(\mathbf{v}) > 1$ for $1 \leq i \leq k-1$.

To prove (b), first note that if $\mathbf{v}$ is admissible of length $k$, then

$$T^{(i)}(n) \geq \frac{3^{v_0 + \ldots + v_{i-1}}}{2^i} n \geq n \text{ for } 1 \leq i \leq k-1 ,$$

and so all elements of $S(\mathbf{v})$ have stopping time at least $k$. Now define $\epsilon_k > 0$ by

(2.12) $$\epsilon_k = 1 - \frac{3^{[k\theta]}}{2^k} ,$$

where $\theta = (\log_2 3)^{-1}$, and note that (2.11) implies that

$$\epsilon_k = 1 - \lambda_k(\mathbf{v}) = 1 - \frac{3^{v_0 + \ldots + v_{k-1}}}{2^k}$$

for all *admissible* $\mathbf{v}$. Now for $n \in S(\mathbf{v})$ for an admissible $\mathbf{v}$, (2.4) may be rewritten as

(2.13) $$T^{(k)}(n) = n + (\rho_k(\mathbf{v}) - \epsilon_k n) .$$

Hence when $\mathbf{v}$ is admissible, those $n$ in $S(\mathbf{v})$ with

(2.14) $$n > \epsilon_k^{-1} \rho_k(\mathbf{v})$$

have stopping time $k$, and $\omega(n) = \sigma(n) = k$ in this case.

Now suppose $\mathbf{v}$ is not admissible. There are two cases, depending on whether or not some initial segment $(v_0, \ldots, v_i)$ of $\mathbf{v}$ is admissible. No initial segment of $\mathbf{v}$ is admissible if and only if

$$(2.15) \qquad (v_0 + \ldots + v_{i-1}) \log 3 > i \log 2 \text{ for } 1 \le i \le k - 1 \;,$$

and when (2.15) holds say that $\mathbf{v}$ is *inflating*. If $\mathbf{v}$ is inflating, $\lambda_k(\mathbf{v}) > 1$ so that $T^{(k)}(n) \ge n$ for all $n$ in $S(\mathbf{v})$ by (2.4), so that no elements of $S(\mathbf{v})$ have stopping time $k$ or less. In the remaining case $\mathbf{v}$ has an initial segment $\mathbf{w} = (v_0, v_1, \ldots, v_i)$ with $i < k - 1$ which is admissible. Now $S(\mathbf{v}) \subseteq S(\mathbf{w})$ and all sufficiently large elements of $S(\mathbf{w})$ have stopping time $i + 1 < k$ by the argument just given. $\quad\square$

Theorem C asserts that the set of integers $I_k$ with a given coefficient stopping time $k$ is a set of arithmetic progressions (mod $2^k$), which has the immediate consequence that $I_k$ has the asymptotic density

$$d(I_k) = \lim_{x \to \infty} \frac{1}{x} \#\{n : n \le x \text{ and } n \in I_k\}$$

which is given by

$$d(I_k) = \frac{1}{2^k} \#\{\mathbf{v} : \mathbf{v} \text{ is admissible and of length } k\} \;.$$

Furthermore Theorem C asserts that the set

$$S_k = \{n : n \text{ has stopping time } k \}$$

differs from $I_k$ by a finite set, so that $S_k$ also has an asymptotic density which is the same as that of $I_k$. Consequently, Theorem C implies the first part of Theorem A, that the set of all integers with stopping time at most $k$ have an asymptotic density $F(k)$ given by

$$(2.16) \qquad F(k) = \sum_{\substack{\mathbf{v} \text{ admissible} \\ \text{length } (\mathbf{v}) \le k}} \text{weight } (\mathbf{v}) \;,$$

where

$$\text{weight } (\mathbf{v}) = 2^{-\text{ length } (\mathbf{v})} \;.$$

Now the formula (2.16) can be used to prove the second part of Theorem A, and in fact to prove the stronger result that $F(k)$ approaches 1 at an exponential rate as $k \to \infty$.

THEOREM D. *For all $k \ge 1$,*

$$(2.17) \qquad 1 - F(k) = \lim_{x \to \infty} \frac{1}{x} \#\{n : n \le x \text{ and } \sigma(n) > k\} \le 2^{-\eta k} \;,$$

*where*

$$(2.18) \qquad \eta = 1 - H(\theta) \approx .05004\ldots \;.$$

*Here $H(x) = -x \log_2 x - (1 - x) \log_2(1 - x)$ is the entropy function and $\theta = (\log_2 3)^{-1}$.*

*Proof.* Let $C = C_1 \cup C_2$, where

$$C_1 = \{\mathbf{v} : \mathbf{v} \text{ is admissible and length } (\mathbf{v}) \le k\}$$

and

$$C_2 = \{\mathbf{v} : \mathbf{v} \text{ is inflating and length } (\mathbf{v}) = k\} \;.$$

Then $C$ has the property that for any binary word $\mathbf{w}$ of length $k$ there is a unique $\mathbf{v} \in C$ with $\mathbf{v}$ a prefix of $\mathbf{w}$. Now for any $\mathbf{v}$ with length $(\mathbf{v}) \leq k$

$$\text{weight } (\mathbf{v}) = \sum \text{weight}(\mathbf{w}) ,$$

where the sum is over all $\mathbf{w}$ of length $k$ for which $\mathbf{v}$ is a prefix of $\mathbf{w}$. Hence

$$\sum_{\mathbf{v} \in C} \text{weight}(\mathbf{v}) = \sum_{\text{length}(\mathbf{w})=k} \text{weight}(\mathbf{w}) = 1 .$$

From (2.16) this implies that

$$\sum_{\mathbf{v} \in C_2} \text{weight}(\mathbf{v}) = 2^{-k}|C_2| = 1 - F(k) ,$$

where $|C_2|$ denotes the number of vectors in $C_2$. The already proved first part of Theorem A shows that

$$1 - F(k) = \lim_{x \to \infty} \frac{1}{x} \#\{n : n \leq x \text{ and } \sigma(n) > k\} ,$$

so that to prove (2.17) it suffices to bound $|C_2|$ from above.

Now the definition (2.15) of an inflating vector implies that

$$C_2 \subseteq \left\{ \mathbf{v} : \sum_{i=0}^{k-1} v_i > k\theta \right\} ,$$

so that

(2.19) $$|C_2| \leq \sum_{j > k\theta} \binom{k}{j} .$$

The right side of (2.19) is just the tail of the binomial distribution. It is easily checked using Stirling's formula that for any constant $\alpha > \frac{1}{2}$ and any $\epsilon > 0$ the bound

$$\sum_{j > k\alpha} \binom{k}{j} \leq k \binom{k}{[ka]} \leq 2^{(H(\alpha)+\epsilon)k}$$

holds for all sufficiently large $k$. With more work one can obtain the more precise estimate (Ash [8], Lemma 4.7.2) that for any $\alpha > \frac{1}{2}$

$$\sum_{j > k\alpha} \binom{k}{j} \leq 2^{H(\alpha)k} ,$$

which used in (2.19) implies (2.17). $\square$

Theorem D cannot be substantially improved; it can be proved that for any $\epsilon > 0$ we have

$$|C_2| \geq 2^{(H(\theta)-\epsilon)k}$$

for all sufficiently large $k$ depending on $\epsilon$. Hence for any $\epsilon > 0$

$$1 - F(k) \geq 2^{-(\eta+\epsilon)k}$$

holds for all sufficiently large $k$ depending on $\epsilon$.

**2.3. What is the relation between the coefficient stopping time and the stopping time?** Theorem C shows that generally they are equal: For any fixed $k$ at most a finite number of those $n$ having coefficient stopping time $\omega(n) \leq k$ have $\sigma(n) \neq \omega(n)$. Terras [67] and later Garner [34] conjecture that this never occurs.

COEFFICIENT STOPPING TIME CONJECTURE. *For all $n \geq 2$, the stopping time $\sigma(n)$ equals the coefficient stopping time $\omega(n)$.*

The Coefficient Stopping Time Conjecture has the aesthetic appeal that if it is true, then the set of positive integers with stopping time $k$ is exactly a collection of congruence classes (mod $2^k$), as described by part (i) of Theorem C. Furthermore, the truth of the Coefficient Stopping Time Conjecture implies that there are no nontrivial cycles. To see this, suppose that there were a nontrivial cycle of period $k$ and let $n_0$ be its smallest element, and note that $\sigma(n_0) = \infty$. Then $T^{(i)}(n_0) > n_0$ for $1 \leq i \leq k - 1$ and

$$(2.20) \qquad T^{(k)}(n_0) = \lambda_k(n_0)n_0 + \rho_k(n_0) = n_0 .$$

Now $\rho_k(n_0) \neq 0$ since $n_0$ isn't a power of 2, so that (2.20) implies that $\lambda_k(n_0) < 1$. Hence $\omega(n_0) \leq k$, so that $\omega(n_0) \neq \sigma(n_0)$.

The following result shows that the Coefficient Stopping Time Conjecture is "nearly true." I will use it later to bound the number of elements not having a finite stopping time.

THEOREM E. *There is an effectively computable constant $k_0$ such that if $\mathbf{v}$ is admissible of length $k \geq k_0$, then all elements of $S(\mathbf{v})$ have stopping time $k$ except possibly the smallest element $n_0(\mathbf{v})$ of $S$.*

*Proof* (sketch). The results of A. Baker and N. I. Feldman on linear forms in logarithms of algebraic numbers ([10], Theorem 3.1) imply that there is an effectively computable absolute constant $c_0 > 0$ such that for all $k, l \geq 1$,

$$|k \log 2 - l \log 3| \geq k^{-c_0} .$$

Consequently there is an effectively computable absolute constant $c_1$ such that for $k, l \geq c_1$ one has

$$|2^k - 3^l| \geq \frac{1}{2}2^k k^{-c_1} ,$$

and (2.12) then yields

$$\epsilon_k \geq k^{-c_1} .$$

Since $\mathbf{v}$ is admissible, $v_0 + \ldots + v_{k-1} \leq \theta k$, where $\theta = (\log_2 3)^{-1}$ by (2.11). Therefore

$$\rho_k(\mathbf{v}) \;=\; \sum_{i=0}^{k-1} v_i \frac{3^{v_{i+1}} + \ldots + v_{k-1}}{2^{k-i}} \leq \left( \sum_{i=0}^{[k\theta]} \frac{3^i}{2^{i+1}} \right) + k(1 - \theta)\left(\frac{3}{2}\right)^{\theta k}$$

$$\leq \;\; k2^{(1-\theta)k} .$$

But all elements of $S(\mathbf{v})$ except $n_0(\mathbf{v})$ exceed $2^k$ and

$$2^k > k^{c_1+1}2^{(l-\theta)k} > \epsilon_k^{-1}\rho_k(\mathbf{v})$$

for all sufficiently large $k$, so the theorem follows by (2.14). $\square$

**2.4. How many elements don't have a finite stopping time?** The results proved so far can be used to obtain an upper bound for the number of elements not having a finite stopping time. Let

$$\pi^*(x) = |\{n : n \le x \text{ and } \sigma(n) < \infty\}| .$$

The following result is the sharpest known result concerning the size of the "exceptional" set of $n$ with $\sigma(n) = \infty$.

THEOREM F. *There is a positive constant $c_1$ such that*

(2.21)                                $$|\pi^*(x) - x| \le c_1 x^{1-\eta} ,$$

*where $\eta \approx .05004\ldots$ is the constant defined in Theorem D.*

*Proof.* Suppose $2^{k-1} \le x \le 2^k$. Then

$$|\{n : n \le x \text{ and } \sigma(n) = \infty\}| = |\pi^*(x) - x| \le S_1 + S_2 ,$$

where $S_1 = \#\{n \le 2^k : \omega(n) \ge k + 1\}$ and $S_2 = \#\{n \le 2^k : \omega(n) \le k \text{ and } \omega(n) \ne \sigma(n)\}$. Now Theorem D shows that

(2.22)                                $$S_1 \le c_2 (2^k)^{1-\eta} \le 2 c_2 x^{1-\eta} ,$$

and Theorem E shows that

$$S_2 \le \#\{\mathbf{v} : \mathbf{v} \text{ admissible and length } (\mathbf{v}) \le k\} + c_3 ,$$

where $c_3 = \#\{n : \omega(n) \le k_0 \text{ and } \omega(n) \ne \sigma(n)\}$ is a constant by Theorem C. Now

$$\#\{\mathbf{v} : \mathbf{v} \text{ admissible and length } (\mathbf{v}) = i\} \le \#\{\mathbf{v} : v_0 + \ldots + v_{i-1} = [i\theta]\}$$

$$= \binom{i}{[i\theta]} \le c_4 2^{(1-n)i}$$

using the binomial theorem and Stirling's formula. Hence

$$S_2 \le c_5 2^{(l-\eta)k} + c_3 \le (2c_5 + c_3) x^{1-\eta} .$$

Then this inequality and (2.22) imply (2.21) with $c_1 = 2c_2 + c_3 + 2c_5$.  □

**2.5. Behavior of the total stopping time function.** Much less is known about the total stopping time function than about the stopping time function. One phenomenon immediately observable from a table of the total stopping times of small integers is the occurrence of many pairs and triples of integers having the same finite total stopping time. From Figure 1 we see that $\sigma_\infty(20) = \sigma_\infty(21) = 6$, $\sigma_\infty(12) = \sigma_\infty(13) = 7$, $\sigma_\infty(84) = \sigma_\infty(85) = 8$, $\sigma_\infty(52) = \sigma_\infty(53) = 9$, and $\sigma_\infty(340) = \sigma_\infty(341) = 10$. Indeed for larger values of $n$, multiple consecutive values occur with the same total stopping time. For example there are 17 consecutive values of $n$ with $\sigma_\infty(n) = 40$ for $7083 \le n \le 7099$. A related phenomenon is that over short ranges of $n$ the function $\sigma_\infty(n)$ tends to assume only a few values (C. W. Dodge [70]). As an example the values of $\sigma_\infty(n)$ for $1000 \le n \le 1099$ are given in Table 3. Only 19 values for $\sigma_\infty(n)$ are observed, for which a frequency count is given in Table 4. Both of these phenomena have a simple explanation; they are caused by coalescence of trajectories of different $n$'s after a few steps. For example the trajectories of $8k + 4$ and $8k + 5$ coalesce after 3 steps, for all $k \ge 0$. More generally, the large number of coalescences of numbers $n_1$ and $n_2$ close together in size can be traced to the trivial cycle $(1,2)$, as follows. Suppose $n_1$ and $n_2$ have $\sigma_\infty(n_1) \equiv \sigma_\infty(n_2) \pmod{2}$, and let $\sigma_\infty(n_1) = r_1 \ge \sigma_\infty(n_2) = r_2$. Then the trajectories of $n_1$ and $n_2$ coalesce after at most $r_1 - 1$ iterations, since $T^{(r_1 - 1)}(n_1) =$

JEFFREY C. LAGARIAS

TABLE 3. Values of the total stopping time $\sigma_\infty(n)$ for $1000 \le n \le 1099$.

|   | 1000 −1009 | 1010 −1019 | 1020 −1029 | 1030 −1039 | 1040 −1049 | 1050 −1059 | 1060 −1069 | 1070 −1079 | 1080 −1089 | 1090 −1099 |
|---|---|---|---|---|---|---|---|---|---|---|
| 0 | 72 | 42 | 34 | 80 | 23 | 23 | 80 | 18 | 31 | 31 |
| 1 | 91 | 42 | 34 | 26 | 80 | 61 | 80 | 107 | 88 | 31 |
| 2 | 72 | 72 | 42 | 80 | 80 | 53 | 80 | 23 | 31 | 23 |
| 3 | 29 | 72 | 42 | 99 | 80 | 53 | 50 | 18 | 88 | 23 |
| 4 | 45 | 26 | 10 | 80 | 23 | 53 | 23 | 18 | 31 | 23 |
| 5 | 45 | 26 | 26 | 80 | 23 | 107 | 50 | 18 | 31 | 50 |
| 6 | 45 | 34 | 26 | 80 | 80 | 23 | 23 | 23 | 88 | 61 |
| 7 | 61 | 99 | 26 | 80 | 80 | 53 | 42 | 23 | 88 | 88 |
| 8 | 72 | 34 | 80 | 42 | 23 | 23 | 18 | 34 | 15 | 61 |
| 9 | 72 | 42 | 80 | 42 | 42 | 23 | 18 | 34 | 31 | 23 |

TABLE 4. Values of $\sigma_\infty(n)$ and their frequencies for $1000 \le n \le 1099$.

| $\sigma_\infty(n)$ | freq. | $\sigma_\infty(n)$ | freq. | $\sigma_\infty(n)$ | freq. | $\sigma_\infty(n)$ | freq. |
|---|---|---|---|---|---|---|---|
| 10 | 1 | 29 | 1 | 50 | 3 | 88 | 5 |
| 15 | 1 | 31 | 7 | 53 | 4 | 91 | 1 |
| 18 | 6 | 34 | 6 | 61 | 4 | 99 | 2 |
| 23 | 17 | 42 | 9 | 72 | 6 | 107 | 2 |
| 26 | 6 | 45 | 3 | 80 | 16 | | |

$T^{(r_1-1)}(n_2) = 2$, since the trajectory of $n_2$ continues to cycle around the trivial cycle. If in addition $\lambda_{r_1-1}(n_1) = \lambda_{r_1-1}(n_2)$, which nearly always happens if $n_1$ and $n_2$ are about the same size, then the trajectories of $2^{r_1-1}k + n_1$, and $2^{r_1-1}k + n_2$ coalesce after at most $r_1 - 1$ iterations, for $k \ge 0$. In particular, $\sigma_\infty(2^{r_1-1}k+n_1) = \sigma_\infty(2^{r_1-1}k+n_2)$ then holds for $k \ge 1$. In this case the original coalescence of $n_1$ and $n_2$ has produced an infinite arithmetic progression (mod $2^{r_1-1}$) of coalescences. The gradual accumulation of all these arithmetic progressions of coalescences of numbers close together in size leads to the phenomena observed in Tables 3 and 4.

Although the $3x + 1$ Conjecture asserts that all integers $n$ have a finite total stopping time, the strongest result proved so far concerning the density of the set of integers with a finite total stopping time is much weaker.

THEOREM G. (Crandall). *Let*

$$\pi_{\text{total}}(x) = |\{n : n \le x \text{ and } \sigma_\infty(n) < \infty\}| \ .$$

*Then there is a positive constant $c_4$ such that*

$$\pi_{\text{total}}(x) > x^{c_4}$$

*for all sufficiently large $x$.*

Assuming that the $3x + 1$ Conjecture is true, one can consider the problem of determining the *expected size* of the total stopping time function $\sigma_\infty(n)$. Crandall [28] and Shanks [63] were guided by probabilistic heuristic arguments (like the one described earlier) to conjecture that the average order of $\sigma_\infty(n)$ should be a constant times $\ln n$; more precisely, that

$$\frac{1}{x} \sum_{n=1}^{X} \sigma_\infty(n) \sim 2 \left( \ln \frac{4}{3} \right)^{-1} \ln x \ .$$

A modest amount of empirical evidence supports these conjectures, see [**28**].

**2.6. Are there non-trivial cycles?** A first observation is that there are other cycles if negative integers are allowed in the domain of the function. There is a cycle of period 1 starting from $n = -1$, and there are cycles of length 3 and 11 starting from $n = -5$ and $n = -17$, respectively. Böhm and Sontacchi [**13**] conjecture that these cycles together with the cycles starting with $n = 0$ and $n = 1$ make up the entire set of cycles occurring under iteration of $T(n)$ applied to the integers $\mathbf{Z}$. Several authors have proposed the following conjecture ([**13**], [**28**], [**41**], [**67**]).

FINITE CYCLES CONJECTURE. *There are only a finite number of distinct cycles for the function $T(n)$ iterated on the domain $\mathbf{Z}$.*

One can easily show that for any given length $k$ there are only a finite number of integers $n$ that are periodic under iteration by $T$ with period $k$, in fact at most $2^k$ such integers, as observed by Böhm and Sontacchi [**13**]. To see this, substitute the equation (2.4) into

$$(2.23) \qquad\qquad T^{(k)}(n) = n \ , \ n \in \mathbf{Z}$$

to obtain the equation

$$(2.24) \qquad \left( 1 - \frac{3^{x_0 + \ldots + x_{k-1}}}{2^k} \right) n = \frac{3^{x_0 + \ldots + x_{k-1}}}{2^k} \sum_{i=0}^{k-1} x_i \frac{2^i}{3^{x_0 + \ldots + x_i}} \ .$$

There are only $2^k$ choices for the $0 - 1$ vector $\mathbf{v} = (x_0, \ldots, x_{k-1})$, and for each choice of $\mathbf{v}$ the equation (2.24) determines a unique rational solution $n = n(\mathbf{v})$. Consequently there are at most $2^k$ solutions to (2.23). Böhm and Sontacchi also noted that this gives an (inefficient) finite procedure for deciding if there are any cycles of a given length $k$, as follows: Determine the rational number $n(\mathbf{v})$ for each of the $2^k$ vectors $\mathbf{v}$, and for each $n(\mathbf{v})$ which is an integer test if (2.23) holds.

The argument of Böhm and Sontacchi is a very general one that makes use only of the fact that the necessary condition (2.24) for a cycle has a unique solution when the values $x_i$ are fixed. In fact, considerably more can be proved about the nonexistence of nontrivial cyclic trajectories using special features of the necessary condition (2.24). For example, several authors have independently found a much more efficient computational procedure for proving the nonexistence of nontrivial cyclic trajectories of period $\leq k$; it essentially makes use of the inequality

$$(1 - \lambda_k(\mathbf{v}))n \leq \rho_k(\mathbf{v}) \ ,$$

which must hold for $\mathbf{v} = (x_0, x_1, \ldots, x_k)$ satisfying (2.24). This approach also allows one to check the truth of the Coefficient Stopping Time Conjecture for all $n$ with $\omega(n) \leq k$. The basic result is as follows.

THEOREM H. (Terras). *For each $k$ there is a finite bound $M(k)$ given by*

$$(2.25) \qquad M(k) = \max\{\epsilon_i^{-1} \rho_i(\mathbf{v}) : \mathbf{v} \ admissible, \ length \ (\mathbf{v}) = i \leq k\}$$

*such that $\omega(n) \leq k$ implies that $\omega(n) = \sigma(n)$ whenever $n \geq M(k)$. Consequently:*

(i) *If $\sigma(n) < \infty$ for all $n \leq M(k)$, then there are no non-trivial cycles of length $\leq k$.*

(ii) *If $\omega(n) = \sigma(n)$ for all $n \leq M(k)$, then $\omega(n) \leq k$ implies $\omega(n) = \sigma(n)$.*

*Proof.* The existence of the bound $M(k)$ follows immediately from (2.14), and (ii) follows immediately from this fact.

To prove (i), suppose a nontrivial cycle of length $\leq k$ exists. We observed earlier that if $n_0$ is the smallest element in a purely periodic nontrivial cycle of length $\leq k$, then $\omega(n_0) = i \leq k$ and $\sigma(n_0) = \infty$. The first part of the theorem then implies that $n_0 \leq M(k)$. This contradicts the hypothesis of (i).  $\square$

Theorem H can be used to show the nonexistence of nontrivial cycles of small period by obtaining upper bounds for the $M(k)$ and checking that condition (i) holds. This approach has been taken by Crandall [28], Garner [34], Schuppar [61] and Terras [67]. In estimating $M(k)$, one can show that the quantities $\rho_i(\mathbf{v})$ are never very large, so that the size of $M(k)$ is essentially determined by how large

$$\epsilon_i^{-1} = \left(1 - \frac{3^{[i\theta]}}{2^i}\right)^{-1}$$

can get. The worst cases occur when $3^{[i\theta]}$ is a very close approximation to $2^i$, i.e., when $i/[i\theta]$ is a very good rational approximation to $\phi = \log_2 3$. The best rational approximations to $\phi$ are given by the convergents $p_k/q_k$ of the continued fraction expansion of $\phi = [1; 1, 1, 2, 2, 3, 1, 5, 2, 23, 2, 2, 1, 1, 55, 1, 4, 3 \ldots]$. Crandall [28] uses general properties of continued fraction convergents to obtain the following quantitative result.

THEOREM I. (Crandall). *Let $n_0$ be the minimal element of a purely periodic trajectory of period $k$. Then*

(2.26) $$k > \frac{3}{2} \min\left(q_j, \frac{2n_0}{q_j + q_{j+1}}\right),$$

*where $p_i/q_j$ is any convergent of the continued fraction expansion of $\log_2 3$ with $j \geq 4$.*

As an application, use Yoneda's bound [2] that $n_0 > 2^{40}$ and choose $j = 13$ in (2.26), noting that $q_{13} = 190737$ and $q_{14} = 10590737$, to conclude that *there are no nontrivial cycles with period length less than 275,000.*

Further information about the nonexistence of nontrivial cyclic trajectories can be obtained by treating the necessary condition (2.24) as an *nonexponential Diophantine equation*. Davidson [29] calls a purely periodic trajectory of period $k$ a *circuit* if there is a value $i$ for which

$$n_0 < T(n_0) < \cdots < T^{(i)}(n_0)$$

and

$$T^{(i)}(n_0) > T^{(i+1)}(n_0) > \cdots > T^{(k)}(n_0) = n_0 ,$$

i.e., the parity vector $\mathbf{v}_k(n_0) = (x_0(n_0), \ldots, x_{k-1}(n_0))$ has the special form

$$x_j(n_0) = \begin{cases} 1 , & \text{when } 0 \leq j \leq [k\theta] - 1 , \\ \\ 0 , & \text{when } [k\theta] \leq j \leq k - 1 , \end{cases}$$

where $\theta = (\log_2 3)^{-1}$. The cycle starting with $n_0 = 1$ is a circuit. Davidson observed that each solution to the exponential Diophantine equation

$$(2.27) \qquad\qquad (2^{a+b} - 3^b)h = 2^a - 1 \ , \ a \geq 1$$

gives rise to a circuit of length $k = a + b$ with $[k\theta] = b$ and $n_0 = 2^b h - 1$, and conversely. (The equation (2.27) is the necessary condition (2.24) specialized to the vector (2.27). R. Steiner [64] showed that $(a, b, h) = 1, 1, 1$ is the only solution of (2.27), thus proving the following result.

THEOREM J. (Steiner). *The only cycle that is a circuit is the trivial cycle.*

*Proof* (sketch). Steiner's method is to show first that any solution of (2.27) with $a \geq 4$ has the property that $(a + b)/b$ is a convergent in the continued fraction expansion of $\log_2 3$ since (2.27) implies that

$$(2.28) \qquad\qquad 0 < \left| \frac{a+b}{b} - \log_2 3 \right| \leq \frac{1}{b \ln 2 (2^b - 1)} \ .$$

He checks that this rational approximation $(a + b)/b$ is so good that it violates the effective estimates of A. Baker [[10], p. 45] for linear forms in logarithms of algebraic numbers if $b > 10^{199}$. Finally he checks that (2.28) fails to hold for all that $b < 10^{199}$ by computing the convergents of the continued fractions of $\log_2 3$ up to $10^{199}$. $\square$

The most remarkable thing about Theorem J is the weakness of its conclusion compared to the strength of the methods used in its proof. The proof of Theorem J does have the merit that it shows that the coefficient Stopping Time Conjecture holds for the *infinite set* of admissible vectors **v** of the form (2.27).

**2.7. Do divergent trajectories exist?** Several authors have observed that heuristic probabilistic arguments suggest that no divergent trajectories occur.

DIVERGENT TRAJECTORIES CONJECTURE. *The function $T : \mathbf{Z} \to \mathbf{Z}$ has no divergent trajectories, i.e., there exists no integer $n_0$ for which*

$$(2.29) \qquad\qquad \lim_{k \to \infty} |T^{(k)}(n_0)| = \infty \ .$$

If a divergent trajectory $\{T^{(k)}(n_0) : 0 \leq k < \infty\}$ exists, it cannot be equidistributed (mod 2). Indeed if one defines

$$N^*(k) = |\{j : j \leq k \text{ and } T^{(j)}(n_0) \equiv 1 \pmod{2}\}| \ ,$$

then it can be proved that the condition (2.29) implies that

$$(2.30) \qquad\qquad \liminf_{k \to \infty} \frac{N^*(k)}{k} \geq (\log_2 3)^{-1} \approx .63097 \ .$$

Theorem F constrains the possible behavior of divergent trajectories. Indeed, associated to any divergent trajectory $D = \{T^{(k)}(n_0) : k \geq 1\}$ is the infinite set $U_D = \{n : n \in D \text{ and } T^{(k)}(n) > n \text{ for all } k \geq 1\}$. Since $\sigma(n) = \infty$ for all $n \in U_D$, Theorem F implies that

$$(2.31) \qquad\qquad |\{n \in U_D : n \leq x\}| \leq c_1 x^{1-\eta} \ ,$$

where $\eta \approx .05004$. Roughly speaking, (2.31) asserts that the elements of a divergent trajectory cannot go to infinity "too slowly."

**2.8. Connections of the $3x+1$ problem to ergodic theory.** The study of the general behavior of the iterates of measure preserving functions on a measure space is called *ergodic theory*. The $3x+1$ problem has some interesting connections to ergodic theory, because the function $T$ extends to a measure-preserving function on the 2-adic integers $\mathbf{Z}_2$ defined with respect to the 2-adic measure. To explain this, I need some basic facts about the 2-adic integers $\mathbf{Z}_2$, cf. [**14**], [**49**]. The 2-adic integers $\mathbf{Z}_2$ consist of all series

$$\alpha = a_0 + a_1 2 + a_2 2^2 + \cdots \ , \quad \text{all } a_i = 0 \text{ or } 1 \ ,$$

where the $\{a_i : 0 \le i < \infty\}$ are called the 2-*adic digits* of $\alpha$. One can define congruences (mod $2^k$) on $\mathbf{Z}_2$ by $\alpha \equiv \beta$ (mod $2^k$) if the first $k$ 2-adic digits of $\alpha$ and $\beta$ agree. Addition and multiplication on $\mathbf{Z}_2$ are given by

$$X \ = \ \alpha + \beta \Leftrightarrow X \text{ (mod } 2^k) \equiv \alpha \text{ (mod } 2^k) + \beta(\text{mod } 2^k) \text{ for all } k \ ,$$

$$X \ = \ \alpha\beta \Leftrightarrow X \text{ (mod } 2^k) \equiv \alpha \text{ (mod } 2^k) \cdot \beta(\text{mod } 2^k) \text{ for all } k \ .$$

The 2-*adic valuation* $||_2$ on $\mathbf{Z}_2$ is given by $|0|_2 = 0$ and for $\alpha \ne 0$ by $|\alpha|_2 = 2^{-k}$, where $a_k$ is the first nonzero 2-adic digit of $\alpha$. The valuation $||_2$ induces a metric $d$ on $\mathbf{Z}_2$ defined by

$$d(\alpha, \beta) = |\alpha - \beta|_2 \ .$$

As a topological space $\mathbf{Z}_2$ is compact and complete with respect to the metric $d$; a basis of open sets for this topology is given by the 2-*adic discs of radius* $2^{-k}$ about $\alpha$:

$$B_k(\alpha) = \{\beta \in \mathbf{Z}_2 : \alpha \equiv \beta \text{ (mod } 2^k)\} \ .$$

Finally one may consistently define the 2-*adic measure* $\mu_2$ on $\mathbf{Z}_2$ so that

$$\mu_2(B_k(\alpha)) = 2^{-k} \ ;$$

in particular $\mu_2(\mathbf{Z}_2) = 1$. The integers $\mathbf{Z}$ are a subset of $\mathbf{Z}_2$; for example

$$-1 = 1 + 1 \cdot 2 + 1 \cdot 2^2 + \cdots \ .$$

Now one can extend the definition of the function $T : \mathbf{Z} \to \mathbf{Z}$ given by (2.1) to $T : \mathbf{Z}_2 \to \mathbf{Z}_2$ by

$$T(\alpha) = \begin{cases} \dfrac{\alpha}{2} \ , & \text{if } \alpha \equiv 0 \text{ (mod } 2) \ , \\[2mm] \dfrac{3\alpha + 1}{2} \ , & \text{if } \alpha \equiv 1 \text{ (mod } 2) \ . \end{cases}$$

Ergodic theory is concerned with the extent to which iterates of a function mix subsets of a measure space. I will use the following basic concepts of ergodic theory specialized to the measure space $\mathbf{Z}_2$ with the measure $\mu_2$. A measure-preserving function $H : \mathbf{Z}_2 \to \mathbf{Z}_2$ is *ergodic* if the only $\mu_2$-measurable sets $E$ for which $H^{-1}(E) = E$ are $\mathbf{Z}_2$ and the empty set, i.e., such a function does such a good job of mixing points in the space that it has no nontrivial $\mu_2$-invariant sets. It can be shown [[**39**], p. 36] that an equivalent condition for ergodicity is that

$$\lim_{N\to\infty} \frac{1}{N} \sum_{j=1}^{N} \mu_2(H^{-j}(B_k(\alpha)) \cap B_l(\beta)) = \mu_2(B_k(\alpha))\mu_2(B_l(\beta)) = 2^{-(k+l)} \ ,$$

for all $\alpha, \beta \in \mathbf{Z}_2$ and all integers $k, l \geq 0$. This condition in turn is equivalent to the assertion that for almost all $\alpha \in \mathbf{Z}_2$ the sequence of iterates

$$\{H^i(\alpha) : i = 0, 1, 2, \ldots\}$$

is uniformly distributed ( mod $2^k$) for all $k \geq 1$. A function $H : \mathbf{Z}_2 \to \mathbf{Z}_2$ is *strongly mixing* if

$$\lim_{N\to\infty} \mu_2(H^{-N}(B_k(\alpha)) \cap B_l(\beta)) = 2^{-(k+l)}$$

for all $\alpha, \beta \in \mathbf{Z}_2$ and all $k, l \geq 0$. Strongly mixing functions are ergodic.

The following result is a special case of a result of K. P. Matthews and A. M. Watts [**51**].

THEOREM K. *The map $T$ is a measure-preserving transformation of $\mathbf{Z}_2$ which is strongly mixing. Consequently it is ergodic, and hence for almost all $\alpha \in \mathbf{Z}_2$ the sequence*

$$\{T^{(i)}(\alpha) : i = 0, 1, 2, \ldots\}$$

*is uniformly distributed (mod $2^k$) for all $k \geq 1$.*

Theorem K implies nothing about the behavior of $T$ on the set of integers $\mathbf{Z}$ because it is a measure 0 subset of $\mathbf{Z}_2$. In fact, the trajectory $\{T^{(i)}(n) : i = 0, 1, 2, \ldots\}$ of any integer $n$ can *never* have the property of the conclusion of Theorem K, for if the trajectory is eventually periodic with period $k$, it cannot be uniformly distributed ( mod $2^{k-1}$), while if it is a divergent trajectory, it cannot even be equidistributed (mod 2) by (2.30). Consequently, this connection of the $3x+1$ problem to ergodic theory does not seem to yield any deep insight into the $3x+1$ problem itself.

There is, however, another connection of the $3x+1$ problem to ergodic theory of $\mathbf{Z}_2$ that may conceivably yield more information on the $3x+1$ problem. For each $\alpha \in \mathbf{Z}_2$ define the 0-1 variables $x_i$ by

$$T^{(i)}(\alpha) \equiv x_i \ (\text{mod } 2) .$$

Now define the function $Q_\infty : \mathbf{Z}_2 \to \mathbf{Z}_2$ by $Q_\infty(\alpha) = \beta$, where

(2.32) $$\beta = x_0 + x_1 2 + x_2 2^2 + \cdots .$$

The value $Q_\infty(\alpha)$ thus encodes the behavior of *all* the iterates of $\alpha$ under $T$.

The following result has been observed by several people, including R. Terras and C. Pomerance, but has not been explicitly stated before.

THEOREM L. *The map $Q_\infty : \mathbf{Z}_2 \to \mathbf{Z}_2$ is a continuous, one-to-one, onto and measure-preserving map on the 2-adic integers $\mathbf{Z}_2$.*

*Proof.* This is essentially a consequence of Theorem B. Use the fact that $Q_\infty(\alpha) \equiv \bar{Q}_n(\alpha) (\text{mod } 2^n)$. For any $\alpha_1, \alpha_2$ in $\mathbf{Z}_2$, if $|\alpha_1 - \alpha_2| \leq 2^{-n}$, then $\alpha_1 \equiv \alpha_2$ (mod $2^n$), so

$$Q_\infty(\alpha_1) \equiv \bar{Q}_n(\alpha_1) \equiv \bar{Q}_n(\alpha_2) \equiv Q_\infty(\alpha_2)(\text{mod } 2^n) ,$$

so that $|Q_\infty(\alpha_1) - Q_\infty(\alpha_2)| \leq 2^{-n}$ and $Q_\infty$ is continuous. If $\alpha_1 \neq \alpha_2$, then $\alpha_1 \not\equiv \alpha_2$ (mod $2^n$) for some $n$, so that

$$Q_\infty(\alpha_1) \equiv \bar{Q}_n(\alpha_1) \not\equiv \bar{Q}_n(\alpha_2) \equiv Q_\infty(\alpha_2) \ (\text{mod } 2^n)$$

and $Q_\infty$ is one-to-one. To see that $Q_\infty$ is onto, given $\alpha$ one can find $\beta_n$ so that

$$\bar{Q}_n(\beta_n) \equiv \alpha \ (\text{mod } 2^n) ,$$

since $\bar{Q}_n$ is a permutation. Then $|Q_\infty(\beta_n) - \alpha|_2 \leq 2^{-n}$. Now $\{\beta_n\}$ forms a Cauchy sequence in the 2-adic metric and $\mathbf{Z}_2$ is compact, hence the limiting value $\beta$ of $\{\beta_n\}$ satisfies $Q_\infty(\beta) = \alpha$. Now $Q_\infty^{-1}$ is defined, and $Q_\infty(\alpha) \equiv \bar{Q}_n^{-1}(\alpha) (\mathrm{mod}\ 2^n)$ implies that $Q_\infty^{-1}$ is continuous. $\square$

The $3x + 1$ Conjecture can be reformulated in terms of the function $Q_\infty$ as follows.

$3x+1$ CONJECTURE (Third form). *Let* $\mathbf{N}^+$ *denote the positive integers. Then* $Q_\infty(\mathbf{N}^+) \subseteq \frac{1}{3}\mathbf{Z}$. *In fact* $Q_\infty(N^+) \subseteq \frac{1}{3}\mathbf{Z} - \mathbf{Z}$.

For example $Q_\infty(1) = \sum_{i=0}^{\infty} 2^{2n} = -1/3$, $Q_\infty(2) = -2/3$, and $Q_\infty(3) = -20/3$.

The behavior of the function $Q_\infty$ under iteration is itself of interest. Let $\mathbf{Q}_2$ denote the set of all rational numbers having odd denominators, so that $\mathbf{Q}_2 \subseteq \mathbf{Z}_2$. The set $\mathbf{Q}_2$ consists of exactly those 2-adic integers whose 2-adic expansion is finite or eventually periodic. The Finite Cycles Conjecture is equivalent to the assertion that there is a finite odd integer $M$ such that

$$Q_\infty(\mathbf{Z}) \subseteq \frac{1}{M}\mathbf{Z} \ .$$

In fact one can take $M = \prod(2^l - 1)$, where the product runs over all integers $l$ for which there is a cycle of minimal length $l$. As a hypothesis for further work I advance the following conjecture.

PERIODICITY CONJECTURE. $Q_\infty(\mathbf{Q}_2) = \mathbf{Q}_2$.

For example, one may calculate that $Q_\infty(10) = -26/3$, $Q_\infty(-26/3) = -54$, $Q_\infty(-54) = -82/7$, $Q_\infty(-82/7) = ?/15$. It can be shown that if $n$ has a divergent trajectory, then the sequence $(x_0(n), x_1(n), x_2(n), \ldots)$ cannot be eventually periodic. As a consequence the truth of the Periodicity Conjecture implies the truth of the Divergent Trajectories Conjecture.

Theorem B has a curious consequence concerning the fixed points of iterates of $Q_\infty$.

THEOREM M. *Suppose the kth iterate* $Q_\infty^{(k)}$ *of* $Q_\infty$ *has a fixed point* $\alpha \in \mathbf{Z}_2$ *which is not a fixed point of any* $Q_\infty^{(l)}$ *for* $1 \leq l < k$. *Then* $k$ *is a power of 2.*

*Proof.* By hypothesis $Q_\infty^{(k)}(\alpha) = \alpha$ and $Q_\infty^{(l)}(\alpha) = \alpha_l \neq \alpha$, for $1 \leq l < k$. All the $\alpha_l$'s are distinct for $0 \leq l \leq k$, since $Q_\infty^{(l_1)}(\alpha) = Q_\infty^{(l_2)}(\alpha)$ implies $Q_\infty^{(l_1 - l_2)}(\alpha) = \alpha$, since $Q_\infty$ is one-one and onto. Consequently one can pick $m$ large enough so that all the residue classes $\alpha_l\ (\mathrm{mod}\ 2^m)$ are distinct, for $0 \leq l \leq k$, where $\alpha_0 = \alpha$. Now the action of $Q_\infty(\mathrm{mod}\ 2^m)$ is exactly that of the permutation $\bar{Q}_m$, hence

$$\bar{Q}_m^{(l)}(\alpha(\mathrm{mod}\ 2^m)) \equiv \alpha_l\ (\mathrm{mod}\ 2^m)$$

for $0 \leq l < k$. In particular $(\alpha_0(\mathrm{mod}\ 2^m),\ \alpha_1\ (\mathrm{mod}\ 2^m), \ldots, \alpha_{k-1}\ (\mathrm{mod}\ 2^m))$ makes up a single cycle of the permutation $\bar{Q}_m$, hence $k$ is a power of 2 by Theorem B. $\square$

## 3. Generalizations of $3x + 1$ problem.

The $3x + 1$ problem can be generalized by considering other functions $U$ : $\mathbf{N} \to \mathbf{N}$ defined on the natural numbers $\mathbf{N}$ that are similar to the function $T$. The functions $I$ consider to be similar to the function $T$ are the *periodically linear functions*, which are those functions $U$ for which there is a finite modulus $d$ such that the function $U$ when restricted to any congruence class $k$ (mod $d$) is linear. Some reasons to study generalizations of the $3x + 1$ problem are that they may uncover new phenomena, they can indicate the limits of validity of known results, and they can lead to simpler, more revealing proofs. Here I discuss three directions of generalizations of the $3x + 1$ problem. These deal with algorithmic decidability questions, with the existence of stopping times for almost all integers, and with the fractional parts of $(3/2)^k$.

**3.1. Algorithmic decidability questions.** J. H. Conway [26] proved the remarkable result that a simple generalization of the $3x + 1$ problem is algorithmically undecidable. He considers the class $\mathbf{F}$ of periodically piecewise linear functions $g : \mathbf{N} \to \mathbf{N}$ having the structure

$$(3.1) \qquad g(n) = \frac{\alpha_k}{(k, d)} n \text{ if } n = k \ (\text{mod} \ d) \ , \text{ for } 0 \leq k \leq d - 1 \ ,$$

specified by the nonnegative integers $(d, \alpha_0, \ldots, \alpha_{d-1})$. These are exactly the functions $g : \mathbf{N} \to \mathbf{N}$ such that $g(n)/n$ is periodic with some finite period $d$.

THEOREM O. (Conway). *For every partial recursive function $f$ defined on a subset $D$ of the natural numbers $\mathbf{N}$ there exists a function $g : \mathbf{N} \to \mathbf{N}$ such that*

(1) *$g(n)/n$ is periodic (mod $d$) for some $d$ and takes rational values.*
(2) *There is some iterate $k \geq 1$ such that $g^{(k)}(2^m) = 2^j$ for some $j$ if and only if $m$ is in $D$.*
(3) *$g^{(k)}(2^m) = 2^{f(m)}$ for the minimal $k \geq 1$ such that $g^{(k)}(2^m)$ is a power of 2.*

Conway's proof actually gives in principle a procedure for explicitly constructing such a function $g$ given a description of a Turing machine[1] that computes $f$. He carried out this procedure to find a function $g$ associated to a particular partial recursive function $f$ having the property that $f(2^{p_n}) = 2^{p_{n+1}}$, where $p_n$ is the $n$th prime; this is described in Guy [37].

By choosing a particular partial recursive function whose domain is not a recursive subset of $\mathbf{N}$, e.g., a function $f_0$ that encodes the halting problem for Turing machines, we obtain the following corollary of Theorem O.

THEOREM P (Conway). *There exists a particular, explicitly constructible function $g_0 : \mathbf{N} \to \mathbf{N}$ such that $g_0(n)/n$ is periodic (mod $d$) for a finite modulus $d$ and takes rational values, for which there is no Turing machine that, when given $n$, always decides in a finite number of steps whether or not there exists some iterate $g_0^{(k)}(n)$ with $k \geq 1$ which is a power of 2.*

---

[1] Conway's proof used Minsky machines, which have the same computational power as Turing machines.

**3.2. Existence of stopping times for almost all integers.** Several authors have investigated the range of validity of the result that $T(n)$ has a finite stopping time for almost all integers $n$ by considering more general classes of periodicity linear functions. One such class **G** consists of all functions $U = U(m, d, R)$ which are given by

$$(3.2) \quad U(n) = \begin{cases} \dfrac{n}{d}, & \text{if } n \equiv 0 \ (\text{mod } d), \\[2mm] \dfrac{mn - r}{d}, & \text{if } n \not\equiv 0 \ (\text{mod } d), \text{ and } r \in R \text{ is such that} \\ & mn \equiv r \ (\text{mod } d), \end{cases}$$

where $m$ and $d$ are positive integers with $(m, d) = 1$ and $R = \{r_i : r_i \equiv i \ (\text{mod } d), 1 \le i \le d - 1\}$ is a fixed set of residue class representatives of the nonzero residue classes ( mod $d$). The $3x+1$ function $T$ is in the class **G**. H. Möller [54] completely characterized the functions $U = U(m, d, R)$ in the set **G** which have a finite stopping time for almost all integers $n$. He showed they are exactly those functions for which

$$(3.3) \qquad\qquad m < d^{d/(d-1)} .$$

E. Heppner [41] proved the following quantitative version of this result, thereby generalizing Theorem D.

THEOREM Q. (Heppner). *Let $U = U(m, d, R)$ be a function in the class* **G**.
(i) *If $m < d^{d/(d-1)}$, then there exist real numbers $\delta_1, \delta_2 > 0$ such that for $N = [\log x / \log d]$ we have $\#\{n : n \le x \text{ and } U^{(N)}(n) > nx^{-\delta_1}\} = O(x^{1-\delta_2})$ as $x \to \infty$.*
(ii) *If $m > d^{d/(d-1)}$, then there exist real numbers $\delta_3, \delta_4 > 0$ such that for $N = [\log x / \log d]$ we have $\#\{n : n \le x \text{ and } U^{(N)}(n) < nx^{\delta_3}\} = O(x^{1-\delta_4})$ as $x \to \infty$.*

J.-P. Allouche [1] has further sharpened Theorem Q and Matthews and Watts [51], [52] have extended it to a larger class of functions.

It is a measure of the difficulty of problems in this area that even the following apparently weak conjecture is unsolved.

EXISTENCE CONJECTURE. *Let $U$ be any function in the class* **G**. *Then:*
(i) *$U$ has at least one purely periodic trajectory if $m < d^{d/(d-1)}$;*
(ii) *$U$ has at least one divergent trajectory if $m > d^{d/(d-1)}$.*

**3.3. Fractional parts of $(3/2)^k$.** Attempts to understand the distribution ( mod 1) of the sequence $\{(3/2)^k : 1 \le k < \infty\}$ have uncovered oblique connections with ergodic-theoretic aspects of a generalization of the $3x + 1$ problem. It is conjectured that the sequence $(3/2)^k$ is uniformly distributed (mod 1). (This conjecture seems intractable at present.)

One approach to this problem is to determine what kinds of (mod 1) distributions can occur for sequences $\{(3/2)^k \xi : 1 \le k < \infty\}$, where $\xi$ is a fixed real number. In this vein K. Mahler [48] considered the problem of whether or not there exist real numbers $\xi$, which he called *Z-numbers*, having the property that

$$(3.4) \qquad\qquad 0 \le \left\{ \left(\frac{3}{2}\right)^k \xi \right\} \le \frac{1}{2} , \quad k = 1, 2, 3, \dots ,$$

where $\{x\} = x - [x]$ is the fractional part of $x$. He showed that the set of $Z$-numbers is countable, by showing that there is at most one $Z$-number in each interval $[n, n+1)$, for $n = 1, 2, 3, \ldots$. He went on to show that a necessary condition for the existence of a $Z$-number in the interval $[n, n+1)$ is that the trajectory $(n, W(n), W^{(2)}(n), \ldots)$ of $n$ produced by the periodically linear function

$$(3.5) \qquad W(n) = \begin{cases} \dfrac{3n}{2}, & \text{if } n \equiv 0 \ (\text{mod } 2), \\[2ex] \dfrac{3n+1}{2}, & \text{if } n \equiv 1 \ (\text{mod } 2), \end{cases}$$

satisfy

$$(3.6) \qquad W^{(k)}(n) \not\equiv 3 \ (\text{mod } 4), \ 1 \leq k < \infty.$$

Mahler concluded from this that is unlikely that any $Z$-numbers exist. This is supported by the following heuristic argument. The function $W$ may be interpreted as acting on the 2-adic integers by (3.5), and it has properties exactly analogous to the properties of $T$ given by Theorem K. In particular, for almost all 2-adic integers $\alpha$ the sequence of iterates $(\alpha, W(\alpha), W^{(2)}(\alpha), \ldots)$ has infinitely many values $k$ with $W^{(k)}(\alpha) \equiv 3 \ (\text{mod } 4)$. Thus if a given $n \in \mathbf{Z}$ behaves like almost all 2-adic integers $\alpha$, then (3.6) will not hold for $n$. Note that it is possible that all the trajectories $(n, W(n), W^{(2)}(n), \ldots)$ for $n \geq 1$ are uniformly distributed $(\text{mod } 2^k)$ for *all* $k$, unlike the behavior of the function $T(n)$.

In passing, I note that the possible distributions $(\text{mod } 1)$ of $\{(3/2)^k \xi; 1 \leq k < \infty\}$ for real $\xi$ have an intricate structure (see G. Choquet [16]–[22] and A. D. Pollington [57], [58]). In particular, Pollington [58] proves that there are uncountably many real numbers $\xi$ such that

$$\frac{1}{25} \leq \left\{ \left( \frac{3}{2} \right)^k \xi \right\} \leq \frac{24}{25} \ ; \ k = 1, 2, 3, \ldots,$$

in contrast to the at most countable number of solutions $\xi$ of (3.4).

## 4. Conclusion.

*Is the $3x+1$ problem intractably hard?* The difficulty of settling the $3x+1$ problem seems connected to the fact that it is a deterministic process that simulates "random" behavior. We face this dilemma: On the one hand, to the extent that the problem has structure, we can analyze it — yet it is precisely this structure that seems to prevent us from proving that it behaves "randomly." On the other hand, to the extent that the problem is structureless and "random," we have nothing to analyze and consequently cannot rigorously prove anything. Of course there remains the possibility that someone will find some hidden regularity in the $3x+1$ problem that allows some of the conjectures about it to be settled. The existing general methods in number theory and ergodic theory do not seem to touch the $3x+1$ problem; in this sense it seems intractable at present. Indeed all the conjectures made in this paper seem currently to be out of reach if they are true; I think there is more chance of disproving those that are false.

*If the $3x+1$ problem is intractable, why should one bother to study it?* One answer is provided by the following aphorism: "No problem is so intractable that something interesting cannot be said about it." Study of the $3x+1$ problem has uncovered a number of interesting phenomena; I believe further study of it may be rewarded by the discovery of other new phenomena. It also serves as a benchmark to measure the progress of general mathematical theories. For example, future developments in solving exponential Diophantine equations may lead to the resolution of the Finite Cycles Conjecture.

*If all the conjectures made in this paper are intractable, where would one begin to do research on this deceptively simple problem?* As a guide to doing research, I ask questions. Here are a few that occur to me: For the $3x + 1$ problem, what restrictions are there on the growth in size of members of a divergent trajectory assuming that one exists? What interesting properties does the function $Q_\infty$ have? Is there some direct characterization of $Q_\infty$ other than the recursive definition (2.32)?

## References

[1] J.-P. Allouche, Sur la conjecture de "Syracuse-Kakutani-Collatz," Séminaire de Théorie des Nombres, 1978–1979. Exp. No. 9, 15, pp., CNRS, Talence (France) 1979.

[2] Shiro Ando, letter to J. C. Lagarias, Feb. 18, 1983. Reports that Prof. Nabuo Yoneda (Dept. of Information Science, Tokyo Univ.) has verified the $3x + 1$ conjecture for all $n < 2^{40} = 1.2 \times 10^{12}$.

[3] R. V. Andree, *Modern Abstract Algebra*, Holt, Reinhart, and Winston, 1971.

[4] Anon., The $3x + 1$ problem. Popular Computing, **1** (April 1973) 1–2.

[5] Anonymous, $3x + 1$ (continued), Popular Computing, **1** (July 1973) 6–7.

[6] Anon., $3x + 1$ strings, Popular Computing, **2** (April 1973) 12–13.

[7] Anon., $3x + 1$ once again, Popular Computing, **3** (April 1975) 4–5.

[8] R. Ash, *Information Theory*, Interscience Publishers, Wiley, New York, 1965.

[9] A. O. L. Atkin, Comment on Problem 63–13, SIAM Review, **8** (1966) 234–236.

[10] A. Baker, *Transcendental Number Theory*, Cambridge Univ. Press, 1975.

[11] M. Beeler, W. Gosper, R. Schroeppel, HAKMEM, Memo 239, Artificial Intelligence Laboratory, M.I.T. 1972, p. 64. Presents some numerical results on cycles for the $3x + 1$ and related problems.

[12] J. Blazewicz and A. Pettorossi, Collatz's Conjecture and Binary Sequences, Institute of Control Engineering, Technical University of Poznan, Poznan, Poland (preprint) March 1983.

[13] C. Böhm and G. Sontacchi, On the existence of cycles of given length in integer sequences like $x_{n+1} = x_n/2$ if $x_n$ even, and $x_{n+1} = 3x + 1$ otherwise, Atti Accad. Naz. Lincei Rend. Cl. Sci. Fis. Mat. Natur., **64** (1978) 260–264,

[14] Z. I. Borevich and I. R. Shafarevich, *Number Theory*, Academic Press, New York, 1966.

[15] M. J. Bruce, Crazy roller coaster, Math. Teacher, **71** (January 1978) 45–49.

[16] G. Choquet, Répartition des nombres $k(3/2)^n$; mesures et ensembles associés, C. R. Acad. Sci. Paris, **290** (31 mars 1980) 575–580.

[17] —————, Algorithmes adaptés aux suites $(k\theta^n)$ et aux chaines associées, C. R. Acad. Sci. Paris, **290** (28 avril 1980) 719–724.

[18] —————, $\theta$-Jeux récursifs et application aux suites $(k\theta^n)$; solénoïdes de $II^Z$, C. R. Acad. Sci. Paris, **290** (19 mai 1980) 863–868.

[19] —————, Construction effective de suites $(k(3/2)^n)$. Etude des mesures (3/2)-stables, C. R. Acad. Sci. Paris, **291** (15 septembre 1980) 69–74.

[20] —————, Les fermés (3/2)-stables de II; structure des fermés dénombrables; applications arithmétiques, C. R. Acad. Sci. Paris, **291** (29 septembre 1980) 239–244.

[21] —————, $\theta$-fermés; $\theta$-chaînes et $\theta$-cycles (pour $\theta = 3/2$), C. R. Acad. Sci. Paris, **292** (5 janvier 1981) 5–10.

[22] —————, $\theta$-fermés et dimension de Hausdorff. Conjecturers de travail. Arithmétique des $\theta$-cycles (où $\theta = 3/2$), C. R. Acad. Sci. Paris, **292** (9 février 1981) 339–344. See especially Conjecture D, which arises from a generalization of the problem considered by Mahler [**48**]; cf. (3.4) in the text.

[23] L. Collatz, letters to R. Terras, 7 May 1976 and 2 Sept. 1976.

[24] —————, letter to L. Garner, 17 March 1980. In none of these letters does L. Collatz actually state that he proposed the $3x + 1$ problem.

[25] —————, Verzweigungsdiagramme und Hypergraphen, International Series for Numerical Mathematics, vol. 38, Birkhäuser, 1977. This paper remarks on the use of graphical representations to study iteration problems, as in Fig. 1 in the text. It does not discuss the $3x + 1$ problem.

[26] J. H. Conway, Unpredictable Iterations, Proc. 1972 Number Theory Conference, University of Colorado, Boulder, Colorado (1972) 49–52.

[27] H. S. M. Coxeter, Cyclic Sequences and Frieze Patterns (The Fourth Felix Behrend Memorial Lecture), Vinculum **8** 1971, 4–7. In this 1970 lecture Coxeter offered a $50 prize for a proof of the $3x + 1$ Conjecture and $100 for a counterexample, according to C. W. Trigg [**70**].

[28] R. E. Crandall, On the "$3x + 1$" problem, Math. Comp., **32** (1978) 1281–1292.

[29] J. L. Davidson, Some Comments on an Iteration Problem, Proc. 6th Manitoba Conf. on Numerical Mathematics (1976) 155–159.

[30] R. Dunn, On Ulam's Problem, Dept. of Computer Science Report #CU-CS-011-73, University of Colorado, Boulder, January 1973. Reports on computer programs to test the $3x + 1$ Conjecture and to calculate the densities $F(k)$ in (2.16) for $k \leq 21$.

[31] C. J. Everett, Iteration of the number theoretic function $f(2n) = n$, $f(2n + 1) = 3n + 2$, Advances in Math.,**25** (1977) 42–45.

[32] A. S. Fraenkel, private communication 1981.

[33] M. Gardner, Mathematical Games, Scientific American, **226** (June 1972) 114–118.

[34] L. E. Garner, On the Collatz $3n + 1$ algorithm, Proc. Amer. Math. Soc., **82** (1981) 19–22.

[35] F. Gruenberger, $3x + 1$ revisited, Popular Computing, **7** (Oct. 1979) 3–12.

[36] R. K. Guy, *Unsolved Problems in Number Theory*, Springer-Verlag, New York, 1981 (Problem E16).

[37] —————, Don't try to solve these Problems!, AMERICAN MATH. MONTHLY, **90** (1983) 35–41.

[38] —————, Conway's prime producing machine, Math. Mag., **56** (1983) 26–33.

[39] P. Halmos, *Lectures on Ergodic Theory*, Math. Soc. of Japan, Tokyo, 1956 (Reprint: Chelsea, New York).

[40] H. Hasse, *Unsolved Problems in Elementary Number Theory*, Lectures at U. Maine (Orono), Spring 1975, dittoed notes, pp. 23–33. He states (p. 23) that Thompson has proved the Finite Cycles Conjecture, a result I am unable to verify.

[41] E. Heppner, Eine Bemerkung zum Hasse-Syracuse Algorithmus, Archiv. Math., **31** (1978) 317–320,

[42] I. N. Herstein and I. Kaplanaky, *Matters Mathematical*, 2nd ed., Chelsea, New York, 1978, 44–45. They call the problem the "Syracuse algorithm."

[43] D. Hofstadter, *Gödel, Escher, Bach,* Basic Books, New York, 1979. Brings the $3x+1$ problem to a wide audience by stating it on pp. 400-402.

[44] S. D. Isard and A. M. Zwicky, Three open questions in the theory of one-symbol Smullyan systems, SIGACT News (1969) 11–14. They propose the problem: Let $f(n) = n/3$ if $n \equiv 0 \pmod{3}$ and let $g(n) = 2n + 1$. Can all numbers $n \equiv 0$ or $1 \pmod{3}$ be reduced to 1 by a series of operations $f$ and $g$? See [**70**], p. 148.

[45] S. Kakutani, private communication 1981.

[46] David Kay, Undergraduate Research Project no. 2, Pi Mu Epsilon J. (1972) 338.

[47] M. S. Klamkin, Problem 63–12, SIAM Review, **5** (1963) 275–276. States the problem: Consider the permutation $f : N^+ \to N^+$ defined by $f(3n) = 2n$, $f(3n - 1) = 4n - 1$, $f(3n - 2) = 4n - 3$. Does $n = 8$ have an infinite trajectory? How many finite cycles are there for $f$? This function was the original one proposed by L. Collatz in 1932.[**24**].

[48] K. Mahler, An unsolved problem on the powers of 3/2, J. Austral. Math. Soc., **8** (1968) 313–321. The problem Mahler studies was originally proposed by Prof. Saburô Uchiyama (Tsukuba Univ.), according to S. Ando [**2**]. It arose indirectly in connection with the function $g(k)$ of Waring's problem, see G. H. Harding and E. M. Wright, An Introduction to the Theory of Numbers (4th ed.), Oxford Univ. Press, 1960, Theorem 393ff.

[49] —————, *P-adic numbers and their functions, 2nd edition* Cambridge Univ. Press, 1976.

[50] K. R. Matthews and G. M. Leigh, A generalization of the Syracuse algorithm to $F_q[x]$, J. Number Thery **25** (1987), 274–278.

[51] K. R. Matthews and A. M. Watts, A generalization of Hasse's generalization of the Syracuse algorithm, Acta Arithmetica, **43** (1983) 75–83.

[52] ⸺, A Markov approach to the generalized Syracuse algorithm, Acta Arithmetica **45** (1985), 29–42.

[53] ⸺, Some invariant probability measures on the polyadic numbers, Colloq. Math. **53** (1992), 191–202.

[54] H. Möller, Über Hasses Verallgemeinerung des Syracuse-Algorithmus (Kakutani's Problem), Acta Arithmetica, **34** (1978) 219–226. This paper claims A. Fraenkel checked that $f^{(k)}(n) = 1$ for all $n \leq 2^{50}$, a result in error [**32**]. In a note added in proof it asserts the proofs of Terras [**67**] are faulty. They seem essentially correct to me, and Terras [**68**] supplies elided details.

[55] J. Nievergelt, J. Farrar, and E. M. Reingold, *Computer Approaches to Mathematical Problems*, Prentice-Hall, Englewood Cliffs, NJ, 1974, 211-217. They discuss the problem of checking the $3x + 1$ Conjecture for a given $n$ and give some interesting empirical observations.

[56] C. Stanley Ogilvy, *Tomorrow's Math*, Oxford Univ. Press, Oxford, 1972, 103.

[57] A. D. Pollington, Intervals Constructions in the Theory of Numbers, Thesis, University of London, 1980. See especially Chapter 5, which contains results related to Mahler's problem [**48**].

[58] ⸺, Progressions arithmétiques géneralisées et le problème des $(3/2)^n$, C. R. Acad. Sci. Paris, **292** (16 février 1981) 383–384.

[59] L. Ratzan, Some work on an unsolved palindromic algorithm, Pi Mu Epsilon J., **5** (Fall 1973) 463–466.

[60] D. A. Rawsthorne, Imitation of an Iteration, Math. Magazine **58** (1985), 172–176.

[61] B. Schuppar, Kettenbrüche und der $3a + 1$ Algorithmus, preprint.

[62] D. Shanks, Comments on Problem 63–13, SIAM Review, **7** (1965) 284–286. Gives the result of a computer search on Collatz' original problem (see [**9**], [**47**]), and a heuristic concerning lengths of finite cycles.

[63] ⸺, Western Number Theory Conference Problem 1975 #5 (R. Guy, ed.) (mimeographed sheets). States a conjecture on the average size of the total stopping time function.

[64] R. P. Steiner, A Theorem on the Syracuse Problem, Proc. 7th Manitoba Conference on Numerical Mathematics–1977, Winnipeg (1978) 553–559.

[65] ⸺, On the "$Qx + 1$ Problem," $Q$ odd, Fibonacci Quarterly, **19** (1981) 285–288.

[66] ⸺, On the "$Qx + 1$ Problem," $Q$ odd II, Fibonacci Quarterly, **19** (1981) 293–296.

[67] R. Terras, A stopping time problem on the positive integers, Acta Arith.metica **30** (1976) 241–252.

[68] ⸺, On the existence of a density, Acta Arithmetica **35** (1979) 101–102.

[69] R. Tijdeman, Note on Mahler's 3/2 problem, Det Kongelige Norske Videnskabers Selskab Skrifter, No. 16, 1972.

[70] C. W. Trigg, C. W. Dodge and L. F. Meyers, Comments on Problem 133, Eureka (now Crux Mathematicorum) **2**, no. 7 (August–September 1976) 144–150. Problem 133 is the $3x + 1$ problem, proposed by K. S. Williams, who said he was shown it by one of his students. C. W. Trigg gives a good history of the problem.

[71] M. Vaughan-Lee, letter to L. Garner, Aug. 10, 1981. Michael Vaughan-Lee (Oxford U.) has verified the $3x + 1$ Conjecture for $n < 2^{32} \cong 4.5 \times 10^9$ on a Sinclair 2X81 Microcomputer.

[72] C. Williams, B. Thwaites, A. van der Poorten, W. Edwards, and L. Williams, Ulam's Conjecture continued again, 1000 pounds for proof. Rumored proof disclaimed. References, PPC Calculator J., **9** (Sept. 1982) 23–24. Gives some history of the problem. B. Thwaites states he originated the $3x + 1$ problem in 1952 and offers 1000 pounds for a proof of the $3x + 1$ Conjecture.

[73] M. Yamada, A convergence proof about an integral sequence, Fibonacci Quarterly, **18** (1980) 231–242. The proofs in this paper are incorrect, see MR 82d:10026.

# PART II.

## Survey Papers

# A $3x + 1$ Survey: Number Theory and Dynamical Systems

Marc Chamberland

## 1. Introduction

The 3x+1 Problem is perhaps today's most enigmatic unsolved mathematical problem: it can be explained to child who has learned how to divide by 2 and multiply by 3, yet there are relatively few strong results toward solving it. Paul Erdös was correct when he stated, "Mathematics is not ready for such problems."

The problem is also referred to as the $3n + 1$ problem and is associated with the names of Collatz, Hasse, Kakutani, Ulam, Syracuse, and Thwaites. It may be stated in a variety of ways. Defining the *Collatz function* as

$$C(x) = \left\{ \begin{array}{ll} 3x + 1 & x \equiv 1 \pmod 2 \\ \frac{x}{2} & x \equiv 0 \pmod 2, \end{array} \right.$$

the conjecture states that for each $m \in \mathbf{Z}^+$, there is a $k \in \mathbf{Z}^+$ such that $C^{(k)}(m) = 1$, that is, any positive integer will eventually iterate to 1. Note that an odd number $m$ iterates to $3m+1$ which then iterates to $(3m+1)/2$. One may therefore "compress" the dynamics by considering the map

$$T(x) = \left\{ \begin{array}{ll} \frac{3x+1}{2} & x \equiv 1 \pmod 2 \\ \frac{x}{2} & x \equiv 0 \pmod 2. \end{array} \right.$$

The map $T$ is usually favored in the literature.

To a much lesser extent some authors work with the most dynamically streamlined $3x + 1$ function, $F : \mathbf{Z}^+_{\text{odd}} \rightarrow \mathbf{Z}^+_{\text{odd}}$, defined by

$$F(x) = \frac{3x + 1}{2^{m(3x+1)}}$$

where $m(x)$ equals the number of factors of 2 contained in $x$. While working with $F$ allows one to work only on the odd positive integers, the variability of $m$ seems to prohibit any substantial analysis.

There have been several earlier surveys on the 3x+1 problem. The paper of Lagarias[**52**](1985) thoroughly catalogued earlier results, made copious connections, and developed many new lines of attack; it has justly become the classical reference for this problem. The book of Wirsching[**102**](1998) studies dynamical systems attached to the 3x+1 problem, and begins with an extensive survey, followed by several chapters of new results. In 2003 Lagarias posted on the math arXiv an

---

2000 *Mathematics Subject Classification.* Primary: 11B83.
*Key words and phrases.* $3x + 1$ problem, Collatz conjecture.

annotated bibliography of work on this problem through 2000, another valuable resource: an updated version of it appears in this volume. He also has a sequel annotated bibliography available [57](2008) as a preprint on the mathematics arXiv.

Here we present a survey that is complementary to the work of Lagarias and Wirsching. It covers results in number theory and in dynamical systems, with special emphasis on the latter. It reflects my views on how work on this problem can be structured. In active areas, I review both new contributions and state some older contributions for the sake of completeness. In other areas which have seen fewer recent developments, such as work on functional equations and cellular automata, the reader may consult other surveys, and other papers in this volume.

This paper is a revised and updated version of my earlier survey article on the 3x+1 problem, officially published in Chamberland[24](2003), in Catalan.

## 2. Numerical Investigations and Stopping Time

The structure of the positive integers forces any orbit of $T$ to iterate to one of the following:

(1) the trivial cycle $\{1, 2\}$
(2) a non-trivial cycle
(3) infinity (the orbit is divergent)

The $3x + 1$ Problem claims that option 1 occurs in all cases. Oliveira e Silva[72, 73](1999,2000) proved that this holds for all numbers $n < 100 \times 2^{50} \approx 1.12 \times 10^{17}$. This was accomplished with two 133MHz and two 266MHz DEC Alpha computers and using 14.4 CPU years. This computation ended in April 2000. Roosendaal[76](2003) claims to have improved this to $n = 195 \times 2^{50} \approx 2.19 \times 10^{17}$. His calculations continue, with the aid of many others in this distributed-computer project. Since 2005 Olivera e Silva has resumed computations on the problem on a larger scale, and results of his computations are reported in this special issue. In particular, he has verified that the 3x+1 Conjecture holds for all integers less than $4.0 \times 10^{18}$.

The record for proving the non-existence of non-trivial cycles is that any such cycle must have length no less than 272,500,658 (see equation (3)). This was derived with the help of numerical results like those in the last paragraph coupled with the theory of continued fractions – see section 5 for more on cycles.

There is a natural algorithm for checking that all numbers up to some $N$ iterate to one. First, check that every positive integer up to $N - 1$ iterates to one, then consider the iterates of $N$. Once the iterates go below $N$, you are done. For this reason, one considers the so-called *stopping time* of $n$, that is, the number of steps needed to iterate below $n$ :

$$\sigma(n) = \inf\{k : T^{(k)}(n) < n\}.$$

Related to this is the *total stopping time*, the number of steps needed to iterate to 1:

$$\sigma_\infty(n) = \inf\{k : T^{(k)}(n) = 1\}.$$

One considers the *height* of $n$, namely, the highest point to which $n$ iterates:

$$h(n) = \sup\{T^{(k)}(n) : k \in \mathbf{Z}^+\}.$$

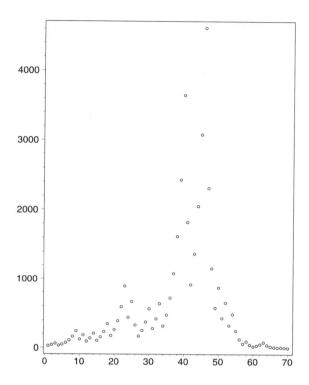

FIGURE 1. Orbit starting at 27.

Note that if $n$ is in a divergent trajectory, then $\sigma_\infty(n) = h(n) = \infty$. These functions may be surprisingly large even for small values of $n$. For example,

$$\sigma(27) = 59, \ \sigma_\infty(27) = 111, \ h(27) = 9232.$$

The orbit of 27 is depicted in Figure 1.

Roosendaal[**76**](2003) has computed various "records" for these functions[1]. Various results about consecutive numbers with the same height are catalogued by Wirsching[**102**, pp.21–22](1998).

**2.1. Stopping Time** The natural algorithm mentioned earlier can be rephrased: the $3x + 1$ problem is true if and only if every positive integer has a finite stopping time. Terras[**91**, **92**](1976,1979) has proven that the set of positive integers with finite stopping time has density one. Specifically, he showed that the

---

[1]Roosdendaal has different names for these functions, and he applies them to the map $C(x)$.

limit

$$F(k) := \lim_{m \to \infty} \frac{1}{m} |\{n \leq m : \sigma(n) \leq k\}|$$

exists for each $k \in \mathbf{Z}^+$ and $\lim_{k \to \infty} F(k) = 1$. A shorter proof was provided by Everett[35](1977). Lagarias[52](1985) proved a result regarding the speed of convergence:

$$F(k) \geq 1 - 2^{-\eta k}$$

for all $k \in \mathbf{Z}^+$, where $\eta = 1 - H(\theta)$, $H(x) = -x\log(x) - (1-x)\log(1-x)$ and $\theta = (\log_2 3)^{-1}$. He uses this to constrain any possible divergent orbits:

$$\left|\{n \in \mathbf{Z}^+ : n \leq x, \sigma(n) = \infty\}\right| \leq c_1 x^{1-\eta}$$

for some positive constant $c_1$. This implies that any divergent trajectory cannot diverge too slowly. Along these lines, Garcia and Tal[38](1999) have recently proven that the density of any divergent orbit is zero. This result holds for more general sets and more general maps.

Along different lines, Venturini[95](1989) showed that for every $\rho > 0$, the set $\{n \in \mathbf{Z}^+ : T^{(k)}(n) < \rho n \text{ for some k}\}$ has density one. Allouche[1](1979) showed that $\{n \in \mathbf{Z}^+ : T^{(k)}(n) < n^c \text{ for some k}\}$ has density one for $c > 3/2 - \log_2 3 \approx 0.869$. This was improved by Korec[46](1994) who obtained the same result with $c > \log_4 3 \approx 0.7924$.

Not surprisingly, results of a probabilistic nature abound concerning these $3x+1$ functions. An assumption often made is that after many iterations, the next iterate has an equal chance of being either even or odd. Wagon[99](1985) argues that the average stopping time for odd $n$ (with the function $C(x)$) approaches a constant, specifically, the number

$$\sum_{i=1}^{\infty} [1 + 2i + i\log_2 3] \frac{c_i}{2^{[i\log_2 3]}} \approx 9.477955$$

where $c_i$ is the number of sequences containing $3/2$ and $1/2$ with exactly $i$ $3/2$'s such that the product of the whole sequence is less than 1, but the product of any initial sequence is greater than 1. Note that this formula bypasses the pesky "+1", perhaps justified asymptotically. This seems to be borne out in Wagon's numerical testing of stopping times for odd numbers up to $n = 10^9$, which matches well with the approximation given above. Slakmon and Macot[84] formally consider $T-$iterations as a random walk and show that almost all starting points approach the trivial cycle.

**2.2. Total Stopping Time** Results regarding the total stopping time are also plenteous. Applegate and Lagarias[9](2002) make use of two new auxiliary functions: the *stopping time ratio*

$$\gamma(n) := \frac{\sigma_\infty(n)}{\log n},$$

and the *ones-ratio* $\rho(n)$ for convergent sequences, defined as the ratio of the number of odd terms in the first $\sigma_\infty(n)$ iterates divided by $\sigma_\infty(n)$. It is easy to see that $\gamma(n) \geq 1/\log 2$ for all $n$, with equality only when $n = 2^k$. Stronger inequalities are

$$\gamma(n) \geq \frac{1}{\log 2 - \rho(n)\log 3}$$

for any convergent trajectory, while if $\rho(n) \leq 0.61$, then for any positive $\epsilon$,

$$\gamma(n) \leq \frac{1}{\log 2 - \rho(n) \log 3} + \epsilon$$

for all sufficiently large $n$. If one assumes that the ones-ratio equals $1/2$ (which is equivalent to the equal probability of encountering an even or an odd after many iterates), we have that the average value of $\gamma(n)$ is $2/\log(4/3) \approx 6.95212$, as has been observed by Shanks[80](1965), Crandall[30](1978), Lagarias[52](1985), Rawsthorne[75](1985), Lagarias and Weiss [56](1992), and Borovkov and Pfeifer [18](2000). Experimental evidence supports this observation. Compare this with the upper bounds suggested by the stochastic models of Lagarias and Weiss [56] (1992):

$$\limsup_{n \to \infty} \gamma(n) \approx 41.677647.$$

Applegate and Lagarias[9](2002) note from records of Roosendaal[76](2003) what is apparently the largest known value of $\gamma$:

$$n = 7, 219, 136, 416, 377, 236, 271, 195$$

produces $\sigma_\infty(n) = 1848$ and $\gamma(n) \approx 36.7169$. Applegate and Lagarias seek a lower bound for $\gamma$ which holds infinitely often. Using the well-known fact that $T^{(k)}(2^k - 1) = 3^k - 1$ (see, for example, Kuttler[51](1994)), they note that

$$\gamma(2^k - 1) \geq \frac{\log 2 + \log 3}{(\log 2)^2} \approx 3.729.$$

They go on to show that there are infinitely many converging $n$ whose ones-ratio is at least $14/29$, hence giving the lower bound

$$\gamma(n) \geq \frac{29}{29 \log 2 - 14 \log 3} \approx 6.14316$$

for infinitely many $n$. Though the proof involves an extensive computational search on the Collatz tree to depth 60, the authors note with surprise that the probabilistic average of $\gamma(n)$ – approximately 6.95212 – was not attained.

Roosendaal[76](2003) defines a function which he calls the *residue* of a number $x \in \mathbf{Z}^+$, denoted Res($x$). Suppose $x$ is a convergent $C$-trajectory with $E$ even terms and $O$ odd terms before reaching one. Then the residue is defined as

$$\text{Res}(x) := \frac{2^E}{x3^O}.$$

Roosendall notes that $\text{Res}(993) = 1.253142\ldots$, and that this is the highest residue attained for all $x < 2^{32}$. He conjectures that this holds for all $x \in \mathbf{Z}^+$.

Zarnowski[104](2001) recasts the $3x + 1$ problem as one with Markov chains. Let $g$ be the slightly altered function

$$g(x) = \begin{cases} \frac{3x+1}{2} & x \equiv 1 \pmod 2, \ x > 1 \\ \frac{x}{2} & x \equiv 0 \pmod 2 \\ 1 & x = 1, \end{cases}$$

$\mathbf{e}_n$ the column vector whose $n^{th}$ entry is one and all other entries zero, and the transition matrix $P$ be defined by

$$P_{ij} = \begin{cases} 1 & \text{if } g(i) = j, \\ 0 & \text{otherwise.} \end{cases}$$

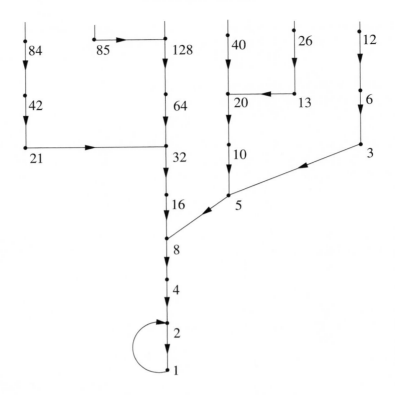

FIGURE 2. The Collatz graph.

This gives a correspondence between $g^{(k)}(n)$ and $\mathbf{e}_n^T P^k$, and the $3x+1$ Problem is true if

$$\lim_{k \to \infty} P^k = [\mathbf{1} \ \mathbf{0} \ \mathbf{0} \ \cdots],$$

where $\mathbf{1}$ and $\mathbf{0}$ represent constant column vectors. Zarnowski goes on to show that the structure of a certain generalized inverse $X$ of $I - P$ encodes the total stopping times of $\mathbf{Z}^+$.

For pictorial representations of stopping times for an extension of $T$, see Dumont and Reiter[33].

**2.3. The Collatz Graph and Predecessor Sets** The topic of stopping times is intricately linked with the *Collatz tree* or *Collatz graph*, the directed graph whose vertices are predecessors of one via the map $T$. It is depicted in Figure 2.3.

The structure of the Collatz graph has attracted some attention. Andaloro [4](2002) studied results about the connectedness of subsets of the Collatz graph. Urvoy [93](2000) proved that the Collatz graph is non-regular, in the sense that it does not have a decidable monadic second order theory; this is related to the work of Conway mentioned in Section 4.

Instead of analyzing how fast iterates approach one, one may consider the set of numbers which approach a given number $a$, that is, the *predecessor set of $a$*:

$$P_T(a) := \{b \in \mathbf{Z}^+ : T^{(k)}(b) = a \text{ for some } k \in \mathbf{Z}^+\}$$

Since such sets are obviously infinitely large, one may measure those terms in the predecessor set not exceeding a given bound $x$, that is,

$$Z_a(x) := |\{n \in P_T(a) : n \leq x\}|.$$

The size of $Z_a(x)$ was first studied by Crandall[30](1979), who proved the existence of some $c > 0$ such that

$$Z_1(x) > x^c, \ x \text{ sufficiently large.}$$

Wirsching[102, p.4](1998) notes that this result extends to $Z_a(x)$ for all $a \not\equiv 0 \bmod 3$. Using the tree-search method of Crandall, Sander[78](1990) gave a specific lower bound, $c = 0.25$, and Applegate and Lagarias[6](1995) extended this to $c = 0.643$. Using functional difference inequalities, Krasikov[48](1989) introduced a different approach and obtained $c = 3/7 \approx 0.42857$. Wirsching[100](1993) used the same approach to obtain $c = 0.48$. Applegate and Lagarias[7](1995) superseded these results with $Z_1(x) > x^{0.81}$, for sufficiently large $x$, by enhancing Krasikov's approach with nonlinear programming. Recently, Krasikov and Lagarias[49](2002) streamlined this approach to obtain

$$Z_1(x) > x^{0.84}, \ x \text{ sufficiently large.}$$

Wirsching[102](1998) has pushed these types of results in a different direction. On a seemingly different course, for $j, k \in \mathbf{Z}^+$ let $R_{j,k}$ denote the set of all sums of the form

$$2^{\alpha_0} + 2^{\alpha_1} 3 + 2^{\alpha_2} 3^2 + \cdots + 2^{\alpha_j} 3^j$$

where $j + k \geq \alpha_0 > \cdots > \alpha_j \geq 0$. The size of $R_{j,k}$ satisfies

$$|R_{j,k}| = \binom{j + k + 1}{j + 1}.$$

To maximize the size, one has $|R_{j-1,j}| = \binom{2j}{j}$. Wirsching posed the following "covering conjecture": there is a constant $K > 0$ such that for every $j, l \in \mathbf{Z}^+$ we have the implication

$$|R_{j-1,j}| \geq K \cdot 2 \cdot 3^{l-1} \implies R_{j-1,j} \text{ covers the prime residue classes to modulus } 3^l.$$

This conjecture implies a stronger version of the earlier inequalities:

$$\liminf_{x \to \infty} \left( \inf_{a \not\equiv 0 \bmod 3} \frac{Z_a(ax)}{x^\delta} \right) > 0 \text{ for any } \delta \in (0, 1).$$

Wirsching argues that, intuitively, the sums of mixed powers can be seen as accumulated non-linearities occuring when iterating the map $T$.

A large part of Wirsching's book [102](1998) concerns a finer study of the predecessor sets using the functions

$$e_l(k, a) = |\{b \in \mathbf{Z}^+ : T^{(k+l)}(b) = a, \ k \text{ even iterates, } l \text{ odd iterates}\}|.$$

Defining the $n^{th}$ *estimating series* as

$$s_n(a) := \sum_{l=1}^{\infty} e_l \left( n + \lfloor l \log_2(3/2) \rfloor, a \right),$$

Wirsching proves the implication

$$\liminf_{n \to \infty} \frac{s_n(a)}{\beta^n} > 0 \implies Z_a(x) \geq C \left( \frac{x}{a} \right)^{\log_2 \beta} \text{ for some constant } C > 0 \text{ and large } x.$$

Since $e_l(k, a)$ depends on $a$ only through its residue class to modulo $3^l$, Wirsching considers the functions $e_l(k, \cdot)$ whose domain is $\mathbf{Z}_3$, the group of 3-adic integers. Since $e_l(k, a) = 0$ whenever $l \geq 1$ and $3|a$, the set $\{e_l(k, \cdot) : l \geq 1, k \geq 0\}$ is a family of functions on the compact topological group $\mathbf{Z}_3^*$ of invertible 3-adic integers. This application of 3-adic integers to the $3x + 1$ Problem was first seen in Wirsching[101](1994), and similar analysis was also done by Applegate and Lagarias[8](1995). The functions $e_l(k, \cdot)$ are integrable with respect to $\mathbf{Z}_3^*$'s unique normalized Haar measure, yielding the 3-adic average

$$\bar{e}_l(k) := \int_{\mathbf{Z}_3^*} e_l(k, a) da = \frac{1}{2 \cdot 3^{l-1}} \binom{k+l}{l}.$$

Several results follow with this approach, including

$$\liminf_{n \to \infty} \frac{1}{2^n} \int_{\mathbf{Z}_3^*} s_n(a) da > 0,$$

implying that every predecessor set $P_T(a)$ with $a \not\equiv 0 \bmod 3$ has positive density.

**2.4. Other Trajectory Statistics** Gluck and Taylor[39](2002) have studied another "global statistic" besides the total stopping time function $\sigma_\infty(n)$. If $\sigma_\infty(a_1) = p$ under the map $C$, the authors define a function[2] $A$ as

$$A(a_1) = \frac{a_1 a_2 + a_2 a_3 + \cdots + a_p a_1}{a_1^2 + a_2^2 + \cdots + a_p^2}.$$

where $a_2, a_3, \ldots, a_p$ are the consecutive $C$-iterates of $a_1$. For any odd $m > 3$, they show that

$$\frac{9}{13} < A(m) < \frac{5}{7}.$$

Moreover, these bounds are sharp since

$$\lim_{n \to \infty} A\left(\frac{4^n - 1}{3}\right) = \frac{9}{13}$$

and

$$\lim_{k \to \infty} A\left(\frac{2^k(2^{3^{k-1}} + 1)}{3^k}\right) = \frac{5}{7}.$$

Gluck and Taylor also produce a normalized histogram of $A$ for values between $2^{20}$ and $2^{20} + 20001$, and its randomized counterpart. These two histograms bear a resemblance, hinting at the stochastic/probabilistic approach to this problem.

Kontorovich and Miller[45] connect the 3x+1 problem to Benford's Law. Let $y_0$ be an odd integer taken randomly from $[1, T]$ and consider the subsequence of odd numbers $\{y_k\}$ of its 3x+1 iterates (use the streamlined function $F$ mentioned at the beginning of this article). Form the sequence $\{\frac{y_k}{y_0}, 1 \leq k \leq n\}$. Then for any base B such that $\log_B 2$ is irrational of finite Diophantine type, the sequence

$$\lim_{n \to \infty} \lim_{T \to \infty} \{\frac{y_k}{y_0}, 1 \leq k \leq n\}$$

approaches the Benford distribution. Lagarias and Soundararajan[55] extend this to give a more precise version of how close the $T$-iterates are to the Benford distribution.

---

[2]Gluck and Taylor used the symbol $C$ for their new function; to avoid confusion with the Collatz function $C$, I am using the symbol $A$.

## 3. Representations of Iterates of a $3x+1$ Map

The oft-cited result in this area is due to Böhm and Sontacchi[17](1978): the $3x+1$ problem is equivalent to showing that each $n \in \mathbf{Z}^+$ may be written as

$$n = \frac{1}{3^m}\left(2^{v_m} - \sum_{k=0}^{m-1} 3^{m-k-1} 2^{v_k}\right)$$

where $m \in \mathbf{Z}^+$ and $0 \le v_0 < v_1 < \cdots v_m$ are integers. Similar results were obtained by Amigó[2, Prop.4.2](2001).

Sinai[83](2003) studies the function $F$ on the set $\sqcap$ defined as

$$\sqcap = \{1\} \cup \sqcap^+ \cup \sqcap^-, \quad \sqcap^{\pm 1} = 6\mathbf{Z}^+ \pm 1.$$

Let $k_1, k_2, \ldots, k_m \in \mathbf{Z}^+$ and $\epsilon = \pm 1$. Then the set of $x \in \sqcap^\epsilon$ to which one can apply $F^{(k_1)}F^{(k_2)}\cdots F^{(k_m)}$ is an arithmetic progression

$$\sigma^{(k_1,k_2,\ldots,k_m,\epsilon)} = \{6 \cdot (2^{k_1+k_2+\cdots+k_m} + q_m) + \epsilon\}$$

for some $q_m$ such that $1 \le q_m \le 2^{k_1+k_2+\cdots+k_m}$. Moreover,

$$F^{(k_1)}F^{(k_2)}\cdots F^{(k_m)}(\sigma^{(k_1,k_2,\ldots,k_m,\epsilon)}) = \{6 \cdot (3^m p + r) + \delta\}$$

for some $r$ and $\delta$ satisfying $1 \le r \le 3^m$, $\delta = \pm 1$. Results in this spirit may be found in Andrei $et\ al.$[5](2000) and Kuttler[51](1994). Sinai uses this result to obtain some distributional information regarding trajectories of $F$. Let $1 \le m \le M$, $x_0 \in \sqcap$ and

$$\omega\left(\frac{m}{M}\right) = \frac{\log(F^{(m)}(x_0)) - \log(x_0) + m(2\log 2 - \log 3)}{\sqrt{M}}.$$

If $0 \le t \le 1$, then $\omega(t)$ behaves like a Wiener trajectory.

Related to these results is the connection made in Blecksmith $et\ al.$[16](1998) between the $3x+1$ problem and 3-smooth representations of numbers. A number $n \in \mathbf{Z}^+$ has a 3-smooth representation if and only if there exist integers $\{a_i\}$ and $\{b_i\}$ such that

$$n = \sum_{i=1}^k 2^{a_i} 3^{b_i}, \quad a_1 > a_2 > \cdots > a_k \ge 0, \ 0 \le b_1 < b_2 < \cdots < b_k.$$

Note the similarity with the terms considered by Wirsching in the last section. These numbers were studied at least as early as Ramanujan. A 3-smooth representation of $n$ is $special\ of\ level\ k$ if

$$n = 3^k + 3^{k-1}2^{a_1} + \cdots + 3 \cdot 2^{a_{k-1}} + 2^{a_k}$$

in which every power of 3 appears up to $3^k$. For a fixed $k$, each $n$ has at most one such representation – see Lagarias[53](1990), who credits the proof to Don Coppersmith. Blecksmith $et\ al.$ then offer the reader to prove that $m \in \mathbf{Z}^+$ iterates to 1 under $C$ if and only if there are integers $e$ and $f$ such that the positive integer $n = 2^e - 3^f m$ has a special 3-smooth representation of level $k = f - 1$. The choice of $e$ and $f$ is not unique, if it exists.

FIGURE 3. Dynamics of Integers (a) *mod* 3 and (b) *mod* 4.

## 4.  Reduction to Residue Classes and Other Sets

Once one is convinced that the $3x + 1$ problem is true, a natural approach is to find subsets $S \subset \mathbf{Z}^+$ such that proving the conjecture on $S$ implies it is true on $\mathbf{Z}^+$. It is obvious this holds if $S = \{x : x \equiv 3 \bmod 4\}$ since numbers in the other residue classes decrease after one or two iterations. Puddu[74](1986) and Cadogan[22](1984) showed that this works if $S = \{x : x \equiv 1 \bmod 4\}$. The work of Böhm and Sontacchi[17](1978) implies that $m$ odd $T$-iterates of $x$ equals $(3/2)^m(x + 1) - 1$, hence odds must eventually become even. Coupling this with the dynamics of integers in $\mathbf{Z}_4$ (see Figure 3b) implies both $S = \{x : x \equiv 1 \bmod 4\}$ and $S = \{x : x \equiv 2 \bmod 4\}$ are sufficient. Similarly, Figure 3a indicates $S = \{x : x \equiv 1 \bmod 3\}$ or $S = \{x : x \equiv 2 \bmod 3\}$ suffice. Andaloro[3](2000) has improved these to $S = \{x : x \equiv 1 \bmod 16\}$. All of these sets have an easily computed positive density, the lowest being Andaloro's set $S$ at $1/16$.

Korec and Znam[47](1987) significantly improve this by showing the sufficiency of the set $S = \{x : x \equiv a \bmod p^n\}$ where $p$ is an odd prime, $a$ is a primitive root mod $p^2$, $p \nmid a$, and $n \in \mathbf{Z}^+$. This set has density $p^{-n}$ which can be made arbitrarily small. In a similar vein, Yang[103](1998) proved the sufficiency of the set

$$\left\{ n : n \equiv 3 + \frac{10}{3}(4^k - 1) \bmod 2^{2k+2} \right\}$$

for any fixed $k \in \mathbf{Z}^+$. Most recently, Monks[68] showed that any set of the form $\{A + Bn : n \geq 0\}$ with $A \geq 0$ and $B \geq 1$ is sufficient.

Korec and Znam also claimed to have a sufficient set with density zero, but details were not provided. To this end, a recent result of Monks[67](2002) is noteworthy. The last section showed how difficult it is to find a usable closed form expression for $T^{(k)}(x)$. If the "+1" was omitted from the iterations, this would yield $T^{(k)}(x) = 3^m x/2^k$, where $m$ is the number of odd terms in the first $k$ iterations of $x$. Monks[67](2002) has proven that there are infinitely many "linear versions" of the $3x+1$ problem. An example given is the map

$$R(n) = \begin{cases} \frac{1}{11}n & \text{if } 11|n \\ \frac{136}{15}n & \text{if } 15|n \text{ and } NOTA \\ \frac{5}{17}n & \text{if } 17|n \text{ and } NOTA \\ \frac{4}{5}n & \text{if } 5|n \text{ and } NOTA \\ \frac{26}{21}n & \text{if } 21|n \text{ and } NOTA \\ \frac{7}{13}n & \text{if } 13|n \text{ and } NOTA \\ \frac{1}{7}n & \text{if } 7|n \text{ and } NOTA \\ \frac{33}{4}n & \text{if } 4|n \text{ and } NOTA \\ \frac{5}{2}n & \text{if } 2|n \text{ and } NOTA \\ 7n & \text{otherwise} \end{cases}$$

where $NOTA$ means "none of the above" conditions hold. Monks shows the $3x+1$ problem is true if and only if for every positive integer $n$ the $R$-orbit of $2^n$ contains 2. The proof uses Conway's *Fractran* language[29](1987) which Conway used [28](1972) to prove the existence of a similar map whose long-term behavior on the integers was algorithmically undecidable. Connecting this material back to sufficient sets, one notes that the density of the set $\{2^n : n \in \mathbf{Z}^+\}$ is zero (albeit, one has a much more complicated map).

## 5. Cycles

Studying the structure of any possible cycles of $T$ has received much attention.

Letting $\Omega$ be a cycle of $T$, and $\Omega_{odd}$, $\Omega_{even}$ denote the odd and even terms in $\Omega$, one may rearrange the equation

$$\sum_{x\in\Omega} x = \sum_{x\in\Omega} T(x)$$

to obtain

$$\sum_{x\in\Omega_{even}} x = \sum_{x\in\Omega_{odd}} x + |\Omega_{odd}|.$$

This was noted by Chamberland[25](1999) and Monks[67](2002).

Using modular arithmetic, Figure 3 indicates the dynamics of integers under $T$ both in $\mathbf{Z}_3$ and $\mathbf{Z}_4$. Since no cycles may have all of its elements of the form 0 *mod* 3, 0 *mod* 4, or 3 *mod* 4, the figures imply that no integer cycles (except $\{0\}$) have elements of the form form 0 *mod* 3. Also, the number of terms in a cycle congruent to 1 *mod* 4 equals the number congruent to 2 *mod* 4. This analysis was presented by Chamberland[25](1999).

An early cycles result concerns a special class of $T$-cycles called circuits. A *circuit* is a cycle which may be written as $k$ odd elements followed by $l$ even elements. Davison[31](1976) showed that there is a one-to-one correspondence between circuits and solutions $(k,l,h)$ in positive integers such that

(1) $$(2^{k+l} - 3^k)h = 2^l - 1.$$

It was later shown using continued fractions and transcendental number theory (see Steiner[85](1977), Rozier[77](1990)) that equation (1) has only the solution $(1,1,1)$. This implies that $\{1,2\}$ is the only circuit. Much later, Simons[81](2005) showed that the 3x+1 problem admits no 2-cycles (2 concatenated circuits). Simons and de Wegner[82](2005) continue this approach for $m$-cycles. They show that there are no nontrivial $m$-cycles for $m \leq 68$.

For general cycles of $T$, Böhm and Sontacchi[17](1978) showed that $x \in \mathbf{Z}^+$ is in an $n$-cycle of $T$ if and only if there are integers $0 \leq v_0 \leq v_1 \leq \cdots \leq v_m = n$ such that

$$x = \frac{1}{2^n - 3^m} \sum_{k=0}^{m-1} 3^{m-k} 2^{v_k}$$

A similar result is derived by B. Seifert[79](1988).

Eliahou[34](1993) has given some strong results concerning any non-trivial cycle $\Omega$ of $T$. Letting $\Omega_0$ denote the odd terms in $\Omega$, he showed

(2) $$\log_2 \left( 3 + \frac{1}{M} \right) \leq \frac{|\Omega|}{|\Omega_o|} \leq \log_2 \left( 3 + \frac{1}{m} \right)$$

where $m$ and $M$ are the smallest and largest terms in $\Omega$. Note that for "large" cycles this implies

$$\frac{|\Omega|}{|\Omega_0|} \approx \log_2 3.$$

Eliahou uses this in conjunction with the Diophantine approximation of $\log_2 3$ and the numerical bound $m > 2^{40}$ to show that

$$|\Omega| = 301994a + 17087915b + 85137581c$$

where $a, b, c$ are nonnegative integers, $b \geq 1$ and $ac = 0$. Similar results were found by Chisala[26](1994) and Halbeisen and Hungerbühler[40](1997). This approach was pushed the farthest by Tempkin and Arteaga[89](1997). They tighten the relations in (2) and use a better lower bound on $m$ to obtain

(3) $$|\Omega| = 187363077a + 272500658b + 357638239c$$

where $a, b, c$ are nonnegative integers, $b \geq 1$ and $ac = 0$. This gives the lower bound 272,500,658 for any nontrivial cycle.

It is not apparent how the even-odd dissection of a cycle used in these results may be extended to a finer dissection of the terms in a cycle, say, mod 4. Related to this, Brox[20](2000) has proven that there are finitely many cycles such that

$$\sigma_1 < 2 \log(\sigma_1 + \sigma_3)$$

where $\sigma_i$ equals the number of terms in a cycle congruent to $i$ mod 4.

## 6. Extending $T$ to Larger Spaces

A common problem-solving technique is to imbed a problem into a larger class of problems and use techniques appropriate to that new space. Much work has been done along these lines for the $3x + 1$ problem. The subsequent subsections detail work done in increasingly larger spaces.

**6.1. The Integers Z** The first natural extension of $T$ is to all of $\mathbf{Z}$. The definition of $T$ suffices to cover this case. One soon finds three new cycles: $\{0\}$, $\{-5, -7, -10\}$, and the long cycle

$$\{-17, -25, -37, -55, -82, -41, -61, -91, -136, -68, -34\}.$$

These new cycles could also be obtained (without minus signs) if one considered the map define $T'(n) = -T(-n)$, which corresponds to the "$3x - 1$" problem. It is conjectured that these cycles are all the cycles of $T$ on $\mathbf{Z}$. This problem was considered by Seifert[79](1988).

**6.2. Rational Numbers with Odd Denominators** Just as the study of the "3x-1" problem of the last subsection is equivalent to the $3x + 1$ problem on $\mathbf{Z}^-$, Lagarias[53](1990) has extended the $3x + 1$ to the rationals by considering the class of maps

$$T_k(x) = \begin{cases} \frac{3x+k}{2} & x \equiv 1 \ (\mathrm{mod}\ 2) \\ \frac{x}{2} & x \equiv 0 \ (\mathrm{mod}\ 2) \end{cases}$$

positive $k \equiv \pm 1 \bmod 6$ with $(x, k) = 1$. He has shown that cycles for $T_k$ correspond to rational cycles $x/k$ for the $3x + 1$ function $T$. Lagarias proves there are integer cycles of $T_k$ for infinitely many $k$ (with estimates on the number of cycles and bounds on their lengths) and conjectures that there are integer cycles for all $k$. Halbeisen and Hungerbühler[40](1997) derived similar bounds as Eliahou[34](1993) on cycle lengths for rational cycles.

**6.3. The Ring of 2-adic Integers $\mathbf{Z}_2$** The next extension is to the ring $\mathbf{Z}_2$ of 2-adic integers consisting of infinite binary sequences of the form

$$a = a_0 + a_1 2 + a_2 2^2 + \cdots = \sum_{j=0}^{\infty} a_j 2^j$$

where $a_j \in \{0, 1\}$ for all non-negative integers $j$. Congruence is defined by $a \equiv a_0 \bmod 2$.

Chisala[26](1994) extends $C$ to $\mathbf{Z}_2$ but then restricts his attention to the rationals. He derives interesting restrictions on rational cycles, for example, if $m$ is the least element of a positive rational cycle, then

$$m > 2^{\frac{\lceil m \log_2 3 \rceil}{m}} - 3.$$

A more developed extension of $T$, initiated by Lagarias[52](1985), defines

$$T : \mathbf{Z}_2 \to \mathbf{Z}_2, \ T(a) := \begin{cases} a/2 & \text{if } a \equiv 0 \bmod 2 \\ (3a + 1)/2 & \text{if } a \equiv 1 \bmod 2 \end{cases}$$

It has been shown (see Mathews and Watts[61](1984) and Müller[70](1991)) that the extended $T$ is surjective, not injective, infinitely many times differentiable, not analytic, measure-preserving with respect to the Haar measure, and strongly mixing. Similar results concerning iterates of $T$ may be found in Lagarias[52](1985), Müller[70](1991),[71](1994) and Terras[91](1976). Defining the shift map $\sigma : \mathbf{Z}_2 \to \mathbf{Z}_2$ as

$$\sigma(x) = \begin{cases} \frac{x-1}{2} & x \equiv 1 \ (\mathrm{mod}\ 2) \\ \frac{x}{2} & x \equiv 0 \ (\mathrm{mod}\ 2) \end{cases},$$

Lagarias[52](1985) proved that $T$ is conjugate to $\sigma$ via the parity vector map $\Phi^{-1} : \mathbf{Z}_2 \to \mathbf{Z}_2$ defined by

$$\Phi^{-1}(x) = \sum_{k=0}^{\infty} 2^k \ (T^k(x) \bmod 2),$$

that is, $\Phi \circ T \circ \Phi^{-1} = \sigma$. Bernstein[14](1994) gives an explicit formula for the inverse conjugacy $\Phi$, namely

$$\Phi(2^{d_0} + 2^{d_1} + 2^{d_2} + \cdots) = -\frac{1}{3}2^{d_0} - \frac{1}{9}2^{d_1} - \frac{1}{27}2^{d_2} - \cdots$$

where $0 \leq d_0 < d_1 < d_2 < \dots$. He also shows that the $3x+1$ problem is equivalent to having $\mathbf{Z}^+ \subset \Phi\left(\frac{1}{3}\mathbf{Z}\right)$. Letting $\mathbb{Q}_{odd}$ denote the set of rationals whose reduced form has an odd denominator, Bernstein and Lagarias[15](1996) proved that the $3x+1$ problem has no divergent orbits if $\Phi^{-1}(\mathbb{Q}_{odd}) \subset (\mathbb{Q}_{odd})$.

Recent developments along these lines have been made by Monks and Yazinski [69](2002). Since Hedlund[41] (1969) proved that the automorphism group of $\sigma$ is simply $\mathrm{Aut}(\sigma) = \{id, V\}$, where $id$ is the identity map and $V(x) = -1 - x$, Monks and Yazinski define a function $\Omega$ as

$$\Omega = \Phi \circ V \circ \Phi^{-1}.$$

We then have that $\Omega$ is the unique nontrivial autoconjugacy of the $3x+1$ map, i.e. $\Omega \circ T = T \circ \Omega$ and $\Omega^2 = id$. Coupled with the results of Bernstein and Lagarias, the authors show that three statements are equivalent: $\Phi^{-1}(\mathbb{Q}_{odd}) \subset (\mathbb{Q}_{odd})$, $\Omega(\mathbb{Q}_{odd}) \subset (\mathbb{Q}_{odd})$, and no rational 2-adic integer has a divergent $T$-trajectory.

Monks and Yazinski also extend the results of Eliahou[34] (1993) and Lagarias [52](1985) concerning the density of "odd" points in an orbit. Let $\kappa_n(x)$ denote the number of ones in the first $n$-digits of the parity vector $x$. If $x \in \mathbb{Q}_{odd}$ eventually enters an $n$-periodic orbit, then

$$\frac{\ln(2)}{\ln(3+1/m)} \leq \lim_{n \to \infty} \frac{\kappa_n(x)}{n} \leq \frac{\ln(2)}{\ln(3+1/M)}$$

where $m, M$ are the least and greatest cyclic elements in the eventual cycle. If $x \in \mathbb{Q}_{odd}$ diverges, then

$$\frac{\ln(2)}{\ln(3)} \leq \liminf_{n \to \infty} \frac{\kappa_n(x)}{n}.$$

Monks and Yazinski define another extension of $T$, namely $\xi : \mathbf{Z}_2 \to \mathbf{Z}_2$ as

$$\xi(x) = \begin{cases} \Omega(x) & x \equiv 1 \pmod 2 \\ \frac{x}{2} & x \equiv 0 \pmod 2 \end{cases}$$

and prove that the $3x+1$ problem is equivalent to having the number one in the $\xi$-orbit of every positive.

### 6.4. The Gaussian Integers and $\mathbf{Z}_2[i]$

Joseph[43](1998) extends $T$ further to $Z_2[i]$, defining $\tilde{T}$ as

$$\tilde{T}(\alpha) = \begin{cases} \alpha/2, & \text{if } \alpha \in [0] \\ (3\alpha + 1)/2, & \text{if } \alpha \in [1] \\ (3\alpha + i)/2, & \text{if } \alpha \in [i] \\ (3\alpha + 1 + i)/2, & \text{if } \alpha \in [1+i] \end{cases}$$

where $[x]$ denotes the equivalence class of $x$ in $\mathbf{Z}_2[i]/2\mathbf{Z}_2[i]$. Joseph shows that $\tilde{T}$ is not conjugate to $T \times T$ via a $\mathbf{Z}_2$-module isomorphism, but is topologically conjugate to $T \times T$. Arguing akin to results of the last subsection, Joseph shows that $\tilde{T}$ is chaotic (in the sense of Devaney). Kucinski[50](2000) studies cycles of Joseph's extension $\tilde{T}$ restricted to the Gaussian integers $\mathbf{Z}[i]$.

### 6.5. The Real Line R

A further extension of $T$ to the real line $\mathbb{R}$ is interesting in that it allows tools from the study of iterating continuous maps. In an unpublished paper, Tempkin[90](1993) studies what is geometrically the simplest such extension: the straight-line extension to the Collatz function $C$:

$$L(x) = \begin{cases} C(x), & x \in \mathbf{Z} \\ C(\lfloor x \rfloor) + (x - \lfloor x \rfloor)(C(\lceil x \rceil) - C(\lfloor x \rfloor)), & x \notin \mathbf{Z} \end{cases}$$

$$= \begin{cases} -(5n - 2)x + n(10n - 3), & x \in [2n - 1, 2n] \\ (5n + 4)x - n(10n + 7), & x \in [2n, 2n + 1] \end{cases}.$$

Tempkin proves that on each interval $[n, n + 1]$, $n \in \mathbf{Z}^+$, $L$ has periodic points of every possible period. Capitalizing on the fact that iterates of piecewise linear functions are piecewise linear, he also shows that every eventually periodic point of $L$ is rational, every rational is either eventually periodic or divergent, and rationals of the form $k/5$ with $k \not\equiv 0 \bmod 5$ are divergent.

Tempkin also mentions a smooth extension to $C$, namely

$$E(x) := \frac{7x + 2}{4} + \frac{5x + 2}{4} \cos(\pi(x + 1))$$

but conducts no specific analysis with it. Chamberland[23](1996) studies a similar extension to $T$:

$$f(x) := \frac{x}{2} \cos^2\left(\frac{\pi x}{2}\right) + \frac{3x + 1}{2} \sin^2\left(\frac{\pi x}{2}\right)$$

$$= x + \frac{1}{4} - \frac{2x + 1}{4} \cos(\pi x).$$

Chamberland shows that any cycle on $\mathbf{Z}^+$ must be locally attractive. By also showing that the Schwarzian derivative of $f$ is negative on $\mathbb{R}^+$, this implies the long-term dynamics of almost all points coincides with the long-term dynamics of the critical points. One quickly finds that there are two attracting cycles,

$$A_1 := \{1, 2\}, \quad A_2 := \{1.192531907\ldots, 2.138656335\ldots\}.$$

Chamberland conjectures that these are the only two attractive cycles of $f$ on $\mathbb{R}^+$. This is equivalent to the $3x + 1$ problem. It is also shown that a monitonically increasing divergent orbit exists. Chamberland compactifies the map via the homeomorphism $\sigma(x) = 1/x$ on $[\mu_1, \infty)$ ($\mu_1$ is the first positive fixed point of the map $f$), yielding a dynamically equivalent map $h$ defined on $[0, \mu_1]$ as

$$h(x) = \begin{cases} \frac{4x}{4 + x - (2 + x)\cos(\pi x)}, & x \in (0, \mu_1] \\ 0, & x = 0 \end{cases}.$$

Lastly, Chamberland makes statements regarding any general extension of $f$: it must have 3-cycle, a homoclinic orbit (snap-back repeller), and a monotonically increasing divergent trajectory. To illustrate how chaos figures into this extension, Ken Monks replaced $x$ with $z$ in Chamberland's map and numerically generated the filled-in Julia set[3]. A portion of the set is indicated in Figure 4.

In a more recent paper, Dumont and Reiter[32](2003) produce similar results for the real extension

$$f(x) := \frac{1}{2}\left(3^{\sin^2(\pi x/2)} x + \sin^2(\pi x/2)\right).$$

---

[3]The Julia set of a map $f$ informally consists of those points whose long-time behavior under repeated iteration of $f$ can change drastically under arbitrarily small perturbations. The filled-in Julia set is indicated in black in Figure 4.

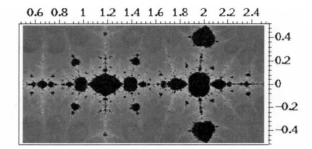

FIGURE 4. Julia set of Chamberland's map extended to the complex plane

The authors have produced several other extensions, as well as figures representing stopping times; see Dumont and Reiter[33](2001).

Borovkov and Pfeifer[18](2000) also show that any continuous extension of $T$ has periodic orbits of every period by arguing that the map is turbulent: there exist compact intervals $A_1$ and $A_2$ such that $A_1 \cup A_2 = f(A_1) \cap f(A_2)$.

Misiurewicz and Rodrigues[65](2004) considered extending $T$ to the positive real line, but with a twist. Let $T_0(x) = x/2$ and $T_1(x) = (3x+1)/2$. For a specified initial point, what happens if one applies $T_0$ or $T_1$ with equal probability? One now considers the semigroup $S$ whose generators are $T_0$ and $T_1$. The authors show that every intial point has a dense orbit, and the set of periodic points (points $x$ where there is some $A \in S$ such that $A(x) = x$) is dense. This type of map and its dynamics are commonly seen in iterated function systems. These results are extended to $T_0(x) = ax$, $T_1(x) = bx + 1$, where $0 < a < 1 < b$ by Bergelson, Misiurewicz and Senti[13](2006).

Lastly for extensions of $T$ to the real line, we mention the work of Konstandinidis [44] who considers the discontinuous funtions

$$T(x) = \begin{cases} \frac{3x+1}{2} & \lfloor x \rfloor \equiv 1 \pmod 2 \\ \frac{x}{2} & \lfloor x \rfloor \equiv 0 \pmod 2. \end{cases}$$

An advantage of this extension over the continuous ones is the main result: all cycles are contained in the integers. Konstandinidis conjectures that all positive $x$ approach $\{1, 2\}$.

**6.6. The Complex Plane $\mathbb{C}$** Letherman, Schleicher and Wood[59](1999) offer the next refinement of this approach: they extend $T$ to the complex plane with

$$f(z) := \frac{z}{2} + \frac{1}{2}\left(1 - \cos(\pi z)\right)\left(z + \frac{1}{2}\right) + \frac{1}{\pi}\left(\frac{1}{2} - \cos(\pi z)\right)\sin(\pi z) + h(z)\sin^2(\pi z).$$

Note that the first two terms (with $z$ replaced by $x$) match Chamberland's extension. The clear dynamic advantage of this new function is that the set of critical points on the real line is exactly $\mathbf{Z}$. They improve Chamberland's result by showing that on $[n, n + 1]$, with $n \in \mathbf{Z}^+$, there is a Cantor set of points that diverge monotonically. On the complex plane, Letherman *et al.* use techniques of complex dynamics to derive results about Fatou components[4]. In particular, they show that

---

[4]The Fatou set of a function $f$ is the complement of the Julia set.

no integer is in a Baker domain (domain at infinity). One concludes then that any integer either belongs to a super-attracting periodic orbit or a wandering domain.

## 7. Generalizations of $3x + 1$ Dynamics

There have been many investigations of maps which have similar dynamics to $T$ but are not extensions of $T$. Here one is usually changing the *function*, as opposed to last section where the *space* was extended.

Belaga and Mignotte[12](1998) considered the "$3x+d$" problem, and conjecture that for any odd $d \geq -1$ and not divisible by three, all integer orbits enter a finite set (hence every orbit is eventually periodic). Note that the "$3x - 1$" problem is equivalent to the $3x+1$ problem on the negative integers, considered in the previous section. Belaga[11](2003) continues this study.

Another obvious generalization is the class of "$qx + 1$" problems. Steiner[86, 87](1981) extended his cycle results and showed that for $q = 5$, there is only one non-trivial circuit ($13 \to 208 \to 13$), while $q = 7$ has no non-trivial circuits. Franco and Pomerance[37](1995) showed that if $q$ is a Wieferich number[5], then some $x \in \mathbf{Z}^+$ never iterates to one. Crandall[30](1978) conjectured that this is true for *any* odd $q \geq 5$. Wirsching[102](1998) offers a heuristic argument using $p$-adic averages that for $q \geq 5$, the $qx + 1$ problem admits either a divergent trajectory or infinitely-many different periodic orbits. Related here is the work of Volkov[98](2006) considers the 5x+1 problem and develops similar results to those seen in Section 2.3.

Mignosi[64](1995) looked at a related generalization, namely

$$T_\beta(n) = \begin{cases} \lceil \beta n \rceil & n \equiv 1 \pmod{2} \\ \frac{x}{2} & n \equiv 0 \pmod{2} \end{cases}$$

for any $\beta > 1$ and $r \in \mathbb{R}$. The case $\beta = 3/2$ is equivalent to $T$. Mignosi conjectures that for any $\beta > 1$, there are finitely many periodic orbits. This is proven for $\beta = \sqrt{2}$ and a heuristic argument is made that this conjecture holds for almost all $\beta \in (1, 2)$. Brocco[19](1995) modifies the map to

$$T_{\alpha,r}(n) = \begin{cases} \lceil \alpha n + r \rceil & n \equiv 1 \pmod{2} \\ \frac{x}{2} & n \equiv 0 \pmod{2} \end{cases}$$

for $1 < \alpha < 2$. He shows for his map that the Mignosi's conjecture is false if the interval $((r-1)/(\alpha-1), r/(\alpha-1))$ contains an odd integer and $\alpha$ is a Salem number or a PV number[6] .

Matthews and Watts[61, 62](1984,1985) deal with multiple-branched maps of the form

$$T(x) := \frac{m_i x - r_i}{d}, \text{ if } x \equiv i \bmod d$$

---

[5] An odd integer $q$ is called a *Wieferich number* if $2^{l(q)} \equiv 1 \bmod q^2$, where $l(q)$ is the order of 2 in the multiplicative group $(\mathbf{Z}/q\mathbf{Z})$. Wieferich numbers have density one in the odd numbers.

[6] A *Salem number* is a real algebraic number greater than 1 all of whose conjugates $z$ satisfy $|z| \leq 1$, with at least one conjugate satisfying $|z| = 1$. A *Pisot-Vijayaraghavan (PV) number* is a real algebraic number greater than 1 all of whose conjugates $z$ satisfy $|z| < 1$.

where $d \geq 2$ is an integer, $m_0, m_1, \ldots, m_{d-1}$ are non-zero integers, and $r_0, r_1, r_{d-1} \in \mathbf{Z}$ such that $r_i \equiv im_i \bmod d$. They mirror their earlier results (seen in the last section) by extending this map to the ring of $d$-adic integers and show it is measure-preserving with respect to the Haar measure. Information about divergent trajectories may be found in Leigh[58](1986) and Buttsworth and Matthews[21](1990). Ergodic properties of these maps has been studied by Venturini [94, 96, 97] (1982,1992,1997). An encapsulating result of [97] (1997) essentially claims that the condition $|m_0 m_1 \cdots m_{d-1}| < 1$ gives "converging" behavior, while divergent orbits may occur if $|m_0 m_1 \cdots m_{d-1}| > 1$. A special class of these maps – known as *Hasse functions*, which are "closer" to the $3x + 1$ map $T$ – have been studied by Allouche [1](1979), Heppner [42](1978), Garcia and Tal [38](1999) and Möller [66](1978), with similar results. A recent survey of these generalized mappings – with many examples – has been written by Matthews[63](2002).

Farkas[36] considered the map

$$F(n) = \begin{cases} n/3, & n \equiv 0 \pmod 3 \\ (3n + 1)/2, & n \equiv 7 \text{ or } 11 \pmod{12} \\ (n + 1)/2, & n \equiv 1 \text{ or } 5 \pmod{12} \end{cases}.$$

and showed that each positive integer iterates to one.

Seemingly farther removed from the $3x+1$ problem are results due to Stolarsky [88](1998) who completely solves a problem of "similar appearance." First, recall Beatty's Theorem which states that if $\alpha, \beta > 1$ are irrational and satisfy $1/\alpha + 1/\beta = 1$, then the sets

$$A = \{\lfloor n\alpha \rfloor : n \in \mathbf{Z}^+\}, \ B = \{\lfloor n\beta \rfloor : n \in \mathbf{Z}^+\}$$

form a partition of $\mathbf{Z}^+$. If $\alpha = \phi := (1 + \sqrt{5})/2$, we have $\beta = \phi^2 = (3 + \sqrt{5})/2$. Stolarsky considers the map

$$f(m) = \begin{cases} \lceil \lceil \frac{m}{\phi^2} \rceil \phi^2 \rceil + 1, & m \in A \\ \lceil \frac{m}{\phi^2} \rceil, & m \in B \end{cases}.$$

He proves that $f$ admits a unique periodic orbit, namely $\{3, 7\}$. Defining $b(n) = \lfloor n\beta \rfloor$, the set $\{b^{(k)}(3) : k \in \mathbf{Z}^+\}$ – which has density zero – characterizes the eventually periodic points. All other positive integers have divergent orbits. The symbolic dynamics are simple: any trajectory has an itinerary of either $B^l(AB)^\infty$ or $B^l(AB)^k A^\infty$ for some integers $k, l \geq 0$.

## 8. Miscellaneous

Margenstern and Matiyasevich[60](1999) have encoded the $3x + 1$ problem as a logical problem using one universal quantifier and several existential quantifiers. Specifically, they showed that $T^{(m)}(2a) = b$ for some $m \in \mathbf{Z}^+$ if and only if there exist $w, p, r, s \in \mathbf{Z}^+$ such that $a, b \leq w$ and

$$\binom{4(w + 1)(p + r) + 1}{p + r} \binom{pw}{s} \binom{rw}{t} \times$$
$$\binom{2w + 1}{w} \binom{2s + 2t + r + b((4w + 3)(p + r) + 1)}{3a + (4w + 4)(3t + 2r + s)} \times$$
$$\binom{p + r}{p} \binom{3a + (4w + 4)(3t + 2r + s)}{2s + 2t + r + b((4w + 3)(p + r) + 1)} \equiv 1 \bmod 2.$$

Farkas[**36**] considers the multiplicative semigroup generated by $\{(2n+1)/(3n+2) : n \geq 1\} \cup \{2\}$. If the 3x+1 Problem is true, then the semigroup contains all the positive integers. Applegate and Lagarias[**10**](2006) proved that this semigroup indeed contains all the positive integers. Similar semigroups and associated problems (such as the so-called "Weak 3x+1 Problem") may be found in Lagarias[**54**](2006) and Applegate and Lagarias[**10**](2006).

**Acknowledgement**: I owe a huge debt to the earlier surveys of Lagarias and Wirsching on this problem, and the annotated bibliography. The author is grateful to Jeff Lagarias and Ken Monks for sharing some the most recent developments on this problem, and to Ken Monks, Toni Guillamon i Grabolosa and the anonymous referee for many helpful suggestions.

## 9.  *

## References

[1] J.-P. Allouche. *Sur la conjecture de "Syracuse-Kakutani-Collatz"*. Séminaire de Théorie des Nombres, 1978–1979, Exp. No. 9, 15 pp., CNRS, Talence, 1979.

[2] J.M. Amigo. *Accelerated Collatz Dynamics*. Centre de Recerca Matemàtica preprint, no. 474, July 2001.

[3] P. Andaloro. *On total stopping times under $3x+1$ iteration*. Fibonacci Quarterly, $\underline{38}$, (2000), 73–78.

[4] P. Andaloro. *The $3x+1$ problem and directed graphs*. Fibonacci Quarterly, $\underline{40}$(1), (2002), 43–54.

[5] S. Andrei, M. Kudlek and R.S. Niculescu. *Some Results on the Collatz Problem*. Acta Informatica , $\underline{37}$(2), (2000), 145–160.

[6] D. Applegate and J. Lagarias. *Density Bounds for the $3x+1$ Problem. I. Tree-search method.* Mathematics of Computation, $\underline{64}$, (1995), 411–426.

[7] D. Applegate and J. Lagarias. *Density Bounds for the $3x+1$ Problem. II. Krasikov-Inequalities.* Mathematics of Computation, $\underline{64}$, (1995), 427–438.

[8] D. Applegate and J. Lagarias. *The Distribution of $3x+1$ Trees*. Experimental Mathematics, $\underline{4}$(3), (1995), 193–209.

[9] D. Applegate and J. Lagarias. *Lower Bounds for the Total Stopping Time of $3x+1$ Iterates*. Mathematics of Computation, $\underline{72}$(242), (2002), 1035–1049.

[10] D. Applegate and J. Lagarias. *The $3x+1$ semigroup*. Journal of Number Theory, $\underline{177}$, (2006), 146–159.

[11] E. Belaga. *Effective polynomial upper bounds to perigees and numbers of (3x+d)- cycles of a given odd length*. Acta Arithmetica, $\underline{106}$(2), (2003), 197–206.

[12] E. Belaga and M. Mignotte. *Embedding the $3x+1$ Conjecture in a $3x+d$ Context*. Experimental Mathematics, $\underline{7}$(2), (1998), 145–151.

[13] V. Bergelson, M. Misiurewicz and S. Senti. *Affine Actions of a Free Semigroup on the Real Line* eprint: arxiv:math.DS/0601473, (2006).

[14] D. Bernstein. *A Non-iterative 2-adic Statement of the $3N+1$ Conjecture* . Proceedings of the American Mathematical Society, $\underline{121}$, (1994), 405–408.

[15] D. Bernstein and J. Lagarias. *The $3x+1$ Conjugacy Map*. Canadian Journal of Mathematics, $\underline{48}$, (1996), 1154–169.

[16] R. Blecksmith, M. McCallum and J. Selfridge. *3-Smooth Representations of Integers*. American Mathematical Monthly, $\underline{105}$(6), (1998), 529–543.

[17] C. Böhm and G. Sontacchi. *On the Existence of Cycles of given Length in Integer Sequences like $x_{n+1} = x_n/2$ if $x_n$ even, and $x_{n+1} = 3x_n+1$ otherwise*. Atti della Accademia Nazionale dei Lincei. Rendiconti. Classe di Scienze Fisiche, Matematiche e Naturali. Serie VIII , $\underline{64}$, (1978), 260–264.

[18] K. Borovkov and D. Pfeifer. *Estimates for the Syracuse Problem via a Probabilistic Model*. Theory Probab. Appl., $\underline{45}$, (2000), 300–310.

[19] S. Brocco. *A note on Mignosi's generalization of the* $(3X + 1)$-*problem*. Journal of Number Theory, 52(2), (1995), 173–178.

[20] T. Brox. *Collatz Cycles with Few Descents*. Acta Arithmetica, 92(2), (2000), 181–188.

[21] R. Buttsworth and K. Matthews. *On some Markov matrices arising from the generalized Collatz mapping* . Acta Arithmetica, 55(1), (1990), 43–57.

[22] C. Cadogan. *A Note on the* $3x + 1$ *Problem*. Caribbean Journal of Mathematics, 3, (1984), 67–72.

[23] M. Chamberland. *A Continuous Extension of the* $3x+1$ *Problem to the Real Line*. Dynamics of Continuous, Discrete and Impulsive Systems, 2, (1996), 495–509.

[24] M. Chamberland. *Una actualizacio del problema 3x+1 (Catalan, translated by Toni Guillamon i Grabolosa) [An Update on the 3x+1 Problem]* . Butlleti de la Societat Catalana de Matematiques, 18, (2003), 19–45.

[25] M. Chamberland. Announced at the "Roundatable Discussion", International Conference on the Collatz Problem and Related Topics, August 5-6, 1999, Katholische Universität Eichstätt, Germany .

[26] B. Chisala. *Cycles in Collatz Sequences*. Publicationes Mathematicae Debrecen, 45, (1994), 35–39.

[27] K. Conrow. http : //www − personal.ksu.edu/~kconrow/gentrees.html.

[28] J. Conway. *Unpredictable Iterations*. Proceedings of the Number Theory Conference (University of Colorado, Boulder), (1972), 49–52.

[29] J. Conway. *FRACTRAN: A Simple Universal Programming Language for Arithmetic*. Open Problems in Communication and Computation (Ed. T.M. Cover and B. Gopinath), Springer, New York, (1987), 4–26.

[30] R.E. Crandall. *On the "*$3x + 1$*" Problem*. Mathematics of Computation, 32, (1978), 1281–1292.

[31] J. Davison. *Some Comments on an Iteration Problem*. Proceedings of the Sixth Manitoba Conference on Numerical Mathematics , (1976), 155–159.

[32] J. Dumont and C. Reiter. *Real Dynamics Of A 3-Power Extension Of The* $3x + 1$ *Function*. Dynamics of Continuous, Discrete and Impulsive Systems, (2003), to appear.

[33] J. Dumont and C. Reiter. *Visualizing Generalized 3x+1 Function Dynamics*. Computers & Graphics, 25(5), (2001), 883–898.

[34] S. Eliahou. *The* $3x + 1$ *Problem: New Lower Bounds on Nontrivial Cycle Lengths*. Discrete Mathematics, 118, (1993), 45–56.

[35] C.J. Everett. *Iteration of the Number-Theoretic Function* $f(2n) = n,\ f(2n + 1) = 3n + 2$. Advances in Mathematics, 25, (1977), 42–45.
H. M. Farkas (2005), Variants of the 3N + 1 problem and multiplicative semigroups, In: Geometry, Spectral Theory, Groups and Dynamics: Proceedings in Memory of Robert Brooks, Contemporary Math., Volume 387, Amer. Math. Soc., Providence, 2005, pp. 121127.

[36] H. M. Farkas. *Variants of the 3N + 1 problem and multiplicative semigroups*. Geometry, Spectral Theory, Groups and Dynamics: Proceedings in Memory of Robert Brooks, Contemporary Mathematics, American Mathematical Society, Providence, 387, (2005), 121–127.

[37] Z. Franco and C. Pomerance. *On a Conjecture of Crandall concerning the* $qx + 1$ *Problem*. Mathematics of Computation, 64(211), (1995), 1333–1336.

[38] M. Garcia and F. Tal. *A Note on the Generalized* $3n + 1$ *Problem*. Acta Arithmetica, 90(3), (1999), 245–250.

[39] D. Gluck and B. Taylor. *A New Statistic for the* $3x+1$ *Problem*. Proceedings of the American Mathematical Society, 130(5), (2002), 1293–1301.

[40] L. Halbeisen and N. Hungerbühler. *Optimal Bounds for the Length of Rational Collatz Cycles*. Acta Arithmetica, 78(3), (1997), 227–239.

[41] G. Hedlund. *Endomorphisms and Automorphisms of the Shift Dynamical System*. Mathematical Systems Theory, 3, (1969), 320–375.

[42] E. Heppner. *Eine Bemerkung zum Hasse-Syracuse Algorithmus*. Archiv der Mathematik, 31(3), (1978), 317–320.

[43] J. Joseph. *A Chaotic Extension of the* $3x + 1$ *Function to* $Z_2[i]$. Fibonacci Quarterly, 36(4), (1998), 309–316.

[44] P. Konstadinidis. *The Real* $3x + 1$ *Problem*. Acta Arithmetica, 122(1), (2006), 35–44.

[45] A. Kontorovich and S. Miller. *Benfords law, values of L-functions and the 3x+1 problem*. Acta Arithmetica, 120(3), (2005), 269–297.

[46] I. Korec. *A Density Estimate for the $3x + 1$ Problem.* Mathematica Slovaca, 44(1), (1994), 85–89.

[47] I. Korec and Š. Znám. *A Note on the $3x + 1$ Problem.* American Mathematical Monthly, 94, (1987), 771–772.

[48] I. Krasikov. *How many numbers satisfy the $3X + 1$ conjecture?* International Journal of Mathematics and Mathematical Sciences , 12, (1989), 791–796.

[49] I. Krasikov and J. Lagarias. *Bounds for the $3x + 1$ Problem using Difference Inequalities.* arXiv:math.NT/0205002 v1, April 30, 2002.

[50] G. Kucinski. *Cycles of the $3x + 1$ Map on the Gaussian Integers.* Preprint dated May, 2000.

[51] J. Kuttler. *On the $3x + 1$ Problem.* Advances in Applied Mathematics, 15, (1994), 183–185.

[52] J. Lagarias. *The $3x + 1$ Problem and its Generalizations.* American Mathematical Monthly, 92, (1985), 1–23. Available online at www.cecm.sfu.ca/organics/papers.

[53] J. Lagarias. *The Set of Rational Cycles for the $3x + 1$ Problem.* Acta Arithmetica, 56, (1990), 33–53.

[54] J. Lagarias. *Wild and Wooley Numbers.* American Mathematical Monthly, 113, (2006), 97–108.

[55] Lagarias and Soundararajan. *Benfords Law for the 3x + 1 Function,* Journal of the London Mathematical Society, 74, (2006), 289–303.

[56] J. Lagarias and A. Weiss. *The $3x + 1$ Problem; Two Stochastic Models.* Annals of Applied Probability, 2, (1992), 229–261.

[57] J. Lagarias. *$3x + 1$ Problem Annotated Bibliography.* http://arxiv.org/abs/math/0608208, April 22, 2008.

[58] G.M. Leigh. *A Markov process underlying the generalized Syracuse algorithm.* Acta Arithmetica, 46(2), (1986), 125–143.

[59] S. Letherman, D. Schleicher, and R. Wood. *The $3n + 1$-Problem and Holomorphic Dynamics.* Experimental Mathematics, 8(3), (1999), 241–251.

[60] M. Margenstern and Y. Matiyasevich. *A binomial representation of the $3x + 1$ problem.* Acta Arithmetica, 91(4), (1999), 367–378.

[61] K. Matthews and A.M. Watts. *A Generalization of Hasse's Generalization of the Syracuse Algorithm.* Acta Arithmetica, 43(2), (1984), 167–175.

[62] K. Matthews and A.M. Watts. *A Markov Approach to the generalized Syracuse Algorithm.* Acta Arithmetica, 45(1), (1985), 29–42.

[63] K. Matthews. http : //www.maths.uq.edu.au/~krm, August 15, 2002.

[64] F. Mignosi. *On a generalization of the $3x + 1$ problem.* Journal of Number Theory, 55(1), (1995), 28–45.

[65] M. Misiurewicz and A. Rodrigues. *Real $3x + 1$* Proceedings of the American Mathematical Society, 133(4), (2004), 1109–1118.

[66] H. Möller. *Über Hasses Verallgemeinerung des Syracuse-Algorithmus (Kakutanis Problem).* Acta Arithmetica, 34(3), (1978), 219–226.

[67] K. Monks. *$3x + 1$ Minus the +.* Discrete Mathematics and Theoretical Computer Science, 5(1), (2002), 47–54.

[68] K. Monks. *The sufficiency of arithmetic progressions for the 3x + 1 conjecture.* Proceedings of the American Mathematical Society, 134(10), (2006), 2861–2872.

[69] K. Monks and J. Yazinski. *The Autoconjugacy of the $3x + 1$ Function.* Discrete Mathematics, to appear.

[70] H. Müller. *Das '$3n + 1$' Problem.* Mitteilungen der Mathematischen Gesellschaft in Hamburg, 12, (1991), 231–251.

[71] H. Müller. *Über eine Klasse 2-adischer Funktionen im Zusammenhang mit dem '$3x + 1$'-Problem.* Abhandlungen aus dem Mathematischen Seminar der Universität Hamburg , 64, (1994), 293–302.

[72] T. Oliveira e Silva. *Maximum excursion and stopping time record-holders for the $3x + 1$ problem: computational results.* Mathematics of Computation, 68(225), (1999), 371–384.

[73] T. Oliveira e Silva. http : //www.ieeta.pt/~tos/3x + 1.html.

[74] S. Puddu. *The Syracuse Problem (spanish).* Notas de la Sociedad de Matemtica de Chile , 5, (1986), 199-200.

[75] D. Rawsthorne. *Imitation of an Iteration.* Mathematics Magazine, 58, (1985), 172–176.

[76] E. Roosendaal. The $3x + 1$ Problem, (web document). http : //personal.computrain.nl/eric/wondrous/.

[77] O. Rozier. *Démonstration de l'absense de cycles d'une certain forme pour le Problème de Syracuse.* Singularité, <u>1</u>, (1990), 9–12.

[78] J. Sander. *On the (3N + 1)-Conjecture.* Acta Arithmetica, <u>55</u>, (1990), 241–248.

[79] B. Seifert. *On the Arithmetic of Cycles for the Collatz-Hasse ('Syracuse') Problem.* Discrete Mathematics, <u>68</u>, (1988), 293–298.

[80] D. Shanks. *Comments on Problem 63-13.* SIAM Review, <u>7</u>, (1965), 284–286.

[81] J. Simons. *On the Nonexistence of 2-Cycles for the 3x+1 Problem* Mathematics of Computation, <u>74</u>251, (2005), 1565–1572.

[82] J. Simons and B. de Weger. *Theoretical and computational bounds for m-cycles of the 3n+1-problem.* Acta Arithmetica, <u>117</u>(1), (2005), 51–70.

[83] Y. Sinai. *Statistical (3x+1) - Problem.* Communications on Pure and Applied Mathematics, <u>56</u>(7), (2003), 1016–1028.

[84] A. Slakmon and L. Macot. *On the Almost Convergence of Syracuse Sequences* Statistics and Probability Letters, <u>76</u>(15), (2006), 1625–1630.

[85] R. Steiner. *A Theorem on the Syracuse Problem.* Proceedings of the Seventh Manitoba Conference on Numerical Mathematics and Computing , (1977), 553–559.

[86] R. Steiner. *On the "QX + 1 problem", Q odd.* Fibonacci Quarterly, <u>19</u>(3), (1981), 285–288.

[87] R. Steiner. *On the "QX+1 problem", Q odd. II.* Fibonacci Quarterly, <u>19</u>(4), (1981), 293–296.

[88] K. Stolarsky. *A Prelude to the 3x + 1 Problem.* Journal of Difference Equations and Applications, <u>4</u>, (1998), 451–461.

[89] J. Tempkin and S. Arteaga. *Inequalities Involving the Period of a Nontrivial Cycle of the 3n + 1 Problem.* Draft of Circa October 3, 1997.

[90] J. Tempkin. *Some Properties of Continuous Extensions of the Collatz Function.* Draft of Circa October 5, 1993.

[91] R. Terras. *A Stopping Time Problem on the Positive Integers.* Acta Arithmetica, <u>30</u>, (1976), 241–252.

[92] R. Terras. *On the Existence of a Density.* Acta Arithmetica, <u>35</u>, (1979), 101–102.

[93] T. Urvoy. *Regularity of congruential graphs.* Mathematical foundations of computer science 2000 (Bratislava), 680–689, Lecture Notes in Computer Science, <u>1893</u>, Springer, Berlin, (2000), see also http : //www.irisa.fr/galion/turvoy/ .

[94] G. Venturini. *Behavior of the iterations of some numerical functions. (Italian).* Istituto Lombardo. Accademia di Scienze e Lettere. Rendiconti. Scienze Matematiche e Applicazioni. A , <u>116</u>, (1982), 115–130.

[95] G. Venturini. *On the 3x + 1 problem.* Advances in Applied Mathematics, <u>10</u>(3), (1989), 344–347.

[96] G. Venturini. *Iterates of number-theoretic functions with periodic rational coefficients (generalization of the 3x + 1 problem).* Studies in Applied Mathematics, <u>86</u>(3), (1992), 185–218.

[97] G. Venturini. *On a generalization of the 3x + 1 problem.* Advances in Applied Mathematics, <u>19</u>(3), (1997), 295–305.

[98] S. Volkov. *A probabilistic model for the 5k + 1 problem amd related problems.* Stochastic Processes and Applications, <u>116</u>(4), (2006), 662-674.

[99] S. Wagon. *The Collatz Problem.* Mathematical Intelligencer, <u>7</u>(1), (1985), 72–76.

[100] G. Wirsching. *An Improved Estimate concerning 3n+1 Predecessor Sets.* Acta Arithmetica, <u>63</u>, (1993), 205–210.

[101] G. Wirsching. *A Markov Chain underlying the Backward Syracuse Algorithm.* Revue Roumaine de Mathématiques Pures et Appliquées, <u>39</u>, (1994), 915–926.

[102] G. Wirsching. *The Dynamical System Generated by the 3n + 1 Function.* Springer, Heidelberg, (1998)

[103] Z.H. Yang. *An Equivalent Set for the 3x + 1 Conjecture.* Journal of South China Normal University, Natural Science Edition, no.2, (1998), 66–68.

[104] R. Zarnowski. *Generalized Inverses and the Total Stopping Times of Collatz Sequences.* Linear and Multilinear Algebra, <u>49</u>(2), (2001), 115–130.

DEPARTMENT OF MATHEMATICS AND STATISTICS, GRINNELL COLLEGE, GRINNELL IA, 50112, USA

*E-mail address:* chamberl@math.grinnell.edu

# Generalized $3x + 1$ Mappings: Markov Chains and Ergodic Theory

## K.R. Matthews

ABSTRACT. This paper reviews connections of the $3x + 1$ mapping and its generalizations with ergodic theory and Markov chains. The work arose out of efforts to describe the observed limiting frequencies of divergent trajectories in the congruence classes $(\bmod\, m)$.

## 1. Introduction

One of the most tantalizing conjectures in number theory is the so–called $3x+1$ conjecture, attributed to Lothar Collatz [**6**]. Let the Collatz mapping $C : \mathbb{Z} \to \mathbb{Z}$ be defined by

$$(1) \qquad C(x) = \begin{cases} x/2 & \text{if } x \equiv 0 \,(\bmod\, 2) \\ 3x + 1 & \text{if } x \equiv 1 \,(\bmod\, 2). \end{cases}$$

Collatz conjectured that if $x \geq 1$, then the trajectory

$$x,\ C(x),\ C^2(x), \ldots$$

eventually reaches the cycle $1, 4, 2, 1$, also denoted by $\langle 1, 4, 2 \rangle$. If $x \in \mathbb{Z}$, it is also conjectured that the trajectory $\{C^k(x)\}$ eventually reaches this cycle or one of the cycles

(a) $\langle -1 \rangle$;
(b) $\langle -5, -14, -7, -20, -10 \rangle$;
(c) $\langle -17, -50, -25, -74, -37, -110, -55, -164, -82, -41, -122, -61,$
$\quad -182, -91, -272, -136, -68, -34 \rangle$;
(d) $\langle 0 \rangle$.

Another version of the problem iterates the $3x + 1$ function, $T(x)$, defined by

$$(2) \qquad T(x) = \begin{cases} x/2 & \text{if } x \equiv 0 \,(\bmod\, 2) \\ (3x + 1)/2 & \text{if } x \equiv 1 \,(\bmod\, 2). \end{cases}$$

The conjecture here is that a positive integer $x$ has iterates

$$x,\ T(x),\ T^2(x), \ldots$$

that eventually reach the cycle $\langle 1, 2 \rangle$.

The Collatz map and the $3x+1$ map are both special cases of generalized Collatz mappings, introduced in Section 2, about which conjectural (heuristic) predictions can be made concerning the behaviour of trajectories. Broadly speaking, there are similar conjectures that can be made about such mappings which seem equally unsolvable and tantalizing.

Since 1981, in a series of papers [**26, 27, 25, 20, 4, 22**], the author and collaborators were led to connections with ergodic theory and Markov chains. This analysis proceeds by studying the behaviour of the iterates modulo $d$. This is modelled by a Markov chain, which if irreducible, predicts that most trajectories spend a fixed fraction of time in each congruence class mod $d$ (for a given number of steps). This permits one to compute an asymptotic exponential growth rate of the size of iterates; if it exceeds 1, one predicts most trajectories will diverge; while if it is between 0 and 1, one predicts almost all trajectories will enter cycles. More generally, one can also analyze the behaviour of iterates modulo $m$ for an arbitrary modulus $m$ (where it simplifies things to require that $d$ divides $m$). It turns out experimentally that divergent trajectories also possess some regularity of distribution of iterates in congruence classes mod $m$ for arbitrary modulus $m$.

It is easy to describe the conjectural picture for mappings of relatively prime type, as defined in section 2, a class which include the $3x+1$ mapping. The Markov chains take a fairly simple form in this case. However the Markov chains are more difficult to describe for maps of non-relatively prime type, and this case leads to many open problems.

There are natural generalizations to other rings of interest to number theorists, namely $GF(q)[x]$ and the ring of integers of an algebraic number field. The conjectural picture is not so clear with $GF(q)[x]$. Also with number fields, it is not just a matter of studying finitely many cycles in the ring of integers. It seems likely there are in addition to finitely many cycles, finitely many "lower dimensional" $T$–invariant subsets within which divergent trajectories move in a regular manner. We illustrate these possibilities with maps in $GF(2)[x], \mathbb{Z}[\sqrt{2}]$ and $\mathbb{Z}[\sqrt{3}]$.

For an account of other work on the $3x+1$ problem, we refer the reader to Lagarias' survey [**18**], Wirsching's book [**33**] and Guy's problem book [**12**, pages 215–218].

## 2. Generalized $3x+1$ mappings

Let $d \geq 2$ be a positive integer and $m_0, \ldots, m_{d-1}$ be non–zero integers. Also for $i = 0, \ldots, d-1$, let $r_i \in \mathbb{Z}$ satisfy $r_i \equiv im_i \,(\mathrm{mod}\, d)$. Then the formula

$$(3) \qquad T(x) = \frac{m_i x - r_i}{d} \qquad \text{if } x \equiv i \,(\mathrm{mod}\ d)$$

defines a mapping $T : \mathbb{Z} \to \mathbb{Z}$ called the *generalized Collatz mapping* or *generalized $3x+1$ mapping*. In this definition we allow $\gcd(m_i, r_i, d) > 1$, so that the denominator in lowest terms of (3) for a given $x \equiv i \,(\mathrm{mod}\ d)$ may be any divisor of $d$.

The minimal integer $d = d(T) \geq 1$ such that the mapping is affine on each residue class (mod $d$) is called the *modulus* of the mapping $T$.

As examples, the Collatz mapping $C(x)$ in (1) corresponds to parameter choices $d = 2$, $m_0 = 1\, m_1 = 6$, $r_0 = 0$, $r_1 = -2$ in (3) while the $3x+1$ mapping corresponds to the choices $d = 2$, $m_0 = 1$, $m_1 = 3$, $r_0 = 0$, $r_1 = -1$.

**EQUIVALENT FORM**: *An equivalent form of* (3) *is:*

$$(4) \qquad T(x) = \left\lfloor \frac{m_i x}{d} \right\rfloor + a_i \qquad \text{if } x \equiv i \ (mod\ d),$$

where $a_0, \ldots, a_{d-1}$ are any given integers and $\lfloor y \rfloor$ denotes the largest integer no larger than $y$. (Here $a_i = \frac{s_i - r_i}{d}$ where $0 \le s_i < d$ has $s_i \equiv i m_i \ (mod \ d)$.)

We say that a generalized $3x+1$ map $T$ is of *relatively prime type* if all multipliers $m_i$ are relatively prime to $d$, i.e., $gcd(m_0 m_1 \cdots m_{d-1}, d) = 1$, and otherwise it is of *non-relatively prime type*. If no restrictions are put on $T$, we call it of *general type*. For example, the $3x+1$ map (2) is of relatively prime type, and the Collatz map (1) is of non-relatively prime type.

The behaviour of generalized $3x+1$ maps (at least conjecturally) is quite well understood in the relatively prime case, as we shall explain in sections 3 and 4. We describe auxiliary models which appear to make accurate predictions how trajectories behave in the relatively prime case. These models continue to work well for many maps in the non-relatively prime case; however many mysteries remain for maps in this case.

Starting in 1982 , Tony Watts and I investigated the frequency of occupation of congruence classes modulo $m$ of trajectories for generalized 3x+1 functions of relatively prime type. Here we consider a general integer modulus $m$, not necessarily related to $d$. The distribution of occupation frequencies in the particular case of congruence classes modulo $d$ is directly relevant to the growth in size of its members of a trajectory, since it determines the number of multipliers $m_i$ occurring in the iteration. In effect we construct auxiliary Markov model problems which lead to predictions of when trajectories should enter cycles or divergent trajectories, as given in Conjecture 3.1 below. These models are described in the next section.

In the remainder of this section we state basic formulas for forward and backwards iteration of a generalized Collatz map.

We first give a basic formula on forward iteration of a generalized Collatz map. Let $T$ be a generalized Collatz mapping and let $T^K(x)$ represent its $K$-th iterate. If $T^K(x) \equiv i \,(mod \ d)$, $0 \le i < d$, we define $m_K(x) = m_i, r_K(x) = r_i$ so that

$$(5) \qquad T^{K+1}(x) = \frac{m_K(x) T^i(x) - r_K(x)}{d}.$$

THEOREM 2.1.  *Let $T$ be a generalized Collatz mapping, and let $T^K(x)$ represent its $K$-th iterate. Then*

(i)  *There holds*

$$(6) \qquad T^K(x) = \frac{m_0(x) \cdots m_{K-1}(x)}{d^K} \left( x - \sum_{i=0}^{K-1} \frac{r_i(x) d^i}{m_0(x) \cdots m_i(x)} \right).$$

(ii)  *If $T^i(x) \ne 0$ for all $i \ge 0$, then*

$$(7) \qquad T^K(x) = \frac{m_0 \cdots m_{K-1}(x)}{d^K} x \prod_{i=0}^{K-1} \left( 1 - \frac{r_i(x)}{m_i(x) T^i(x)} \right).$$

Proof. Both parts follow by induction on $K$, using equation (5).  □

The following result gives basic facts on allowable sequences of congruence classes modulo $m$, which underlies the analysis of backwards iteration.

Let $B(j, m)$ denote the congruence class consisting of integers $j$ modulo $m$, i.e.,

$$B(j, m) = \{x \in \mathbb{Z} : x \equiv j \,(mod \ m)\}.$$

THEOREM 2.2. *Let $T$ be a generalized Collatz map with modulus $d$. Let $B(j, m)$ denote the congruence class consisting of integers $x \equiv j \pmod{m}$. Then*

(i) *The set $T^{-1}(B(j, m))$ is a disjoint union of $N_{j,m} \geq 0$ congruence classes $\pmod{md}$, where*

$$N_{j,m} = \sum_{\substack{i = 0 \\ \gcd(m_i, m) | j - T(i)}}^{d-1} \gcd(m_i, m),$$

*where $T(i) = \frac{m_i i - r_i}{d}$. In particular, in the relatively prime case, where $\gcd(m_i, m) = 1$ for $i = 0, \ldots, d - 1$, all $N_{j,m} = d$. In the general case it is possible that some $N_{i,m} = 0$.*

(ii) *In the relatively prime case, the $d^\alpha$ cylinders*

(8)     $$C(i_0, i_1, \ldots, i_{\alpha-1}; d) = B(i_0, d) \cap T^{-1}(B(i_1, d)) \cap \cdots \cap T^{-(\alpha-1)}(B(i_{\alpha-1}, d))$$

*comprise the complete set of $d^\alpha$ congruence classes modulo $d^\alpha$.*

(iii) *In the relatively prime case, if $A = B(j, d^\alpha)$ and $B = B(k, d^\beta)$, then $T^{-K}(A) \cap B$ is a disjoint union of $d^{K-\beta}$ congruence classes mod $d^{K+\alpha}$, if $K \geq \beta$.*

Proof. Properties (i) and (ii) are proved in [**26**]. Property (ii) has been the basis of many papers on the subject. Property (iii) then follows from Property (ii) by expressing $A$ and $B$ as disjoint unions of cylinders. □

In 4.2 we associate to this data a *Markov matrix* $Q_T(m)$.

## 3. Relatively Prime Case: Growth Rate of Iterates

In this section we treat generalized Collatz maps of relatively prime type. In section 3.1 we formulate a general conjecture on the existence of cycles and divergent trajectories. In section 3.2 we describe proved results supporting this conjecture.

### 3.1. Conjecture on Cycles and Divergent Trajectories. Matthews and Watts [**26**] analyzed the relatively prime case and formulated the following conjectures, and supported them with computer evidence.

CONJECTURE 3.1. *For generalized Collatz mappings (3) of relatively prime type,*

(i) *If $|m_0 \cdots m_{d-1}| < d^d$, then all trajectories $\{T^K(n)\}$, $n \in \mathbb{Z}$, eventually cycle. In particular there always must be at least one cycle.*

(ii) *If $|m_0 \cdots m_{d-1}| > d^d$, then almost all trajectories $\{T^K(n)\}$, $n \in \mathbb{Z}$ are divergent (that is, $T^K(n) \to \pm\infty$, except for an exceptional set $S$ of integers $n$ satisfying $\#\{n \in S| - X \leq n \leq X\} = o(X)$.) In particular, there must exist at least one divergent trajectory.*

(iii) *The number of cycles is finite.*

(iv) *If the trajectory $\{T^K(n)\}$, $n \in \mathbb{Z}$ is not eventually cyclic, then the iterates are uniformly distributed $\pmod{d^\alpha}$ for each $\alpha \geq 1$:*

$$\lim_{N \to \infty} \tfrac{1}{N+1} card\{K \leq N | T^K(n) \equiv j \pmod{d^\alpha}\} = \frac{1}{d^\alpha},$$

*for $0 \leq j \leq d^\alpha - 1$.*

We note that the equality case

$$(9) \qquad\qquad |m_0 \cdots m_{d-1}| = d^d$$

which is not covered in (i) and (ii) can never occur due to the relative primality condition $gcd(m_0 m_1 \cdots m_{d-1}, d) = 1$. Thus Conjecture 3.1 covers all generalized Collatz maps $T$ of relatively prime type.

The first three parts of Conjecture 3.1 are generalizations of earlier conjectures of Möller [28], who studied a special subclass of mappings (3). Also the last part of (i) generalizes a conjecture of Lagarias [19] concerning the existence of at least one cycle.

Note that in these conjectures the prediction of overall cycling or divergence is independent of the remainder values $r_i$ in (3), or equivalently of the integers $a_i$ in (1)). This is a nice feature of relatively prime type mappings; in the non-relatively prime case, for fixed $(d, m_0, m_1, ..., m_{d-1})$ the choice of the integers $a_i$ may affect the cycling or divergence behaviour of trajectories. Another difference with the general case is that the condition (9) may occur for some $T$ of non-relatively prime type.

These conjectures (i)-(iv) all appear intractable for general maps of relatively prime type. However apart from (iv) they can be proved in some very special cases, namely where the map $T$ is strictly increasing (resp. strictly decreasing) on all sufficiently large $|n|$.

In Section 5 we present various examples of such maps, which give supporting empirical evidence in favor of parts (i)-(iv) of Conjecture 3.1. In the next two subsections we present theorems which also support parts of this conjecture.

**3.2. Growth Rate Heuristic for Forward Iteration.** It follows from Theorem 2.1 that the size of iterates is determined by the fraction of the time the iterate spends in the class $i$ (mod $d$). Suppose over the first $K$ steps a trajectory spends time $f_i$ in the class $i$ (mod $d$). Then formula Theorem 2.1(2) suggests that

$$(10) \qquad\qquad T^K(x) \sim \prod_{i=0}^{d-1} \left(\frac{m_i}{d}\right)^{f_i} x = \frac{1}{d} \left(\prod_{i=0}^{d-1} m_i^{f_i}\right) x,$$

For mappings of relatively prime type, it turns out that the distribution describing an "average" input is uniform distribution, where all $f_i = \frac{1}{d}$.

Assuming that the $T^K(x)$ are uniformly distributed mod $d$, we would conclude from (10) that the iterates $|T^K(x)|$ grow geometrically at some rate (either expanding or contracting). For, on taking logarithms on the right side of Theorem 2.1 (2), we obtain

$$\log|T^K(x)| = \sum_{i=0}^{K-1} \log|m_i(x)| + \log x - K \log d + \sum_{i=0}^{K-1} \log\left|1 - \frac{r_i(x)}{m_i(x)T^i(x)}\right|.$$

Then as $a_i = r_i(x)/(m_i(x)T^i(x)) \to 0$, we have $b_i = \log|1 - a_i| \to 0$ and hence $\frac{1}{K}(b_0 + \cdots + b_{K-1}) \to 0$. Consequently

$$\frac{1}{K} \log|T^K(x)| = \frac{1}{K} \sum_{i=0}^{K-1} \log|m_i(x)| - \log d + o(1).$$

Hence if the $T^K(x)$ are uniformly distributed mod $d$, we deduce

$$\frac{1}{K} \log |T^K(x)| \to \frac{1}{d} \sum_{i=0}^{d-1} \log |m_i| - \log d,$$

and hence

$$|T^K(x)|^{1/K} \to \frac{(|m_0 \cdots m_{d-1}|)^{1/d}}{d}.$$

(See [**27**, Theorem 1.1(b)] for a more general statement.)

This supports parts (i) and (ii) of Conjecture 3.1. One can also see it as justifying part (iii) in the case that the growth rate constant is smaller than one. In this case one expects that the iterates will contract to a finite basin, in which there can be only finitely many periodic orbits.

In fact in the forward iteration approach one can look at the growth of size of iterates, for a fixed number $K$ of steps, averaged over all inputs $-N \le x \le N$, as $N \to \infty$, and precisely justify this heuristic in this limiting situation.

THEOREM 3.1. *Let $T$ be a generalized Collatz mapping of relatively prime type with modulus $d$. Then the first $K$ steps of the iteration are periodic for all $x$ (mod $d^K$), and the values of iterates*

$$(x, T(x), T^2(x), \cdots, T^{K-1}(x)) \; (mod \; d)$$

*take on all $d^K$ possible values (mod $d$). That is, the values of the first $K$ classes are uniformly distributed (mod $d$).*

Proof. This follows from Theorem 2.2, part (2).

Theorem 3.1 does not rigorously establish the heuristic, which is concerned with behaviour in a different limiting case, where $x$ is fixed and $K \to \infty$. To switch between the two limiting cases, requires establishing ergodic behaviour, which we discuss further in section 6.

## 4. Relatively Prime Case: Markov Models and Ergodic Sets

In this section we consider the behaviour modulo $m$ of the iterates of a generalized Collatz map of relatively prime type. We can consider this problem in two ways: one is for a fixed number of iterations, averaging over all inputs, the second is on all iterates of a divergent trajectory.

### 4.1. Conjecture on Distribution Modulo $m$. 
We considered many examples of generalized Collatz maps of relatively prime type which conjecturally have divergent trajectories. Computer experiments indicated that for each $m > 1$, every divergent trajectory eventually occupies certain congruence classes mod $m$ with positive limiting frequencies, depending on $T$. We make the following conjecture.

CONJECTURE 4.1. *Let $T$ be a generalized Collatz mapping of relatively prime type. Then for each integer $m \ge 2$, and each trajectory viewed modulo $m$ will have a limiting frequency of iterates $f_i$ in each residue class $i$ (mod $m$), which depends on the initial starting point of the trajectory.*

Stated this way, the conjecture is obviously true for any trajectory that enters a periodic orbit; the limiting frequencies are completely determined by the periodic orbit. The interesting case is that of divergent trajectories.

A simple example is given by the $5x + 1$ mapping. Here divergent trajectories are not known to exist, but conjecturally most trajectories will be divergent (by Conjecture 3.1), and we can experimentally sample the distribution modulo $m$ of apparently divergent trajectories. Now consider the case of modulus $m = 5$. One can show that any trajectory starting from a non–zero integer will eventually visit the class $3 \, (\text{mod} \, 5)$ and thereafter remain in the $T$–invariant set $\mathbb{Z} - B(0, 5)$. Here we find experimentally the elements in each divergent trajectory appear to occupy the congruence classes $0, 1, 2, 3, 4 \, (\text{mod} \, 5)$ with limiting frequencies $\mathbf{v} = (0, \frac{1}{15}, \frac{2}{15}, \frac{8}{15}, \frac{4}{15})$ respectively.

We explain this conjecture below, using Markov chain models for iteration $(\text{mod} \ m)$. We use such models to formulate a more precise version of this conjecture below, which gives a way to predict the limiting frequencies in any particular case.

For the $5x + 1$ problem these rational numbers are the limits

$$(11) \qquad \rho_j = \lim_{K \to \infty} \mu\{B(i, 5) \cap T^{-K}(B(j, 5))\} / \mu\{B(i, 5)\}, \quad 0 \le j \le 4,$$

where $\rho_j$ is independent of $i$ and we define $\mu(S) = \frac{r}{m}$ if $S$ is a disjoint union of $r$ congruence classes mod $m$. (Here $\mu$ is a finitely additive measure on $\mathbb{Z}$.) Also $\mu\{B(i, 5) \cap T^{-K}(B(j, 5))\} / \mu\{B(i, 5)\}$ is the $(i, j)$ element of the $K$–th power of a Markov matrix $Q_T(5)$ defined by the recipe in section 4.2.

**4.2. Markov chains: maps of relatively prime type.** Now we describe Markov chain models $Q_T(m)$ for describing the iterates modulo $m$ of a generalized Collatz mapping $T$ of relatively prime type.

It was implicit in the proof of [**27**, Lemma 2.8] that if $B$ is the cylinder:

$$B = B(i_0, m) \cap T^{-1}(B(i_1, m)) \cap \cdots \cap T^{-K}(B(i_K, m)),$$

then

$$(12) \qquad \mu(B) = q_{i_0 i_1}(m) \cdots q_{i_{K-1} i_K}(m) \mu\{B(i_0, m)\},$$

where

$$(13) \qquad q_{ij}(m) = \mu\{B(i, m) \cap T^{-1}(B(j, m))\} / \mu\{B(i, m)\} \ 0 \le i, j \le m - 1,$$

Then (see [**27**, Lemma 2.9]), the matrix

$$Q_T(m) = [q_{ij}(m)]$$

is an $m \times m$ *Markov matrix*, i.e., a matrix whose elements are non–negative and whose rows sum to unity. That is, it is the transition matrix of a finite Markov chain, see [**14**] or [**29**]. (Here we are using the transpose of the matrix used in [**27**].) If $d | m$, a simple formula exists for $q_{ij}(m)$:

$$q_{ij}(m) = \begin{cases} \frac{1}{d} & \text{if } T(i) \equiv j \, (\text{mod} \, \frac{m}{d}) \\ 0 & \text{otherwise.} \end{cases}$$

If $d$ does not divide $m$, the formula for $q_{ij}(m)$ becomes more complicated.

With $p_{ij}(m) = dq_{ij}(m)$, we have $[p_{ij}(m)]^K = [p_{Kjk}(m)]$, where $p_{Kjk}(m)$ is the number of congruence classes $(\text{mod} \, md^K)$ that constitute $B(i, m) \cap T^{-K}(B(j, m))$. (See [**27**, Lemma 2.8]). This result also holds under the more general condition

$$\gcd(m_i, d^2) = \gcd(m_i, d), \ 0 \le i < d,$$

provided $d$ divides $m$.

Then for example, Theorem 2.2 (3) tells us that with $m = d^\alpha$, $\{Q(m)\}^K = \frac{1}{d^\alpha}H$ if $K \geq \alpha$, where $H$ has all its entries 1.

For information on Markov chains, the reader may consult Grimmett [11] or Kemeny, Snell and Knapp [15].

To introduce Markov chains, we need a probability space, which we take to be the Prüfer ring $\hat{\mathbb{Z}}$. (See [30] or [10, pages 7–11].) Like the $d$–adic integers, this ring can be defined as a completion of $\mathbb{Z}$. The congruence class $\{x \in \hat{\mathbb{Z}} | x \equiv j \,(\mathrm{mod}\, m)\}$ is also denoted by $B(j, m)$. Then our finitely additive measure $\mu$ on $\mathbb{Z}$ extends to a probability Haar measure on $\hat{\mathbb{Z}}$.

Equation (12) can then be interpreted as showing that the sequence of random set–valued functions $Y_K(x) = B(T^K(x), m)$, $x \in \hat{\mathbb{Z}}$, forms a Markov chain with states $B(0, m), \ldots, B(m-1, m)$, with transition probabilities $q_{ij}(m)$, given by equation (13). For equation (12) can be rewritten as

$$Pr\left(Y_0(x) = B(i_0, m), \ldots Y_K(x) = B(i_K, m)\right) = q_{i_0 i_1}(m) \cdots q_{i_{K-1} i_K}(m)/m.$$

Then from an ergodic theorem for Markov chains (Durrett [7, Example 2.2, page 341]) we have the following result:

PROPOSITION 4.1. *Let $\mathcal{C}$ be a positive recurrent class and for each $B \in \mathcal{C}$, let $\rho_B$ be the component of the unique stationary distribution over $\mathcal{C}$.*

$$Pr\left(\lim_{K\to\infty} \tfrac{1}{K+1} card\{n; n \leq K,\ Y_n(x) = B\} = \rho_B | Y_n(x) \text{ enters } \mathcal{C}\right) = 1.$$

*In other words, if $\mathcal{S}_C$ is the union of the congruence classes of $\mathcal{C}$,*

$$Pr\left(\lim_{K\to\infty} \tfrac{1}{K+1} card\{n; n \leq K,\ T^n(x) \in B\} = \rho_B | T^n(x) \text{ enters } \mathcal{S}_C\right) = 1.$$

This, together with computer evidence, leads to formulation of the following sharper form of Conjecture 4.1.

CONJECTURE 4.2. *Let $T$ be a generalized Collatz mapping of relatively prime type, having a divergent trajectory. Then for each integer $m \geq 2$, this divergent trajectory viewed modulo $m$ will eventually enter some positive recurrent ergodic component $\mathcal{S}_C$ of the associated Markov chain $Q_T(m)$, and will occupy each congruence class $B$ of $\mathcal{C}$ with positive limiting frequency $\rho_B$, given by the invariant probability distribution on $\mathcal{C}$.*

We mention two other asymptotic density conjectures:

CONJECTURE 4.3. *If $\{T^K(n)\}$ is a divergent trajectory starting in $\mathcal{S}_C$, then*

$$(14) \quad \lim_{N\to\infty} \tfrac{1}{N+1} card\{K \leq N | T^K(n) \equiv i_0 \,(mod\ m), \ldots, T^{K+k}(n) \equiv i_k \,(mod\ m)\}$$
$$= q_{i_0 i_1}(m) \cdots q_{i_{k-1} i_k}(m)/m.$$

$$(15) \quad \lim_{N\to\infty} \tfrac{1}{N+1} card\{K \leq N | T^K(n) \equiv j \,(mod\ md)\}$$
$$= \frac{1}{d} \lim_{N\to\infty} \tfrac{1}{N+1} card\{K \leq N | T^K(n) \equiv j \,(mod\ m)\}.$$

Now we present an example.

EXAMPLE 4.1. The $3x + 1$ mapping with $m = 3$. Here

$$Q_T(3) = \begin{bmatrix} \frac{1}{2} & 0 & \frac{1}{2} \\ 0 & 0 & 1 \\ 0 & \frac{1}{2} & \frac{1}{2} \end{bmatrix}.$$

Then

$$\{Q_T(3)\}^k = \begin{bmatrix} \frac{1}{2^k} & \frac{1}{3}(1 + \frac{(-1)^k}{2^{k+1}}) - \frac{1}{2^{k+1}} & \frac{1}{3}(2 - \frac{(-1)^k}{2^{k+1}}) - \frac{1}{2^{k+1}} \\ 0 & \frac{1}{3}(1 + \frac{(-1)^k}{2^{k-1}}) & \frac{1}{3}(2 - \frac{(-1)^k}{2^{k-1}}) \\ 0 & \frac{1}{3}(1 - \frac{(-1)^k}{2^k}) & \frac{1}{3}(2 + \frac{(-1)^k}{2^k}) \end{bmatrix}.$$

Also $\mathcal{C} = \{B(1,3), B(2,3)\}$ and $\{Q_T(3)\}^k \to \begin{bmatrix} 0 & 1/3 & 2/3 \\ 0 & 1/3 & 2/3 \\ 0 & 1/3 & 2/3 \end{bmatrix}.$

We remark that in papers [**4, 22, 27**], the matrices $Q_T(m)$ and sets $\mathcal{S}_\mathcal{C}$ were studied in some detail for mappings $T$ of relatively prime type. The structure of these sets can be quite complicated.

The positive recurrent classes can be determined numerically using an algorithm from [**9**]. Also see implementation [**24**].

It seems likely that there are finitely many sets $\mathcal{S}_\mathcal{C}$ as $m$ varies, if and only if $T(\mathbb{Z}) = \mathbb{Z}$.

## 5. Relatively Prime Case: Examples

Our first example is a strict generalization of the $3x + 1$ problem, usually called the $3x + k$ problem.

EXAMPLE 5.1.

$$T_k(x) = \begin{cases} x/2 & \text{if } x \equiv 0 \,(\mathrm{mod}\, 2) \\ (3x + k)/2 & \text{if } x \equiv 1 \,(\mathrm{mod}\, 2), \end{cases}$$

where $k$ is an odd integer.

Here, as with the $3x + 1$ problem, Conjecture 3.1 predicts that all trajectories appear to reach one of finitely many cycles. Here we observe empirically that the number of cycles for $T_{5^t}$ appears to strictly increase with $t$. (See Fig. 1.) On the other hand, in [**19**] it is observed that the only cycles for $T_{3^t}$ are exactly $3^t$ times those of $T_1$.

Cycles for the $3x + k$ problem were studied in Lagarias [**19**]. He showed that there exist $k$ having an arbitrarily large (finite) number of cycles. He noted that the number of cycles as a function of $k$ empirically appeared to be a complicated function of $k$, and that numbers $k$ of the form $k = 2^a - 3^b$ for $a \geq 1$ appeared to have an unusually large number of cycles. In fact, as pointed out by Alan S. Jones in 2002, if $k = 2^{2c+1} - 9$, the numbers $2^n + 3, 1 \leq n \leq c$, generate $c$ different cycles of length $2c + 1$. Actually the paper of Lagarias [**19**] studied an equivalent problem, concerning rational cycles of the $3x + 1$ problem, and noted that rational cycles with denominator $k$ correspond in a one-to-one fashion with integral cycles of the $3x + k$ probem.

For $k = 371$, we believe there are 9 cycles (lengths in parentheses):

$$0\,(1), -371\,(1), 371\,(2), -1855\,(3), -6307\,(11), 25\,(222), 265\,(4), 721\,(29), -563\,(14).$$

| t | # of cycles for $T_{5^t}$ | max cycle–length |
|---|---|---|
| 0 | 5 | 11 |
| 1 | 10 | 27 |
| 2 | 13 | 34 |
| 3 | 17 | 118 |
| 4 | 19 | 118 |
| 5 | 21 | 165 |
| 6 | 23 | 433 |

FIGURE 1. Observed number of cycles for $T_{5^t}$, $0 \le t \le 6$.

We next consider cycles in a further generalization of the $3x + 1$ mapping.

EXAMPLE 5.2.

$$T(x) = \begin{cases} (x + a)/2 & \text{if } x \equiv 0 \,(\text{mod}\, 2) \\ (3x + b)/2 & \text{if } x \equiv 1 \,(\text{mod}\, 2), \end{cases}$$

where $a$ is even, $b$ odd.

We expect all iterates to eventually cycle, with finitely many cycles including the following:

(i) $\langle a \rangle$;
(ii) $\langle -b \rangle$;
(iii) $\langle b + 2a, 2b + 3a \rangle$;
(iv) $\langle -5b - 4a, -7b - 6a, -10b - 9a \rangle$;
(v) $\langle -17b - 16a, -25b - 24a, -37b - 36a, -55b - 54a, -82b - 81a,$
$-41b - 40a, -61b - 60a, -91b - 90a, -136b - 135a, -68b - 67a, -34b - 33a \rangle$.

The reader is referred to the author's CALC number theory calculator program [**23**] for a cycle-finding program called `cycle`.

By choosing $m_0, \ldots, m_{d-1}$ so that their product is close to, but less than $d^d$ in absolute value, one expects some trajectories to take many iterations to reach a cycle and for some cycles to be rather long. For example:

EXAMPLE 5.3.

$$T(x) = \begin{cases} x/4 & \text{if } x \equiv 0 \,(\text{mod}\, 4) \\ (3x - 3)/4 & \text{if } x \equiv 1 \,(\text{mod}\, 4) \\ (5x - 2)/4 & \text{if } x \equiv 2 \,(\text{mod}\, 4) \\ (17x - 3)/4 & \text{if } x \equiv 3 \,(\text{mod}\, 4). \end{cases}$$

Here $1 \cdot 3 \cdot 5 \cdot 17 = 255 = 4^4 - 1 < 4^4$, so this map falls in case (i) of Conjecture 3.1. We have found 17 cycles, starting at values $0, -3, 2, 3, 6,$ (period 1747), $-18,$ $-46, -122, -330, -117, -137, -186, -513$ (period 1426), $-261, -333, 5127, -5205$.

Regarding the divergent trajectory part of conjecture 3.1(ii), the simplest example where things are evident numerically, but defy proof, is the $5x + 1$ mapping:

EXAMPLE 5.4. The $(5x + 1)/2$ Problem.

$$T(x) = \begin{cases} x/2 & \text{if } x \equiv 0 \,(\text{mod}\, 2) \\ (5x + 1)/2 & \text{if } x \equiv 1 \,(\text{mod}\, 2). \end{cases}$$

Probabilistic models for behaviour of the $5x + 1$ problem indicate that most trajectories should be divergent. Some models are detailed at length in Volkov [32]. The trajectory $\{T^K(7)\}$ appears to be divergent. However $T$ is known to have 5 cycles, with starting values 0, 1, 13, 17, −1, and infinitely many integers have trajectories entering these cycles.

We next consider the original map attributed to Collatz (see [18]), which is a permutation. It first appeared in print in Klamkin [16] in 1963.

EXAMPLE 5.5. (Collatz-Klamkin)

$$T(x) = \begin{cases} 2x/3 & \text{if } x \equiv 0 \,(\text{mod}\,3) \\ (4x - 1)/3 & \text{if } x \equiv 1 \,(\text{mod}\,3) \\ (4x + 1)/3 & \text{if } x \equiv 2 \,(\text{mod}\,3). \end{cases}$$

The map $T$ is a 1–1 mapping and its inverse is the 4–branched mapping

$$T^{-1}(x) = \begin{cases} 3x/2 & \text{if } x \equiv 0 \,(\text{mod}\,4) \\ (3x + 1)/4 & \text{if } x \equiv 1 \,(\text{mod}\,4) \\ 3x/2 & \text{if } x \equiv 2 \,(\text{mod}\,4) \\ (3x - 1)/4 & \text{if } x \equiv 3 \,(\text{mod}\,4). \end{cases}$$

The trajectory $\{T^K(8)\}$ appears to be divergent. There are 9 known cycles, with starting values: $-44, -4, -2, -1, 0, 1, 2, 4, 44$.

The set of generalized $3x + 1$ mappings is closed under composition: If $T_1, T_2$ have $d_1, d_2$ branches respectively, then $T = T_2 T_1$ has $d_1 d_2$ branches. For example if we take $T_1, T_2$ to be the $3x + 1, 5x + 1$, mappings, then we obtain the following map.

EXAMPLE 5.6.

$$T(x) = \begin{cases} x/4 & \text{if } x \equiv 0 \,(\text{mod}\,4) \\ (3x + 1)/4 & \text{if } x \equiv 1 \,(\text{mod}\,4) \\ (5x + 2)/4 & \text{if } x \equiv 2 \,(\text{mod}\,4) \\ (15x + 7)/4 & \text{if } x \equiv 3 \,(\text{mod}\,4). \end{cases}$$

Here all trajectories appear to enter cycles, with 7 known cycles, having starting values: $-749, -2, 0, 1, 7, 10, 514$.

Now we consider examples which conjecturally have divergent trajectories, and we consider the behaviour of iterates modulo $m$ for various $m$.

EXAMPLE 5.7. The $(5x+1)/2$ mapping with $m = 5$. The Markov chain formed by states 0, 1, 2, 3, 4 mod 5 has transition matrix

$$Q_T(5) = \begin{bmatrix} 1/2 & 0 & 0 & 1/2 & 0 \\ 0 & 0 & 0 & 1 & 0 \\ 0 & 1/2 & 0 & 1/2 & 0 \\ 0 & 0 & 0 & 1/2 & 1/2 \\ 0 & 0 & 1/2 & 1/2 & 0 \end{bmatrix}.$$

The set of states $\mathcal{C} = \{1, 2, 3, 4 \,(\text{mod } 5)\}$ form a positive recurrent class with limiting probabilities $\mathbf{v} = (\frac{1}{15}, \frac{2}{15}, \frac{8}{15}, \frac{4}{15})$.

EXAMPLE 5.8. The $(5x - 3)/2$ mapping with $3|m$. This example, discovered by Tony Watts, was a very interesting one. Empirical data suggested there exist two measures on $\hat{\mathbb{Z}}$ with respect to which $T$ is ergodic and whose values on congruence classes give the observable frequencies of occupation of integer congruence classes.

With $m = 15$, the Markov chain formed by states $0, \ldots, 14 \pmod{15}$ has two positive classes:

$$\mathcal{C}_1 = \{1, 2, 4, 7, 8, 11, 13, 14 \,(\mathrm{mod}\,15)\}, \quad \mathcal{C}_2 = \{3, 6, 9, 12 \,(\mathrm{mod}\,15)\},$$

with limiting probabilities

$$\mathbf{v}_1 = \left( \frac{4}{15}, \frac{1}{30}, \frac{1}{15}, \frac{1}{30}, \frac{2}{15}, \frac{4}{15}, \frac{2}{15}, \frac{1}{15} \right), \quad \mathbf{v}_2 = \left( \frac{4}{15}, \frac{8}{15}, \frac{2}{15}, \frac{1}{15} \right),$$

respectively.

## 6. Ergodic Theory

One innovation in Matthews and Watts [26] was the introduction of ergodic theory (see [3]). It uses an extension of mapping $T$ to a mapping of the $d$–adic integers $\mathbb{Z}_d$ into itself. The $d$-adic integers $\mathbb{Z}_d$ can be regarded as a completion of $\mathbb{Z}$, consisting of all formal sums

$$x = \sum_{i=0}^{\infty} a_i d^i, \quad a_i \in \{0, 1, \ldots, d-1\},$$

with addition and multiplication done as with ordinary positive integers, by "carrying the digit". (See [17] or [21].) Here $\mathbb{Z}_d$ is a complete metric space, under the $d$-adic metric

$$d(x, y) = |x - y|_d,$$

in which $|\cdot|_d$ is the $d$-adic norm, given by $|0|_d = 0$ and $|x|_d = d^{-k}$, where $k = \min_{a_j \neq 0}(j)$, if $x \neq 0$. We note that the integers $\mathbb{Z}$ are dense in $\mathbb{Z}_d$ in its metric topology.

THEOREM 6.1. *Let $T : \mathbb{Z} \to \mathbb{Z}$ be a generalized Collatz mapping, with modulus $d$. Then there is a unique continuous extension $\hat{T} : \mathbb{Z}_d \to \mathbb{Z}_d$, where $\mathbb{Z}_d$ denotes the $d$-adic integers.*

Proof. The behaviour of the map $\hat{T}(x)$ for $x \in \mathbb{Z}_d$ is determined by its first $d$-adic digit $x \equiv a_0 \pmod{d}$, which determines its congruence class $a_0 \equiv i \pmod{d}$, and $\hat{T}(x) = \frac{m_i x - r_i}{d}$ is a $d$-adic integer. This defines a continuous map of $\mathbb{Z}_d$ to itself, because the sets $\{\alpha \in \mathbb{Z}_d : \alpha \equiv i \pmod{d}$ are both open and closed in the $d$-adic topology. It obviously agrees with the map $T$ on $\mathbb{Z}$, and since $\mathbb{Z}$ is dense in $\mathbb{Z}_d$ it is unique. $\square$

This extension makes sense for an arbitrary generalized Collatz mapping. For such mappings of relatively prime type, one can say much more. The space $\mathbb{Z}_d$ is a compact group under addition, and as such has a canonical Haar measure, which may be normalized to have volume 1. This measure is a (Borel) measure called the *$d$-adic measure* $\mu_d$, which is completely determined by its values

$$\mu_d(B(j, d^\alpha)) = \mu_d(\{x \in \mathbb{Z}_d : \ x \equiv j \,(\mathrm{mod}\ d^\alpha)\}) = \frac{1}{d^\alpha}.$$

Here we note that the cylinders $B(j, d^\alpha)$, originally defined on $\mathbb{Z}$, now make sense on $\mathbb{Z}_d$.

THEOREM 6.2. *Let $T : \mathbb{Z} \to \mathbb{Z}$ be a generalized Collatz mapping, with modulus $d$, that is of relatively prime type, and let $\hat{T} : \mathbb{Z}_d \to \mathbb{Z}_d$ denote its continuous extension to $\mathbb{Z}_d$.*

*(1) The mapping $\hat{T}$ is measure-preserving for the d-adic measure. That is,*

$$\mu(\hat{T}^{-1}(A)) = \mu(A),$$

*where $A$ is a Haar–measurable set in $\mathbb{Z}_d$.*

*(2) The mapping $\hat{T}$ is strongly mixing. That is,*

$$\lim_{K \to \infty} \mu(\hat{T}^{-K}(A) \cap B) = \mu(A)\mu(B)$$

*for all Haar–measurable sets $A$ and $B$ in $\mathbb{Z}_d$.*

*(3) The map $\hat{T}$ is ergodic. That is, $\hat{T}^{-1}(A) = A \Rightarrow \mu(A) = 0$ or $1$. In particular, there holds*

$$\lim_{N \to \infty} \tfrac{1}{N+1} \, card \{K \le N | \hat{T}^K(x) \equiv j \,(mod \; d^\alpha)\} = \frac{1}{d^\alpha}$$

*for almost all $x \in \mathbb{Z}_d$.*

Proof. (1) Let us assume that $T$ is a generalized Collatz mapping of relatively prime type. Then by Theorem 2.2, Property (ii), the inverse image of a congruence class mod $d^\alpha$ is the disjoint union of $d$ classes mod $d^{\alpha+1}$. This gives the measure-preserving property on cylinders, i.e., $\hat{T}^{-1}(B(j, d^\alpha))$ has $\mu_d$-measure equal to $\frac{1}{d^\alpha}$. The measure preserving property holds in general since the cylinders generate the Borel sets.

(2) Theorem 2.2, Property (iii) implies the strongly mixing property on cylinders $B(j, d^\alpha)$, whence it holds for general measurable sets.

(3) The strong mixing property for a measure-preserving map implies ergodicity of $\hat{T}$. Now the Ergodic Theorem [**3**, page 12] applied to the measurable set $B(j, d^\alpha)$ gives the required property

$$\lim_{N \to \infty} \tfrac{1}{N+1} \, card \{K \le N | \hat{T}^K(x) \equiv j \,(\mathrm{mod} \; d^\alpha)\} = \frac{1}{d^\alpha},$$

for almost all $x \in \mathbb{Z}_d$. $\square$

For a map of relatively prime type, applying the ergodic theorem to the cylinder (8) gives a similar result for the $d$–adic integers $x$ which satisfy

$$\hat{T}^K(x) \equiv i_0 \,(\mathrm{mod}\, d), \ldots, \hat{T}^{K+\alpha-1}(x) \equiv i_{\alpha-1} \,(\mathrm{mod}\, d),$$

namely, for almost all $x \in \mathbb{Z}_d$ these have a limiting density $\frac{1}{d^\alpha}$.

Theorem 6.2 implies that, on the level of ergodic theory, for generalized Collatz maps of relatively prime type, the predictions of limiting densities of iterates (mod $d^\alpha$) of trajectories are correct, if one takes a generic input $x \in \mathbb{Z}_d$.

A much stronger ergodic property can be proved for some generalized Collatz maps of relatively prime type.

THEOREM 6.3. *Let $T$ be the $3x + 1$ map. There is a $3x + 1$ conjugacy map $C_T : \mathbb{Z}_2 \to \mathbb{Z}_2$, which is an automorphism such that*

$$C_T \circ \hat{T} = \hat{S}_2 \circ C_T,$$

*where $\hat{S}_2$ is the 2-adic shift map*

$$S_2(x) = \frac{x - a_0}{2}, \; for \; x \equiv a_0 \;(\mathrm{mod}\; 2).$$

Here the 2-adic shift mapping simply chops off the first 2-adic digit and shifts the 2-adic expansion one unit to the right; it has the maximal amount of mixing possible. This result is proved in Lagarias [18] and is studied further in Bernstein [1] and Bernstein and Lagarias [2]. The $3x+1$ conjecture can be encoded in properties of this conjugacy map. An analogous result should hold for all generalized Collatz mappings of relatively prime type (by a similar proof): they are conjugate to the $d$-adic shift map.

Finally we note another nice feature of the $d$-adic extension for maps of relatively prime type.

THEOREM 6.4. (Möller) *Let $T$ be a generalized Collatz mapping of relatively prime type. Then every $d$–adic integer $x \in \mathbb{Z}_d$ possesses an $d$–adically convergent series given by:*

$$(16) \qquad x = \sum_{i=0}^{\infty} \frac{r_i(x)d^i}{m_0(x) \cdots m_i(x)} \ \ \textit{if } x \in \mathbb{Z}_d.$$

Proof. Möller [28, p. 221] proves this result for a related mapping $H$. Equation (6) holds for $d$-adic integers and gives

$$m_0(x) \cdots m_{K-1}(x)x \equiv \sum_{i=0}^{K-1} r_i(x)d^i m_{i+1}(x) \cdots m_{K-1}(x) \, (\mathrm{mod} \, d^K)$$

and this is equivalent to (16), as $m_0(x), \ldots, m_{K-1}(x)$ are units in $\mathbb{Z}_d$.

Theorem 6.4 tells us that the congruence classes mod $d$ occupied by the iterates of $x \in \mathbb{Z}_d$ in fact completely determine $x$. A corresponding expansion is used to advantage later in Section 8 for a mapping $T : GF(2)[x] \to GF(2)[x]$. (See mapping (27) below.)

What happens from the ergodic theory viewpoint for generalized Collatz maps not of relatively prime type? Here we enter mysterious terrain, which is much more complicated. The first issue is that the $d$-adic measure need not be an invariant measure for the map $T : \mathbb{Z}_d \to \mathbb{Z}_d$. There is the issue of finding an invariant Borel measure for this map. We do not know of a general construction of an invariant measure on $\mathbb{Z}_d$, although it can be done for special classes of maps. Also, it appears that there are cases where there may be more than one invariant measure that is absolutely continuous with respect to the $d$-adic measure. When an invariant measure is constructed, there then remains the issue of determining ergodicity of the maps for such measures.

## 7. General Case: Leigh and Venturini Markov chains

In 1983, George Leigh, then a 4th year mathematics student at the University of Queensland, suggested that to predict the frequencies mod $m$ of divergent trajectories for the generalized Collatz mapping (3), we should restrict $m$ to be a multiple of $d$ and allow the states of the Markov chain to be all congruence classes in $\hat{\mathbb{Z}}$ of the form $mk$, where $k$ divides some power of $d$. The idea is *to keep track of how much information we have on the congruence classes to which an iterate belongs.*

Leigh's viewpoint leads to chains with states with the startling feature that the congruence classes for different states may sometimes have one included inside another. For example, in the mapping of Example 8.5 below, which has modulus $d = 8$, if we start off in state $B(4, 8)$, then we know $T(x)$ is in $B(0, 32)$ (here we keep track only of powers of 2 in the modulus) Then $B(0, 32)$ is all the information we have about where $T(x)$ is located. All the information is in the current state, regardless of the previous states of the trajectory. It turns out that this chain also has a state $B(0, 8)$, which must be kept separate from $B(0, 32)$.

Leigh introduced two related Markov chains, denoted by $\{X_n\}$ and $\{Y_n\}$. We shall consider here only the latter one. (Leigh proved the two chains contain equivalent information.) To define his Markov chain $\{Y_n\}$, we need some definitions: Let $m_i = b_i d_i$, where $b_i \in \mathbb{Z}$, $d_i \in \mathbb{N}$ and $\gcd(d, b_i) = 1$, where $d_i$ divides some power of $d$, $0 \le i < d$. Let $d | m$. We define a sequence of random functions on $\hat{\mathbb{Z}} : x \to Y_n(x) \in \mathcal{B}$, where $\mathcal{B}$ is the collection of congruence classes of the form $B(j, mk)$, $k$ dividing some power of $d$:

(1) The set of states $\{\mathcal{B}\}$ that the chain can reach, are defined recursively by the following recipe:

    (a) $Y_0(x) = B(x, m)$;
    (b) $Y_{n+1}(x) = B(T^{n+1}(x), mk_{n+1})$, where

$$(17) \qquad k_0 = 1 \text{ and } k_{n+1} = \frac{d_j k_n}{\gcd(d_j k_n, d)},$$

and where $j$ is determined by $T^n(x) \equiv j \pmod{d}$, $0 \le j \le d - 1$.

(2) Transition probabilities $q_{BB'}$ are defined for $B$, $B' \in \mathcal{B}$, as follows: Let $B = B(j, mk)$, $B' = B(j', mk')$. Then

$$(18)$$
$$q_{BB'} = \begin{cases} \frac{kd_j}{k'd} = \frac{\gcd(kd_j, d)}{d} & \text{if } k' = \frac{kd_j}{\gcd(kd_j, d)} \text{ and } T(j) \equiv j' \pmod{mkd_j/d}, \\ 0 & \text{otherwise.} \end{cases}$$

Then (see [**20**, page 133]), the set–valued functions $\{Y_n(x)\}$ form a Markov chain with transition probabilities $q_{BB'}$, by virtue of the equation

$$\Pr(Y_0(x) = B_0, \dots, Y_n(x) = B_n) = q_{B_0 B_1} q_{B_1 B_2} \cdots q_{B_{n-1} B_n}/m.$$

Leigh gives a recursive scheme for constructing the states reached. This shows that at stage $n$, there are $d/\gcd(d_j k_n, d))$ possibilities for $Y_{n+1}(x)$ and that all these possibilities arise for suitable initial values $x$. The scheme depends on the following result:

LEMMA 7.1. *If $Y_n(x) = B(j, M), 0 \le j < M$ and $M' = Md_j/d$, then*

$$(19) \qquad Y_{n+1}(x) = B(T(j) + tM', M''),$$

*where $M'' = \operatorname{lcm}(M', m)$ and $0 \le t < \frac{M''}{M'} = d/\gcd(d_j k_n, d)$.*

*Conversely if $0 \le t < \frac{M''}{M}$, there exists a $y$ such that (19) holds with $x$ replaced by $y$.*

REMARK 7.1. The converse follows from a fundamental equation of Leigh ([**20**, (42) p. 132]), which asserts that there exist $U, w$, with $\gcd(w, d) = 1$, such that for all $a$

$$(20) \qquad T^{n+1}(x + aU) = T^{n+1}(x) + awM'$$

and the fact that $\frac{M''}{M'}$ divides a power of $d$.

The recursive step is depicted as follows, in Leigh's notation:

$$B(j, M) \overset{H}{\to} B(T(j), M') \overset{G}{\to} B(T(j), M'')$$
$$\to B(T(j) + M', M'')$$
$$\cdots$$
$$\to B(T(j) + (\tfrac{M''}{M'} - 1)M', M'').$$

Now let $\mathcal{C}$ be a positive recurrent class (assuming that one exists) and for each $B \in \mathcal{C}$, let $\rho_B$ be the corresponding limiting probability. Then from the result of Durrett [7] mentioned earlier, we have

$$Pr\left(\lim_{K \to \infty} \tfrac{1}{K+1} \operatorname{card} \{n; n \leq K, \ Y_n(x) = B\} = \rho_B | Y_n(x) \text{ enters } \mathcal{C}\right) = 1.$$

To find the limiting frequency $p_i$ of occupancy of a particular congruence class $B(i, m)$, we must sum the contributions of each congruence class in $\mathcal{C}$ which is contained in $B(i, m)$, obtaining

(21) $$Pr\left(f_i = p_i | Y_n(x) \text{ enters } \mathcal{C}\right) = 1,$$

where

(22) $$f_j = \lim_{K \to \infty} \tfrac{1}{K+1} \operatorname{card} \{n; n \leq K, \ T^n(x) \equiv j \, (\operatorname{mod} m)\}$$

and

(23) $$p_j = \sum_{\substack{B \,\in\, \mathcal{C} \\ B \,\subseteq\, B(j,\, m)}} \rho_B.$$

(See [20, Theorem 5].)

For mappings $T$ of relatively prime type, or when

$$\gcd(m_i, d^2) = \gcd(m_i, d), \ 0 \leq i < d,$$

(equivalently $d_i | d$ for all $i$) this Markov chain reduces to the one implicitly studied by Matthews and Watts. However in the general case the chain can be infinite.

Leigh ([20, equations (15, 23)]) also proved that if

$$p_{nij} = Pr(T^n(x) \equiv j \, (\operatorname{mod} m) | x \equiv i \, (\operatorname{mod} m)),$$

then

(24) $$p_{nij} = \sum_{B \subseteq B(j,m)} q^{(n)}_{B(i,m)B}$$

(25) $$\lim_{N \to \infty} \frac{1}{N} \sum_{n \leq N} p_{nij} = \sum_{\mathcal{C}} p_j f_{B(i,m)\mathcal{C}},$$

where $f_{B(i,m)\mathcal{C}} = Pr(Y_n(x) \in \mathcal{C} \text{ for some } n | Y_0 = B(i, m))$.

In the case where there is just one positive recurrent class $\mathcal{C}$ which is aperiodic and $B(i, m) \in \mathcal{C}$, then (25) reduces to

(26) $$\lim_{N \to \infty} \sum_{n \leq N} p_{nij} = p_j.$$

We finish this section with some generalizations of conjectures 3.1 (i),(ii) and (iv) due to Leigh. Here we define $\mathcal{S}_C$ to be the union of the states in a positive class $\mathcal{C}$, i.e.,

$$S_C = \bigcup_{B \in C} B.$$

CONJECTURE 7.1. (Leigh) *Suppose that the Markov chain $\{Y_n\}$ above has finitely many states. Then:*

*(a) Every divergent trajectory will eventually enter some positive class $\mathcal{S}_C$ and will occupy each class $B$ of $\mathcal{C}$ with limiting frequency $\rho_B$.*

*(b) Let $\mathcal{C}$ be a positive recurrent class for the Markov chain mod $d$ and let $p_i$ be defined by (23). Then if*

$$\prod_{i=0}^{d-1} \left| \frac{m_i}{d} \right|^{p_i} < 1,$$

*all trajectories starting in $\mathcal{S}_C$ will eventually cycle.*

*However if*

$$\prod_{i=0}^{d-1} \left| \frac{m_i}{d} \right|^{p_i} > 1,$$

*almost all trajectories starting in $\mathcal{S}_C$ diverge.*

## 8. General Case: Examples

The general case exhibits a much wider variety of behaviours under iteration than the relatively prime case.

We begin with a family of maps including the original form of Collatz version for the $3x + 1$ problem.

EXAMPLE 8.1. (Collatz family) For an integer $a$,

$$C_a(x) = \begin{cases} x/2 & \text{if } x \equiv 0 \,(\text{mod}\,2) \\ 3x + a & \text{if } x \equiv 1 \,(\text{mod}\,2), \end{cases}$$

The original version of Collatz [**6**] for the $3x+1$ problem is the case $a = 1$. The $3x + 1$ map is given by

$$T(x) = \begin{cases} C_1(x) = x/2 & \text{if } x \equiv 0 \,(\text{mod}\,2) \\ C_1^2(x) = (3x + 1)/2 & \text{if } x \equiv 1 \,(\text{mod}\,2). \end{cases}$$

so in this case we expect all trajectories of $C_1$ should enter cycles. However for $a = 0$ every trajectory of $C_0$ diverges (except $x = 0$), for once an iterate becomes odd, it increases in size monotonically. In fact whenever $a$ is an even integer there are divergent orbits, by similar reasoning. This example illustrates that the value of integers $a_i$ in (1) matter in determining the behaviour of orbits in the general case.

EXAMPLE 8.2.

$$T(x) = \begin{cases} x/3 - 1 & \text{if } x \equiv 0 \,(\text{mod}\,3) \\ (x + 5)/3 & \text{if } x \equiv 1 \,(\text{mod}\,3) \\ 10x - 5 & \text{if } x \equiv 2 \,(\text{mod}\,3) \end{cases}$$

There appear to be five cycles, with starting values 0, 5, 17, −1, −4. However this example becomes less mysterious if we consider the related mapping

$$T'(x) = \begin{cases} T(x) & \text{if } x \equiv 0 \text{ or } 1 \, (\text{mod } 3) \\ T^2(x) = (10x - 8)/3 & \text{if } x \equiv 2 \, (\text{mod } 3). \end{cases}$$

For $T'$ is a mapping of relatively prime type and Conjecture 3.1 predicts that all trajectories eventually cycle.

Our next example contains provably divergent trajectories.

EXAMPLE 8.3.

$$T(x) = \begin{cases} 2x & \text{if } x \equiv 0 \, (\text{mod } 3) \\ (7x + 2)/3 & \text{if } x \equiv 1 \, (\text{mod } 3) \\ (x - 2)/3 & \text{if } x \equiv 2 \, (\text{mod } 3). \end{cases}$$

Using an abbreviated notation, the transition probabilities scheme is:

$$
\begin{array}{ccccc}
B(0,3) & \rightarrow & B(0,3) & \rightarrow & B(0,3) \\
B(1,3) & \rightarrow & B(0,1) & \rightarrow & B(0;1;2,3) \\
B(2,4) & \rightarrow & B(0,1) & \rightarrow & B(0;1;2,3).
\end{array}
$$

The states are $B(0,3), B(1,3)$ and $B(2,3)$ with

$$Q_T(3) = \begin{bmatrix} 1 & 0 & 0 \\ \frac{1}{3} & \frac{1}{3} & \frac{1}{3} \\ \frac{1}{3} & \frac{1}{3} & \frac{1}{3} \end{bmatrix}.$$

There is one positive recurrent class $\mathcal{C}_1 = \{B(0,3)\}$ and transient states $B(1,3)$ and $B(2,3)$.

Here $3|x$ implies $3|T(x)$; so once a trajectory enters the zero residue class mod 3, it remains there and diverges. Experimental evidence strikingly suggests that if a trajectory takes values $T^k(x) \equiv \pm 1 \, (\text{mod } 3)$ for all $k \geq 0$, then the trajectory must eventually enter one of the cycles $-1, -1$ or $-2, -4, -2$. The author offers a \$100 (Australian) prize for a proof. This problem seems just as intractable as the $3x + 1$ problem and is a simple example of the more general Conjecture (4.2) on divergent trajectories.

EXAMPLE 8.4.

$$T(x) = \begin{cases} 12x - 1 & \text{if } x \equiv 0 \, (\text{mod } 4) \\ 20x & \text{if } x \equiv 1 \, (\text{mod } 4) \\ (3x - 6)/4 & \text{if } x \equiv 2 \, (\text{mod } 4) \\ (x - 3)/4 & \text{if } x \equiv 3 \, (\text{mod } 4). \end{cases}$$

To predict the frequency distribution mod 4 of divergent trajectories, we find there are 8 states in the Markov chain:

$\mathcal{S} = \{B(0,4), B(1,4), B(2,4), B(3,4), B(15,16), B(4,16), B(47,64), B(11,16)\}$ with corresponding transition probability scheme and matrix $Q(4)$:

$$
\begin{array}{ccccc}
B(0,4) & \rightarrow & B(15,16) & \rightarrow & B(15,16) \\
B(1,4) & \rightarrow & B(4,16) & \rightarrow & B(4,16) \\
B(2,4) & \rightarrow & B(0,1) & \rightarrow & B(0;1;2;3,4) \\
B(3,4) & \rightarrow & B(0,1) & \rightarrow & B(0;1;2;3,4) \\
B(15,16) & \rightarrow & B(3,4) & \rightarrow & B(3,4) \\
B(4,16) & \rightarrow & B(47,64) & \rightarrow & B(47,64) \\
B(47,64) & \rightarrow & B(11,16) & \rightarrow & B(11,16) \\
B(11,16) & \rightarrow & B(2,4) & \rightarrow & B(2,4)
\end{array}
$$

$$Q_T(4) = \begin{bmatrix} 0 & 0 & 0 & 0 & 1 & 0 & 0 & 0 \\ 0 & 0 & 0 & 0 & 0 & 1 & 0 & 0 \\ 1/4 & 1/4 & 1/4 & 1/4 & 0 & 0 & 0 & 0 \\ 1/4 & 1/4 & 1/4 & 1/4 & 0 & 0 & 0 & 0 \\ 0 & 0 & 0 & 1 & 0 & 0 & 0 & 0 \\ 0 & 0 & 0 & 0 & 0 & 0 & 1 & 0 \\ 0 & 0 & 0 & 0 & 0 & 0 & 0 & 1 \\ 0 & 0 & 1 & 0 & 0 & 0 & 0 & 0 \end{bmatrix}.$$

The Markov matrix is primitive (its 8th power is positive) and the limiting probabilities are $(\frac{1}{10}, \frac{1}{10}, \frac{2}{10}, \frac{2}{10}, \frac{1}{10}, \frac{1}{10}, \frac{1}{10}, \frac{1}{10})$ To find the predicted frequency $p_0$ for the congruence class $0 \,(\mathrm{mod}\,4)$, we must sum the contributions arising from the states $B(0, 4)$ and $B(4, 16)$, namely $p_0 = \frac{1}{10} + \frac{1}{10} = \frac{2}{10}$. Similarly the other frequencies are $p_1 = \frac{1}{10}, p_2 = \frac{2}{10}, p_3 = \frac{2}{10} + \frac{1}{10} + \frac{1}{10} + \frac{1}{10} = \frac{5}{10}$. Then as

$$12^{2/10} 20^{1/10} (\tfrac{3}{4})^{2/10} (\tfrac{1}{4})^{5/10} > 1,$$

we expect most trajectories to be divergent. The trajectory starting with 21 appears to be divergent.

We have found 6 cycles, with starting values $-1$, $5$, $-19$, $-4$, $-6$, $19133$.

EXAMPLE 8.5. (Leigh [20, page 140])

$$T(x) = \begin{cases} x/4 & \text{if } x \equiv 0 \,(\mathrm{mod}\,8) \\ (x+1)/2 & \text{if } x \equiv 1 \,(\mathrm{mod}\,8) \\ 20x - 40 & \text{if } x \equiv 2 \,(\mathrm{mod}\,8) \\ (x-3)/8 & \text{if } x \equiv 3 \,(\mathrm{mod}\,8) \\ 20x + 48 & \text{if } x \equiv 4 \,(\mathrm{mod}\,8) \\ (3x - 13)/2 & \text{if } x \equiv 5 \,(\mathrm{mod}\,8) \\ (11x - 2)/4 & \text{if } x \equiv 6 \,(\mathrm{mod}\,8) \\ (x+1)/8 & \text{if } x \equiv 7 \,(\mathrm{mod}\,8) \end{cases}$$

We find there are 9 states in the Markov chain mod 8:
$B(0, 8)$, $B(1, 8)$, $B(2, 8)$, $B(3, 8)$, $B(4, 8)$, $B(5, 8)$, $B(6, 8)$, $B(7, 8)$, $B(0, 32)$, with corresponding transition probability scheme:

$$
\begin{array}{lcl cl}
B(0, 8) & \to & B(0, 2) & \to & B(0; 2; 4; 6, 8) \\
B(1, 8) & \to & B(1, 4) & \to & B(1; 5, 8) \\
B(2, 8) & \to & B(0, 32) & \to & B(0, 32) \\
B(3, 8) & \to & B(0, 1) & \to & B(0; 1; 2; 3; 4; 5; 6; 7, 8) \\
B(4, 8) & \to & B(0, 32) & \to & B(0, 32) \\
B(5, 8) & \to & B(1, 4) & \to & B(1; 5, 8) \\
B(6, 8) & \to & B(0, 2) & \to & B(0; 2; 4; 6, 8) \\
B(7, 8) & \to & B(0, 1) & \to & B(0; 1; 2; 3; 4; 5; 6; 7, 8) \\
B(0, 32) & \to & B(0, 8) & \to & B(0, 8)
\end{array}
$$

The corresponding transition matrix is

$$Q_T(8) = \begin{bmatrix} 1/4 & 0 & 1/4 & 0 & 1/4 & 0 & 1/4 & 0 & 0 \\ 0 & 1/2 & 0 & 0 & 0 & 1/2 & 0 & 0 & 0 \\ 0 & 0 & 0 & 0 & 0 & 0 & 0 & 0 & 1 \\ 1/8 & 1/8 & 1/8 & 1/8 & 1/8 & 1/8 & 1/8 & 1/8 & 0 \\ 0 & 0 & 0 & 0 & 0 & 0 & 0 & 0 & 1 \\ 0 & 1/2 & 0 & 0 & 0 & 1/2 & 0 & 0 & 0 \\ 1/4 & 0 & 1/4 & 0 & 1/4 & 0 & 1/4 & 0 & 0 \\ 1/8 & 1/8 & 1/8 & 1/8 & 1/8 & 1/8 & 1/8 & 1/8 & 0 \\ 1 & 0 & 0 & 0 & 0 & 0 & 0 & 0 & 0 \end{bmatrix}.$$

There are two positive recurrent classes: $\mathcal{C}_1 = \{B(1, 8),\ B(5, 8)\}$ and

$$\mathcal{C}_2 = \{B(0, 8),\ B(0, 32),\ B(2, 8),\ B(4, 8),\ B(6, 8)\},$$

with transient states $B(3, 8)$ and $B(7, 8)$.

The limiting probabilities are $\mathbf{v}_1 = (\frac{1}{2}, \frac{1}{2})$ and $\mathbf{v}_2 = (\frac{3}{8}, \frac{1}{4}, \frac{1}{8}, \frac{1}{8}, \frac{1}{8})$, respectively.

We have $p_1 = p_5 = \frac{1}{2}$ and as

$$\left(\tfrac{1}{2}\right)^{1/2}\left(\tfrac{3}{2}\right)^{1/2} < 1,$$

we expect every trajectory starting in $\mathcal{S}_{\mathcal{C}_1} = B(1,8) \cup B(5,8)$ to cycle.

Also $p_0 = \frac{3}{8} + \frac{1}{4} = \frac{5}{8}$ and $p_2 = p_4 = p_6 = \frac{1}{8}$. Then as

$$\left(\tfrac{1}{4}\right)^{5/8} 20^{1/8} 20^{1/8} \left(\tfrac{11}{4}\right)^{1/8} > 1,$$

we expect most trajectories starting in $\mathcal{S}_{\mathcal{C}_2} = B(0,2)$ to diverge, displaying frequencies $\mathbf{v} = (\frac{5}{8}, \frac{1}{8}, \frac{1}{8}, \frac{1}{8})$ in the respective component congruence classes. (Leigh uses another Markov chain $\{X_n\}$ and arrives at $\mathbf{v} = (\frac{4}{7}, \frac{1}{7}, \frac{1}{7}, \frac{1}{7})$, which is erroneous, as one discovers on observing the apparently divergent trajectory starting with 46.)

We have found 13 cycles, with starting values

$$\{0,\ 1,\ 10,\ 13,\ 61,\ 158,\ 205,\ 3292,\ 4244,\ -2,\ -11,\ -12,\ -18\}.$$

G. Venturini (see [**31**]) further expanded on Leigh's ideas and gave many interesting examples. One spectacular example is worth mentioning.

EXAMPLE 8.6. (Venturini)

$$T(x) = \begin{cases} 2500x/6 + 1 & \text{if } x \equiv 0\,(\mathrm{mod}\,6) \\ (21x - 9)/6 & \text{if } x \equiv 1\,(\mathrm{mod}\,6) \\ (x + 16)/6 & \text{if } x \equiv 2\,(\mathrm{mod}\,6) \\ (21x - 51)/6 & \text{if } x \equiv 3\,(\mathrm{mod}\,6) \\ (21x - 72)/6 & \text{if } x \equiv 4\,(\mathrm{mod}\,6) \\ (x + 13)/6 & \text{if } x \equiv 5\,(\mathrm{mod}\,6). \end{cases}$$

There are 9 states in the Markov chain mod 6:
$B(0,6), B(1,6), B(2,6), B(3,6), B(4,6), B(5,6), B(1,12), B(5,12), B(9,12)$.
These form a recurrent class with limiting probabilities

$$\mathbf{v} = (\tfrac{18}{202}, \tfrac{20}{202}, \tfrac{53}{202}, \tfrac{20}{202}, \tfrac{18}{202}, \tfrac{55}{202}, \tfrac{6}{202}, \tfrac{6}{202}, \tfrac{6}{202}).$$

Noting that $B(1,12) \subseteq B(1,6)$, $B(5,12) \subseteq B(5,6)$, $B(9,12) \subseteq B(3,6)$, we get
$p_0 = \frac{9}{101}$, $p_1 = \frac{20}{202} + \frac{6}{202} = \frac{13}{101}$, $p_2 = \frac{53}{202}$,
$p_3 = \frac{20}{202} + \frac{6}{202} = \frac{13}{101}$, $p_4 = \frac{9}{101}$, $p_5 = \frac{55}{202} + \frac{6}{202} = \frac{61}{202}$.

Then $\left(\frac{2500}{6}\right)^{9/101} \left(\frac{21}{6}\right)^{13/101} \left(\frac{1}{6}\right)^{53/202} \left(\frac{21}{6}\right)^{13/101} \left(\frac{21}{6}\right)^{9/101} \left(\frac{1}{6}\right)^{61/202} < 1$ and we expect all trajectories to eventually cycle. in fact, there appear to be two cycles, with starting values 2 and 6.

EXAMPLE 8.7.

$$T(x) = \begin{cases} 2x - 2 & \text{if } x \equiv 0\,(\mathrm{mod}\,6) \\ (x - 3)/2 & \text{if } x \equiv 1\,(\mathrm{mod}\,6) \\ (2x - 1)/3 & \text{if } x \equiv 2\,(\mathrm{mod}\,6) \\ 7x/3 & \text{if } x \equiv 3\,(\mathrm{mod}\,6) \\ (5x - 2)/6 & \text{if } x \equiv 4\,(\mathrm{mod}\,6) \\ 9x & \text{if } x \equiv 5\,(\mathrm{mod}\,6). \end{cases}$$

There are 14 Markov states, with two periodic positive recurrent classes:
$$\mathcal{C}_1 = \{B(5,6),\ B(45,54),\ B(15,18)\},$$
$$\mathcal{C}_2 = \{B(5,12),\ B(45,108),\ B(33,36)\},$$

where $\mathcal{S}_{\mathcal{C}_2} \subseteq \mathcal{S}_{\mathcal{C}_1}$ and corresponding limiting probabilities $\mathbf{v}_1 = \mathbf{v}_2 = (\frac{1}{3}, \frac{1}{3}, \frac{1}{3})$.
We also have transient states
$B(0,6), B(1,6), B(2,6), B(3,6), B(4,6), B(10,12), B(1,12), B(9,12)$.

For both $\mathcal{C}_1$ and $\mathcal{C}_2$, we have $p_3 = \frac{1}{3} + \frac{1}{3} = \frac{2}{3}$, $p_5 = \frac{1}{3}$. Also $(\frac{7}{3})^{2/3} 9^{1/3} > 1$.
Hence we expect most trajectories starting in $S_{\mathcal{C}_1}$ and $S_{\mathcal{C}_2}$ to diverge.

The trajectories $\{T^k(-1)\}$ and $\{T^k(5)\}$ apparently diverge. Experimentally, $\{T^k(-1)\}$ meets each of $B(11,12), B(99,108), B(15,36)$ with limiting frequencies $1/3$ and does not meet $S_{\mathcal{C}_2}$; $\{T^k(5)\}$ meets each of $B(5,12), B(45,108), B(33,36)$ with limiting frequencies $1/3$. Also $B(11,12) \cup B(99,108) \cup B(15,36) = S_{\mathcal{C}_1} - S_{\mathcal{C}_2}$ is $T$-invariant. There appears to be one cycle $\langle -2 \rangle$.

Venturini gave the following sufficient condition for the $\{Y_n\}$ chain to be finite:

THEOREM 8.1. ([**31**, Theorem 7, p. 196]). *Let* $D_n(x) = d_{j_0} \cdots d_{j_{n-1}}$, *where* $T^i(x) \equiv j_i \,(\mathrm{mod}\,d)$ *for* $0 \le i \le n-1$. *Suppose* $D_\nu(x)$ *divides* $d^\nu$ *for all* $x, 0 \le x < d^\nu$. *Then* $K_n(x) = mK'_n(x)$, *where* $K'_n(x)$ *divides* $d^{\nu-1}$ *for all* $n \ge 0$.

We finish this section with an example where there are infinitely many states in the Markov chain. The example was suggested by Chris Smyth in 1993.

EXAMPLE 8.8.

$$T(x) = \left\{ \begin{array}{ll} \lfloor 3x/2 \rfloor & \text{if } x \equiv 0 \,(\mathrm{mod}\,2) \\ \lfloor 2x/3 \rfloor & \text{if } x \equiv 1 \,(\mathrm{mod}\,2). \end{array} \right.$$

This can be regarded as a 6–branched mapping. The integer trajectories are much simpler to describe than the Markov chain: Non–zero even integers are successively multiplied by $3/2$ until one gets an odd integer: if $k$ is even, say $k = 2^r(2c+1), r \ge 1$, then $k$ eventually reaches $3^{r+2}(2c+1) = 6k' + 3$. Also $6k + 1 \to 4k \to 6k \to 9k \to 6k$, while $6k + 3 \to 4k + 2 \to 6k + 3$.

Finally $6k + 5 \to 4k + 3$ and unless we encounter 0 or $-1$ (fixed points), we must eventually reach $B(1,6)$ or $B(3,6)$.

With $m = 6$, $Y_n(0) = B(0, 2 \cdot 3^{n+1})$ for $n \ge 0$.

## 9. Other rings: Finite fields

It is natural to investigate analogous mappings $T$ for rings other than $\mathbb{Z}$, where division is meaningful, namely the ring of integers of a global field. Here global fields are algebraic number fields or function fields in one variable over a finite field, cf. Cassels and Fröhlich [**5**, Chapter 2].

The author and George Leigh experimented with the ring $GF(2)[x]$ of polynomials over the field with two elements. Here the conjectural picture for trajectories is not so clear. In [**25**] we examined the following mapping:

EXAMPLE 9.1. (Matthews and Leigh [**25**]) Over $GF(2)[x]$,

(27) $$T(f) = \left\{ \begin{array}{ll} \dfrac{f}{x} & \text{if } f \equiv 0 \,(\mathrm{mod}\,x) \\[2ex] \dfrac{(x^2+1)f+1}{x} & \text{if } f \equiv 1 \,(\mathrm{mod}\,x) \end{array} \right.$$

This is an example of relatively prime type and $|m_0 \cdots m_{|d|-1}| = |d|^{|d|}$, where $|f| = 2^{\deg f}$. Most trajectories appear to cycle. However the trajectory starting from $1 + x + x^3$ exhibits a regularity which enabled its divergence to be proved (see Figure 2). If $L_n = 5(2^n - 1)$, then

$$T^{L_n}(1 + x + x^3) = \frac{1 + x^{3 \cdot 2^n + 1} + x^{3 \cdot 2^n + 2}}{1 + x + x^2}.$$

Figure 2 shows the first 37 iterates.

```
1101
11001
111101
1001001
01101101
1101101
11011001
110111101
1101001001
11001101101
111111011001
1000010111101
01001001001001
1001001001001
01101101101101
110110110110101
110101011011001
110110110111101
1101101101001001
1101101001101101
110110111111011001
1101101000010111101
1101100100100100100
110111101101101101101
110100101101101101001
110011001101101101101
11111111110110110101001001
1000000001011011001101101
01000000100110111111011001
10000001001101111111011001
0100001011110100001011101
1000010111101000010111101
010010010010010010010010001001
1001001001001001001001001001
011011011011011011011101101
110110110110110110110110101
110110110110110110110110101
110110110110110110110110001
```

FIGURE 2. The first 37 iterates for Example 9.1.

There are infinitely many cycles. In particular, the trajectories starting with $g_n = (1+x^{2^n-1})/(1+x)$ are purely periodic, with period–length $2^n$. Figure 3 shows the cycle printout for $g_4$.

```
111111111111111
100000000000011
0100000000001111
100000000001111
01000000000110011
1000000000110011
010000001111111
1000000011111111
01000001100000011
100001100000011
0100011110001111
100011110001111
01011001100110011
1011001100110011
001111111111111111
01111111111111111
111111111111111
```

FIGURE 3. Cycle $g_4$.

EXAMPLE 9.2. (Hicks, Mullen, Yucas and Zavislak [13]) Over $GF(2)[x]$,

$$(28) \qquad T(f) = \begin{cases} \dfrac{f}{x} & \text{if } f \equiv 0 \pmod{x} \\[2mm] \dfrac{(x+1)f+1}{x} & \text{if } f \equiv 1 \pmod{x} \end{cases}$$

The paper [13] shows that all orbits converge to one of the fixed points $0, 1$.

## 10. Other rings: algebraic number fields

Let $d$ be a non–unit in the ring $O_K$ of integers of an algebraic number field $K$. Again, the conjectural picture for trajectories is not entirely clear. Here are three examples.

EXAMPLE 10.1. $T : \mathbb{Z}[\sqrt{2}] \to \mathbb{Z}[\sqrt{2}]$ be defined by

$$T(\alpha) = \begin{cases} \alpha/\sqrt{2} & \text{if } \alpha \equiv 0 \,(\text{mod } \sqrt{2}) \\ (3\alpha + 1)/\sqrt{2} & \text{if } \alpha \equiv 1 \,(\text{mod } \sqrt{2}). \end{cases}$$

Writing $\alpha = x + y\sqrt{2}$, where $x, y \in \mathbb{Z}$, we have equivalently

$$T(x, y) = \begin{cases} (y, x/2) & \text{if } x \equiv 0 \,(\text{mod } 2) \\ (3y, (3x + 1)/2) & \text{if } x \equiv 1 \,(\text{mod } 2). \end{cases}$$

There appear to be finitely many cycles with starting values

$$\{0, 1, -1, -5, -17, -2 - 3\sqrt{2}, \quad -3 - 2\sqrt{2}, \quad 9 + 10\sqrt{2}\}.$$

An interesting feature is the presence of at least three one–dimensional $T$–invariant sets $S_1$, $S_2$, $S_3$ in $\mathbb{Z} \times \mathbb{Z}$:

$$S_1 : \ x = 0 \text{ or } y = 0, \quad S_2 : \ 2x + y + 1 = 0 \text{ or } x + 4y + 1 = 0,$$
$$S_3 : \ x + y + 1 = 0 \text{ or } x + 2y + 1 = 0 \text{ or } x + 2y + 2 = 0.$$

Trajectories starting in $S_1$ or $S_2$ oscillate from one line to the other, while those starting in $S_3$ oscillate between the first and either of the second and third. Trajectories starting in $S_1$ will cycle, as $T^2(x, 0) = (C(x), 0)$, where $C$ denotes the $3x + 1$ mapping.

Most divergent trajectories appear to be evenly distributed mod $(\sqrt{2})^\alpha$. However divergent trajectories starting in $S_2$ or $S_3$ present what at first sight appear to be anomalous frequency distributions mod 2. By considering $T^2$, these are explicable in terms of the predicted "one–dimensional" uniform distribution. For example, if $2x + y + 1 = 0$, then

$$T^2(x, y) = \begin{cases} (\frac{3x}{2}, -3x - 1) & \text{if } x \equiv 0 \,(\text{mod } 2) \\ (\frac{9x+3}{2}, -9x - 4) & \text{if } x \equiv 1 \,(\text{mod } 2). \end{cases}$$

EXAMPLE 10.2. $T : \mathbb{Z}[\sqrt{3}] \to \mathbb{Z}[\sqrt{3}]$ is defined by

$$T(x) = \begin{cases} x/\sqrt{3} & \text{if } x \equiv 0 \,(\text{mod } \sqrt{3}) \\ (x - 1)/\sqrt{3} & \text{if } x \equiv 1 \,(\text{mod } \sqrt{3}) \\ (4x + 1)/\sqrt{3} & \text{if } x \equiv 2 \,(\text{mod } \sqrt{3}) \end{cases}$$

We have found 103 cycles. The trajectory starting with $-1 + 7\sqrt{3}$ appears to be divergent. Divergent trajectories produce limiting frequencies approximating $(\cdot 27, \cdot 32, \cdot 40)$ in the residue classes $0, 1, 2 \pmod{\sqrt{3}}$.

Finally it should be pointed out that divergent trajectories need not tend to infinity in absolute value:

EXAMPLE 10.3. $T : \mathbb{Z}[\sqrt{2}] \to \mathbb{Z}[\sqrt{2}]$ be defined by

$$T(\alpha) = \begin{cases} (1 - \sqrt{2})\alpha/\sqrt{2} & \text{if } \alpha \equiv 0 \,(\text{mod } \sqrt{2}) \\ (3\alpha + 1)/\sqrt{2} & \text{if } \alpha \equiv 1 \,(\text{mod } \sqrt{2}). \end{cases}$$

If $\{T^K(x)\}$ is a divergent trajectory and $T^K(x) = x_K + y_K\sqrt{2}, x_K, y_K \in \mathbb{Z}$, then apparently $x_K/y_K \to -\sqrt{2}$ and $T^K(x) \to 0$.

## 11. Concluding Remarks

This paper is intended to draw attention to the Markov chains that arise from generalized Collatz maps $T : \mathbb{Z} \to \mathbb{Z}$. When the number of states is finite, these chains give confident predictions regarding cycling and almost everywhere divergence of trajectories, as well as predicting the frequencies of occupation of certain congruence classes by divergent trajectories.

In the non-relatively prime case, the question of when there are only finitely many cycles remains open.

The conjectural situation for rings other than $\mathbb{Z}$ also invites systematic investigation.

**Acknowledgements.** The author is indebted to the reviewer for vastly improving the presentation of the paper and for suggesting the inclusion of Theorem 6.3. The author also thanks George Leigh for answering email queries in December 1993 and providing a computer program which constructs the transition matrix (18) for the $\{Y_n\}$ chain. The author also thanks Owen Jones, Richard Pinch and Anthony Quas for helpful conversations when the author was on study leave at the Department of Pure Mathematics, University of Cambridge, July–December 1993.

## References

[1] D. J. Bernstein, *A non-iterative 2-adic statement of the $3x+1$ conjecture*, Proc. Amer. Math. Soc. **121** (1994), 405–408.

[2] D. J. Bernstein and J. C. Lagarias, *The $3x+1$ conjugacy map*, Canadian J. Math. **48** (1996), 1154–1169.

[3] P. Billingsley, *Ergodic theory and information*, John Wiley, New York 1965.

[4] R. N. Buttsworth and K. R. Matthews, *On some Markov matrices arising from the generalized Collatz mapping*, Acta Arith. 55 (1990), 43–57.

[5] J. W. S. Cassels and A. Fröhlich, *Algebraic Number Theory*, Academic Press, London 1967.

[6] L. Collatz, *On the origin of the $3x + 1$ problem (Chinese)*, J. of Qufu Normal University, Natural Science Edition **12** (1986) No. 3, 9–11.

[7] R. Durrett, *Probability: Theory and Examples*, Third Edition. Thomson-Brooks/Cole (2005).

[8] H. M. Farkas, *Variants of the $3N + 1$ problem and multiplicative semigroups*, in: Geometry, Spectral Theory, Groups and Dynamics: Proceedings in Memory of Robert Brooks, Contemporary Math., Volume 387, Amer. Math. Soc., Providence, 2005, pp. 121–127.

[9] B. L. Fox and D. M. Landi, *An algorithm for identifying the ergodic subchains and transient states of a stochastic matrix*, Communications of the ACM, 11 (1968), 619–621.

[10] M. D. Fried and M. Jarden, *Field Arithmetic*, Ergebnisse der Mathematik und ihrer Grenzgebiete, Band 11, Springer, 1986.

[11] G. R. Grimmett and D. R. Stirzaker, *Probability and random processes*, Oxford University Press, 1992.

[12] R. K. Guy, *Unsolved problems in number theory*, Second Edition, Vol. 1, Problem books in Mathematics, Springer, Berlin 1994.

[13] K. Hicks, G. L. Mullen, J. L. Yucas and R. Zavislak, *A Polynomial Analogue of the $3N + 1$ Problem?*, American Math. Monthly **115** (2008), No. 7, 615-622.

[14] A. Kaufmann and R. Cruon, *Dynamic Programming*, Academic Press, New York 1967.

[15] J. R. Kemeny, J. L. Snell and A. W. Knapp, *Denumerable Markov Chains*, Second Edition. With a chapter of Markov random fields, by David Griffeath. Graduate Texts in Mathematics, No. 40, Springer-Verlag: New York 1976.

[16] M. S. Klamkin, *Problem $63 - 13$*, SIAM Review 5 (1963), 275–276.

[17] N. Koblitz, *p–adic numbers, p–adic analysis and zeta–functions*, Graduate Text 58, Springer, 1981.

[18] J. C. Lagarias, *The $3x + 1$ problem*, Amer. Math. Monthly (1985), 3–23.

[19] J. C. Lagarias, *The set of rational cycles for the $3x+1$ problem*, Acta Arith. 56 (1990), 33–53.

[20] G. M. Leigh, *A Markov process underlying the generalized Syracuse algorithm*, Acta Arith. 46 (1985), 125–143.

[21] K. Mahler, *p−adic numbers and their functions*, Second Edition, Cambridge University Press, Cambridge, 1981.

[22] K. R. Matthews, *Some Borel measures asssociated with the generalized Collatz mapping*, Colloquium Math. 63 (1992), 191–202.

[23] K. R. Matthews, `http://www.numbertheory.org/calc/krm_calc.html`, Number theory program CALC.

[24] K. R. Matthews, `http://www.numbertheory.org/php/fox_landi0.php`, Implementation of the Fox-Landi algorithm.

[25] K. R. Matthews and G. M. Leigh, *A generalization of the Syracuse algorithm in $F_q[x]$*, J. Number Theory, 25 (1987), 274–278.

[26] K. R. Matthews and A. M. Watts, *A generalization of Hasse's generalization of the Syracuse algorithm*, Acta Arith. 43 (1984), 167–175.

[27] K. R. Matthews and A. M. Watts, *A Markov approach to the generalized Syracuse algorithm*, ibid. 45 (1985), 29–42.

[28] H. Möller, *Über Hasses Verallgemeinerung des Syracuse-Algorithmus (Kakutanis Problem)*, ibid. 34 (1978), 219–226.

[29] M. Pearl, *Matrix theory and finite mathematics*, McGraw–Hill, New York 1973.

[30] A. G. Postnikov. *Introduction to analytic number theory*, Amer. Math. Soc., Providence R.I. 1988.

[31] G. Venturini. *Iterates of number-theoretic functions with periodic rational coefficients (generalization of the $3x + 1$ problem)*, Stud. Appl. Math. 86 (1992), no.3, 185–218.

[32] S. Volkov, *A probabilistic model for the $5k + 1$ problem and related problems*, Stochastic Processes and Applications 116 (2006), 662–674.

[33] G. Wirsching. *The Dynamical System Generated by the $3n + 1$ Function*, Lecture Notes in Mathematics 1682, Springer 1998.

DEPARTMENT OF MATHEMATICS, UNIVERSITY OF QUEENSLAND, BRISBANE, QUEENSLAND, AUSTRALIA 4072

*E-mail address*: `keithmatt@gmail.com`

# Generalized 3x+1 functions and the theory of computation

## Pascal Michel and Maurice Margenstern

ABSTRACT. This survey displays some links between the theory of computation and functions that generalize the 3x+1 function $T$ defined by $T(n) = n/2$ if $n$ is even and $T(n) = (3n + 1)/2$ if $n$ is odd. While the behavior of function $T$, when it is iterated on integers, is an open problem, the behavior of some of its generalizations, when iterated, was proved by Conway to be undecidable. The idea of the proof is that iteration steps simulate computation steps. The 3x+1 function and its generalizations are shown to be useful in the classifications of some well known models of computation such as Turing machines, tag systems and cellular automata. They are also related to Diophantine representations, grammars and rewriting systems.

## 1. Introduction

The 3x+1 *function* is the function $T$ on integers defined by

$$T(n) = \begin{cases} n/2 & \text{if } n \text{ is even,} \\ (3n + 1)/2 & \text{if } n \text{ is odd.} \end{cases}$$

Let us iterate this function, starting for example from number 7, as follows.

$$7, 11, 17, 26, 13, 20, 10, 5, 8, 4, 2, 1, 2, 1, \ldots.$$

We see that the values reach the loop $2, 1, 2, 1, \ldots$. What will happen if we start from another positive number? Do we always reach the loop $2, 1, 2, 1, \ldots$? We do not know the answer to this question. It is an open problem in mathematics. The *Collatz conjecture* claims that, when function $T$ is iterated on a positive integer, its values always reach this loop (see Lagarias (1985) for a survey).

When a mathematician faces a problem stubbornly resisting a proof, a natural way out is to study variants or generalizations of the initial problem. The condition on the parity of $n$ in the definition of $T(n)$ can be replaced by a condition on $n$ modulo $d$ for any integer $d \geq 2$. Then, the functions $n/2$ and $(3n + 1)/2$ can be replaced by functions $(an+b)/d$. So we get generalized Collatz functions, hoping that studying them will lead to ideas useful to solve Collatz conjecture. The outcome turns out to be both unexpected and fruitful. The bad news is that there exist generalized Collatz functions with an *intrinsically* undecidable behavior. The good news is that generalized Collatz functions have risen to the status of all-purpose model of computation. Roughly, the idea is that the lines of the definition of a generalized Collatz function are equivalent to the lines of a program, iteration is equivalent to computation, and the successive values of the function are equivalent to the successive states of the computing device, with initial and final values corresponding to input and output.

In Section 2, we display the ideas that were used by Conway (1972) to prove that generalized Collatz functions can have an undecidable behavior when they are iterated, and we give some consequences of this seminal property. Then we present some links between generalized Collatz functions and some other models of computation: Turing machines in Section 3, tag systems in Section 4, and cellular automata in Section 5. In Section 6, we give a Diophantine representation of the $3x+1$ problem by means of binomial coefficients. In Section 7, we define congruential systems, that is a further generalization of generalized Collatz functions. In Section 8, we see some other encodings of the $3x+1$ problem.

## 2. Undecidability of predicting iterations

We define the notions of generalized Collatz function, periodically linear function and Fractan program, and we state Conway's theorem: when these functions are iterated, there are no algorithms to predict their behaviors, or, equivalently, their behaviors are undecidable. Then, these iterated functions have the properties of all well known computation models. They have universal models, they can compute some fashionable functions, such as the $3x+1$ function, and they can be used to prove the undecidability of new problems. They also provide a new example of $\Pi_2^0$-complete problem. Finally, we display another generalization of the $3x+1$ function with iterates that have an undecidable behavior.

### 2.1. Conway's theorem.

DEFINITION 2.1. *A mapping $g : \mathbb{Z} \to \mathbb{Z}$ is a* generalized Collatz function *if there exist an integer $d \geq 2$ and rational numbers $a_0, b_0, \ldots, a_{d-1}, b_{d-1}$ such that, for all $i$, $0 \leq i \leq d-1$, and all $n \in \mathbb{Z}$,*

$$g(n) = a_i n + b_i \quad if \quad n \equiv i \pmod{d}.$$

In this definition, rational numbers $a_i$, $b_i$ are chosen such that $a_i n + b_i$ is always integral when $n \equiv i \pmod d$.

The iterates of $g$ are defined by $g^0(n) = n$, and, for all $k \geq 0$, $g^{k+1}(n) = g(g^k(n))$. The problem is to know whether the sequence of iterates $n, g(n), g^2(n), \ldots$ reaches a given number, 1 for example. Conway (1972) proved the following theorem.

THEOREM 2.2 (Conway, 1972). *There is no algorithm which, given a generalized Collatz function $g$ and an integer $n$, determines whether or not there exists a positive integer $k$ such that $g^k(n) = 1$.*

Of course, the integer 1 to be reached can be replaced by any other integer.

Note that computing $n, g(n), g^2(n), \ldots$, waiting for the value 1, is not a relevant algorithm. With this procedure, if 1 is reached, we will know, but if 1 is never reached, we will never know. And we want to know within a finite time.

What is an algorithm? In 1936, many authors were aiming at a formal definition of the notion of algorithm. Church had defined $\lambda$-calculus, Kleene had defined $\mu$-recursive functions, but it was Alan Turing who gave the most fruitful definition, by defining what is now called a *Turing machine* (see definition in Section 3). Then a function on the set of integers is *computable* (or *recursive*) if it is computable by a Turing machine. The other definitions were proved to be equivalent to Turing's one. Some classical references for the theory of computation are Rogers (1967), Soare (1987) and Odifreddi (1989).

Theorem 2.2 states that there is no computable function which, given the numerators and denominators of the rational numbers defining $g$, and an integer $n$, has value 1 if there is a positive integer $k$ such that $g^k(n) = 1$, and has value 0 otherwise. When there is no algorithm to solve a problem, this problem is said to be *undecidable*. So the behavior of iterating generalized Collatz functions is undecidable.

To prove Theorem 2.2, Conway proved that it is true for the special case of generalized Collatz functions such that $b_i = 0$ for all $i$.

DEFINITION 2.3. *A mapping $g : \mathbb{Z} \to \mathbb{Z}$ is a* periodically linear function *if it satisfies the following condition:*

*(i) There is an integer $d \geq 2$ and rational numbers $a_0, \dots, a_{d-1}$ such that, for all $i$, $0 \leq i \leq d-1$, and all $n \in \mathbb{Z}$,*

$$g(n) = a_i n \quad if \quad n \equiv i \pmod{d},$$

*or, equivalently, the following condition:*

*(ii) The function with rational values $g(n)/n$ is periodic with smallest positive period $d \geq 2$.*

In this definition, rational numbers $a_i = p_i/q_i$ are such that $q_i$ divides $\gcd(i, d)$. It is very easy to prove that Conditions (i) and (ii) are equivalent: integer $d \geq 2$ in Condition (i) is the period of $g(n)/n$ in Condition (ii).

THEOREM 2.4 (Conway, 1972). *There is no algorithm which, given a periodically linear function $g$ and an integer $n$, determines whether or not there exists a positive integer $k$ such that $g^k(n) = 1$.*

To prove this theorem, Conway proved that iterating periodically linear functions enable us to simulate any computable function.

THEOREM 2.5 (Conway, 1972). *(i) For any partial computable function $f$, there exists a periodically linear function $g$ such that, for all nonnegative integer $n$,*

$$2^{f(n)} = g^k(2^n),$$

*where $k$ is the smallest positive integer such that $g^k(2^n)$ is a power of 2.*

*(ii) Moreover, there exists a computable function which maps a program for $f$ to the parameters $d, a_0, \dots, a_{d-1}$ of the definition of $g$.*

This theorem deals with *partial computable* functions, which are not necessarily defined everywhere, but only on the set of integers on which the Turing machine that defines the function stops. If $f(n)$ is undefined, then there is no positive integer $k$ such that the iterate $g^k(2^n)$ is a power of 2.

Let us see why Theorem 2.5 implies Theorem 2.4. Suppose that there exists an algorithm which, given a periodically linear function $g$ and an integer $n$, determines whether or not $g^k(n) = 1$ for a positive integer $k$. Then, by Theorem 2.5, there exists an algorithm which, given a program for a computable function $f$, and an integer $n$, determines whether or not $f(n) = 0$. But it is known that no such algorithm exists (This is a special case of Rice's theorem: see Rogers (1967), pp. 33–34).

Theorem 2.5 is true for special periodically linear functions, those that are definable by a Fractan program.

DEFINITION 2.6 (Conway, 1987).          (a) *A Fractan program is a finite sequence of positive rational numbers $r = [r_1, \ldots, r_t]$.*
(b) *A Fractan program $r = [r_1, \ldots, r_t]$ defines a partial function $\varphi_r$ on positive integers by $\varphi_r(n) = r_i n$, if $i$ is the smallest number such that $r_i n$ is integral, and $\varphi_r(n)$ is undefined if such a smallest $i$ does not exist.*
(c) *Iterating a Fractan program $r = [r_1, \ldots, r_t]$ defines a partial function $\psi_r$ on positive integers by $\psi_r(n) = \varphi_r^k(n)$, where $\varphi_r(n), \varphi_r^2(n), \ldots, \varphi_r^k(n)$ are defined and $\varphi_r^{k+1}(n)$ is undefined. If $\varphi_r(n)$ is undefined, or if $\varphi_r^k(n)$ is defined for all $k$, then $\psi_r(n)$ is undefined.*

Note that, by adding to a Fractan program $[r_1, \ldots, r_t]$ an integer $r_{t+1}$ that is a prime number larger than any prime number dividing a denominator of an $r_i$, we get a function $\varphi_r$ that is always defined, and function $\psi_r$ is no longer relevant. Then iterating function $\varphi_r$ can still define a partial function $\psi_r'$ as follows. We examine the sequence $\varphi_r(n), \varphi_r^2(n), \ldots$, until a number with a special form occurs, and then $\psi_r'(n)$ is extracted from this number. That is the way Conway proceeds in his theorem stated below as Theorem 2.7.

**Example.** If $r = [\frac{21}{10}, \frac{5}{7}]$, then $\varphi_r(10) = 21$, $\varphi_r(21) = 15$, and $\varphi_r(15)$ is undefined, so $\psi_r(10) = 15$.

Functions $\varphi_r$ defined by Fractan programs are special cases of partial periodically linear functions. If $r = [r_1, \ldots, r_t]$, $r_i = \frac{p_i}{q_i}$ is an irreducible fraction, and $d = \text{lcm}(q_1, \ldots, q_t)$, then there exist rational numbers $a_0, \ldots, a_{d-1}$ such that, for all positive integer $n$, $\varphi_r(n) = a_i n$ or is undefined when $n \equiv i \pmod{d}$.

**Example.** $r = [\frac{21}{10}, \frac{5}{7}]$, $d = \text{lcm}(10,7) = 70$, and for a positive integer $n$,
$\varphi_r(n) = \frac{21}{10} n$ if $n \equiv 0, 10, 20, 30, 40, 50, 60 \pmod{70}$,
$\varphi_r(n) = \frac{5}{7} n$ if $n \equiv 7, 14, 21, 28, 35, 42, 49, 56, 63 \pmod{70}$,
$\varphi_r(n)$ undefined otherwise.

THEOREM 2.7 (Conway, 1987). *(i) For any partial computable function $f$, there exists a Fractan program $r$ such that, for all nonnegative integer $n$,*

$$2^{f(n)} = \varphi_r^k(2^n),$$

*where $k$ is the smallest positive integer such that $\varphi_r^k(2^n)$ is a power of 2.*

*(ii) Moreover, there exists a computable function which maps a program for $f$ to the Fractan program $r$.*

**Sketch of proof.** A computable function can be computed by a Minsky machine (Minsky, 1967). A Minsky machine has registers $a_1, \ldots, a_m$, that contain nonnegative integers, and a program made of a finite sequence of numbered instructions, each one being of one among the following three types.

- add 1 to register $a_i$, and go to the next instruction,
- if $a_i > 0$, decrease it by 1 and go to the next instruction, and if $a_i = 0$, go to instruction No. $k$,
- stop the computation.

For any partial computable function $f$, there exists a Minsky machine starting with $n, 0, \ldots, 0$ in its registers, stopping with $f(n), 0, \ldots, 0$ in its registers if $f(n)$ is defined, and never stopping if $f(n)$ is undefined.

To prove that any partial computable function can be computed by iterating a Fractan program, it is sufficient to prove that a Minsky machine can be simulated by iterating a Fractan program. If the Minsky machine has $m$ registers, with nonnegative integers $a_1, \ldots, a_m$, these registers are coded by the number $2^{a_1} \times 3^{a_2} \times \cdots \times p_m^{a_m}$, where $p_m$ is the $m$th prime number. If the program of the Minsky machine has $r$ instructions, numbered from 1 to $r$, then prime numbers $p_{m+1}, \ldots, p_{m+r}$ are used to code these instructions. The machine starts with $n$ in its first register, 0 elsewhere, and is reading the first instruction, so the initial code is $2^n \times p_{m+1}$. If the $k$-th instruction is "add 1 to register $a_i$ and go to the next instruction", then it is coded by multiplying by $p_i \frac{p_{m+k+1}}{p_{m+k}}$. If the $k$th instruction is "if $a_i > 0$, decrease it by 1 and go to the next instruction, and if $a_i = 0$, go to instruction No. $j$", then it is coded by two successive fractions, the first one being $\frac{p_{m+k+1}}{p_i p_{m+k}}$, and the second one being $\frac{p_{m+j}}{p_{m+k}}$.

We will not specify in detail how iterating a Fractan program can simulate a Minsky machine. For more detailed accounts, see Conway (1987), Burckel (1994) or Kurtz and Simon (2007). The following two examples can be easily checked.

(1) $r = [\frac{21}{10}, \frac{5}{7}]$. Then $\psi_r(2^a \times 5) = 3^a \times 5$ if $a > 0$, which corresponds to moving a number from the first register to the second one.

(2) $r' = [\frac{110}{21}, \frac{7}{11}]$. Then $\psi_{r'}(2^a \times 3^b \times 7) = 2^{a+b} \times 5^b \times 7$ if $b > 0$, which corresponds to adding to the first register the number contained in the second one, and moving this number from the second register to the third one.

## 2.2. Universal functions.

Turing machines, that will be seen in Section 3, Minsky machines, and iteration of Fractan programs are *computation models* that are able to compute all partial computable functions. A computation model endowed with this property has *universal* members. A universal member can simulate single-handed the computation of all partial computable functions. For example, a universal Turing machine can simulate any Turing machine on any input. Likewise, there exist universal Fractan programs.

THEOREM 2.8 (Conway, 1987). *There exists a Fractan program $u$ that is universal in the following sense.*

*Let $f_c$ be defined by*

$$f_c(n) = m \quad if \quad \psi_u(c 2^{2^n}) = 2^{2^m},$$

*and $f_c(n)$ is undefined if $\psi_u(c 2^{2^n})$ is undefined or is not of the form $2^{2^m}$.*

*Then every partial computable function appears in the sequence of functions $f_0, f_1, f_2, \ldots$.*

Conway (1987) gave an explicit example of universal Fractan program, with 24 fractions, and stated that John Rickard had found a universal Fractan program with 7 fractions, but with very complicated fractions.

Kaščák (1992) gave a universal generalized Collatz function with modulus $d = 396$.

**2.3. Fractan programs as a computation model.** Once we have a computation model that can compute all partial computable functions, it is tempting to use it to program some fashionable functions. So Guy (1983) and Conway (1987) gave Fractan programs which produce by iteration the list of prime numbers. Conway (1987) gave a Fractan program which produces the list of decimal digits of the number $\pi$. Monks (2002) gave a Fractan program which produces the $3x+1$ function, as follows.

THEOREM 2.9 (Monks, 2002). *Let $T$ be the 3x+1 function, defined by $T(n) = n/2$ if $n$ is even, $T(n) = (3n+1)/2$ if $n$ is odd. When the Fractan program*

$$r = \left[ \frac{1}{11}, \frac{136}{15}, \frac{5}{17}, \frac{4}{5}, \frac{26}{21}, \frac{7}{13}, \frac{1}{7}, \frac{33}{4}, \frac{5}{2}, 7 \right]$$

*starts at $2^n$, then the powers of 2 that appears are exactly $2^n, 2^{T(n)}, 2^{T^2(n)}, \ldots$.*

**2.4. Conway's theorem as a tool to prove undecidability.** The usual way to prove that a new problem is undecidable is the following one: a problem known to be undecidable is chosen, and this problem is reduced to a special case of the new problem. Then, were the new problem decidable, the known one would be too, which is impossible. So the new problem is undecidable. This is what was done above to prove that generalized Collatz functions have undecidable behavior. The behavior of Minsky machines, known to be undecidable, was reduced to a special case of the behavior of iterated generalized Collatz functions.

Once a problem is known to be undecidable, it can be added to the list of problems that can be used to prove that a new problem is undecidable. So, now, the undecidable problem of the behavior of iterated generalized Collatz functions can be used to prove that new problems are undecidable. This is what was done by Devienne et al. (1993) and by Marcinkowski (1999).

Devienne et al. (1993) extended Theorem 2.4 to the following one.

THEOREM 2.10 (Devienne et al., 1993). *There is no algorithm which, given a periodically linear function $g$ such that there exists a positive integer $p$ such that $g^p(1) = 1$, and given an integer $n$, determines whether or not there is $k$ such that $g^k(n) = 1$.*

These authors used this theorem to prove the undecidability of the halting problem for programs with exactly one right-linear binary Horn clause.

Marcinkowski (1999) wrote that he was inspired by the work of Devienne et al. (1993) when he proved the undecidability of some properties of DATALOG programs by using generalized Collatz functions.

**2.5. A $\Pi_2^0$-complete problem.** We can define a hierarchy of problems, all of them undecidable, but "more and more undecidable", as follows. A subset $A$ of $\mathbb{N}$ is *recursive* if its characteristic function $\chi_A$ (such that $\chi_A(n) = 1$ if $n \in A$, and $\chi_A(n) = 0$ if $n \notin A$) is computable. Likewise, recursive subsets of $\mathbb{N}^k$ can be defined for $k \geq 2$. Then a subset $B$ of $\mathbb{N}$ is $\Pi_k^0$ if there exists a recursive subset $A$ of $\mathbb{N}^{k+1}$, such that, for all $n \in \mathbb{N}$,

$$n \in B \iff (\forall n_1)(\exists n_2) \ldots (Q n_k)(n, n_1, n_2, \ldots, n_k) \in A,$$

where quantifiers $\forall$ and $\exists$ are alternating, so $Q$ is $\forall$ if $k$ is odd and $Q$ is $\exists$ if $k$ is even.

There exist, among $\Pi_k^0$ sets, sets called $\Pi_k^0$-complete, that are the hardest $\Pi_k^0$ sets, because any $\Pi_k^0$ set can be reduced to a $\Pi_k^0$-complete set. For example, the set {program for $f : f$ is total} is $\Pi_2^0$-complete, where a partial computable function is *total* if it is defined everywhere.

Kurtz and Simon (2007) proved that the set of generalized Collatz functions $g$ such that $(\forall n)(\exists k)g^k(n) = 1$ is a $\Pi_2^0$-complete set.

**2.6. Undecidability of another variant of the $3x+1$ function.** Lehtonen (2008) proved the following theorem.

THEOREM 2.11 (Lehtonen, 2008). *There exist recursive sets $A_t$, with $t \in \{-9, -8, \ldots, 8, 9\}$, that form a partition of the set of odd positive integers, such that, if function $f$ is defined by*

$$f(n) = \begin{cases} 3n + t & \text{if } n \in A_t \\ n/2 & \text{if } n \text{ is even,} \end{cases}$$

*then there is no algorithm to decide, given a positive integer $n$, whether or not there exists a positive integer $k$ such that $f^k(n) = 1$.*

The idea of the proof is the following one. In a first stage, as it is usual, the first $p$ steps of a computation of a Turing machine can be coded by a string that contains the description of the Turing machine and, successively for each step $0, 1, \ldots, p$, the description of the tape and the state at this step. In a second stage, each of these strings is coded by string made of a 1, followed by a string on alphabet $\{0, 2\}$, providing an integer written in ternary representation. Then, not only integers coding for a finite number of steps of a computation of a Turing machine are considered, but also arbitrary prefixes of such integers written in ternary representation. If $n$ codes for a prefix of an ongoing computation of a Turing machine, then $f(n)$ codes for a prefix of the same computation, with one digit added: $f(n) = 3n + 0$ or $f(n) = 3n + 2$. Else, if $n$ codes for a stopped computation or does not code validly for a prefix of a computation, then $f(n)$ is defined such that the iterates $f^k(n)$ go down to 1. So only a never halting computation is coded by a sequence of iterates $f^k(n)$ going to infinity, and in all other cases the sequence of iterates $f^k(n)$ has limit 1. If the simulated Turing machine is universal, an algorithm which decides, given $n$, whether or not there exists $k$ such that $f^k(n) = 1$ would decide the halting problem for this Turing machine, which is impossible.

## 3. Classification of Turing machines

Turing machines are one of the standard models of computation, and the halting problem for Turing machines is the paradigmatic example of an undecidable problem. Of course, given two natural models of computation that have universal members, such as Turing machines and generalized Collatz functions, it is always possible to code either of them into the other one. But we present two unexpected links between them. First, the $3x+1$ function and other generalized Collatz functions enable us to refine the classification of Turing machines according to their numbers of states and symbols. Second, the behaviors of the successive record holders in the busy beaver competition turn out to involve generalized Collatz functions, or a further generalization of these functions.

|   | 0 | 1 |
|---|---|---|
| A | 1R$B$ | 1R$H$ |
| B | 0R$C$ | 1R$B$ |
| C | 1L$C$ | 1L$A$ |

TABLE 1. An example of Turing machine

**3.1. Turing machines.** There are many types of Turing machines. Here, we will consider Turing machines with a single tape, made of cells, infinite in both directions. Each cell contains a *symbol* from a finite set $S = \{s_1, \ldots, s_n\}$. The machine can be in a *state* from a finite set $Q = \{q_1, \ldots, q_k\}$. A single read–write tape head moves on the tape, reading and writing symbols on the cells. These moves are determined by the *transition function*, which is a mapping

$$\delta : Q \times S \longrightarrow S \times \{L, R\} \times (Q \cup \{H\}).$$

If $\delta(q, s) = (s', D, q')$, then, when the machine is in state $q$ and is reading symbol $s$ on a cell, it replaces $s$ by $s'$ on this cell, moves one cell left or right according to $D = L$ or $D = R$, and leaves state $q$ to enter state $q'$.

Initially, the tape contains an *input*, which is a word made of a finite sequence of symbols, while the other cells of the tape contain the *blank symbol* $s_1$. Initially, the tape head reads the leftmost symbol of the input, and the state is the initial state $q_1$. A *computation step* consists in an application of the transition function $\delta$. A special state $H$, outside the set $Q$, is called the *halting state*. If the halting state $H$ is reached, then the computation stops, and then the *output* of the computation is what is written on the tape. If state $H$ is never reached, the computation never stops. A *configuration* of the machine consists in the description of the symbols written on the tape, the state of the machine, and the place of the tape head.

**Example.** Consider the following Turing machine with set of symbols $S = \{0, 1\}$ and set of states $Q = \{A, B, C\}$. The blank symbol is 0, the initial state is $A$, and the transition function is given by Table 1.

When this machine is launched on the blank tape (where all symbols are blank ones), the initial configuration is $\ldots 0(A0)0 \ldots$, where the place of the tape head is shown by putting both the state and the scanned symbol between parentheses, and the dots mean an infinite string of 0s. After 14 computation steps, the configuration is $\ldots 0111(H1)110 \ldots$ and the computation stops, since halting state $H$ is reached.

Even with the simple type of Turing machine we have defined, any computable function on integers can be computed. That is, for any partial computable function $f$, there exists a Turing machine as it is defined above such that, when it is launched on an input coding integer $n$, it stops with an output coding integer $f(n)$, and never stops if $f(n)$ is undefined.

The *halting problem* for a Turing machine $M$ is the following one: given an input $x$ for $M$, does the computation of $M$ on $x$ stop?

There exists a Turing machine $U$ with inputs of the form $(M, x)$, where $M$ codes for a Turing machine and $x$ codes for an input, such that $U$ simulates the computation of Turing machine $M$ on input $x$. Such a Turing machine is called *universal*.

The halting problem for a universal Turing machine $U$ is undecidable, that is, there is no algorithm to decide whether or not machine $U$ stops on an input. It

| symbols | 2 | 3 | 4 | 5 | 6 | 7 | 8 | 9 | 10 | ⋯ | 18 |
|---|---|---|---|---|---|---|---|---|---|---|---|
| 18 | $U$ | | | | | | | | | | |
| ⋮ | ⋮ | | | | | | | | | | |
| 9 | | $U$ | | | | | | | | | |
| 8 | $T$ | | | | | | | | | | |
| 7 | | | | | | | | | | | |
| 6 | | | $U$ | | | | | | | | |
| 5 | | $T$ | | $U$ | | | | | | | |
| 4 | $O$ | | $T$ | | $U$ | | | | | | |
| 3 | $D$ | $O$ | | $T$ | | | | $U$ | | | |
| 2 | $D$ | $D$ | $O$ | | | | | | $T$ | ⋯ | $U$ |

states

TABLE 2. Type of Turing machine in $TM$(states, symbols) : $U$ = Universal, $T$ = Three–$x$–plus–one, $O$ = Open Collatz-like problem, $D$ = all Decidable

can be proved that there exist Turing machines that are not universal but have an undecidable halting problem. But the proof is indirect, and presently all proofs of undecidability for the halting problem of a particular Turing machine are made by proving that the machine can simulate a universal Turing machine, and so is itself universal.

**3.2. Classification according to numbers of states and symbols.** Let $TM(k,n)$ be the set of Turing machines that have $k$ states and $n$ symbols. This is a finite set, and with the definition given above, it has $(2(k+1)n)^{kn}$ members. Note also that $TM(k,n)$ can be embedded in $TM(k',n')$ if $k \leq k'$, $n \leq n'$. A natural question is: from which values of $k$ and $n$ can we find universal Turing machines in $TM(k,n)$? We refer to Margenstern (2000) for a survey of this topic and we give the presently known results.

There are universal Turing machines in the following sets.

- $TM(2,18)$: Rogozhin (1996).
- $TM(3,9)$: Kudlek and Rogozhin (2002).
- $TM(4,6)$: Rogozhin (1982).
- $TM(5,5)$: Rogozhin (1982).
- $TM(6,4)$: Neary and Woods (2007).
- $TM(9,3)$: Neary (2006).
- $TM(18,2)$: Neary (2006).

These sets are labelled with a $U$ in Table 2.

Note that these universal Turing machines should not be confused with universal Turing machines from Cook (2004), that are never halting Turing machines that simulate universal cellular automata on a tape with a well designed infinite initial configuration.

On the other hand, all the Turing machines in the following classes are decidable.

- $TM(k,1)$: trivial.
- $TM(1,n)$: Hermann (1968).

|   | $b$ | $0$ | $1$ |
|---|-----|-----|-----|
| $A$ |      | $b\mathrm{R}A$ | $b\mathrm{R}D$ |
| $B$ |      | $0\mathrm{R}B$ | $1\mathrm{R}C$ |
| $C$ | $1\mathrm{L}E$ | $1\mathrm{R}B$ | $0\mathrm{R}D$ |
| $D$ | $0\mathrm{R}C$ | $0\mathrm{R}C$ | $1\mathrm{R}D$ |
| $E$ | $b\mathrm{R}A$ | $0\mathrm{L}E$ | $1\mathrm{L}E$ |

TABLE 3. Machine $M_0$ in $TM(5,3)$

- $TM(3,2)$: Pavlotskaya (1978).
- $TM(2,3)$: Pavlotskaya (unpublished).

These sets are labelled with a $D$ in Table 2.

This leaves 42 sets $TM(k,n)$ for which the problem is unsettled. We can find another criterion to make a distinction between these remaining sets. In some of these sets, there are Turing machines that simulate the iteration of the $3x+1$ function defined by $T(n) = n/2$ if $n$ is even, $T(n) = (3n+1)/2$ if $n$ is odd. Michel (1993) found a machine in $TM(6,3)$ which, on an input $n$ written in binary, computes the iterates $T^k(n)$ until it reaches 1, and then stops. So the halting problem for this machine depends on the conjecture on the iterates of $T$. Margenstern (2000) and Baiocchi (1998) found machines that simulate function $T$ without halting, so they compute successively all iterates $T^k(n)$ for $k \geq 1$. Then, reaching the value 1 is attested by the machines reaching some special configurations. Such machines were found in the following sets : $TM(2,8)$, $TM(3,5)$, $TM(4,4)$, $TM(5,3)$ and $TM(10,2)$. These sets are labelled with a $T$ in Table 2. Table 3 gives the transition function for Turing machine $M_0$ in $TM(5,3)$, where $b$ is the blank symbol. For this machine, we have the following theorem.

THEOREM 3.1. *The following two conditions are equivalent.*

(i) *(Collatz conjecture) For all positive integer $n$, there exists a positive integer $k$ such that $T^k(n) = 1$.*

(ii) *For all positive integer $n$, written $n_p \ldots n_1 n_0$ in binary, there exists a positive integer $t$ such that the initial configuration $\ldots b(An_0)n_1 \ldots n_p b \ldots$ of machine $M_0$ leads in $t$ computation steps to the configuration $\ldots b(A1)b\ldots$.*

For many generalized Collatz functions, the behavior of their iterates is an open problem. So simulating these iterations by Turing machines give halting problems for these Turing machines that are open problems. Michel (1993, 2004) found such machines in the following sets: $TM(2,4)$, $TM(3,3)$ and $TM(5,2)$. These sets are labelled with an $O$ in Table 2. We will see below how the machines in these sets were found.

**3.3. Generalized Collatz functions from busy beaver competition.** The busy beaver competition was defined by Rado (1962) to provide a noncomputable function with a simple and explicit definition. In the set $TM(k,n)$ of Turing machines with $k$ states and $n$ symbols, we consider the subset $HTM(k,n)$ of machines that halt when they are launched on a blank tape, that is a tape full of blank symbols. For a Turing machine $M$ in $HTM(k,n)$, we denote by $s(M)$ the number of steps made by $M$ before it stops, and we denote by $\sigma(M)$ the number of non-blank symbols left on the tape when $M$ stops. Then the busy beaver functions are

$$S(k,n) = \max\{s(M) : M \in HTM(k,n)\},$$

$$\Sigma(k,n) = \max\{\sigma(M) : M \in HTM(k,n)\}.$$

That is, $S(k,n)$ is the greatest number of steps made by a Turing machine with $k$ states and $n$ symbols that stops when it starts from a blank tape. And $\Sigma(k,n)$ is the greatest number of non-blank symbols left on the tape by such a machine when it stops.

Rado (1962) considered only machines with $n = 2$ symbols. He proved that functions $S(k) = S(k,2)$ and $\Sigma(k) = \Sigma(k,2)$ are not computable (they grow faster than any computable function), and he asked for the values of $S(k)$ and $\Sigma(k)$ for small $k$. Brady (1988) considered machines with $n \geq 3$ symbols. The following values are known.

- $S(2,2) = 6$, $\Sigma(2,2) = 4$: Rado (1962).
- $S(3,2) = 21$, $\Sigma(3,2) = 6$: Lin and Rado (1965).
- $S(4,2) = 107$, $\Sigma(4,2) = 13$: Brady (1983), and Kopp, cited by Machlin and Stout (1990).

For all other values of $k, n \geq 2$, only lower bounds are known.

- $S(5,2) \geq 47,176,870$, $\Sigma(5,2) \geq 4098$: Marxen and Buntrock (1990).
- $S(6,2) \geq 2.5 \times 10^{2879}$, $\Sigma(6,2) \geq 4.6 \times 10^{1439}$: Terry and Shawn Ligocki in 2007.
- $S(2,3) \geq 38$, $\Sigma(2,3) \geq 9$: Brady (1988) and Michel (2004).
- $S(3,3) \geq 1.1 \times 10^{17}$, $\Sigma(3,3) \geq 374,676,383$: Terry and Shawn Ligocki in 2007.
- $S(4,3) \geq 1.0 \times 10^{14072}$, $\Sigma(4,3) \geq 1.3 \times 10^{7036}$: Terry and Shawn Ligocki in 2008.
- $S(2,4) \geq 3,932,964$, $\Sigma(2,4) \geq 2050$: Terry and Shawn Ligocki in 2005.
- $S(3,4) \geq 5.2 \times 10^{13036}$, $\Sigma(3,4) \geq 3.7 \times 10^{6518}$: Terry and Shawn Ligocki in 2007.
- $S(2,5) \geq 1.9 \times 10^{704}$, $\Sigma(2,5) \geq 1.7 \times 10^{352}$: Terry and Shawn Ligocki in 2007.
- $S(2,6) \geq 2.4 \times 10^{9866}$, $\Sigma(2,6) \geq 1.9 \times 10^{4933}$: Terry and Shawn Ligocki in 2008.

Accounts of successive record holders for functions $S(k,n)$ and $\Sigma(k,n)$ have been given in Marxen's and Michel's websites (see References). When we study the behaviors of these record holders to find out how simply definable machines can achieve big performances, we face a surprising fact. Nearly all these machines simulate generalized Collatz functions or variants of such functions. Let us give two examples. In the following, if $C$ and $D$ are two configurations of a Turing machine, we write $C \vdash (t) \, D$ if configuration $C$ leads to configuration $D$ in $t$ computation steps.

**Example 1.** Table 4 gives the transition function for Turing machine $M_1$ with 5 states and 2 symbols discovered in September 1989 by Marxen and Buntrock (1990). It stops in 47,176,870 steps, and leaves 4098 symbols 1 on the tape. It is the present record holder for both $S(5,2)$ and $\Sigma(5,2)$. Let $C(n) = \ldots 0(A0)1^n 0 \ldots,$

|   | 0 | 1 |
|---|---|---|
| $A$ | 1R$B$ | 1L$C$ |
| $B$ | 1R$C$ | 1R$B$ |
| $C$ | 1R$D$ | 0L$E$ |
| $D$ | 1L$A$ | 1L$D$ |
| $E$ | 1R$H$ | 0L$A$ |

TABLE 4. Machine $M_1$ in $TM(5,2)$

|   | 0 | 1 | 2 | 3 |
|---|---|---|---|---|
| $A$ | 1R$B$ | 2L$A$ | 1R$A$ | 1R$A$ |
| $B$ | 1L$B$ | 1L$A$ | 3R$B$ | 1R$H$ |

TABLE 5. Machine $M_2$ in $TM(2,4)$

where $1^n$ denotes the string of $n$ symbols 1. Then we have, for all $k \geq 0$,

$$
\begin{array}{lll}
C(3k) & \vdash (5k^2 + 19k + 15) & C(5k+6) \\
C(3k+1) & \vdash (5k^2 + 25k + 27) & C(5k+9) \\
C(3k+2) & \vdash (6k+12) & \ldots 01(H0)1(001)^{k+1}10\ldots
\end{array}
$$

So the behavior of machine $M_1$ depends on the iterates of the partial generalized Collatz function $g_1$ defined by

$$
\begin{aligned}
g_1(3k) &= 5k + 6 \\
g_1(3k+1) &= 5k + 9 \\
g_1(3k+2) &\text{ undefined.}
\end{aligned}
$$

Michel (1993) proved that this machine halts on all inputs if and only if for all nonnegative integers $n$ there is a positive integer $k$ such that $g_1^k(n)$ is undefined. So the halting problem for machine $M_1$ is presently an open problem.

**Example 2.** Table 5 gives the transition function for Turing machine $M_2$ with 2 states and 4 symbols discovered in February 2005 by Terry and Shawn Ligocki. It stops in 3,932,964 steps and leaves 2050 non-blank symbols on the tape. It is the present record holder for both $S(2,4)$ and $\Sigma(2,4)$. Let $C(n,1) = \ldots 0(A0)2^n 10 \ldots$, and $C(n,2) = \ldots 0(A0)2^n 110 \ldots$. Then we have $\ldots 0(A0)0 \cdots \vdash (6)\ C(1,2)$, and, for all $k \geq 0$,

$$
\begin{array}{lll}
C(3k,1) & \vdash (15k^2 + 9k + 3) & C(5k+1,1) \\
C(3k+1,1) & \vdash (15k^2 + 24k + 13) & \ldots 013^{5k+2}1(H1)0\ldots \\
C(3k+2,1) & \vdash (15k^2 + 29k + 17) & C(5k+4,2) \\
C(3k,2) & \vdash (15k^2 + 11k + 3) & C(5k+1,2) \\
C(3k+1,2) & \vdash (15k^2 + 21k + 7) & C(5k+3,1) \\
C(3k+2,2) & \vdash (15k^2 + 36k + 21) & \ldots 013^{5k+4}1(H1)0\ldots
\end{array}
$$

So the behavior of machine $M_2$ depends on the iterates of the following partial function $g_2 : \mathbb{N} \times \{1,2\} \to \mathbb{N} \times \{1,2\}$.

$$g_2(3k,1) = (5k+1,1)$$
$$g_2(3k+1,1) \text{ undefined}$$
$$g_2(3k+2,1) = (5k+4,2)$$
$$g_2(3k,2) = (5k+1,2)$$
$$g_2(3k+1,2) = (5k+3,1)$$
$$g_2(3k+2,2) \text{ undefined.}$$

We have here a further generalization of generalized Collatz functions, that Michel (1993) called *Collatz-like functions*. Because the halting problem for machine $M_2$ depends on the iterates of function $g_2$, it is presently an open problem.

## 4. Classification of tag systems

**4.1. Tag systems.** Tag systems have a very simple definition, first given by Emil Post in 1921. A tag system $S$ has $m$ *symbols* from a finite alphabet $A = \{a_1, \ldots, a_m\}$, a *shift number* $n$ which is a positive integer, and *productions* $P(a_1), \ldots, P(a_m)$ in the set $A^*$ of words of finite length over the alphabet $A$. The productions are often written as follows.

$$a_1 \quad \to \quad P(a_1)$$
$$\cdots$$
$$a_m \quad \to \quad P(a_m).$$

The *computation* of tag system $S$ on a word $\alpha \in A^*$ is the sequence of words $\alpha = \alpha_0$, $\alpha_1, \ldots$, where $\alpha_{k+1}$ is obtained from $\alpha_k$ by deleting the first $n$ symbols of $\alpha_k$ and appending $P(a_i)$ if $a_i$ is the first symbol of $\alpha_k$. The computation stops if the length of $\alpha_k$ is less than $n$. See Minsky (1967) for more on tag systems.

For example, let $S_1$ be the tag system defined by $m = 2$, $A = \{0,1\}$, $n = 3$, and the productions

$$0 \quad \to \quad 00$$
$$1 \quad \to \quad 1101.$$

This tag system was first defined by Post. It has the following computation on the word $\alpha = 100$.

$$100, 1101, 11101, 011101, 10100, 001101, 10100, \ldots$$

This computation becomes periodic and never stops.

The *halting problem* for a tag system $S$ is the following one: given a word $\alpha$, does the computation of $S$ on $\alpha$ stop? If there is an algorithm which, on input $\alpha$, decides whether or not the computation of tag system $S$ on $\alpha$ stops, then $S$ is said to be *decidable*. Otherwise, $S$ is *undecidable*. The decidability of tag system $S_1$ defined above is still an open problem, despite the simplicity of its definition.

Like for Turing machines, an undecidability result for a tag system is gotten by proving it to be *universal*, that is capable of simulating a universal Turing machine.

**4.2. Classification according to number of symbols and shift number.** Let $TS(m,n)$ be the set of tag systems with $m$ symbols in the alphabet, and shift number $n$. All the tag systems in the following sets are decidable.

- $TS(1,n)$: trivial.
- $TS(m,1)$: Wang (1963).

- $TS(2, 2)$: De Mol (2007).

On the other hand, there are universal tag systems in $TS(576, 2)$ (see De Mol (2007, 2008); the existence of a universal tag system in $TS(288, 2)$ is not yet proved (De Mol, personal communication)).

We have seen that Collatz function and Collatz-like problems enable us to display Turing machines with an open halting problem in sets $TM(k, n)$ with small values of $k$ and $n$. Likewise, they enable De Mol (2008) to get a tag system with an open halting problem in the set $TS(3, 2)$. This tag system $S_2$ has alphabet $A = \{0, 1, 2\}$, shift number $n = 2$, and productions

$$
\begin{array}{ccc}
0 & \rightarrow & 12 \\
1 & \rightarrow & 0 \\
2 & \rightarrow & 000.
\end{array}
$$

The computation of $S_2$ on $0^n$ yields $0^{n/2}$ in $n$ computation steps if $n$ is even, and it yields $0^{(3n+1)/2}$ in $n+1$ computation steps if $n$ is odd. The computation of $S_2$ stops when it reaches the word 0. So the halting problem for tag system $S_2$ depends on the conjecture for the $3x+1$ function.

Thus, tag system $S_1$ in $TS(2, 3)$ and tag system $S_2$ in $TS(3, 2)$ show that there are open decidability problems for tag systems in sets $TS(m, n)$ with small values of $m$ and $n$.

## 5. Classification of cellular automata

Cellular automata are another widespread model of computation. By making time and space discrete, they can be used in physics to simulate evolution of systems where the future state of each point depends on the current states of its neighbor points. We give the definition of cellular automata and a simulation of the $3x+1$ function by a 1-dimensional cellular automaton.

**5.1. Cellular automata.** A *cellular automaton* $\mathcal{A}$ has a $d$-dimensional *grid* $\mathbb{Z}^d$, a finite set of *states* $S = \{q_1, \ldots, q_n\}$, a finite set $N = \{v_1, \ldots, v_m\}$ of vectors of $\mathbb{Z}^d$, called the *neighborhood* of $\mathcal{A}$, and a *transition function* $\delta : S^m \rightarrow S$. A *configuration* is a mapping $c : \mathbb{Z}^d \rightarrow S$ which assigns a state in $S$ to each cell of the grid $\mathbb{Z}^d$.

The *computation* of the cellular automaton $\mathcal{A}$ on a configuration $c$ is the sequence of configuration $c = c_0, c_1, \ldots$, where $c_{k+1}$ is obtained from $c_k$ by replacing the state in every cell $x \in \mathbb{Z}^d$ by the state given by function $\delta$ from the states in the $m$ cells in the neighborhood $\{x + v_1, \ldots, x + v_m\}$ of cell $x$, that is, for all $x \in \mathbb{Z}^d$,

$$
c_{k+1}(x) = \delta(c_k(x + v_1), \ldots, c_k(x + v_m)).
$$

The most studied cellular automata are the 2-dimensional ones with von Neumann neighborhood (each cell has 5 neighbors) or Moore neighborhood (each cell has 9 neighbors), and the 1-dimensional ones, where the neighborhood is $N = \{-1, 0, 1\}$, so each cell has 3 neighbors, itself and the adjacent cells.

For each dimension $d \geq 1$ and each choice of neighborhood $N$, one can ask for the smallest number of states that allows a cellular automaton to be universal, or to simulate the $3x+1$ function.

Results are known for various notions of universality in the 2-dimensional and 1-dimensional cases. Ollinger (2002) gave an intrinsically universal 1-dimensional cellular automaton with neighborhood $N = \{-1, 0, 1\}$ and 6 states. We give below a 1-dimensional cellular automaton that simulates the $3x+1$ function with neighborhood $N = \{-1, 0, 1\}$ and 5 states.

**5.2. Simulation of the $3x+1$ function.** Several papers on the simulation of the $3x+1$ function were written, by Korec (1992), Goles et al. (1998), Baiocchi and Margenstern (2001) and Bostick (2004), where the cellular automaton with the smallest number of states have 7, 7, 5 and 7 states respectively. Here, we shall construct a cellular automaton according to the guidelines of Baiocchi and Margenstern (2001). The construction relies on the simulation of the Turing machine which we indicate in Table 1.

|     |       | $-$   | $z$   | $u$    |
|-----|-------|-------|-------|--------|
| $i$ |       |       | $\_L$ | $\_L2$ |
| $0$ |       |       | $L$   | $L1$   |
| $1$ | $uRf$ |       | $uL0$ | $zL2$  |
| $2$ | $zL1$ |       | $L1$  | $L$    |
| $f$ | $Li$  |       | $R$   | $R$    |

Table 1. *Machine $M_3$ in $TM(5,3)$ simulating the function $3x+1$. This machine can be obtained from machine $M_0$ in $TM(5,3)$ by exchanging $R$ and $L$ in the table of the machine, see Table 3.*

The principle of working of the machine is easy to understand. The machine is based on the representation of natural numbers in base 6 in a notation which makes use of both base 2 and base 3. We shall call this the *mixed base*. If we write
$$n = \sum_{j=0}^{k} a_j 6^j,$$
we get that $a_j \in \{0, \ldots, 5\}$ and we shall write $a_j = 3X_j + x_j$ with $X_j \in \{0, 1\}$ and $x_j \in \{0, 1, 2\}$. In machine $M_3$, 0 and 1 are replaced by $z$ and $u$ respectively. Accordingly, a digit in base 6 is replaced by 2 symbols, one in base 2 and the other in base 3. The interest of this representation lies in the following remark. Consider the Euclidean division of $n$ by 2. This provides us with two numbers $m$ and $r$ such that $n = 2m + r$ with $r \in \{0, 1\}$. Writing $m$ in the mixed base as $m = \sum_{j=0}^{k} (3Y_j + x_j) 6^j$, as $m$ requires no more digits than $n$, the identification of the coefficients gives us that:

$(i)$         $Y_k = 0$
$(ii)$       $y_j = (3X_j + x_j) \operatorname{div} 2$    for   $j = 1, \ldots, k$
$(iii)$     $Y_j = (3X_{j+1} + x_{j+1}) \bmod 2$   for   $j = 1, \ldots, k$,

setting $Y_{k+1} = y_{k+1} = 0$.

Now, $3n+1 = 3(2m) + 3r + 1 = 6m + 3r + 1$, so that the main part of the representation of $3n+1$ in the mixed base is obtained by a shift to the left from that of $m$, followed by a simple tuning at the last step. Indeed, $3r+1$ is at most 4 so that it is the lowest digit of the representation of $3n+1$ in base 6. In the mixed base, this last value 4 will be represented by $u$ $\_$. Accordingly, we have to define rules to compute the quotient in the division of $n$ by 2. These rules are easily defined by the above equalities $(i)$, $(ii)$ and $(iii)$.

| R | x |
|---|---|
| x' | R' |

The transformation of these rules into rules of a one-dimensional cellular automaton are straightforward. This gives us a cellular automaton with six states, $z$, $u$, 0, 1, 2 and the blank _.

It is interesting to notice that the representation in the mixed base allows to give the data in the initial configuration of the automaton either in base 2, in base 3 or in the mixed base. In the case of an initial configuration in base 2 or in base 3, there is a first step of the computation of the iteration of the $3x+1$ function which transforms the data in its representation in the mixed base and then, the simulation goes on in this representation. However the simulation can also be read in base 2 and in base 3 if we look at the space-time diagram of the computation. Such a space-time diagram is a two-dimensional table where the rows represent the successive configurations of the cellular automaton and the columns represent the evolution of a cell in time. We consider the diagonals of the space-time diagram. Say that the diagonals from top right to bottom left are of the first kind and that those from top left to bottom right are of the second kind. Then, as explained by Baiocchi and Margenstern (2001), the successive values of the iterations can be read in base 2 on each second diagonal of the first kind and in base 3 on each second diagonal of the second kind.

Below, we give the table of this six-state automaton in a condensed representation.

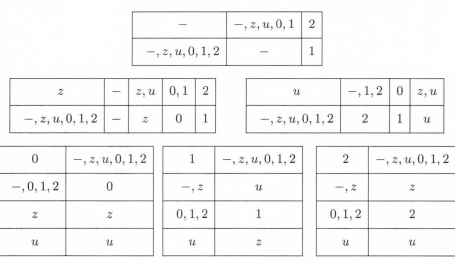

| − | −, z, u, 0, 1 | 2 |
|---|---|---|
| −, z, u, 0, 1, 2 | − | 1 |

| z | − | z, u | 0, 1 | 2 |
|---|---|---|---|---|
| −, z, u, 0, 1, 2 | − | z | 0 | 1 |

| u | −, 1, 2 | 0 | z, u |
|---|---|---|---|
| −, z, u, 0, 1, 2 | 2 | 1 | u |

| 0 | −, z, u, 0, 1, 2 |
|---|---|
| −, 0, 1, 2 | 0 |
| z | z |
| u | u |

| 1 | −, z, u, 0, 1, 2 |
|---|---|
| −, z | u |
| 0, 1, 2 | 1 |
| u | z |

| 2 | −, z, u, 0, 1, 2 |
|---|---|
| −, z | z |
| 0, 1, 2 | 2 |
| u | u |

Table 2. *The cellular automaton with 6 states simulating the computation of $3x+1$.*

Now, we get the announced five-state cellular automaton by requiring that the initial data is given in base 3 and we apply the same table as in Table 2 in which $z$ is identified with the blank. This reduction can be performed by taking into account that a representation in base 3 and in the mixed base implies that the neighbors of $z$ and $u$ are in $\{0, 1, 2\}$ unless the letter is at the border of the word representing the number. And so, the true blank has at least an other blank neighbor.

This gives us the following result:

THEOREM 5.1 (Baiocchi and Margenstern, 2001). *The iterations of the function $3x+1$ can be simulated by a five-state one-dimensional cellular automaton.*

Baiocchi and Margenstern (2001) said that the implementation of the six-state cellular automaton could provide a very fast computation of the iterations of $3x+1$.

To conclude this section about the computation of $3x+1$ with cellular automata, we mention another seven-state cellular automaton simulating these iterations by Bostick (2004).

## 6. Diophantine representations by binomial coefficients

The role of binomial coefficients in computational problems appeared with the pioneer work of Martin Davis, Harry Putnam and Julia Robinson on Hilbert's Tenth Problem, see Davis et al. (1976) and a detailed account by Matiyasevich (1993). We remind the reader that Hilbert's Tenth Problem asks whether there is an algorithm which, being given a polynomial $P(X_1, \ldots, X_k)$ with integer coefficients as a list of the coefficients and the corresponding degrees would say whether or not the equation $P(X_1, \ldots, X_k) = 0$ has solutions among the integers. As the problem turned out to be undecidable, the negative solution could not be imagined before the foundational works of the 30's of the last century on the formalization of the concept of algorithm.

Consider any formula of the form $Q_1 x_1 \ldots Q_k x_k \, G(x_1, \ldots, x_k)$ where $Q_i$'s are quantifiers and $G$ is a quantifier-free formula defined by an elementary function, where the term *elementary* can be interpreted either unformally or formally, *i.e.* as belonging to Grzegorczyk's hierarchy for instance. Then, it is important to remark that binomial coefficients played a key role in the transformation of such a formula into a formula of the form $\exists x_1 \ldots \exists x_k \, P(x_1, \ldots, x_k) = 0$, where $P$ is a polynomial with integer coefficients. Several properties are at work in this transformation. In particular, an important point is an old lemma proved by Kummer in the middle of the $19^e$ century which was several time re-discovered:

LEMMA 6.1 (Kummer, 1852). *Let $a$ and $n$ be natural numbers and let their binary representations be defined by $a = \sum_{i=0}^{m} \alpha_i 2^i$ and $b = \sum_{i=0}^{m} \beta_i 2^i$, using the same number $m$ with possible leading $0$'s as digits. Then we have $\alpha_i \leq \beta_i$ for all $i \in \{0, \ldots, m-1\}$ if and only if $\dbinom{a}{b} \equiv 1 \bmod 2$.*

With the notations of Lemma 6.1, the relation $\alpha_i \leq \beta_i$ for all $i \in \{0, \ldots, m-1\}$ is denoted by $a \preceq b$. This relation is very powerful. As an example, with the same notations as in Lemma 6.1 we have that $\sum_{i=0}^{m} \alpha_i \beta_i = 0$ if and only if $a \preceq (a+b)$.

These relations, with the help of appropriate *masks* can be used to encode finite sequences of numbers of a given length. We have no room to give the details of these techniques and we refer the interested reader to Margenstern and Matiyasevich (1999). Now, this encoding property enables us to view that binomial coefficients can be used to encode the whole history of the computational steps of a Turing machine. They can also be used to state simpler problems, in particular, the conjecture about the iterations of the $3x+1$ function.

Indeed, in Margenstern and Matiyasevich (1999), it is proved that:

THEOREM 6.2 (Margenstern and Matiyasevich, 1999). *The Collatz conjecture is true if and only if for every positive integer $a$ there are natural numbers $w$ and $v$ such that $a \leq w$ and:*

$$\binom{2w+1}{w}\binom{4(w+1)v+1}{v}\sum_{r=0}^{\infty}\sum_{s=0}^{\infty}\sum_{t=0}^{\infty}\binom{v}{r}$$

$$\times \binom{w(v-r)}{s}\binom{wr}{t}$$

$$\times \binom{2s+2t+r+(4w+3)v+1}{3((4w+4)t+a)+2(4w+4)r+(4w+4)s}$$

$$\times \binom{3((4w+4)t+a)+2(4w+4)r+(4w+4)s}{2s+2t+r+(4w+3)v+1} \equiv 1 \ (\mathrm{mod}\ 2).$$

Although impressive, the formula stated in the theorem is a rather simple encoding of properties of sequences of digits 0 and 1. In order that the reader can understand the particularities of this encoding, let us remark the following. Two coefficients which occur under the summations are of the form $\binom{a}{b} \cdot \binom{b}{a}$. In such a summation where $a$ and $b$ are positive integers, $\binom{a}{b} = 0$ when $a > b$. As the summation must be equal to 1 modulo 2, at least one term is not even. This means that the summation replaces existential quantifiers on the corresponding variables. Moreover, all the factors contained in a term under the summations must be equal to 1 modulo 2 when the sum is not even. This means that each factor is equal to 1 modulo 2. Now, when such a term is not 0 modulo 2, necessarily $\binom{a}{b} = \binom{b}{a} = 1$ which is equivalent to $a = b$. At last, note that the summation in the theorem is finite as for large values of the variables on which the summation is performed, one of the binomial coefficients under the summation or in front of it vanishes. And so, connecting this feature with the applications of masks revealed by Kummer's Lemma, we can better understand that binomial coefficients can be used to formalize computations and also conjectures about such computations.

We also think that this example illustrates Yuri Matiyasevich's point of view that binomial coefficients can be used for compact representation of conjectures. But, as Yuri Matiyasevich emphasizes, such a representation can be useful to attack conjectures: there are now efficient softwares to deal with closed formulas involving sums of products of binomial coefficients. See Petrovšek, Wilf and Zeilberger (1996) on automated hypergeometric and binomial coefficient identities, and see Margenstern and Matiyasevich (1999) for other references.

## 7. Congruential systems

A *congruential system* $C$ has, for each letter $a$ from a finite alphabet $A$, a rule of the form

$$(p, r) \overset{a}{\to} (q, s),$$

where $p$, $q$, $r$, $s$ are nonnegative integers, with $r < p$ and $s < q$. The *graph* $G(C)$ of a congruential system $C$ is the directed graph with labelled edges of the form

$$pn + r \overset{a}{\to} qn + s,$$

where $(p, r) \overset{a}{\to} (q, s)$ is a rule of $C$, and $n$ is a nonnegative integer. So the set of vertices of this graph is a subset of $\mathbb{N}$. The graph can have loops, and several edges, with distinct labels, can link two given integers. To be more readable, the rule $(p, r) \overset{a}{\to} (q, s)$ will be written $pn + r \overset{a}{\to} qn + s$.

Generalized Collatz functions are special cases of congruential systems. For example, the $3x+1$ function yields the following congruential system with alphabet $A = \{a, b\}$.

$$2n \quad \overset{a}{\to} \quad n$$
$$2n + 1 \quad \overset{b}{\to} \quad 3n + 2.$$

The graph associated to this congruential system is called the Collatz's graph. It has for set of vertices the set of positive integers, and the Collatz conjecture becomes: for all positive integer $n$, there is a path from $n$ to 1 in the Collatz's graph.

Since congruential systems are a further generalization of generalized Collatz functions, they can have undecidable behaviors, and they give rise to open problems.

Isard and Zwicky (1970) studied the following congruential system

$$n \quad \overset{a}{\to} \quad 3n$$
$$2n + 1 \quad \overset{b}{\to} \quad n,$$

and asked whether there exists, in the associated graph, a path from 1 to every positive integer congruent to either 0 or 1 modulo 3.

Urvoy (2000) made a link between congruential systems and properties of infinite graphs arising from rewriting systems. He proved that the Collatz's graph is not regular and so has an undecidable monadic second order theory.

## 8. Other representations

In this section, we present some other results on the $3x+1$ function, notably some representations by grammars and rewriting systems.

### 8.1. Andrei, Chin and Nguyen's grammar.
A *phrase-structure grammar* $G = (V, T, P, S)$ has a set $V$ of *variables*, a set $T$ of *terminals*, such that $V$ and $T$ are disjoint, a *start symbol* $S \in V$, and a set $P$ of *productions*. The productions in set $P$ are of the form $\alpha \to \beta$, with $\alpha, \beta \in (V \cup T)^*$, that is $\alpha$ and $\beta$ are strings of symbols from the alphabet $V \cup T$, and moreover $\alpha$ contains at least one symbol from $V$.

String $\gamma_0 \in (V \cup T)^*$ *derives directly* string $\gamma_1 \in (V \cup T)^*$ in grammar $G$ if there is a production $\alpha \to \beta$ in $P$ such that $\gamma_0 = u\alpha v$ and $\gamma_1 = u\beta v$. Then we write $u\alpha v \Longrightarrow u\beta v$. String $\gamma_0$ *derives in $n$ steps* string $\gamma_n$ if $\gamma_0 \Longrightarrow \gamma_1 \Longrightarrow \cdots \Longrightarrow \gamma_n$, and then we write $\gamma_0 \overset{n}{\Longrightarrow} \gamma_n$. String $\gamma_0$ *derives* string $\gamma$ if there exists a nonnegative integer $n$ such that $\gamma_0$ derives in $n$ steps string $\gamma$, and then we write $\gamma_0 \overset{*}{\Longrightarrow} \gamma$.

The *language generated* by grammar $G$ is $L(G) = \{w \in T^* : S \overset{*}{\Longrightarrow} w\}$. That is, it is the set of strings of terminals that can be derived from the start symbol $S$.

It is known that a language can be generated by a phrase-structure grammar if and only if it is computably enumerable, so phrase-structure grammars are as powerful as Turing machines.

Andrei et al. (2007) gives the following phrase-structure grammar $G$, with set of variables $V = \{S, A, X, Y\}$, set of terminals $T = \{a, \#\}$, start symbol $S$ and set of productions $P$ as follows.

|     |     |     |     |
|-----|-----|-----|-----|
| (1) | $S \to \#A\#$ | (5) | $AX\# \to Y\#$ |
| (2) | $\#A \to \#XA$ | (6) | $AAAY \to YA$ |
| (3) | $XA \to AAX$ | (7) | $\#Y \to \#$ |
| (4) | $X\# \to \#$ | (8) | $A \to a$ |

Informally, given a string $\#A^n\#$, an $X$ is created by production (2), moves from left to right through the $A$s and doubles their number by production (3), and then vanished by production (4), or is transformed by production (5) into a $Y$, which moves from right to left through the $A$s and divides their number by three by production (6), and vanished by production (7). When no $X$s or $Y$s are present, all the $A$s can be transformed in $a$s by production (8) and the derivations stop.

It is easy to see that we have

$$\#A^k\# \quad \overset{k+2}{\Longrightarrow} \quad \#A^{2k}\#,$$

$$\#A^{3k+2}\# \quad \overset{5k+6}{\Longrightarrow} \quad \#A^{2k+1}\#.$$

Recall that the definition of the $3x+1$ function $T$ can be written as $T(2k) = k$ and $T(2k+1) = 3k+2$. So the derivations of this grammar go back up the Collatz's graph from 1 to all possible numbers that give 1 by iteration of the $3x+1$ function. Thus, the language generated by this grammar is $L(G) = \{\#a^n\# : \exists k \quad T^k(n) = 1\}$, and the Collatz conjecture is true if and only if $L(G) = \{\#a^n\# : n \geq 1\}$.

Andrei et al. (2007) were aiming to use properties of their grammar to study the Collatz's graph.

**8.2. Kaufmann's string arithmetic.** Kaufmann (1995) considered strings with three symbols: the star $*$ and two brackets $\langle$, and $\rangle$. A nonnegative integer $n$ is represented by $n$ stars, and $\langle n \rangle$ stands for twice $n$. Then addition is given by concatenation, and the following rules hold.

(1) $\rangle\langle \leftrightarrow$ blank
(2) $** \leftrightarrow \langle * \rangle$
(3) $*x \leftrightarrow x*$ if $x$ represents a nonnegative integer.

Each nonnegative integer has a normal form where there are no consecutive stars, and which corresponds directly to the binary writing of the integer. For example, 26 is 11010 in binary writing, and has normal form $\langle\langle\langle\langle * \rangle * \rangle\rangle * \rangle$. The normal form can be obtained by applications of rules (1) and (2).

Multiplication of $x$ and $y$ is obtained by substituting $y$ for every occurrence of $*$ in $x$.

Since this system encodes addition and multiplication of nonnegative integers, it has the full power of arithmetic. The $3x+1$ function has a very pleasant form in this system:

$$\begin{aligned} T(\langle x \rangle) &= x \\ T(\langle x \rangle *) &= \langle x* \rangle x. \end{aligned}$$

**8.3. Błażewicz and Pettorossi's rewriting system.** Błażewicz and Pettorossi (1983) considered the $3x+1$ function as a rewriting system. The terms are the binary writings of positive integers, and the rewriting rules are the obvious ones: rewriting of an even integer by suppression of the final 0, and rewriting of an odd integer $n$ as the binary writing of $(3n+1)/2$.

They considered *alternations* in the binary writings, where symbols 0 and 1 are alternating. They proved that, from a term $1x1$, such alternations in $x$ are rewritten as *runs* of 0s or 1s, that is as sequences with only 0s or only 1s. They also studied the patterns that give runs in the rewritten term. From these theorems on the future of alternations and the past of runs, they derived corollaries on the lengths of maximal runs in a term and in the rewritten term.

They were aiming at a measure from binary writings to nonnegative integers, may be based on both the length and the number of runs and alternations, that could well behave when the $3x+1$ function is iterated.

### 8.4. Extensions to infinite words.
The $3x+1$ function on integers can be easily extended to the ring of 2-adic integers $\mathbb{Z}_2$. Canales Chacón and Vielhaber (2004) used this extension to give an isometry of this ring with an exponential structural complexity.

The members of the ring $\mathbb{Z}_2$ can be written as words of infinite lengths on the alphabet $\{0,1\}$. Scollo (2005) considered the extension of the $3x+1$ function to $\mathbb{Z}_2$ as a rewriting system on these infinite words.

### 8.5. Shallit and Wilson's theorem.
For all nonnegative integer $i$, let $S_i$ be the set of positive integers that are the lengths of a trajectory of the $3x+1$ function from an integer to 1. Let $r(S_i)$ be the set of words on the alphabet $\{0,1\}$ that are binary writings of integers in $S_i$. Shallit and Wilson (1992) proved that, for all nonnegative integer $i$, $r(S_i)$ is regular.

## 9. Conclusion

Conway's theorem on the undecidability of behavior for some generalized Collatz functions has deep consequences. By this theorem, these functions are risen to the rank of general model of computation, as were already Turing machines, tag systems, etc. So, since the theorems of mathematics are computably enumerable, any theorem can be coded into the behavior of a generalized Collatz function. On the other hand, this powerful status kills any attempt to solve the Collatz conjecture by embedding it in a well bounded theory of generalized Collatz functions.

We have seen in this survey that the $3x+1$ function and its generalizations are mainly used to give new open problems about Turing machines, tag systems, cellular automata, etc. It can always be hoped that moving a difficult open problem from its original background to another domain will help to find a new approach. However, it seems that the encodings of the $3x+1$ problem and its generalizations into the new domains are too simple to produce a breakthrough.

**Acknowledgement.** We thank Jeffrey Lagarias for giving us the opportunity to write this article.

## References

[1] Ştefan Andrei, Wei-Ngan Chin and Huu Hai Nguyen, A functional view over the Collatz's problem, 2007+, preprint.
[2] Claudio Baiocchi, 3N+1, UTM e Tag-systems (Italian), Dipartimento di Matematica dell'Università "La Sapienza" di Roma **98/38**, 1998.

[3] Claudio Baiocchi and Maurice Margenstern, Cellular automata about the $3x+1$ problem, in: Proc. LCCS 2001, Université Paris 12, 2001, 37–45, available on the website http://lacl.univ-paris12.fr/LCCS2001/.

[4] Jacek Błażewicz and Alberto Pettorossi, Some properties of binary sequences useful for proving Collatz's conjecture, *Foundations of Control Engineering* **8**, 1983, 53–63.

[5] Brenton Bostick, An NKS Approach to the 3n+1 Problem, *NKS Conference 2004*, 2004.

[6] Allen H. Brady, The determination of the value of Rado's noncomputable function $\Sigma(k)$ for four-state Turing machines, *Mathematics of Computation* **40**, 1983, 647–665.

[7] Allen H. Brady, The busy beaver game and the meaning of life, in: The Universal Turing Machine: a Half-Century Survey, R. Herken (Ed.), Oxford University Press, 1988, 259–277.

[8] S. Burckel, Functional equations associated with congruential functions, *Theoret. Comput. Sci.* **123**, 1994, 397–406.

[9] Mónica del Pilar Canales Chacón and Michael Vielhaber, Structural and computational complexity of isometries and their shift commutators, *Electronic Colloquium on Computational Complexity*, Report No. 57, 2004, 24 pp. (electronic).

[10] John H. Conway, Unpredictable iterations, in: Proc. 1972 Number Theory Conference, University of Colorado, Boulder, CO, 1972, 49–52.

[11] John H. Conway, FRACTAN: a simple universal programming language for arithmetic, in: Open Problems in Communication and Computation (T. M. Cover and B. Gopinath, Eds.), Springer-Verlag, 1987, 4–26.

[12] Matthew Cook, Universality in elementary cellular automata, *Complex Systems* **15**, 2004, 1–40.

[13] Martin Davis, Yuri Matiyasevich and Julia Robinson, Hilbert's tenth problem. Diophantine equations: positive aspects of a negative solution, in: Proc. Sympos. Pure Math. **28**, Amer. Math. Soc., 1976, 323-378.

[14] Liesbeth De Mol, Study of limits of solvability in tag systems, in: Proc. MCU 2007, Lecture Notes in Computer Science No. 4664, Springer-Verlag, 2007, 170–181.

[15] Liesbeth De Mol, Tag systems and Collatz-like functions, *Theoret. Comput. Sci.* **390**, 2008, 92–101.

[16] Philippe Devienne, Patrick Lebègue and Jean-Christophe Routier, Halting problem of one binary Horn clause is undecidable, in: Proc. STACS 1993, Lecture Notes in Computer Science No. 665, Springer-Verlag, 1993, 48–57.

[17] Eric Goles, Maurice Margenstern, Martín Matamala, The $3x+1$ problem for cellular automata on line, IFIPCA'98, Santiago de Chile, Dec. 1998.

[18] Richard K. Guy, Conway's prime producing machine, *Math. Magazine* **56**, 1983, 26–33.

[19] G. T. Hermann, The uniform halting problem for generalized one state Turing machines, in: Proc. 9th Annual Symposium on Switching and Automata Theory, IEEE Computer Society Press, 1968, 368–372.

[20] Stephen D. Isard and Arnold M. Zwicky, Three open questions in the theory of one-symbol Smullyan systems, *ACM SIGACT News* No. 7, 1970, 11–19.

[21] František Kaščák, Small universal one-state linear operator algorithm, in: Proc. MFCS'92, Lecture Notes in Computer Science No. 629, Springer-Verlag, 1992, 327–335.

[22] Louis H. Kauffman, Arithmetic in the form, *Cybernetics and Systems* **26**, 1995, 1–57.

[23] Ivan Korec, The $3x+1$ problem, generalized Pascal triangles and cellular automata, *Math. Slovaca* **42**, 1992, 547–563.

[24] Manfred Kudlek and Yurii Rogozhin, A universal Turing machine with 3 states and 9 symbols, in: Proc. DLT'01, Lecture Notes in Computer Science No. 2295, Springer-Verlag, 2002, 311–318.

[25] Stuart A. Kurtz and Janos Simon, The undecidability of the generalized Collatz problem, in: Proc. TAMC 2007, Lecture Notes in Computer Science No. 4484, Springer-Verlag, 2007, 542–553.

[26] Jeffrey C. Lagarias, The $3x+1$ problem and its generalizations, *The American Mathematical Monthly* **92**, 1985, 3–23. See also: The $3x+1$ problem: An annotated bibliography, (1963–1999) at arXiv:math/0309224, and (2000–) at arXiv:math/0608208.

[27] Eero Lehtonen, Two undecidable variants of Collatz's problem, *Theoret. Comput. Sci.* **407**, 2008, 596–600.

[28] Shen Lin and Tibor Rado, Computer studies of Turing machine problems, *J. ACM* **12**, 1965, 196–212.

[29] Rona Machlin and Quentin F. Stout, The complex behavior of simple machines, *Physica D* **42**, 1990, 85–98.

[30] Jerzy Marcinkowski, Achilles, Turtle, and undecidable boundedness problems for small DAT-ALOG programs, *SIAM J. Comput.* **29**, 1999, 231–257.

[31] Maurice Margenstern, Frontier between decidability and undecidability: a survey, *Theoret. Comput. Sci.* **231**, 2000, 217–251.

[32] Maurice Margenstern and Yuri Matiyasevich, A binomial representation of the $3x+1$ problem, *Acta Arithmetica* **91**, 1999, 367–378.

[33] Heiner Marxen, Busy Beaver, http://www.drb.insel.de/~heiner/BB.

[34] Heiner Marxen and Jürgen Buntrock, Attacking the Busy beaver 5, *Bulletin of the EATCS* No. 40, 1990, 247–251.

[35] Yuri Matiyasevich, *Hilbert's Tenth Problem*, Moscow, Fizmatlit, 1993 (in Russian). English transl.: MIT Press, 1993. French transl.: Masson, 1995. Also see at: http://logic.pdmi.ras.ru/~yumat/Journal/4cc, mirrored at: http://www.informatik.uni-stuttgart.de/ifi/ti/personen/Matiyasevich/H10Pbook.

[36] Pascal Michel, Busy beaver competition and Collatz-like problems, *Archive Math. Logic* **32**, 1993, 351–367.

[37] Pascal Michel, Small Turing machines and generalized busy beaver competition, *Theoret. Comput. Sci.* **326**, 2004, 45–56.

[38] Pascal Michel, Historical survey of busy beavers, http://www.logique.jussieu.fr/~michel/ha.html

[39] Marvin L. Minsky, Computation: Finite and Infinite Machines, Prentice-Hall, Englewood Cliffs, N. J., 1967.

[40] Kenneth G. Monks, $3x+1$ minus the $+$, *Discrete Math. Theoret. Comput. Sci.* **5**, 2002, 47–53.

[41] Turlough Neary, Small polynomial time universal Turing machines, in: Proc. 4th Irish Conference on the Mathematical Foundations of Computer Science and Information Technology, T. Hurley et al. (Eds.), 2006, 325–329.

[42] Turlough Neary and Damien Woods, Four small universal Turing machines, in: Proc. MCU 2007, Lecture Notes in Computer Science No. 4664, Springer-Verlag, 2007, 242–254.

[43] Piergiorgio Odifreddi, Classical Recursion Theory, North-Holland, Amsterdam, 1989.

[44] Nicolas Ollinger, The quest for small universal cellular automata, in: Proc. ICALP 2002, Lecture Notes in Computer Science No. 2380, Springer-Verlag, 2002, 318–329.

[45] Liudmila M. Pavlotskaya, Sufficient conditions for halting problem decidability of Turing machines (Russian), *Avtomaty i Mashiny*, 1978, 91–118.

[46] Marko Petrovšek, Herbert S. Wilf and Doron Zeilberger, A = B. With a foreword by Donald E. Knuth, A. K. Peters, Ltd.: Wellesley, Mass., 1996.

[47] Tibor Rado, On non-computable functions, *Bell System Technical Journal* **41**, 1962, 877–884.

[48] Hartley Rogers, Jr., Theory of Recursive Functions and Effective Computability, McGraw-Hill, New-York, 1967.

[49] Yurii Rogozhin, Seven universal Turing machines (Russian), *Mat. Issledovaniya* **69**, 1982, 76–90.

[50] Yurii Rogozhin, Small universal Turing machines, *Theoret. Comput. Sci.* **168**, 1996, 215–240.

[51] Giuseppe Scollo, $\omega$-rewriting the Collatz problem, *Fundamenta Informaticae* **64**, 2005, 405–416.

[52] Jeffrey O. Shallit and David W. Wilson, The "$3x+1$" problem and finite automata, *Bulletin of the EATCS* No. 46, 1992, 182–185.

[53] Robert I. Soare, Recursively Enumerable Sets and Degrees, Springer-Verlag, Berlin, 1987.

[54] Tanguy Urvoy, Regularity of congruential graphs, in: Proc. MFCS 2000, Lecture Notes in Computer Science No. 1893, Springer-Verlag, 2000, 680–689.

[55] Hao Wang, Tag systems and lag systems, *Mathematische Annalen* **152**, 1963, 65–74.

ÉQUIPE DE LOGIQUE DE L'UNIVERSITÉ PARIS 7, CASE 7012, 75251 PARIS CEDEX 05, FRANCE AND UNIVERSITÉ DE CERGY-PONTOISE, IUFM, F-95000 CERGY-PONTOISE, FRANCE;, CORRESPONDING ADDRESS: 59 RUE DU CARDINAL LEMOINE, 75005 PARIS, FRANCE.

*E-mail address*: michel@logique.jussieu.fr

LABORATOIRE D'INFORMATIQUE THÉORIQUE ET APPLIQUÉE, EA 3097, UNIVERSITÉ PAUL VER-
LAINE - METZ, IUT DE METZ, CAMPUS DU SAULCY, 57045 METZ CEDEX, FRANCE
    *E-mail address*: margens@univ-metz.fr

# PART III.

## Stochastic Modelling and Computation Papers

# Stochastic Models for the $3x + 1$ and $5x + 1$ Problems and Related Problems

## Alex V. Kontorovich and Jeffrey C. Lagarias

ABSTRACT. This paper discusses stochastic models for predicting the long-time behavior of the trajectories of orbits of the $3x + 1$ problem and, for comparison, the $5x + 1$ problem. The stochastic models are rigorously analyzable, and yield heuristic predictions (conjectures) for the behavior of $3x + 1$ orbits and $5x + 1$ orbits.

## 1. Introduction

The $3x + 1$ problem concerns the following operation on integers: if an integer is odd "multiply by three and add one," while if it is even "divide by two." This operation is given by the *Collatz function*

$$(1.1) \qquad C(n) = \begin{cases} 3n + 1 & \text{if } n \equiv 1 \ (\text{mod } 2)\,, \\ \dfrac{n}{2} & \text{if } n \equiv 0 \ (\text{mod } 2)\,. \end{cases}$$

The $3x + 1$ problem concerns what happens if one iterates this operation starting from a given positive integer $n$. The unsolved $3x + 1$ *Problem* or *Collatz problem* is to prove (or disprove) that such iterations always eventually reach the number 1 (and therefter cycle, taking values $1, 4, 2, 1$). This problem goes under many other names, including: *Syracuse Problem, Hasse's Algorithm, Kakutani's Problem* and *Ulam's Problem.*

The $3x+1$ Conjecture has now been verified for all $n \leq 5.67 \times 10^{18}$ by computer experiments [**31**].

**1.1. $3x + 1$ Function.** There are a number of different functions that encode the $3x + 1$ problem, which proceed through the iteration at different speeds. The following two functions prove to be more convenient for probabilistic analysis than

---

AVK received support from an NSF Postdoc, grant DMS 0802998.
JCL received support from NSF Grants DMS-0500555 and DMS-0801029.

the Collatz function. The first of these is the $3x+1$ *function* $T(n)$ (or $3x+1$ *map*)

(1.2)
$$T(n) = \begin{cases} \dfrac{3n+1}{2} & \text{if } n \equiv 1 \ (\text{mod } 2) \,, \\[2mm] \dfrac{n}{2} & \text{if } n \equiv 0 \ (\text{mod } 2) \,. \end{cases}$$

This function divides out one power of 2, after an odd input is encountered; it is defined on the domain of all integers.

The second function, the *accelerated* $3x+1$ *function* $U(n)$, is defined on the domain of all odd integers, and removes all powers of 2 at each step. It is given by

(1.3)
$$U(n) = \frac{3n+1}{2^{\text{ord}_2(3n+1)}},$$

in which $\text{ord}_2(n)$ counts the number of powers of 2 dividing $n$. The function $U(n)$ was studied by Crandall [14] in 1978.

The long-term dynamics under iteration of the $3x+1$ map has proved resistant to rigorous analysis. It is conjectured that there is a finite positive constant $C$ so that all trajectories eventually enter and stay in the region $-C \leq n \leq C$. In particular, there are finitely many periodic orbits and all trajectories eventually enter one of these periodic orbits. On the domain of positive integers it is conjectured there is is a single periodic orbit $\{1, 2\}$; this is part of the $3x+1$ Conjecture. On the domain of negative integers, the known periodic orbits are the three orbits $\{-1\}$, $\{-5, -7, -10\}$ and $\{-17, -25, -37, -55, -82, -41, -61, -91, -136, -68, -34\}$.

**1.2. $5x+1$ Problem.** For comparison purposes, we also consider the $5x+1$ *problem*, which concerns iterates of the *Collatz $5x+1$ function*

(1.4)
$$C_5(n) = \begin{cases} 5n+1 & \text{if } n \equiv 1 \ (\text{mod } 2) \,, \\[2mm] \dfrac{n}{2} & \text{if } n \equiv 0 \ (\text{mod } 2) \,. \end{cases}$$

For this function we also have analogues of the other two functions above. We define the $5x+1$ *function* $T_5(n)$ (or $5x+1$ *map*), given by

(1.5)
$$T_5(n) = \begin{cases} \dfrac{5n+1}{2} & \text{if } n \equiv 1 \ (\text{mod } 2) \,, \\[2mm] \dfrac{n}{2} & \text{if } n \equiv 0 \ (\text{mod } 2) \,. \end{cases}$$

It is defined on the set of all integers.

The second function, the *accelerated* $5x+1$ *function* $U_5(n)$, is defined on the domain of all odd integers, and removes all powers of 2 at each step. It is given by

(1.6)
$$U_5(n) = \frac{5n+1}{2^{ord_2(5n+1)}},$$

in which $ord_2(n)$ counts the number of powers of 2 dividing $n$.

The long-term dynamics under iteration of the $5x+1$ map on the integers is conjecturally quite different from the $3x+1$ map. It is conjectured that a density one set of integers belong to divergent trajectories, ones with $|T^{(k)}(n)| \to \infty$. It is also conjectured that there are a finite number of periodic orbits, which include the orbits $\{1, 3, 8, 4, 2\}$ and $\{13, 33, 83, 208, 104, 52, 26\}$ on the positive integers and the orbit $\{-1, -2\}$ on the negative integers. An infinite number of trajectories eventually enter one of these orbits, but the set of all integers entering each of these orbits is believed to have density zero.

**1.3. Stochastic models.** This paper is concerned with probabilistic models for the behavior of the $3x + 1$ function iterates, and for comparison, the $5x + 1$ function iterates. The absence of rigorous analysis of the long-term behavior under iteration of these functions provides one motivation to formulate probabilistic models of the behavior of the $3x + 1$ map and $5x + 1$ map. These models can make predictions that can be compared to empirical data, which, by uncovering discrepancies, may lead to the discovery of new hidden regularities in their behavior under iterations. Note that both the $3x + 1$ map and the $5x + 1$ map have the positive integers and negative integers as invariant subsets; thus their dynamics can be studied separately on these domains. The original problems concern their dynamics restricted to the positive integers.

Here we survey what is known about iteration of these maps, in frameworks which have a probabilistic interpretation. A great deal is known about the initial behavior of the iteration of the $3x+1$ map and $5x+1$ map; such results are summarized in §2 and §7, respectively. Here some models for the $5x + 1$ problem are new, developed in parallel with models in Lagarias and Weiss [**23**]. The major unsolved questions have to do with the behavior of long term aspects of the iterations. It is here that stochastic models have an important role to play. We present models for forward iteration of the map which are of random walk or Markov process type, and models for backwards iteration of the map, which are branching processes or branching random walks. Such models can address how the iteration behaves for a randomly selected input value $n$. More sophisticated models address behavior of "extremal" input values. Analysis of these latter models typically uses some variant of the theory of large deviations.

We are interested in using these stochastic models to explore similarities and differences between the iteration behavior of the $3x+1$ and $5x+1$ functions. There are many similarities which are exact parallels, listed in the concluding §11. The main differences are: in short term iteration on the integers $\mathbb{Z}$, $3x + 1$ iterates tend to get smaller, while $5x + 1$ iterates tend to get larger (in absolute value). For long term iteration it is conjectured that all $3x + 1$ trajectories eventually enter finite cycles; it is conjectured that almost all $5x + 1$ trajectories diverge. Stochastic models permit making some quantitative versions of this behavior. These include the following (conjectural) predictions.

(1) The number of integers $1 \leq n \leq x$ whose $3x + 1$ forward orbit reaches 1 is about $x^{\eta_3 + o(1)}$, where $\eta_3 = 1$.

(2) Restricting to those integers $1 \leq n \leq x$ whose $3x + 1$ map forward orbit includes 1, the trajectories of most such $n$ reach 1 after about $6.95212 \log n$ steps.

(3) Only finitely many $3x + 1$ map trajectories starting at $x$ reach 1 after more than $(\gamma_3 + \epsilon) \log x$ steps, while infinitely many positive $x$ reach 1 after more than $(\gamma_3 - \epsilon) \log x$ steps, where $\gamma_3 \approx 41.67765$.

(4) The number of integers $1 \leq n \leq x$ whose $5x+1$ map forward orbit includes 1 is about $x^{\eta_5 + o(1)}$, where $\eta_5 \approx 0.65049$.

(5) Restricting to those integers $1 \leq n \leq x$ whose $5x + 1$ map forward orbit includes 1, the trajectories of most such $n$ reach 1 after about $9.19963 \log n$ steps.

(6) Only finitely many $5x + 1$ map trajectories starting at $x$ reach 1 after more than $(\gamma_5 + \epsilon) \log x$ steps, while infinitely many positive $x$ reach 1 after more than $(\gamma_5 - \epsilon) \log x$ steps, where $\gamma_5 \approx 84.76012$.

In the case of the $3x + 1$ map, extensive numerical evidence supports these predictions. There has been much less computational testing of the $5x + 1$ map, and the predictions above are less tested in these cases.

We also survey a number of rigorous results that fit in this framework: these results describe aspects of the initial part of the iteration. These include symbolic dynamics for accelerated iteration, given in §6, which were used by Kontorovich and Sinai [18] to show that suitably scaled versions of initial trajectories converge in a limit to geometric Brownian motion. These also include results on Benford's law for the initial base B digits of the initial iterates of the functions above, given in §9.

**1.4. Contents of the paper.** In §2 through §6 we first consider the $3x + 1$ function. Then in §7 and §8 we give comparison results for the $5x + 1$ problem. In §9 and §10 we give results on Benford's law and for 2-adic generalizations, in parallel for both the $3x + 1$ function and $5x + 1$ function.

In §2 we discuss the iteration of the $3x + 1$ map. We describe its symbolic dynamics, and formulate several statistics of orbits, which will be studied via stochastic models in later sections. We state various rigorously proved results about these statistics. For a given starting value $n$, these statistics include the $\lambda$-*stopping time* $\sigma_\lambda(n)$, the *total stopping time* $\sigma_\infty(n)$, the *maximum excursion value* $t(n)$, and *counting functions* $N_k(n)$ and $N_k^*(n)$, for the number of backward iterates at depth $k$ of a given integer $a$, with the latter only counting iterates that are not divisible by 3. We also review what has been rigorously proved about these statistics, and give tables of empirical results known about these statistics, found by large scale computations. Further data appears in the paper of Oliveira e Silva [31] (in this volume).

In §3 we discuss stochastic models for a single orbit under forward iteration of the $3x + 1$ map. These include a multiplicative random product model (MRP model) and a logarithmic rescaling giving an additive random walk model taking unequal steps (BRW model), which has a negative drift. These models predict that all orbits converge to a bounded set, and that the total stopping time $\sigma_\infty(n)$ for the $3x + 1$ map of a random starting point $n$ should be about $6.95212 \log n$ steps, and as $n \to \infty$ have a Gaussian distribution around this value, with standard deviation proportional to $\sqrt{\log n}$.

In §4 we discuss models for extreme values of the total stopping time of the $3x + 1$ map. We introduce a repeated random walk model (RRW model) which produces a random trajectory separately for each integer $n$. We present results obtained using the theory of large deviations which rigorously determine behavior in this model of a statistic which is an analogue of the scaled total stopping time $\gamma(n) := \frac{\sigma_\infty(n)}{\log n}$. The model predicts that the limit superior of these values should be a constant $\gamma_{RRW} \approx 41.67765$, which is larger than the average value $6.95212$

this variable takes. This prediction agrees fairly well with the empirical data given in §2.

In §5 we survey results concerning forward iteration of the accelerated $3x + 1$ map. These include a complete description of its symbolic dynamics. We also show that a suitable scaling limit of these trajectories is a geometric Brownian motion, and discuss the equidistribution of various images via entropy.

In §6 we describe stochastic models simulating backward iteration of the $3x + 1$ function. These models grow random labelled trees, whose levels describe branching random walks. These models give exact answers for the expected number of leaves at a given depth $k$, analogous to the number of integers having total stopping time $k$, and also predict the extremal behavior of the scaled total stopping time function $\gamma(n) := \frac{\sigma_\infty(n)}{\log n}$. It yields a prediction for the limit superior of these values to be $\gamma_{BP} \approx 41.677647$, the same value as for the repeated random walk process above.

In §7 and §8 we present analogous results for the $5x + 1$ map. Much less empirical study has been made for iteration of the $5x + 1$ function, so there is less empirical data available for comparison.

In §7 we define $5x + 1$ statistics of orbits. These are analogues of the $3x + 1$ statistics given in §3, but some require modification to reflect the fact that $5x + 1$ orbits grow on average. We also review what is known rigorously about the behavior of this function; in particular the symbolic dynamics of the forward iteration of the $5x + 1$ map is exactly the same as that for the $3x + 1$ map. The statistics introduced include a reverse analogue of the stopping time, *the $\lambda^+$-stopping time $\sigma_\lambda^+(n)$*, and also the *total stopping time $\sigma_\infty(n; T_5)$*. Since most trajectories are believed to be unbounded, the total stopping time is believed to take the value $+\infty$ for almost all initial conditons. In place of the maximum excursion value, we consider the *minimum excursion value $t^-(n)$*!

In §8 we present results on stochastic models for the $5x + 1$ iteration. These include repeated random walk models for the forward iteration of this function, paralleling results of §4; the convergence to Brownian motion of appropriately scaled trajectories, paralleling results of §5; and branching random walk models for inverse iteration, paralleling results of §6. In the latter case we present some new results. The most interesting results of the analysis of these models is the prediction that the number of integers below $x$ which iterate under the $5x + 1$ to 1 should be about $x^{\delta_5 + o(1)}$ with $\delta_5 \approx 0.65041$, and that all integers below $x$ that eventually iterate to 1 necessarily do it in at most $(\gamma_{5,BP} + o(1)) \log x$ steps, where $\gamma_{5,BP} \approx 84.76012$.

In §9 we discuss another property of $3x+1$ iterates and $5x+1$ iterates: Benford's law. In this context "Benford's law" asserts that the distribution of the initial decimal digits of numbers in a trajectory $\{T^{(k)}(n) : 1 \le k \le m\}$ approaches a particular non-uniform probability distribution, the Benford distribution, in which an initial digit less than $k$ occurs with probability $\log_{10} k$, so that 1 is the most likely initial digit. We summarize results showing that most initial starting values of both the $3x + 1$ map and the $5x + 1$ map have initial iterates exhibiting Benford-like behavior; this property holds for any fixed finite set of initial iterates.

In §10 we review results on the extensions to the domain of 2-adic integers $\mathbb{Z}_2$ of the functions $T_3(n)$ and $T_5(n)$. These functions have the pleasant property that their definition makes sense 2-adically, and each function has a unique continuous

2-adic extension, which we denote $\tilde{T}_3 : \mathbb{Z}_2 \to \mathbb{Z}_2$ and $\tilde{T}_5 : \mathbb{Z}_2 \to \mathbb{Z}_2$, respectively. These extended maps are measure-preserving for the 2-adic Haar measure, and are ergodic in a very strong sense. The interesting feature is that at the level of 2-adic extensions the $3x+1$ map and $5x+1$ map are identical maps from the perspective of measure theory. They are both topologically and measurably conjugate to the full shift on the 2-adic integers, hence they are topologically and measurably conjugate to each other! Thus their dynamics are "the same." This contrasts with the great difference between these maps view on the domain of integers.

In §11 we present concluding remarks, summarizing this paper, comparing properties under iteration of the $3x+1$ map and $5x+1$ map. The short-run behavior under iteration of these maps have some strong similarities. However all evidence indicates that the long-run behavior of iteration for the $3x+1$ map and the $5x+1$ map on the integers $\mathbb{Z}$ is very different. We also list a set of insights and topics for further investigation.

**Notation.** For convenience, when comparing the $3x+1$ maps with the corresponding $5x+1$ maps, we may write $C_3(n), T_3(n), U_3(n)$ in place of $C(n), T(n), U(n)$ above.

**Acknowledgments.** The authors thank Steven J. Miller and Kevin Ford for a careful reading of and many corrections to an earlier draft of this manuscript. AVK wishes to thank the hospitality of Dorian Goldfeld and Columbia University during this project.

## 2. The $3x+1$ Function: Symbolic Dynamics and Orbit Statistics

In this section we consider the $3x+1$ map $T(n)$. We recall basic properties of its symbolic dynamics. We also define several different statistics for describing its behavior on individual trajectories, and summarize what is rigorously proved about these statistics. In later sections we will present probabilistic models which are intended to model the behavior of these statistics.

### 2.1. $3x+1$ Symbolic Dynamics: Parity Sequence.
The behavior of the map $T(n)$ under iteration is completely described by the parities of the successive iterates.

DEFINITION 2.1. *(i) For a function $T : \mathbb{Z} \to \mathbb{Z}$ and input value $n \in \mathbb{Z}$ define the* parity sequence *of $n$ to be*

$$(2.7) \qquad S(n) := (n \ (\mathrm{mod} \ 2), T(n) \ (\mathrm{mod} \ 2), T^{(2)}(n) \ (\mathrm{mod} \ 2), ...)$$

*in which $T^{(k)}(n)$ denotes the k-th iterate, so that $T^{(2)}(n) := T(T(n))$. This is an infinite vector of zeros and ones.*

*(ii) For $k \geq 1$ its $k$-truncated parity sequence is a vector giving the initial segment of $k$ terms of $S(n)$, i.e.*
(2.8)
$$S^{[k]}(n) := (n \ (\ \mathrm{mod} \ 2), T(n) \ (\ \mathrm{mod} \ 2), T^{(2)}(n) \ (\ \mathrm{mod} \ 2), \cdots, T^{(k-1)}(n) \ (\ \mathrm{mod} \ 2)).$$

A basic result on the iteration is as follows.

THEOREM 2.1. *(Parity Sequence Symbolic Dynamics) The $k$-truncated parity sequence $S^{[k]}(n)$ of the first $k$ iterates of the $3x+1$ map $T(x)$ is periodic in $n$ with period $2^k$. Each of the $2^k$ possible $0-1$ vectors occurs exactly once in the initial segment $1 \leq n \leq 2^k$.*

PROOF. This result is due to Terras [38] in 1976 and Everett [16] in 1977. A proof is given as Theorem B in Lagarias [21].    □

An immediate consequence is that an integer $n$ is uniquely determined by the parity sequence $S(n)$ of its forward orbit. To see this, note that any two distinct integers fall in different residue classes (mod $2^k$) for large enough $k$, so will have different parity sequences. The parity sequence thus provides a *symbolic dynamics* which keeps track of the orbit. Taken on the integers, only countably many different parity sequences occur out of the uncountably many possible infinite $0-1$ sequences.

**2.2. $3x + 1$ Stopping Time Statistics: $\lambda$-stopping times.** The initial statistic we consider is the number of iteration steps needed to observe a fixed amount of decrease of size in the iterate.

DEFINITION 2.2. *For fixed $\lambda > 0$ the $\lambda$-stopping time $\sigma_\lambda(n)$ of a map $T : \mathbb{Z} \to \mathbb{Z}$ from input $n$ is the minimal value of $k \geq 0$ such that $T^{(k)}(n) < \lambda n$, e.g.*

$$(2.9) \qquad \sigma_\lambda(n) := \inf \left\{ k \geq 0 : \frac{T^{(k)}(n)}{n} < \lambda \right\}.$$

*If no such value $k$ exists, we set $\sigma_\lambda(n) = +\infty$.*

This notion for $\lambda = 1$ was introduced in 1976 by Terras [38] who called it the *stopping time*, and denoted it $\sigma(n)$. The more general $\lambda$-stopping time is interesting in the range $0 < \lambda \leq 1$; it satisfies $\sigma_\lambda(n) = 0$ for all $\lambda > 1$.

Terras [38] studied the set of numbers having stopping time at most $k$, denoted

$$(2.10) \qquad S_1(k) := \{n : \ \sigma_1(n) \leq k\}.$$

He used Theorem 2.1 to show ([38], [39]) that this set of integers has a natural density, as defined below, and that this density approaches 1 as $k \to \infty$.

Later this result was generalized. Rawsthorne [32] in 1985 introduced the case of general $\lambda$, and Borovkov and Pfeifer [10, Theorem 2] in 2000 considered criteria with several stopping time conditions.

There are several notions of density of a set $\Sigma$ of the natural numbers $\mathbb{N} = \{1, 2, 3, ...\}$. The *lower asymptotic density* $\underline{\mathbb{D}}(\Sigma)$ is defined for all infinite sets $\Sigma$, and is given by

$$(2.11) \qquad \underline{\mathbb{D}}(\Sigma) := \liminf_{t \to \infty} \frac{1}{t} |\{n \in \Sigma : \ n \leq t\}|.$$

The assertion that an infinite set $\Sigma \subset \mathbb{N}$ of natural numbers has a *natural density* $\mathbb{D}(\Sigma)$ is the assertion that the following limit exists:

$$(2.12) \qquad \mathbb{D}(\Sigma) := \lim_{t \to \infty} \frac{1}{t} |\{n \in \Sigma : \ n \leq t\}|.$$

Sets with a natural density automatically have $\mathbb{D}(\Sigma) = \underline{\mathbb{D}}(\Sigma)$.

THEOREM 2.2.    ($\lambda$-Stopping Time Natural Density)

*(i) For the $3x + 1$ map $T(n)$, and any fixed $0 < \lambda \leq 1$ and $k \geq 1$, the set $S_\lambda(k)$ of integers having $\lambda$-stopping time at most $k$ has a well-defined natural density $\mathbb{D}(S_\lambda(k))$.*

*(ii) For $\lambda$ fixed and $k \to \infty$, this natural density satisfies*

$$(2.13) \qquad \mathbb{D}(S_\lambda(k)) \to 1.$$

*In particular, the set of numbers with finite $\lambda$-stopping time has natural density 1.*

PROOF. For the special case $\lambda = 1$, that is the stopping time, this is the basic result of Terras [**38**], [**39**], obtained also by Everett [**16**]. A proof for $\lambda = 1$ is given as Theorem A in Lagarias [**21**]. The idea is that by Theorem 2.1, each arithmetic progression (mod $2^k$) has iterates that multiply by a certain pattern of $\frac{1}{2}$ or $\frac{3}{2}$ for the first $k$ steps. A certain subset of the $2^k$-arithmetic progressions (mod $2^k$) will have the product of these numbers fall below $\lambda$, and these arithmetic progressions give the density. To see that the density goes to 1 as $k \to \infty$, one must show that most arithmetic progressions (mod $2^k$) have a product smaller than one. Theorem 2.1 says that all products occur equally likely, and since the geometric mean of these products is $\left(\frac{3}{4}\right)^{\frac{1}{2}} < 1$, one can establish that such a decrease occurs for all but an exponentially small set of patterns, of size $O(2^{0.94995k})$ out of $2^k$ possible patterns. One can show a similar result for decrease by a factor of any fixed $\lambda$, and a proof of natural density for general $\lambda > 0$ is given in Borovkov and Pfeifer [**10**, Theorem 3]. $\qquad\square$

The results above are rigorous results, and therefore we have no compelling need to find stochastic models to model the behavior of stopping times. Nevertheless stochastic models intended to analyze other statistics produce in passing models for stopping time distributions. In §3.1 we present such a model, which gives an interpretation of these stopping time densities as exact probabilities of certain events.

REMARK. The analysis in Theorem 2.2 treats $\lambda$ as fixed. In fact one can also prove rigorous results which allow $\lambda$ to vary slowly (as a function of $n$), under the restriction that $\lambda \leq \log_2 n$.

**2.3. $3x + 1$ Stopping Time Statistics: Total Stopping Times.** The following concept concerns the speed at which positive integers iterate to 1 under the map $T$, assuming they eventually get there.

DEFINITION 2.3. *The total stopping time $\sigma_\infty(n)$ for iteration of the $3x+1$ map $T(n)$ is defined for positive integers $n$ by*

$$\sigma_\infty(n) := \inf\{k \geq 0 : T^{(k)}(n) = 1\}.$$

*We set $\sigma_\infty(n) = +\infty$ if no finite $k$ has this property.*

The $3x + 1$ Conjecture asserts that all positive integers have a finite total stopping time.

Concerning lower bounds for this statistic, there are some rigorous results. First, since each step decreases $n$ by at most a factor of 2, we trivially have

$$\sigma_\infty(n) \geq \frac{\log n}{\log 2} \approx 1.4426 \log n.$$

The strongest result on the existence of integers having a large total stopping time is the following result of Applegate and Lagarias [**5**, Theorem 1.1].

THEOREM 2.3. (Lower Bound for $3x + 1$ Total Stopping Times) *There are infinitely many $n$ whose total stopping time satisfies*

$$(2.14) \qquad \sigma_\infty(n) \geq \left(\frac{29}{29\log 2 - 14\log 3}\right)\log n \approx 6.14316 \log n.$$

Nothing has been rigorously proved about either the average size of the total stopping time, or about upper bounds for the total stopping time (since such would imply the main conjecture!). This provides motivation to study stochastic models for this statistic, to make guesses how it may behave.

The various stochastic models discussed in §3, as well as empirical evidence given below, suggest that the size of this statistic will always be proportional to $\log n$. This motivates the following definition.

DEFINITION 2.4.  *For $n \geq 1$ the scaled total stopping time $\gamma_\infty(n)$ of the $3x + 1$ function is given by*

$$(2.15) \qquad\qquad \gamma_\infty(n) := \frac{\sigma_\infty(n)}{\log n}.$$

*This value will be finite for all positive $n$ only if the $3x + 1$ conjecture is true.*

A stochastic model in §3 makes strong predictions about the distribution of scaled total stopping times: they should have a Gaussian distribution with mean

$$\mu := \left( \frac{1}{2} \log \frac{4}{3} \right)^{-1} \approx 6.95212$$

and variance

$$\sigma := \frac{1}{2} \log 3 \left( \frac{1}{2} \log \frac{4}{3} \right)^{\frac{3}{2}},$$

cf. Theorem 3.2. In particular, half of all integers ought to have a total stopping time $\sigma_\infty(n) \geq \mu \log n \approx 6.95212 \log n$. It seems **scandalous** that there is no unconditional proof that infinitely many $n$ have a stopping time at least this large, compared to the bound (2.14) in Theorem 2.3 above!

We next define a limiting constant associated with extremal values of the scaled total stopping time for the $3x + 1$ map.

DEFINITION 2.5.  *The $3x + 1$ scaled stopping constant is the quantity*

$$(2.16) \qquad\qquad \gamma = \gamma_3 := \limsup_{n \to \infty} \gamma_\infty(n) = \limsup_{n \to \infty} \frac{\sigma_\infty(n)}{\log n}.$$

We now give empirical data about these extremal values. Table 1 presents empirical data on record holders for the function $\gamma_\infty(n)$, compiled by Roosendaal [**33**]. This table also includes data on another statistic called the *ones-ratio* (or *completeness*), taken from Roosendaal [**33**, Completeness and Gamma Records]. The function ones$(n)$ counts the number of odd iterates of the $3x + 1$ function to reach 1 starting from $n$ (including 1), and

$$(2.17) \qquad\qquad \text{ones-ratio}(n) := \text{ones}(n)/\sigma_\infty(n).$$

Table 1 shows that the function $\gamma(n)$ is not a monotone increasing function of the ones-ratio, compare rows 9 and 10. The values with question marks mean that all intermediate values have not been searched, so these values are not known to be record holders.

In §4 we present a stochastic model which makes a prediction for the extremal value of $\gamma$. A quite different model is discussed in §7, which makes exactly the same prediction! For both models the analogue of the constant $\gamma := \limsup \gamma_\infty(n)$ exists and equals a constant which numerically is approximately 41.677647, with

TABLE 1. Record Values for $\gamma_\infty(n)$ and for ones-ratio(n).

| $k$ | #k-th record $n_k$ | $\sigma_\infty(n_k)$ | $ones(n_k)$ | $ones - ratio$ | $\gamma_\infty(n_k)$ |
|---|---|---|---|---|---|
| 1 | 3 | 5 | 2 | 0.400000 | 4.551196 |
| 2 | 7 | 11 | 5 | 0.454545 | 5.652882 |
| 3 | 9 | 13 | 6 | 0.461358 | 5.916555 |
| 4 | 27 | 70 | 41 | 0.585714 | 21.238915 |
| 5 | 230 631 | 278 | 164 | 0.589928 | 22.512720 |
| 6 | 626 331 | 319 | 189 | 0.592476 | 23.899366 |
| 7 | 837 799 | 329 | 195 | 0.592705 | 24.122828 |
| 8 | 1 723 519 | 349 | 207 | 0.593123 | 24.303826 |
| 9 | 3 732 423 | 374 | 222 | 0.593583 | 24.714906 |
| 10 | 5 649 499 | 384 | 228 | 0.593750 | 24.699176 |
| 11 | 6 649 279 | 416 | 248 | 0.596154 | 26.479917 |
| 12 | 8 400 511 | 429 | 256 | 0.596737 | 26.907006 |
| 13 | 63 728 127 | 592 | 357 | 0.603041 | 32.943545 |
| 14 | 3 743 559 068 799 | 966 | 583 | 0.603520 | 33.366656 |
| 15 | 100 759 293 214 567 | 1134 | 686 | 0.604938 | 35.169600 |
| ?16 | 104 899 295 810 901 231 | 1404 | 850 | 0.605413 | 35.823841 |
| ?17 | 268 360 655 214 719 480 367 | 1688 | 1022 | 0.605450 | 35.885221 |
| ?18 | 6 852 539 645 233 079 741 799 | 1840 | 1115 | 0.605978 | 36.595864 |
| ?19 | 7 219 136 416 377 236 271 195 | 1848 | 1120 | 0.606061 | 36.716918 |

corresponding ones-ratio of about 0.609091. Compare these predictions with the data in Table 1.

**2.4. $3x + 1$ Size Statistics: Maximum Excursion Values.** Another interesting statistic is the maximum value attained in a trajectory, which we call the maximum excursion value.

DEFINITION 2.6. *The* maximum excursion value $t(n)$ *is the maximum value occurring in the forward iteration of the integer $n$, i.e.*

$$(2.18) \qquad t(n) := \max(T^{(k)}(n) : \ k \geq 0),$$

*with $t(n) = +\infty$ if the trajectory is divergent.*

The quantity $t(n)$ will be finite for all $n$ if and only if there are no divergent trajectories for the $3x + 1$ problem (but does not exclude the possibility of as yet unknown loops) .

We define the following extremal statistic for maximum excursions.

DEFINITION 2.7. *Let the $3x + 1$* maximum excursion ratio *be given by*

$$(2.19) \qquad \rho(n) := \frac{\log t(n)}{\log n}.$$

*Then the $3x + 1$* maximum excursion constant *is the quantity*

$$(2.20) \qquad \rho := \limsup_{n \to \infty} \rho(n) = \limsup_{n \to \infty} \frac{\log t(n)}{\log n}.$$

The maximal excursion constant will be $+\infty$ if there is a divergent trajectory. The fact that the logarithmic scaling used in defining this constant is the "correct" scaling is justified by empirical data given in Oliveira e Silva [**31**] (in this volume) and by the predictions of the stochastic model given in §3. As explained in §4.3 , the stochastic model prediction for the maximum excursion constant is $\rho = 2$.

TABLE 2. Seeds $n$ giving record heights for $3x + 1$ maximum excursion value $t(n)$.

| $k$ | #k-th record $n_k^*$ | $t(n_k^*)$ | $r(n_k^*)$ | $\rho(n_k^*)$ |
|---|---|---|---|---|
| 1 | 2 | 2 | 0.500 | 1.000 |
| 2 | 3 | 8 | 0.889 | 1.893 |
| 3 | 7 | 26 | 0.531 | 1.674 |
| 4 | 15 | 80 | 0.356 | 1.618 |
| 5 | 27 | 4 616 | 6.332 | 2.560 |
| 6 | 255 | 6 560 | 0.101 | 1.586 |
| 7 | 447 | 19 682 | 0.099 | 1.620 |
| 8 | 639 | 20 782 | 0.051 | 1.539 |
| 9 | 703 | 125 252 | 0.253 | 1.792 |
| 10 | 1 819 | 638 468 | 0.193 | 1.781 |
| 11 | 4 255 | 3 405 068 | 0.188 | 1.800 |
| 12 | 4 591 | 4 076 810 | 0.193 | 1.805 |
| 13 | 9 663 | 13 557 212 | 0.145 | 1.790 |
| 14 | 20 895 | 25 071 632 | 0.057 | 1.712 |
| 15 | 26 623 | 53 179 010 | 0.075 | 1.746 |
| 16 | 31 911 | 60 506 432 | 0.059 | 1.728 |
| 17 | 60 975 | 296 639 576 | 0.080 | 1.771 |
| 18 | 77 671 | 785 412 368 | 0.130 | 1.819 |
| 19 | 113 383 | 1 241 055 674 | 0.097 | 1.799 |
| 20 | 138 367 | 1 399 161 680 | 0.073 | 1.779 |
| 21 | 159 487 | 8 601 188 876 | 0.338 | 1.861 |
| 22 | 270 271 | 12 324 038 948 | 0.169 | 1.858 |
| 23 | 665 215 | 26 241 642 656 | 0.059 | 1.789 |
| 24 | 704 511 | 28 495 741 760 | 0.057 | 1.788 |
| 25 | 1 042 431 | 45 119 577 824 | 0.042 | 1.770 |

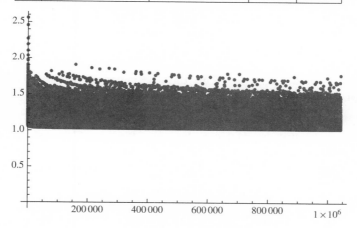

FIGURE 2.1. A plot of $n$ versus the maximal excursion ratio $\rho(n)$ for $3 \leq n \leq 1\,042\,431$ and odd, cf. (2.19). The only seeds $n$ in this range with $\rho(n) > 2$ are $n = 27$, 31, 41, 47, 55, and 63 (which all look at this scale as if they are on the $y$-axis).

TABLE 3. Values of $n$ for which the maximal excursion ratio $\rho(n) = \frac{\log t(n)}{\log n} > 2$ (equivalently, $r(n) = t(n)/n^2 > 1$), culled from Oliveira e Silva's [31, Table 8] record $t(n)$ values.

| $n$ | $t(n)$ | $r(n)$ | $\rho(n)$ |
|---|---|---|---|
| 27 | 4 616 | 6.332 | 2.560 |
| 319 804 831 | 707 118 223 359 971 240 | 6.914 | 2.099 |
| 1 410 123 943 | 3 562 942 561 397 226 080 | 1.792 | 2.028 |
| 3 716 509 988 199 | 103 968 231 672 274 974 522 437 732 | 7.527 | 2.070 |
| 9 016 346 070 511 | 126 114 763 591 721 667 597 212 096 | 1.551 | 2.015 |
| 1 254 251 874 774 375 | 1 823 036 311 464 280 263 720 932 141 024 | 1.159 | 2.004 |
| 1 980 976 057 694 848 447 | $3.2012\ldots \times 10^{36}$ | 8.158 | 2.050 |

In Table 2 we give the set of initial champion values for the maximum excursion, extracted from data of Oliveira e Silva [30]. For comparison we give for each the ratio $r(n) := \frac{t(n)}{n^2}$ and the value of the maximal excursion ratio $\rho(n) = \frac{\log t(n)}{\log n}$. It is also useful to examine the larger table to $10^{18}$ given in Oliveira e Silva [31].

While record values of $t(n)$ have received tremendous computational attention, there has not been a substantial amount of effort put into congregating those $n$ with large $\rho(n)$ (the difference being that the former seeks seeds $n$ with large values of $t(n)$, whereas the latter seeks large values of $t(n)$ *relative* to the size of $n$). We have computed that the only seeds $n < 10^6$ for which $\rho(n) > 2$ are: $n \in \{27, 31, 41, 47, 55, 63\}$, cf. Figure 2.1.

Nevertheless, some "large" values of $\rho(n)$ already appear in tables of large $t(n)$'s. In Table 3 we extract from a table of $t(n)$ champions computed by Oliveira e Silva [31] the subset of seeds $n$ for which $\rho(n) > 2$, i.e.

$$r(n) = \frac{t(n)}{n^2} > 1.$$

Only seven such values appear. This data seems to (however weakly) support Conjecture 4.2.

**2.5. $3x + 1$ Count Statistics: Inverse Iterate Counts.** In considering backwards iteration of the $3x + 1$ function, we can ask: given an integer $a$ how many numbers $n$ have $T^{(k)}(n) = a$, that is, iterate forward to $a$ after exactly $k$ iterations?

The set of backwards iterates of a given number $a$ can be pictured as a tree; we call these $3x + 1$ *trees* and describe their structure in §6. Here $N_k(a)$ counts the number of leaves at depth $k$ of a tree with root node $a$, and $N_k^*(a)$ counts the number of leaves in a *pruned $3x + 1$ tree*,

in which all nodes with label $n \equiv 0 \pmod 3$ have been removed. The definitions are as follows.

DEFINITION 2.8.  *(1) Let $N_k(a)$ count the number of integers that forward iterate under the $3x + 1$ map $T(n)$ to $a$ after exactly $k$ iterations, i.e.*

(2.21)                    $$N_k(a) := |\{n : \ T^{(k)}(n) = a\}|.$$

*(2) Let $N_k^*(a)$ count the number of integers not divisible by 3 that forward iterate under the $3x+1$ map $T(n)$ to $a$ after exactly $k$ iterations, i.e.*

$$(2.22) \qquad N_k^*(a) := |\{n : \ T^{(k)}(n) = a, \ n \not\equiv 0(\text{mod } 3)\}|.$$

The case $a = 1$ is of particular interest, since the quantities then count integers that iterate to 1 . We set

$$N_k := N_k(1), \ \ N_k^* := N_k^*(1).$$

The secondary quantity $N_k^*(a)$ is introduced because it is somewhat more convenient for analysis. It satisfies the monotonicity properties $N_k^*(a) \le N_{k+1}^*(a)$ and

$$N_k^*(m) \le N_k(a) \le \sum_{j=0}^{k} N_j^*(m) \le (k+1) N_k^*(a).$$

We have the trivial exponential upper bound

$$(2.23) \qquad N_k(a) \le 2^k.$$

since each number has at most 2 preimages. We are interested in the exponential growth rate of $N_k(a)$.

DEFINITION 2.9. *(1) For a given $a$ the $3x+1$ tree growth constant $\delta_3(a)$ is given by*

$$(2.24) \qquad \delta_3(a) := \limsup_{k\to\infty} \frac{1}{k} \left(\log N_k(a)\right).$$

*(2) The $3x+1$ universal tree growth constant is $\delta = \delta_3 = \delta_3(1)$.*

The constant $\delta_3(a)$ exists and is finite, as follows from the upper bound (2.23). It is easy to prove unconditionally that $\delta_3(3a) = 0$, because the only preimages of a number $3a$ are $2^k 3a$ and $N_k(3a) = 1$ for all $k \ge 1$. The interesting case is when $a \not\equiv 0 \pmod{3}$.

Applegate and Lagarias [2] determined by computer the maximal and minimal number of leaves in pruned $3x+1$ trees of depth $k$ for $k \le 30$. The maximal and minimal number of leaves in such trees at level $k$ is given by

$$N_k^+ := \max\{N_k^*(a) : \ a \ (\text{mod } 3^{k+1}) \text{ with } a \not\equiv 0 \ (\text{mod } 3)\}$$

and

$$N_k^- := \min\{N_k^*(a) : \ a \ (\text{mod } 3^{k+1}) \text{ with } a \not\equiv 0 \ (\text{mod } 3)\},$$

respectively. Counts for the number of leaves in maximum and minimum size trees of various depths $k$ are given in the following table, taken from Applegate and Lagarias ([2], [4]). It is known that the average number of leaves at depth $k$ (averaged over $a$) is proportional to $\left(\frac{4}{3}\right)^k$, therefore in Table 4 below we include the value $\left(\frac{4}{3}\right)^k$ and the scaled statistics

$$D_k^\pm := N_k^\pm \left(\frac{4}{3}\right)^{-k}.$$

This table also gives the number of distinct types of trees of each depth (there are some symmetries which speed up the calculation) .

Applegate and Lagarias [2, Theorem 1.1] proved the following result by an easy induction using this table.

TABLE 4. Normalized extreme values for $3x + 1$ trees of depth $k$

| $k$ | # tree types | $N_k^-$ | $N_k^+$ | $\left(\frac{4}{3}\right)^k$ | $D_k^-$ | $D_k^+$ |
|---|---|---|---|---|---|---|
| 1 | 4 | 1 | 2 | 1.33 | 0.750 | 1.500 |
| 2 | 8 | 1 | 3 | 1.78 | 0.562 | 1.688 |
| 3 | 14 | 1 | 4 | 2.37 | 0.422 | 1.688 |
| 4 | 24 | 2 | 6 | 3.16 | 0.633 | 1.898 |
| 5 | 42 | 2 | 8 | 4.21 | 0.475 | 1.898 |
| 6 | 76 | 3 | 10 | 5.62 | 0.534 | 1.780 |
| 7 | 138 | 4 | 14 | 7.49 | 0.534 | 1.869 |
| 8 | 254 | 5 | 18 | 9.99 | 0.501 | 1.802 |
| 9 | 470 | 6 | 24 | 13.32 | 0.451 | 1.802 |
| 10 | 876 | 9 | 32 | 17.76 | 0.507 | 1.802 |
| 11 | 1638 | 11 | 42 | 23.68 | 0.465 | 1.774 |
| 12 | 3070 | 16 | 55 | 31.57 | 0.507 | 1.742 |
| 13 | 5766 | 20 | 74 | 42.09 | 0.475 | 1.758 |
| 14 | 10850 | 27 | 100 | 56.12 | 0.481 | 1.782 |
| 15 | 20436 | 36 | 134 | 74.83 | 0.481 | 1.791 |
| 16 | 38550 | 48 | 178 | 99.77 | 0.481 | 1.784 |
| 17 | 72806 | 64 | 237 | 133.03 | 0.481 | 1.782 |
| 18 | 137670 | 87 | 311 | 177.38 | 0.490 | 1.753 |
| 19 | 260612 | 114 | 413 | 236.50 | 0.482 | 1.746 |
| 20 | 493824 | 154 | 548 | 315.34 | 0.488 | 1.738 |
| 21 | 936690 | 206 | 736 | 420.45 | 0.490 | 1.751 |
| 22 | 1778360 | 274 | 988 | 560.60 | 0.489 | 1.762 |
| 23 | 3379372 | 363 | 1314 | 747.47 | 0.486 | 1.758 |
| 24 | 6427190 | 484 | 1744 | 996.62 | 0.486 | 1.750 |
| 25 | 12232928 | 649 | 2309 | 1328.83 | 0.488 | 1.738 |
| 26 | 23300652 | 868 | 3084 | 1771.77 | 0.490 | 1.741 |
| 27 | 44414366 | 1159 | 4130 | 2362.36 | 0.491 | 1.748 |
| 28 | 84713872 | 1549 | 5500 | 3149.81 | 0.492 | 1.746 |
| 29 | 161686324 | 2052 | 7336 | 4199.75 | 0.489 | 1.747 |
| 30 | 308780220 | 2747 | 9788 | 5599.67 | 0.491 | 1.748 |

THEOREM 2.4. ($3x + 1$ Tree Sizes) *For any fixed $a \not\equiv 0$ (mod 3) and for all sufficiently large $k$,*

$$(2.25) \qquad (1.302053)^k \leq N_k^*(a) \leq (1.358386)^k.$$

*In consequence, for any $a \not\equiv 0 (\mathrm{mod}\ 3)$,*

$$(2.26) \qquad \log(1.302053) \leq \delta_3(a) \leq \log(1.358386).$$

We describe probabilistic models for $3x + 1$ inverse iterates in §6. The models are Galton-Watson processes for the number of leaves in the tree, and branching random walks for the sizes of the labels in the tree. The model prediction is that $\delta_3(a) = \log\left(\frac{4}{3}\right)$ for all $a \not\equiv 0$ (mod 3).

**2.6. $3x + 1$ Count Statistics: Total Inverse Iterate Counts.** In considering backwards iteration of the $3x + 1$ function from an integer $a$, complete data is the set of integers that contain $a$ in their forward orbit. The $3x + 1$ problem concerns exactly this question for $a = 1$. The following function describes this set.

DEFINITION 2.10. *Given an integer $a$, the* inverse iterate counting function $\pi_a(x)$ *counts the number of integers $n$ with $|n| \leq x$ that contain $a$ in their forward* *orbit under the $3x+1$ function. That is,*

$$(2.27) \qquad \pi_a(x) := \#\{n : \ |n| \le x \ and \ T^{(k)}(n) = a \ for \ some \ k \ge 0\}.$$

It is possible to obtain rigorous lower bounds for this counting function. For $a \equiv 0 \pmod{3}$ the set of inverse iterates is exactly $\{2^k a : \ k \ge 0\}$ and $\pi_a(x) = \lfloor \log_2(\frac{2x}{|a|}) \rfloor$ grows logarithmically. If $a \not\equiv 0 \pmod{3}$ then $\pi_a(x)$ satisfies a bound $\pi_a(x) > x^c$ for some positive $c$, as was first shown by Crandall [14] in 1978. The strongest method currently known to obtain lower bounds on $\pi_a(x)$ was initiated by Krasikov [19] in 1989, and extended in [3], [20]. It gives the following result.

THEOREM 2.5. (Inverse Iterate Lower Bound) *For each $a \not\equiv 0 \pmod{3}$, there is a positive constant $x_0(a)$ such that for all $x \ge x_0(a)$,*

$$(2.28) \qquad \pi_a(x) \ge x^{0.84}.$$

PROOF. This is proved in Krasikov and Lagarias [20]. The proof uses systems of difference inequalities $(\bmod \ 3^k)$, analyzed in Applegate and Lagarias [3], and by increasing $k$ one gets better exponents. The exponent above was obtained by computer calculation using $k = 9$. $\qquad\square$

The following statistics measure the size of the inverse iterate set in the sense of fractional dimension.

DEFINITION 2.11. *Given an integer $a$, the* upper and lower $3x+1$ growth exponents *for $a$ are given by*

$$\eta_3^+(a) := \limsup_{x\to\infty} \frac{\log \pi_a(x)}{\log x},$$

*and*

$$\eta_3^-(a) := \liminf_{x\to\infty} \frac{\log \pi_a(x)}{\log x}.$$

*If these quantities are equal, we define the $3x+1$* growth exponent $\eta_3(a)$ *to be* $\eta_3(a) = \eta_3^+(a) = \eta_3^-(a)$.

We clearly have $\eta_3(a) = 0$ if $a \equiv 0 \pmod{3}$. For the remaining values Applegate and Lagarias made the following conjecture.

CONJECTURE 2.1. ($3x+1$ Growth Exponent Conjecture) *For all integers $a \not\equiv 0 \pmod{3}$, the $3x+1$ growth exponent $\eta_3(a)$ exists, with*

$$(2.29) \qquad \eta_3(a) = 1.$$

The truth of the $3x+1$ Conjecture would imply that $\eta_3(1) = 1$; however it does not seem to determine $\eta_3(a)$ for all such $a$. Applegate and Lagarias [2, Conjecture A] made the stronger conjecture that for each $a \not\equiv 0 \pmod{3}$ $\pi_a(x)$ grows linearly, i.e. there is a constant $c_a > 0$ such that $\pi_a(x) > c_a x$ holds for all $x \ge 1$.

Note that Theorem 2.5 shows that $\eta_3^-(a) \ge 0.84$ when $a \not\equiv 0 \pmod{3}$. Thus the lower bound in Conjecture 2.1 thus seems approachable. A stochastic model in §6.5 makes the prediction that $\eta_3(a) = 1$.

## 3. $3x + 1$ Forward Iteration: Random Product and Random Walk Models

In this section we formulate stochastic models intended to predict the behavior of iterations of the $3x + 1$ map $T(n)$ on a "random" starting value $n$. These models are exactly analyzable. We describe results obtained for these models, which can be viewed as predictions for the "average" behavior of the $3x + 1$ function.

### 3.1. Multiplicative Random Product Model and $\lambda$-stopping times.
Recall that the $\lambda$-stopping time is defined (see (2.9)) by

$$\sigma_\lambda(n) := \inf\{k \geq 0 : \frac{T^{(k)}(n)}{n} < \lambda\}.$$

Rawsthorne [32] and Borovkov and Pfeifer [10] obtained a probabilistic interpretation of the $\lambda$-stopping time, as follows. They consider a stochastic model which studies the random products

$$Y_k := X_1 X_2 \cdots X_k,$$

in which the $X_i$ are each independent identically distributed (i.i.d.) random variables $X_i$ having the discrete distribution

$$X_i = \begin{cases} \dfrac{3}{2} & \text{with probability } \frac{1}{2}, \\[2mm] \dfrac{1}{2} & \text{with probability } \frac{1}{2}. \end{cases}$$

We call this the $3x + 1$ *multiplicative random product* ($3x + 1$ MRP) model.

This model does not include the choice of the starting value of the iteration, which would correspond to $X_0$; the random variable $Y_k$ really models the *ratio* $\frac{T^{(k)}(X_0)}{X_0}$. They define for $\lambda > 0$ the $\lambda$-*stopping time random variable*

$$(3.30) \qquad V_\lambda(\omega) := \inf\{k : Y_k \leq \lambda\},$$

where $\omega = (X_1, X_2, X_3, \dots)$ denotes a sequence of random variables as above. This random vector $\omega$ will model the effect of choosing a random starting value $n = X_0$ in iteration of the $3x + 1$ map.

This stochastic model can be used to exactly describe the density of $\lambda$-stopping times, as follows. Let $\mathbb{P}[E]$ denote the probability of an event $E$.

THEOREM 3.1.  ($\lambda$-Stopping Time Density Formula) *For the $3x + 1$ function $T(n)$ the natural density $\mathbb{D}(S_\lambda(k))$ for integers having $\lambda$-stopping time at most $k$ is given exactly by the formula*

$$(3.31) \qquad \mathbb{D}(S_\lambda(k)) = \mathbb{P}[V_\lambda(\omega) \leq k],$$

*in which $V_\lambda$ is the $\lambda$-stopping time random variable in the $3x + 1$ multiplicative random product (MRP) model.*

PROOF. In 1985 Rawsthorne [32, Theorem 1] proved a weaker version of this result, with $\mathbb{D}(S_\lambda(k))$ replaced by the lower asymptotic density $\underline{\mathbb{D}}(S_\lambda(k))$. The result, using natural density, is a special case of Borovkov and Pfeifer [10, Theorem 3].  $\square$

It is natural to apply the $3x + 1$ MPR model with an initial condition added, which is a proxy for the expected behavior of the total stopping time. To do this we must allow variable $\lambda$ (as a function of $n$), in a range of parameters where there is no rigorous proof that the model behavior agrees with that of iteration of the map $T(n)$, namely for $\lambda = \alpha \log n$ with various $\alpha > 1$. What is missing is a result saying that it accurately matches the behavior of iteration of the $3x + 1$ map.

The behavior of the resulting probabilistic model is rigorously analyzable, as we discuss in the next subsection, cf. Theorem 3.2 below.

**3.2. Additive Random Walk Model and Total Stopping Times.** The $3x + 1$ iteration takes $x_0 = n$ and $x_k = T^{(k)}(n)$. In studying the iteration, it is often more convenient to use a logarithmic scale and set $y_k = \log x_k$ (natural logarithm) so that

$$y_k = \log x_k := \log T^{(k)}(n).$$

Then we have

$$(3.32) \qquad y_{k+1} = \begin{cases} y_k + \log \frac{3}{2} + e_k & \text{if } x \equiv 1 \ (\text{mod } 2) \,, \\[2mm] y_k + \log \frac{1}{2} & \text{if } x \equiv 0 \ (\text{mod } 2) \,, \end{cases}$$

with

$$(3.33) \qquad e_k := \log \left( 1 + \frac{1}{3x_k} \right).$$

Here $e_k$ is small as long as $|x_k|$ is large.

Theorem 2.1 implies that if an integer is drawn at random from $[1, 2^k]$ then its $k$-truncated parity sequence will be uniformly distributed in $\{0, 1\}^k$. In consequence, equations (3.32) and (3.33) show that the quantities $\log T^{(k)}(n)$ (natural logarithm) can be modeled by a random walk starting at initial position $y_0 = \log n$ and taking steps of size $\log \frac{3}{2}$ if the parity value is odd, and $\log \frac{1}{2}$ if it is even.

The MRP model considered before is converted to an additive model by making a logarithmic change of variable, taking new random variables $W_k := \log X_k$. The additive model considers the random variables $Z_k$ which are a sum of random variables

$$Z_k := Z_0 + \log Y_k = Z_0 + W_1 + W_2 + \cdots + W_k.$$

Here $Z_0$ is a specified initial starting point, and $Z_k$ is the result of a (biased) random walk, taking steps of size either $\log \frac{3}{2}$ or $\log \frac{1}{2}$ with equal probability. In terms of these variables, the $\lambda$-stopping time random variable above is

$$V_\lambda(\omega) = \inf\{k : Z_k - Z_0 \leq \log \lambda\}.$$

We consider the approximation of this iteration process by the following stochastic model, which we term the $3x + 1$ *Biased Random Walk Model* ($3x + 1$ *BRW Model*). For an integer $n \geq 1$ it separately makes a random walk which takes steps of size $\log \frac{1}{2}$ half the time and $\log \frac{3}{2}$ half the time. We can write such a random variable as

$$\xi_k := -\log 2 + \delta_k \log 3,$$

in which $\delta_k$ are independent Bernoulli zero-one random variables. The random walk positions $\{Z_k : k \geq 0\}$, are then random variables having starting value $Z_0 = \log n$, and with

$$Z_k := Z_0 + \xi_1 + \xi_2 + \cdots + \xi_k.$$

The $Z_k$ define a biased random walk, whose expected drift $\mu$ is given by

$$(3.34) \qquad \mu := E[\xi_k)] = -\log 2 + \frac{1}{2}\log 3 = \frac{1}{2}\log\left(\frac{3}{4}\right) \approx -0.14384.$$

The variance $\sigma$ of each step is given by

$$\sigma := \mathrm{Var}[\xi_k] = \frac{1}{2}\log 3 \approx 0.54930.$$

In the addive model we associate to a random walk a *total stopping time random variable*

$$S_\infty(n) := \min\{k > 0 : Z_k \le 0, \text{ given } Z_0 = \log n\},$$

which detects when the walk first crosses 0 (this corresponds in the multiplicative model to reaching 1). The expected number of steps to reach a nonpositive value starting from $Z_0 = \log n$ is

$$E[S_\infty(n)] = \frac{1}{|\mu|}\log a = \frac{1}{\frac{1}{2}\log(\frac{4}{3})}\log n \approx 6.95212\log n.$$

As noted in §2, Borovkov and Pfeifer [10] consider the multiplicative stochastic model obtained by exponentiation of the positions of the biased random walk above, from a given starting value $X_0 = e^{n_0}$. They conclude the following result [10, Theorem 5].

THEOREM 3.2. (3X + 1 BRW Gaussian Limit Distribution) *In the Biased Random Walk Model, for each fixed $n \ge 2$ define the normalized random variable*

$$Z_\infty(n) := \frac{S_\infty(n) - \frac{1}{\mu}\log n}{\mu^{-\frac{3}{2}}\sigma\sqrt{\log n}},$$

*which has cumulative distribution function $P_n(x) := \mathrm{Prob}[Z_\infty(n) < x]$. Here $\mu = |\frac{1}{2}\log\frac{3}{4}|$, and $\sigma = \frac{1}{2}\log 3$. Then for each fixed real $x$, allowing $n$ to vary, one has*

$$P_n(x) := \mathrm{Prob}[Z_\infty(n) < x] \longrightarrow \Phi(x), \text{ as } n \to \infty,$$

*where $\Phi(x) = \frac{1}{\sqrt{2\pi}}\int_{-\infty}^x e^{-\frac{1}{2}t^2}dt$ is the cumulative distribution function of the standard normal distribution $N(0,1)$.*

Borovkov and Pfeifer note further that the rate of convergence of the normalized distribution $P_n(x)$ with fixed $n$ to the limiting normal distribution as $n \to \infty$ is uniform in $x$, but is quite slow. They assert that for all $n \ge 2$ and all $-\infty < x < \infty$,

$$(3.35) \qquad |P_n(x) - \Phi(x)| = O\left((\log n)^{-\frac{1}{2}}\right),$$

where the implied constant in the O-symbol is absolute.

They also propose a better approximation to the distribution of the total stopping time of a random integer of size near $n$, reflecting the fact that it is nonnegative random variable. They assert that the rescaled variable

$$Y_\infty(n) := \frac{S_\infty(n)}{\log n}$$

should have a good second order approximation given by the nonnegative random variable $\tilde{Y}(n)$ having the distribution function

$$\Psi_n(x) = C_n\frac{\sqrt{\log n}}{\sigma}\int_0^x \frac{1}{\sqrt{2\pi t^3}}e^{-\frac{(\mu t-1)^2\log n}{2\sigma^2 t}}dt, \; x > 0.$$

in which $C_n$ is a normalizing constant ([10, eqn. (25)]).

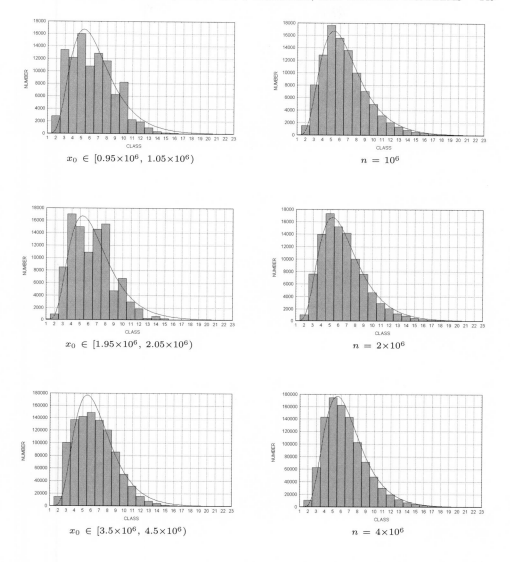

$x_0 \in [0.95 \times 10^6, \ 1.05 \times 10^6)$

$n = 10^6$

$x_0 \in [1.95 \times 10^6, \ 2.05 \times 10^6)$

$n = 2 \times 10^6$

$x_0 \in [3.5 \times 10^6, \ 4.5 \times 10^6)$

$n = 4 \times 10^6$

FIGURE 3.2. Histograms for $\sigma_\infty(x_0)/\ln x_0$ and its stochastic analog $T(n)/\ln n$ with fitted density. Taken from Borovkov-Pfeifer [**10**].

They view the random variable $S_\infty(n)$ as providing a model for the total stopping time $\sigma_\infty(n)$ of the $3x+1$ function, where one compares the ensemble of values $\{\sigma_\infty(n) : x \le n \le c_1 x\}$ with $c_1 > 1$ fixed with independent samples of values $S_\infty(n)$. The result above (with error term $O\left(\frac{1}{\sqrt{\log n}}\right)$) predicts that for any $\epsilon > 0$

the number of values that do not satisfy

$$\left(\frac{1}{\mu} - \frac{1}{(\log x)^{\frac{1}{2}-\epsilon}}\right)\log x \le \sigma_\infty(n) \le \left(\frac{1}{\mu} + \frac{1}{(\log x)^{\frac{1}{2}-\epsilon}}\right)\log x$$

is $o(x)$, as $x \to \infty$. They compare the distribution of $\tilde{Y}(n)$ with numerical data $\frac{\sigma_\infty(n)}{\log n}$ for the $3x+1$ function for $n \approx 10^6$ and find fairly good agreement.

## 4. $3x+1$ Forward Iteration: Large Deviations and Extremal Trajectories

Lagarias and Weiss [23] formulated and studied stochastic models which are intended to give predictions for the extremal behavior of iteration of the $3x+1$ map $T(n)$.

### 4.1. $3X+1$ Repeated Random Walk Model.
Lagarias and Weiss studied the following *Repeated Random Walk Model* ($3x+1$ *RRW Model*). For each integer $n \ge 1$, independently run a $3x+1$ biased random walk model trial with starting value $Z_{0,n} = \log n$. That is, generate an infinite sequence of independent random walks $\{Z_{k,n} : k \ge 0\}$, with one walk generated for each value of $n$. The model data is the countable set of random variables

(4.36) $$\omega := \{Z_{k,n} : n \ge 1, \ k \ge 0\},$$

in which the initial starting points $Z_{0,n} := \log n$ are deterministic, and all other random variables stochastic. From this data, one can form random variables that are functions of $\omega$, corresponding to the total stopping times and the maximum excursion values above.

The $3x+1$ RRW model is exactly analyzable, and makes predictions for the value of the scaled stopping time constant, and for the maximum excursion constant. A subtlety of the RRW model is the fact that there are exponentially many trials with inputs of a given length $j$, namely for those $n$ with $e^j \le n < e^{j+1}$, which have initial condition $j \le Z_{0,n} < j+1$, so that the theory of large deviations becomes relevant to the analysis.

### 4.2. $3x+1$ RRW Model Prediction: Extremal Total Stopping Times.
The $3x+1$ RRW model can be used to produce statistics analogous to the scaled total stopping time $\gamma_\infty(n)$ and the $3x+1$ scaled stopping time constant $\gamma$ , cf. (2.15) and (2.16).

For a given trial $\omega$ it yields an infinite sequence of total stopping time random variables

$$S_\infty(\omega) := (S_\infty(1), S_\infty(2), S_\infty(3), \ldots, S_\infty(n), \ldots),$$

where $S_\infty(n)$ is computed using the individual random walk $\mathcal{R}_n$. Thus we can compute the scaled statistics $\frac{S_\infty(n)}{\log n}$ for $n \ge 2$, and set

$$\gamma(\omega) := \limsup_{n\to\infty} \frac{S_\infty(n)}{\log n}.$$

as a stochastic analogue of the quantity $\gamma$.

The $3x + 1$ RRW model has the following asymptotic limiting behavior for this statistic, given by Lagarias and Weiss [**23**, Theorem 2.1].

THEOREM 4.1. ($3x+1$ RRW Scaled Stopping Time Constant) *For the $3x+1$ RRW model, with probability one the scaled stopping time*

$$\gamma(\omega) := \limsup_{n \to \infty} \frac{S_\infty(n)}{\log n}$$

*is finite and equals a constant*

$$\gamma_{RRW} \approx 41.677647,$$

*which is the unique real number $\gamma > \left(\frac{1}{2} \log \frac{4}{3}\right)^{-1} \approx 6.952$ of the fixed point equation*

$$(4.37) \qquad \gamma \, g\left(\frac{1}{\gamma}\right) = 1.$$

*Here the rate function $g(a)$ is given by*

$$(4.38) \qquad g(a) := \sup_{\theta \in \mathbb{R}} \left(\theta a - \log M_{RRW}(\theta)\right),$$

*in which*

$$(4.39) \qquad M_{RRW}(\theta) := \frac{1}{2}\left(2^\theta + \left(\frac{2}{3}\right)^\theta\right)$$

*is a moment generating function associated to the random walk.*

Lagarias and Weiss also obtain a density result on the number of $n$ getting values close to the extremal constant, as follows ([**23**, Theorem 2.2]).

THEOREM 4.2. ($3x + 1$ RRW Scaled Stopping Time Distribution) *For the $3x + 1$ RRW model, and for any constant $\alpha$ satisfying*

$$(4.40) \qquad \left(\frac{1}{2} \log \frac{4}{3}\right)^{-1} < \alpha < \gamma_{RRW},$$

*one has the bound*

$$(4.41) \qquad E\left[|\{n \le x : \frac{S_\infty(n)}{\log n} \ge \alpha\}|\right] \le \left(1 - \alpha \, g\left(\frac{1}{\alpha}\right)\right)^{-1} x^{1 - \alpha g(1/\alpha)}.$$

*In the reverse direction, for any $\epsilon > 0$ this expected value satisfies*

$$(4.42) \qquad E\left[|\{n \le x : \frac{S_\infty(n)}{\log n} \ge \alpha\}|\right] \ge x^{1 - \alpha g(1/\alpha) - \epsilon}$$

*for all sufficiently large $x \ge x_0(\epsilon)$.*

This theorem says that not only is there an upper bound $\gamma_{RRW}$ on the asymptotic limiting value of the stopping ratio, but the set of $n$ for which one gets a value above $\alpha$ becomes very sparse (in the logarithmic sense) as $\alpha$ approaches $\gamma_{RRW}$ from below. Theorem 4.2 is analogous to obtaining a multifractal spectrum for this problem. This result is well-suited for comparison with experimental data on $3x + 1$ iterates.

This analysis suggest the following prediction, which we state as a conjecture.

CONJECTURE 4.1. ($3x + 1$ Scaled Stopping Constant Conjecture) *The $3x + 1$ scaled stopping constant $\gamma$ is finite and is given by*

$$(4.43) \qquad \gamma = \gamma_{RRW} \approx 41.677647.$$

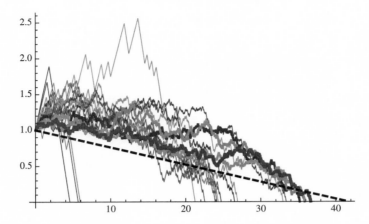

FIGURE 4.3. Scaled trajectories of $n_k$ maximizing $\gamma(n)$ for record values from Table 1 (thin for $1 \leq k \leq 10$; regular for $11 \leq k \leq 15$; thick for $16 \leq k \leq 19$), plotted against the predicted trajectory.

The large deviations model does more than predict an extremal value, it also predicts that the numbers that approach the extremal value must have a trajectory of iterates whose graph have a specified shape, which is a linear function when properly scaled. In Figure 4.3 we graph the set of scaled points

$$\left\{ \left( \frac{k}{\log n}, \frac{\log T^{(k)}(n)}{\log n} \right) : 0 \leq k \leq \sigma_\infty(n) \right\}.$$

The predicted large deviations extremal trajectory in this scaling has graph a straight line connecting the points $(0, 1)$ and $(\gamma_{RRW}, 0)$. Figure 4.3 shows the scaled trajectories with starting seeds $n_k$ taken from Table 1, i.e. those with record values for $\gamma_\infty(n)$. Compare to Lagarias and Weiss [**23**, Figure 3].

### 4.3. $3x + 1$ RRW Model Prediction: Maximum Excursion Constant.

For the $3x + 1$ RRW Stochastic Model, an appropriate statistic for a single trial that corresponds to the maximum excursion value is

$$t(n; \omega) := \sup(e^{Z_{k,n}} : k \geq 0).$$

The $3x + 1$ RRW model behavior for extremal behavior of maximum excursions $t(n; \omega)$ is given in the following result [**23**, Theorem 2.3].

THEOREM 4.3. *(3x + 1 RRW Maximum Excursion Constant) For the $3x + 1$ RRW model, with probability one the quantities $t(n, \omega)$ are finite for every $n \geq 1$. In addition, with probability one the random quantity*

$$(4.44) \qquad \rho(\omega) := \limsup_{n \to \infty} \frac{\log t(n; \omega)}{\log n} = \limsup_{n \to \infty} \left( \sup_{k \geq 0} \frac{Z_{k,n}}{\log n} \right)$$

*equals the constant*

$$(4.45) \qquad \rho_{RRW} = 2.$$

Lagarias and Weiss also prove [**23**, Theorem 2.4] a result permitting a quantitative comparison with data.

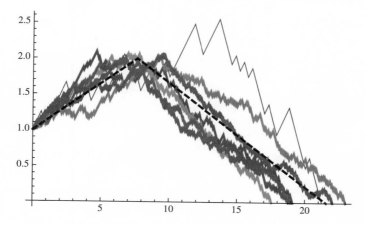

FIGURE 4.4. Scaled trajectories of seeds $n$ from Table 3, plotted against the predicted trajectory. The trajectory of $n = 27$ is thin, while the others are thick.

THEOREM 4.4. $(3x + 1$ RRW Maximum Excursion Density Function$)$ *For the* $3x + 1$ *RRW model, for any fixed* $0 < \alpha < 1$, *the expected value*

$$(4.46) \qquad E\left[|\{n \le x : \frac{\log t(n;\omega)}{\log n} \ge 2 - \alpha\}|\right] = x^{\alpha(1 - o(1))},$$

*as* $x \to \infty$.

These theorems suggest formulating the following conjecture.

CONJECTURE 4.2. *The* $3x + 1$ *maximum excursion constant* $\rho$ *defined in* $(2.20)$ *is finite and is given by*

$$(4.47) \qquad\qquad\qquad \rho = 2.$$

The large deviations model also makes a prediction on the graphs of the trajectories achieving maximum excursion, when plotted as the scaled data points

$$\left\{\left(\frac{k}{\log n}, \frac{\log T^{(k)}(n)}{\log n}\right) : 0 \le k \le \sigma_\infty(n)\right\}.$$

It asserts that extremal large deviation trajectories should approximate two line segments, the first with vertices $(0, 1)$ to $(7.645, 2)$ and then from this vertex to $(21.55, 0)$. The slope of the first line segment is $\frac{3}{4} \log 3 - \log 2 \approx 0.1308$ and that of the second line segment is $(\frac{1}{2} \log \frac{3}{4})^{-1} \approx -0.1453$. This prediction is shown as a dotted black line in Figure 4.4; it shows substantial agreement with the empirical evidence.

**4.4. $3x + 1$ RRW Model: Critique.** The $3x + 1$ repeated random walk model has the feature that random walks for different $n$ are *independent*. However the actual $3x + 1$ map certainly has a great deal of dependency built in, due to the fact that trajectories coalesce under forward iteration. For example, trajectories of numbers $8n + 4$ and $8n + 5$ always coalesce after 3 iterations of $T$. After coalescence, the trajectories are completely correlated. In fact, the $3x + 1$ Conjecture predicts that all trajectories of positive integers $n$ reach the orbit $\{1, 2\}$ and then cycle, whence they all should coalesce into exactly two classes, namely those that reach 1 in an odd number of iterations of $T$, and those that reach this orbit under an even number of iterations.

For this reason, it is not apparent a priori whether the prediction in Conjecture 4.1 above of the constant $\gamma = \gamma_{RRW}$ is reasonable. Our faith in Conjecture 4.1 relies on the fact that first, the same prediction is made using a branching random walk model that incorporates dependency in the model, see Theorem 6.4 in §6, and second, on comparison with empirical data in Table 1 .

## 5. $3x + 1$ **Accelerated Forward Iteration : Brownian Motion**

Now we consider the accelerated $3x + 1$ function $U$. Recall that $U$ is defined on odd integers, and removes all powers of 2 in one fell swoop. Iterates of the accelerated function $U$ are of course equivalent (from the point of view of the main conjecture) to those of $T$, but there are some subtle differences which make studying both points of view appealing.

For an odd integer $n$, we let $\mathfrak{o}(n)$ count the number of powers of 2 dividing $3n + 1$, so that

$$(5.48) \qquad \mathfrak{o}(n) := \mathrm{ord}_2(3n + 1).$$

Then the accelerated $3x + 1$ function $U$ is given by:

$$(5.49) \qquad U(n) := \frac{3n + 1}{2^{\mathfrak{o}(n)}}.$$

In analogy with the (truncated) parity sequence, cf. Definition 2.1, we make the following definition, giving a symbolic dynamics for the accelerated $3x+1$ map.

DEFINITION 5.1. *(i) For an odd integer $n$, define the $\mathfrak{o}$-sequence of $n$ to be*

$$(5.50) \qquad V(n) := (\mathfrak{o}_1(n), \mathfrak{o}_2(n), \mathfrak{o}_3(n), \dots)$$

*where*

$$\mathfrak{o}_k(n) := \mathfrak{o}(U^{(k)}(n)),$$

*and $U^{(k)}(n)$ denotes the $k$-th iterate of $U$, as usual. This is an infinite vector of positive integers.*

*(ii) For $k \geq 1$ the $k$-truncated $\mathfrak{o}$-sequence of $n$ is:*

$$(5.51) \qquad V^{[k]}(n) := (\mathfrak{o}_1(n), \mathfrak{o}_2(n), \dots, \mathfrak{o}_k(n))$$

*i.e. a vector giving the initial segment of $k$ terms of $V(n)$.*

DEFINITION 5.2. *For an odd integer $n$ and $k \geq 1$, let the $k$-size $\mathfrak{s}_k(n)$ be the sum of the entries in $V^{[k]}(n)$, that is*

$$\mathfrak{s}_k(n) := \mathfrak{o}_1(n) + \mathfrak{o}_2(n) + \cdots + \mathfrak{o}_k(n).$$

**5.1. The Structure Theorem.** Notice that $U(n)$ is not only odd, but is also relatively prime to 3. Hence we lose no generality by restricting the domain for $U$ from $\mathbb{Z}$ to the (more natural) set $\Pi$ of positive integers prime to 2 and 3, i.e.

$$(5.52) \qquad \Pi := \{n \in \mathbb{Z} : \gcd(n, 6) = 1\}.$$

Moreover, $\Pi$ is the disjoint union of $\Pi^{(1)}$ and $\Pi^{(5)}$, where $\Pi^{(\varepsilon)}$ consists of positive integers congruent to $\varepsilon \pmod{6}$, $\varepsilon = 1$ or 5.

DEFINITION 5.3. *Given $\varepsilon = 1$ or $5$, $k \geq 1$, and a vector $(\mathfrak{o}_1, \ldots, \mathfrak{o}_k)$ of positive integers, let*

$$\Sigma^{(\varepsilon)}(\mathfrak{o}_1, \ldots, \mathfrak{o}_k)$$

*be the set of all $n \in \Pi^{(\varepsilon)}$ with $V^{[k]}(n) = (\mathfrak{o}_1, \ldots, \mathfrak{o}_k)$.*

The result analogous to Theorem 2.1 is given by Sinai [**35**] and Kontorovich-Sinai [**18**].

THEOREM 5.1. (Structure Theorem for $\mathfrak{o}$ Symbolic Dynamics) *Fix $\varepsilon = 1$ or $5$, and let $n \in \Pi^{(\varepsilon)}$.*

*(i) The $k$-truncated $\mathfrak{o}$-sequence $V^{[k]}(n)$ of the first $k$ iterates of the accelerated map $U(n)$ is periodic in $n$. Its period is $6 \cdot 2^{\mathfrak{s}}$, where*

$$\mathfrak{s} = \mathfrak{s}_k(n) = \mathfrak{o}_1(n) + \mathfrak{o}_2(n) + \cdots + \mathfrak{o}_k(n).$$

*(ii) For any $k \geq 1$ and $\mathfrak{s} \geq k$, each of the $\binom{\mathfrak{s}-1}{k-1}$ possible vectors $(\mathfrak{o}_1, \cdots, \mathfrak{o}_k)$ with $\mathfrak{o}_j \geq 1$ and $\mathfrak{o}_1 + \cdots + \mathfrak{o}_k = \mathfrak{s}$ occurs exactly once as $V^{[k]}(n)$ for some $n \in \Pi^{(\varepsilon)}$ in the               initial                   segment $1 \leq n < 6 \cdot 2^{\mathfrak{s}}$.*

*(iii) The least element $n_0 \in \Sigma^{(\varepsilon)}(\mathfrak{o}_1, \ldots, \mathfrak{o}_k)$ satisfies $n_0 < 6 \cdot 2^{\mathfrak{s}}$; moreover*

$$\Sigma^{(\varepsilon)}(\mathfrak{o}_1, \ldots, \mathfrak{o}_k) = \left\{ 6 \cdot 2^{\mathfrak{s}} \cdot m + n_0 \right\}_{m=0}^{\infty}.$$

PROOF. This is proved as part one of the Structure Theorem in Kontorovich-Sinai [**18**]. Here (iii) follows immediately from (i) and (ii).  □

Again one easily shows that an integer $n$ is uniquely determined by the $\mathfrak{o}$-sequence $V(n)$ of its forward $U$-orbit.

Moreover, the following result shows that the image under the iterated map $U^{(k)}$ of $n \in \Sigma^{(\varepsilon)}(\mathfrak{o}_1, \ldots, \mathfrak{o}_k)$ is also a nice arithmetic progression!

THEOREM 5.2. (Iterated Structure Theorem) *Fix $\varepsilon = 1$ or $5$, $k \geq 1$, a vector $(\mathfrak{o}_1, \cdots, \mathfrak{o}_k)$, and let $\mathfrak{s} = \mathfrak{o}_1 + \cdots + \mathfrak{o}_k$. Suppose $1 \leq n_0 < 6 \cdot 2^{\mathfrak{s}}$ is the least element of $\Sigma^{(\varepsilon)}(\mathfrak{o}_1, \ldots, \mathfrak{o}_k)$. Then there is a $\delta_k = 1$ or $5$ and an $r_k \in \{0, 1, 2, \ldots, 3^k - 1\}$, both depending only on $\varepsilon$ and $(\mathfrak{o}_1, \ldots, \mathfrak{o}_k)$, such that, for each positive integer $m$,*

$$(5.53) \qquad U^{(k)}(6 \cdot 2^{\mathfrak{s}} \cdot m + n_0) = 6(3^k \cdot m + r_k) + \delta_k.$$

*Moreover, $\delta_k$ is determined by the congruence*

$$(5.54) \qquad \delta_k \equiv 2^{\mathfrak{o}_k} (\text{mod } 3).$$

PROOF. This is part two of the Structure Theorem in Kontorovich-Sinai [**18**]. Note that $m$ is the same number on both sides of (5.53); this equation says that an arithmetic progression with common difference $6 \cdot 2^{\mathfrak{s}}$ mapped under $U^{(k)}$ to one with common difference $6 \cdot 3^k$.  □

**5.2. Probability Densities.** We first tweak the notion of natural density defined in (2.12) on subsets of the natural numbers, by restricting to just elements of our domain $\Pi$. For a subset $\Sigma \subset \Pi$, let the $\Pi$-*natural density* be

$$\mathbb{D}_{\Pi}(\Sigma) := \lim_{t \to \infty} \frac{3}{t} \left| \left\{ n \in \Sigma : n \leq t \right\} \right| = \lim_{t \to \infty} \frac{\left| \left\{ n \in \Sigma : n \leq t \right\} \right|}{\left| \left\{ n \in \Pi : n \leq t \right\} \right|},$$

provided that the limit exists. (The factor 3 appears because $\Pi$ contains two residue classes modulo 6.)

For a vector $(\mathfrak{o}_1, \ldots, \mathfrak{o}_k)$ , let

$$\Sigma(\mathfrak{o}_1, \ldots, \mathfrak{o}_k) := \Sigma^{(1)}(\mathfrak{o}_1, \ldots, \mathfrak{o}_k) \ \cup \ \Sigma^{(5)}(\mathfrak{o}_1, \ldots, \mathfrak{o}_k).$$

Recall that a random variable $X$ is *geometrically distributed* with parameter $0 < \rho < 1$ if

$$\mathbb{P}[X = m] = \rho^{m-1}(1 - \rho) \qquad \text{for } m = 1, 2, 3, \ldots$$

THEOREM 5.3.    *(Geometric Distribution)*

*(1) The sets $\Sigma(\mathfrak{o}_1, \ldots, \mathfrak{o}_k)$ have a $\Pi$-natural density given by*

$$(5.55) \qquad \mathbb{D}_\Pi \left( \Sigma(\mathfrak{o}_1, \ldots, \mathfrak{o}_k) \right) = 2^{-\mathfrak{s}} = 2^{-\mathfrak{o}_1} \cdot 2^{-\mathfrak{o}_2} \cdots 2^{-\mathfrak{o}_k}.$$

*(2) This natural density matches the probability density of the distribution for independent geometrically distributed random variables $(\mathfrak{p}_1, \ldots, \mathfrak{p}_k)$ with parameter $\rho = \frac{1}{2}$, which have*

$$\mu_\mathfrak{o} := \mathbb{E}[\mathfrak{p}_j] = 2, \qquad and \qquad \sigma_\mathfrak{o} := Var[\mathfrak{p}_j] = 2.$$

*That is,*

$$(5.56) \qquad \mathbb{P}[(\mathfrak{p}_1 = \mathfrak{o}_1, \ldots, \mathfrak{p}_k = \mathfrak{o}_k)] = \mathbb{D}_\Pi \left( \Sigma(\mathfrak{o}_1, \ldots, \mathfrak{o}_k) \right).$$

PROOF. (1) The existence of a natural density is automatic, since these sets are finite unions of arithmetic progressions. For $\varepsilon = 1$ or $5$, we easily compute from Theorem 5.1 that

$$\mathbb{D}_\Pi \left( \Sigma^{(\varepsilon)}(\mathfrak{o}_1, \ldots, \mathfrak{o}_k) \right) = 3 \cdot \frac{1}{6 \cdot 2^{\mathfrak{o}_1 + \cdots + \mathfrak{o}_k}} = \frac{1}{2} \cdot 2^{-\mathfrak{s}},$$

and hence (5.55) follows.

(2) The identity (5.56) is immediate from independence and (5.55).        $\square$

We now deduce the following result.

THEOREM 5.4.    (Central Limit Theorem) *For the accelerated $3x + 1$ map $U$, with symbolic iterates $(\mathfrak{o}_1, \mathfrak{o}_2, \ldots)$, the scaled ordinates satisfy*

$$\lim_{k \to \infty} \mathbb{D}_\Pi \left[ n : \frac{\mathfrak{o}_1(n) + \cdots + \mathfrak{o}_k(n) - \mu_\mathfrak{o} k}{\sqrt{\sigma_\mathfrak{o} k}} < a \right] = \frac{1}{\sqrt{2\pi}} \int_{-\infty}^{a} e^{-u^2/2} du.$$

PROOF. This follows immediately from the argument above and the Central Limit Theorem for geometrically distributed random variables.        $\square$

Compare the above to Theorem 3.2. The rate of convergence is again quite slow (this feature is shared by Borovkov-Pfeifer; see (3.35)).

**5.3. Brownian Motion.** Consider some starting value $x_0 = n \in \Pi$, denote its iterates by $x_k := U^{(k)}(x_0)$, and take logarithms $y_k := \log x_k$. As in (3.32) , the multiplicative behavior of $U$ is converted via logarithms to an additive behavior. Normalize the above by

$$(5.57) \qquad \omega_k := \frac{y_k - y_0 - k \log(\frac{3}{4})}{\sqrt{2k} \log 2}.$$

Then we have the following scaling limits for "random" accelerated trajectories, chosen in the sense of density.

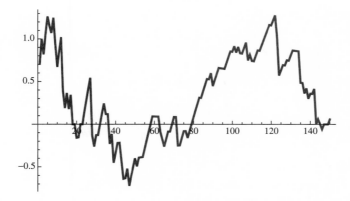

FIGURE 5.5. A sample path of the $3x+1$ map. Here we took the starting value $x_0 = 123\,456\,789\,135\,791\,113\,151\,719$, computed 150 iterates of $U$, and plotted $\omega_k$.

THEOREM 5.5. (Geometric Brownian Motion Increments) *Fix a partition of the interval $[0,1]$ as $0 = t_0 < t_1 < \cdots < t_r = 1$. Given an integer $k$, set $k_j = \lfloor t_j k \rfloor$, with $j = 1, \ldots, r$. Then for any $a_j < b_j$,*

$$\mathbb{D}_\Pi \left[ x_0 : a_j < \omega_{k_j} - \omega_{k_{j-1}} < b_j, \text{ for all } j = 1, 2, \ldots, r \right] \quad \rightarrow \quad \prod_{j=1}^{r} \Big( \Phi(b_j) - \Phi(a_j) \Big),$$

*as $k \to \infty$, where recall that $\Phi(a)$ is the cumulative distribution function for the standard normal distribution:*

$$\Phi(a) = \frac{1}{\sqrt{2\pi}} \int_{-\infty}^{a} e^{-u^2/2} du.$$

PROOF. This appears as Theorem 5 in Kontorovich-Sinai [**18**]. See Figure 5.5. □

The interpretation of the above result is that the paths of the accelerated $3x+1$ map, when properly scaled, approach those of a geometric Brownian motion, that is, a stochastic process whose logarithm is a Brownian motion, or a Weiner process.

REMARK. There are in fact **two** limits taken in the above theorem, whose order is highly non-interchangeable! The first limit is hidden inside the definition of density, that is, first we take the limit as $x \to \infty$ of the set of all $x_0 < x$ satisfying the given condition with the number $k$ of iterates of $U$ fixed, and only then do we let $k \to \infty$. If $x_0$ were to be fixed and $k$ allowed to grow, there would be nothing stochastic at all happening, since we believe the $3x+1$ Conjecture!

REMARK. The drift, as given in (5.57), is $\log(\frac{3}{4}) \approx -0.28768$. Compare this to (3.34), where the drift of the Biased Random Walk model is computed to be $\frac{1}{2} \log(\frac{3}{4}) \approx -0.14384$. While it is not surprising that the accelerated map $U$ should have a more aggressive pull to the origin, it is curious that it is exactly twice as fast (on an exponential scale) as the $3x+1$ function $T$.

REMARK. Given that the drift of the (logarithm of the) accelerated $3x+1$ function $U$ is $\mu = \log(\frac{3}{4})$, one expects that the typical total stopping time of a seed

$n$ is roughly

$$\frac{1}{|\mu|}\log n \approx 3.476 \log n.$$

## 5.4. Entropy.

DEFINITION 5.4.   *The entropy of a random variable* $X$ *taking values in* $[M] :=$ $\{1, 2, \ldots, M\}$ *is given by*

$$H := -\sum_{m=1}^{M} \mathbb{P}[X = m] \log \mathbb{P}[X = m].$$

The following facts are classical:

  (i) If $X$ is distributed uniformly in $[M]$ then $H = \log M$.
  (ii) The entropy $H$ is maximized by the uniform distribution.

The first is an elementary exercise, while the second is proved easily using, e.g., Lagrange's multiplier method.

In light of Theorem 5.2, for any fixed $k \geq 1$, the value $0 \leq r_k \leq 3^k - 1$ is a function of the values $\varepsilon$ and $(\mathfrak{o}_1, \ldots, \mathfrak{o}_k)$, and hence has a natural density. For a fixed $\mathfrak{r} \in [0, 3^k - 1]$ we write

$$\mathbb{D}_\Pi[x_0 : r_k(x_0) = \mathfrak{r}] \qquad \text{to mean} \qquad \sum_{\substack{(\mathfrak{o}_1, \ldots, \mathfrak{o}_k),\ \varepsilon \in \{1,5\} \\ r_k(\varepsilon, \mathfrak{o}_1, \ldots, \mathfrak{o}_k) = \mathfrak{r}}} \mathbb{D}_\Pi[\Sigma^{(\varepsilon)}(\mathfrak{o}_1, \ldots, \mathfrak{o}_k)].$$

One might hope that $r_k$ (which is a deterministic function but can be thought of as a "random variable") is close to being uniformly distributed in $\{0, 1, \ldots, 3^k - 1\}$; then one could attempt to "bootstrap" iterations of $U$ to one-another to have better quantitative control on various asymptotic densities with $k$ not too large. Were this to be the case, the entropy (defined for this using $\mathbb{D}_\Pi$ in place of $\mathbb{P}$) would be $\log 3^k = k \log 3$.

THEOREM 5.6.   (Entropy of $r_k$) *There is some constant* $c > 0$ *such that the entropy* $H$ *of* $r_k$ *satisfies:*

$$H \geq k \log 3 - c \log k.$$

PROOF.   This statement is Theorem 5.1 in Sinai [**35**].          □

The function $r_k$ in Theorem 5.2 is accompanied by the residue class $\delta_k \in \{1, 5\}$, which satisfies, cf. (5.54),

$$\delta_k \equiv 2^{\mathfrak{o}_k} \pmod{3}.$$

It follows immediately from the fact that $\mathfrak{o}_k$ is geometrically distributed with parameter $1/2$, that

$$\mathbb{D}_\Pi[x_0 : \delta_k(x_0) = 1] = \mathbb{D}_\Pi[x_0 : \mathfrak{o}_k \text{ is even }] = \frac{1}{3},$$

and hence of course, $\mathbb{D}_\Pi[x_0 : \delta_k(x_0) = 5] = \frac{2}{3}$.

Moreover, if $r_k$ is uniformly distributed, then so are the digits $h_k(j) \in \{0, 1, 2\}$ in its 3-adic expansion:

$$r_k = h_k(k-1) \cdot 3^{k-1} + h_k(k-2) \cdot 3^{k-2} + \cdots + h_k(1) \cdot 3 + h_k(0).$$

Note that only the first few leading digits $h_k(k-1), h_k(k-2), \ldots, h_k(k-t)$ are needed to specify that location of $r_k/3^k$, to within an error of $1/3^t$.

THEOREM 5.7. (Joint Uniform Distribution) *The joint distributions of $(r_k/3^k, \delta_k)$ converge weakly to the uniform distribution, that is, for any fixed $t \geq 1$ and $\mathfrak{h}_1, \ldots, \mathfrak{h}_t \in \{0,1,2\}$, as $k \to \infty$,*

$$\mathbb{D}_\Pi[x_0 : h_k(k-1) = \mathfrak{h}_1, h_k(k-2) = \mathfrak{h}_2, \ldots, h_k(k-t) = \mathfrak{h}_t, \delta_k(x_0) = 1] \to \frac{1}{3^t} \cdot \frac{1}{3},$$

*and*

$$\mathbb{D}_\Pi[x_0 : h_k(k-1) = \mathfrak{h}_1, h_k(k-2) = \mathfrak{h}_2, \ldots, h_k(k-t) = \mathfrak{h}_t, \delta_k(x_0) = 5] \to \frac{1}{3^t} \cdot \frac{2}{3}.$$

PROOF. This appears as Theorem 1 in Sinai [**36**]. See also [**37**].  □

## 6. $3x+1$ Backwards Iteration: $3x+1$ Trees

One can also model backwards iteration of the $3x+1$ map $T(x)$.

Backwards iteration is described by a tree of inverse iterates, and there are either one or two inverse iterates. Here

$$T^{(-1)}(n) = \begin{cases} \{2n\} & \text{if } n \equiv 0, 1 \ (\text{mod } 3), \\ \{2n, \frac{2n-1}{3}\} & \text{if } n \equiv 2 (\text{mod } 3). \end{cases}$$

Starting from a root node labelled $a$ we can grow an infinite tree $\mathcal{T}(a)$ of all the inverse iterates of $a$. Each node in the tree is labelled by its associated $3x+1$ function value. To a node labelled $n$ we add either one or two (directed) edges from the elements of $T^{-1}(n)$ to $n$, and we label these two edges by the value of this element.

**6.1. Pruned $3x+1$ Trees.** Next we note that any $a \equiv 0 \ (\text{mod } 3)$ has exactly one inverse iterate, which itself is 0 (mod 3). Thus if $a \equiv 0 \ (\text{mod } 3)$ the set of inverse iterates forms a single branch that never divides. However if $a \not\equiv 0 \ (\text{mod } 3)$ then the tree grows exponentially in size. It is convenient therefore to restrict to numbers $a \not\equiv 0 \ (\text{mod } 3)$ and furthermore to prune such a tree to remove all nodes $n \equiv 0 \ (\text{mod } 3)$. This produces an (infinite depth) *pruned tree* $\mathcal{T}^*(a)$ which is described by inverse iterates of the modified map

(6.58) $$\tilde{T}^{(-1)}(n) = \begin{cases} \{2n\} & \text{if } n \equiv 1, 4, 5 \text{ or } 7 \ (\text{mod } 9), \\ \{2n, \frac{2n-1}{3}\} & \text{if } n \equiv 2 \text{ or } 8 \ (\text{mod } 9), \end{cases}$$

applied starting with root node labelled $n_0 := a$. The pruning operation is depicted in Figure 6.6, with root node assigned depth 0.

We obtain a reduced tree $\overline{\mathcal{T}}^*(a)$ obtained by labelling each node with the (mod 2) residue class of the $3x+1$ value assigned to that node. (One may also think of this as labelling the directed edge leaving this node, with the exception of the root node.)

We let $\mathcal{T}_k^*(a)$ denote the pruned tree with root node $n_0 = a$, cut off at depth $k$, and we let $\overline{\mathcal{T}}_k^*(a)$ denote the same tree, keeping only the node labels (mod 2), for all nodes except the root node, where no data is kept. Let $N^*(k; a)$ count the number of depth $k$ leaves in this tree. Then we have

(6.59) $$N^*(k, a) := |\{n : n \not\equiv 0 \ (\text{mod } 3) \text{ and } T^{(k)}(n) = a\}|.$$

We have $N^*(k, a) \leq 2^k$ as a consequence of the fact that each $3x + 1$ tree has at most two upward branches at each node.

The following result gives information on the sizes of depth $k$ trees over all possible tree types ([**23**, Theorem 3.1]).

THEOREM 6.1.   (Structure of Pruned $3x + 1$ Trees)

*(1) For $k \geq 1$ and $a \not\equiv 0 \pmod{3}$, the structure of the pruned level $k$ tree $\overline{\mathcal{T}}_k^*(a)$, and hence the number $N^*(k, a)$, is completely determined by $a \pmod{3^{k+1}}$.*

*(2) There are $2 \cdot 3^k$ residue classes $a \pmod{3^{k+1}}$ with $a \not\equiv 0 \pmod{3}$. For these*

$$(6.60) \qquad \sum_{\substack{a \ (\mathrm{mod} \ 3^{k+1}) \\ m \not\equiv 0 \ (\mathrm{mod} \ 3)}} N^*(k, a) = 2 \cdot 4^k.$$

*It follows that if a residue class $a \pmod{3^{k+1}}$ with $a \not\equiv 0 \pmod{3}$ is picked with the uniform distribution, the expected number of leaves in the random tree $\overline{\mathcal{T}}_k^*(a)$ is exactly $\left(\frac{4}{3}\right)^k$.*

We now consider the complete set of numbers having total stopping time $k$. Set

$$(6.61) \qquad N_k := |\{n : \ \sigma_\infty(n) = k\}|.$$

Recall from §2.5 that $N_k = N_k(1)$, where $N_k(a)$ counts the number of integers that iterate to $a$ after exactly $k$ iterations of the $3x + 1$ map $T$. We defined there the $3x + 1$ tree growth constants

$$\delta_3(a) := \limsup_{k \to \infty} \frac{1}{k} \log N_k(a).$$

Theorem 6.1 suggests the following conjecture for these tree growth constants, made by Lagarias and Weiss [**23**].

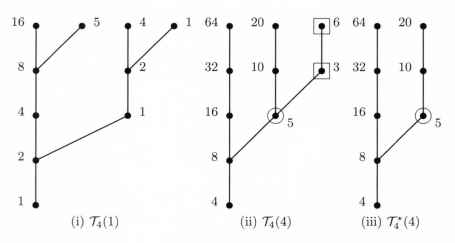

FIGURE 6.6.   $3x + 1$ trees $\mathcal{T}_k(a)$ and "pruned" $3x + 1$ tree $\mathcal{T}_k^*(a)$, with $k = 4$.

CONJECTURE 6.1. *For each $a \not\equiv 0 \pmod 3$, the $3x+1$ tree growth constant $\delta_3(a)$ is given by*

$$(6.62) \qquad \delta_3(a) = \log\left(\frac{4}{3}\right).$$

Applegate and Lagarias [**2**] determined by computer the maximal and minimal number of leaves in $3x+1$ trees of depth $k$ for $k \geq 30$. The maximal and minimal number of leaves in such trees at level $k$ is given by

$$N_k^+ := \max\{N_k^*(a): \ a \pmod{3^{k+1}} \text{ with } a \not\equiv 0 \pmod 3\}.$$

and

$$N_k^- = \min\{N_k^*(a): \ a \pmod{3^{k+1}} \text{ with } a \not\equiv 0 \pmod 3\},$$

respectively. Figure 6.7 pictures maximal and minimal trees for depth $k = 5$. (Circled nodes indicate an omitted inverse iterate under $T^{-1}$ that is $\equiv 0 \pmod 3$.)

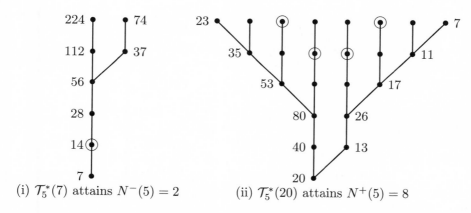

(i) $\mathcal{T}_5^*(7)$ attains $N^-(5) = 2$    (ii) $\mathcal{T}_5^*(20)$ attains $N^+(5) = 8$

FIGURE 6.7. Maximal and Minimal depth 5 pruned $3x+1$ Trees

The data on these counts $N_\pm(k)$ was presented already in §2.5, cf. Table 4. Based on this data, Applegate and Lagarias [**4**, Conjecture C] formulated the following strengthened conjecture, which implies Conjecture 6.1.

CONJECTURE 6.2. *The maximal and minimal number of leaves of $3x+1$ trees satisfy, as $k \to \infty$,*

$$(6.63) \qquad N_k^- = \left(\frac{3}{4}\right)^{k+o(k)}$$

$$(6.64) \qquad N_k^+ = \left(\frac{3}{4}\right)^{k+o(k)}.$$

**6.2. $3x+1$ Backwards Stochastic Models: Branching Random Walks.** Lagarias and Weiss [**23**] formulated stochastic models for the growth of $3x+1$ trees that were *multi-type branching processes*. Such models grow a random tree, with nodes marked as several different kinds of individuals. In this case the number of nodes of each type at each depth $k$ (also called generation $k$) can be viewed as the output of the branching process. The particular branching processes they used are *multi-type Galton-Watson processes*.

Lagarias and Weiss also modeled the size of preimages of elements in a (pruned) $3x+1$ tree. This size is specified by a real number attached to each node. Branching process models which attach to each node in the tree a real number giving the position of those individuals on a line, according to some (possibly random) rule, are models called *multi-type branching random walks*. Here the location of the individuals on the line give the random walk aspect; offspring nodes at level $k$ are shifted in position from their parent ancestor at level $k-1$ by a point process. The process starts with a root node giving a single progenitor at level 0 (generation 0).

Lagarias and Weiss defined a hierarchy of branching random walk models, which they denoted $\mathcal{B}[3^j]$, for each $j \geq 0$. These branching random walk models, having several kinds of individuals, model the backwards iteration viewed ($\mod 3^j$). The model for $j = 0$ is simpler than the other models.

$3x + 1$ *Branching Random walk* $\mathcal{B}[3^0]$. There is one type of individual. With probability $\frac{2}{3}$ an individual has a single offspring located at a position shifted by $\log 2$ on the line from its progenitor, and with probability $\frac{1}{3}$ it has two offspring located at positions shifted $\log 2$ and $\log \frac{2}{3}$ on the line from their progenitor. If the progenitor is in generation (or depth) $k-1$, the offspring are in generation $k$. The tree is grown from a single individual at generation 0, with specified location $\log a$.

The more general models for $j \geq 1$ are given as follows.

$3x + 1$ *Branching Random walk* $\mathcal{B}[3^j], (j \geq 1)$. There are $p = 2 \cdot 3^{j-1}$ types of individuals, indexed by residue classes $a$ ($\mod 3^j$) with $a \not\equiv 0$ ($\mod 3$). The distribution of offspring of an individual of type $a$ ($\mod 3^j$), at any given depth $k$ in the branching, is determined as follows: Regard $a$ ($\mod 3^j$) labelling a node at depth $d-1$. Regard it as being, with probability $\frac{1}{3}$ each, one of the three possible residue classes $\tilde{a}$ ($\mod 3^{j+1}$) consistent with it. The tree (of depth 1) with $\tilde{a}$ as root node, given by $(T^*)^{-1}(\tilde{a})$ has either one or two progeny, at depth 1 and their node labels are well-defined classes ($\mod 3^j$), either $2\tilde{a}$ or, if it legally occurs, $\frac{2\tilde{a}-1}{3}$ ($\mod 3^j$). The branching random walk then produces an individual of type $2\tilde{a}$ at generation $k+1$ whose position is additively shifted by $\log 2$ from that of the generation $k$ progenitor node, plus, if legal, another labelled $\frac{2\tilde{a}-1}{3}$, which is shifted in position by $\log(\frac{2}{3})$ on the line from that of the generation $k$-node. The tree is grown from a single individual at depth 0, with specified type and location $\log a$.

In these models, the behavior of the random walk part of the model can be completely reconstructed from knowing the type of each node. This is a very special property of these branching random walk models, which does not hold for general branching random walks.

In such models, one may think of the nodes as representing individuals, with individuals at level $k$ being children of a particular individual at level $k-1$; the random walk aspect indicates position in space of these individuals.

Let $\omega$ denote a single realization of such a branching random walk $\mathcal{B}[3^j]$ which starts from a single individual $\omega_{0,1}$ of type 1 ($\mod 3^j$) at depth 0, with initial position labeled $\log a$. Here $\omega$ describes a particular infinite tree. We let $N_k(\omega)$ denote the number of individuals at level $k$ of the tree. We let $S(\omega_k, j)$ denote the position of the $j$-th individual at level $k$ in the tree, for $1 \leq j \leq N_k(\omega)$.

These models are all *supercritical branching processes* in the following very strong sense. In *every* random realization $\omega$, the number of nodes at level $d$ grows exponentially in $d$, and there are no extinction events.

Lagarias and Weiss [23] observed that the predictions of these models stabilized for all $j \geq 1$, as far as the behavior of asymptotic statistics related to $3x + 1$ trees is concerned. This is illustrated in the following theorems.

**6.3. $3x + 1$ Backwards Model Prediction: Tree Sizes.** Concerning the number of nodes $N_k(\omega)$ in a realized tree at depth $k$, Lagarias and Weiss proved the following result [23, Corollary 3.1].

THEOREM 6.2. *($3x + 1$ Stochastic Tree Size) For all $j \geq 0$, a realization $\omega$ of a tree grown in the $3x + 1$ branching random walk model $\mathcal{B}[3^j]$ has*

$$(6.65) \qquad \lim_{k \to \infty} \frac{1}{k} \left( \log N_k(\omega) \right) = \log \left( \frac{4}{3} \right), \quad \text{for almost every } \omega.$$

This result only uses the Galton-Watson structure built into the process $\mathcal{B}[3^j]$. Its prediction is consistent with the rigourous results on average tree size for pruned $3x + 1$ trees given in Theorem 6.1, and it also supports Conjecture 6.1.

**6.4. $3x + 1$ Backwards Model Prediction: Extremal Total Stopping Times.** Next, as a statistic that corresponds to an extremal trajectory, consider the *first birth in generation $k$*, which is the leftmost individual on the line at depth $k$ in the branching random walk. Denote the location of this individual by $L_k^*(\omega)$, for a given realization $\omega$ of the random walk. Lagarias and Weiss [23, Theorem 3.4] proved the following result.

THEOREM 6.3. *(Asymptotic First Birth Location) For any $3x + 1$ branching random walk model $\mathcal{B}[3^j]$ with $j \geq 2$, there is a constant $\beta_{BP}$ such that for all $j \geq 0$, the branching random walk $\mathcal{B}[3^j]$ has asymptotic first birth (leftmost birth)*

$$(6.66) \qquad \lim_{k \to \infty} \frac{L_k^*(\omega)}{k} = \beta_{BP} \quad \text{for almost every } \omega.$$

*This constant $\beta_{BP} \approx 0.02399$ is determined uniquely by the properties that it is the unique $\beta > 0$ that satisfies*

$$(6.67) \qquad \tilde{g}(\beta) = 0$$

*where*

$$(6.68) \qquad \tilde{g}(a) := -\sup_{\theta \leq 0} \left( a\theta - \log M_{BP}(\theta) \right).$$

*Here $M_{BP}(\theta)$ is the branching process moment generating function*

$$(6.69) \qquad M_{BP}(\theta) := 2^\theta + \frac{1}{3} \left( \frac{2}{3} \right)^\theta.$$

Since the first birth individual at depth $k$ corresponds to taking $k$ iterations to reach the root node, we can define a *branching process scaled stopping limit* $\gamma_{BP}(\omega)$. This is the BP model's prediction for the scaled stopping constant $\gamma$ from (2.16), defined by

$$\gamma_{BP}(\omega) := \limsup_{k \to \infty} \frac{k}{L_k^*(\omega)}.$$

Theorem 6.3 implies that this value is constant (almost surely independent of $\omega$), and takes the value

$$(6.70) \qquad\qquad \gamma_{BP} = (\beta_{BP})^{-1}.$$

Note that since $\beta_{BP} \approx 0.02399$, we have $1/\beta_{BP} \approx 41.7$. At this point we have two completely different predictions for the scaled stopping constant $\gamma$, one from the RRW model (cf. Theorem 4.1) which approximates forward iterations, and another from the BP models which estimate backwards iterations. Applegate and Lagarias then prove [**23**, Theorem 4.1] the following striking identity.

THEOREM 6.4. ($3x + 1$ Random Walk-Branching Random Walk Duality) *The* $3x + 1$ *repeated random walk (RRW) stochastic model scaled stopping time limit* $\gamma_{RRW}$ *and the* $3x + 1$ *branching random walk (BP) model* $\mathcal{B}[3^j]$ *with* $j \geq 0$, *scaled stopping time limit* $\gamma_{BP}$ *are identical! I.e.,*

$$(6.71) \qquad\qquad \gamma_{RRW} = \gamma_{BP}.$$

PROOF. This is a consequence [**23**] of an identity relating the moment generating functions associated to the two models, which is $M_{BP}(\theta) = M_{RRW}(\theta + 1)$; compare (4.39) and (6.69).                                          □

REMARK. Recall the critique of the RRW model given in §4.4, that various trajectories coalesce in their forward iterates. But the BP models, by their tree construction, completely take into account the dependence caused by coalescing trajectories! Since both models predict the same exact value for $\gamma$, it appears the critique has been thwarted off.

## 6.5. $3x+1$ Backwards Model Prediction: Total Preimage Counts.

We next consider what the branching process models have to say about the number of integers below $x$ that eventually iterate to a given integer $a$.

The following result gives, for the simplest branching random walk model, an almost sure asymptotic of the number of inverse iterates of size below a given bound ([**23**, Theorem 4.2]).

THEOREM 6.5. (Stochastic Inverse Iterate Counts) *For a realization* $\omega$ *of the branching random walk* $\mathcal{B}[1]$, *let* $I^*(t; \omega)$ *count the number of progeny located at positions* $S(\omega_{k,j}) \leq x$, *i.e.*

$$(6.72) \qquad I^*(x; \omega) := \#\{\omega_{k,j} : S(\omega_{k,j}) \leq x, \text{for any } k \geq 1, \ 1 \leq j \leq N_k(\omega)\}.$$

*Then the asymptotic estimate*

$$(6.73) \qquad\qquad I^*(x; \omega) = x^{1+o(1)} \ as \ x \to \infty$$

*holds almost surely.*

The model statistic $I^*(x; \omega)$ functions as a proxy for the function $\pi_a(x)$, where $\log a$ gives the position of the root node of the branching random walk. This result is the stochastic analogue of Conjecture 2.1 about the $3x + 1$ growth exponent.

## 7. The $5x + 1$ Function: Symbolic Dynamics and Orbit Statistics

We now turn for comparison to the $5x + 1$ iteration. Some features of the dynamics of this iteration are similar to that of the $3x + 1$ problem, and some are different. Here the dynamics of iteration in the long run are expected to be quite different globally from the $3x+1$ problem; most trajectories are expected to diverge.

In this section we formulate several orbit statistics for this map, some the same as for the $3x + 1$ map, and some changed. We review basic results on them.

**7.1.  $5x + 1$ Forward Iteration: Symbolic Dynamics.** The basic features of the $5x + 1$ problem are similar to the $3x + 1$ problem. We introduce the parity sequence

$$(7.74) \qquad S_5(n) := (n \ (\text{mod} \ 2), T_5(n) \ (\text{mod} \ 2), T_5^{(2)}(n) \ (\text{mod} \ 2), ...).$$

The symbolic dynamics is similar to the $3x + 1$ map: all finite initial symbol sequences of length $k$ occur, each one for a single residue class (mod $2^k$).

THEOREM 7.1.  ($5x + 1$ Parity Sequence Symbolic Dynamics) *The $k$-truncated parity sequence $S_5^{[k]}(n)$ of the first $k$ iterates of the $5x + 1$ map $T(x)$ is periodic in $n$ with period $2^k$. Each of the $2^k$ possible $0 - 1$ vectors occurs exactly once in the initial segment $1 \le n \le 2^k$.*

PROOF.  The proof of this result exactly parallels that of Theorem 2.1.    □

As before, the parity sequence of an orbit of $x_0$ uniquely determines $x_0$.

Analysis of this recursion, assuming even and odd iterates are equally likely, as prescribed by Theorem 7.1, we find the logarithms of iterates grow in size on the average.

**7.2.  $5x+1$ Forward Iteration: $\lambda^+$- Stopping Times.** Most $5x+1$ iteration sequences grow on average, rather than shrinking on average. An appropriate notion of stopping time for this situation is as follows.

DEFINITION 7.1.  *For fixed $\lambda \ge 1$, the $\lambda^+$-stopping time $\sigma_\lambda^+(n)$ of a map $T_5 : \mathbb{Z} \to \mathbb{Z}$ for input $n$ is the minimal value of $k \ge 0$ such that $T_5^{(k)}n > \lambda n$, e.g.*

$$(7.75) \qquad \sigma_\lambda^+(n) := \inf\{k \ge 0 : \frac{T_5^{(k)}(n)}{n} > \lambda\}.$$

*If no such value $k$ exists, we set $\sigma_\lambda^+(n) = +\infty$.*

One now has the following result, which parallels Theorem 2.2 for the $3x + 1$ map, except that here iterates grow in size rather than shrink in size.

THEOREM 7.2.  ($\lambda^+$-Stopping Time Natural Density)

*(i) For the $5x + 1$ map $T_5(n)$, and fixed $\lambda \ge 1$ and $k \ge 1$ , the set $S_\lambda^+(k)$ of integers having $\lambda^+$-stopping time at most $k$ has a well-defined natural density $\mathbb{D}(S_\lambda^+(k))$.*

*(ii) This natural density satisfies*

$$(7.76) \qquad \lim_{k \to \infty} \mathbb{D}(S_\lambda^+(k)) = 1.$$

*In particular, the set of numbers with finite $\lambda^+$-stopping time has natural density 1.*

PROOF.  Claim (i) follows using the Parity Sequence Theorem 7.1. Here the set is a finite union of arithmetic progressions (mod $2^k$), except a finite number of initial elements may be omitted from each such progression.

The result (ii) can be established by a similar argument to that used for the $3x + 1$ problem in Theorem 2.2.    □

Here we note a surprise: there are infinitely many exceptional integers $n$ that have $\lambda^+$-stopping time equal to $+\infty$! This occurs because the $5x+1$ problem has a periodic orbit $\{1, 3, 8, 4, 2\}$, and infinitely many positive seeds $n_0$ eventually enter this orbit, e.g. $n_0 = \frac{2^{4k}-1}{5}$ for any $k \geq 2$. All of these integers have $\sigma_\lambda^+(n_0) = +\infty$. Nevertheless Theorem 7.2 asserts that such integers have natural density zero.

**7.3. $5x+1$ Stopping Time Statistics: Total Stopping Times.** The $5x+1$ problem has a finite orbit containing 1, and we may define total stopping time as for the $3x+1$ function.

DEFINITION 7.2. *For $n \geq 1$ the* total stopping time $\sigma_\infty(n; T_5)$ *of the $5x+1$ function is given by*

$$(7.77) \qquad \sigma_\infty(n; T_5) := \inf\{k \geq 1 : T_5^{(k)}(n) = 1\}.$$

*We set $\sigma_\infty(n; T_5) = +\infty$ if no finite $k$ has this property.*

Here we expect that the vast majority of positive $n$ will belong to divergent trajectories, and only a small minority of $n$ have a well-defined finite value $\sigma_\infty(n; T_5) < \infty$. It is an open problem to prove that *even a single trajectory* (such as that emanating from the starting seed $n_0 = 7$) is divergent!

The best we can currently show unconditionally is a lower bound on the size of the extremal total stopping time that grows proportionally to $\log n$.

THEOREM 7.3. (Lower Bound for $5x + 1$ Total Stopping Times) *There are infinitely many $n$ whose total stopping time satisfies*

$$(7.78) \qquad \sigma_\infty(n, T_5) \geq \left(\frac{\log 2 + \log 5}{(\log 2)^2}\right) \log n \approx 4.79253 \log n.$$

PROOF. The Parity Sequence Theorem 6.2 implies there is at least one odd number $n_k$ with $1 \leq n_k < 2^k$ whose first $k - 1$ iterates are also odd, so that $T_5^{(k)}(n_k) \geq (\frac{5}{2})^k n_k$. Since a single step can divide by at most 2, we necessarily have (using $\log n_k \leq k \log 2$),

$$\frac{\sigma_\infty(n_k, T_5)}{\log n_k} \geq \frac{k}{\log n_k} + \left(\frac{k \log \frac{5}{2} + \log n_k}{\log 2}\right) \frac{1}{\log n_k} \geq \frac{2}{\log 2} + \frac{\log \frac{5}{2}}{(\log 2)^2} \approx 4.79253.$$

We do not know if these numbers $n_k$ have a finite total stopping time.    □

The methods of Applegate and Lagarias [4] for $3x + 1$ trees can potentially be applied to this problem, to further improve this lower bound, and to establish it for numbers $n$ having a finite total stopping time.

An interesting challenge is whether one can show for each $c > 0$ that *only a density zero set of $n$ have $\frac{\sigma_\infty(n; T_5)}{\log n} < c$*. A stochastic model in §8.9 predicts that all but finitely many trajectories having $\sigma_\infty(n) > 85 \log n$ will necessarily have $\sigma_\infty(n) = +\infty$, so establishing this for $c = 85$ would be consistent with the prediction that only a density zero set of $n$ have 1 in their forward orbit under $T_5$.

**7.4. $5x + 1$ Size Statistics: Minimum Excursion Values.** In the topsy-turvy world of the $5x + 1$ problem, since most trajectories get large, our substitute for the maximum excursion constant is the following reversed notion.

DEFINITION 7.3. *For an integer $n$ the* minimal excursion value $t^-(n)$ *of the $5x + 1$ function is given by*

$$(7.79) \qquad t^-(n) := \inf\{|T_5^{(k)}(n)| : k \geq 0\}.$$

We have $t^-(0) = 0$, while infinitely many $n$ will have minimum excursion value equal to 1.

DEFINITION 7.4. *For $n \geq 1$ the* minimal excursion constant $\rho_5^-(n)$ *of the $5x+1$ function is given by*

$$(7.80) \qquad \rho_5^-(n) := \liminf_{n \to \infty} \frac{\log t^-(n)}{\log n}.$$

We now immediately have the following result.

THEOREM 7.4. *The $5x + 1$ minimum excursion constant is given by*

$$(7.81) \qquad \rho_5^- = 0.$$

PROOF. The inverse orbit of $n = 1$ for $T_5$ contains $\{2^j : j \geq 1\}$, whence $t^-(2^j) = 1$. □

We state this easy result as a theorem, because it has the remarkable feature, among all the constants associated to these $3x + 1$ and $5x + 1$ maps, of being unconditionally proved! It also has the interesting feature that the stochastic models below make an incorrect prediction in this case, cf. Theorem 8.4.

**7.5. $5x + 1$ Count Statistics: $5x + 1$ Tree Sizes.** In considering backwards iteration of the $5x + 1$ function, we can ask: given an integer $a$ how many numbers $n$ iterate forward to $a$ after exactly $k$ iterations, that is, $T_5^{(k)}(n) = a$?

The set of backwards iterates of a given number $a$ can again be pictured as a tree; we call these $5x + 1$ *trees*. Now $N_k(a)$ counts the number of leaves at depth $k$ of the tree with root node $a$, and $N_k^*$ counts the number of leaves in a *pruned* $5x + 1$ *tree*, which is one from which all nodes with label $n \equiv 0 \pmod 5$ have been removed. The definitions are as follows.

DEFINITION 7.5. *(1) Let $N_k(a; T_5)$ count the number of integers that forward iterate under the $5x + 1$ map $T_5(n)$ to $a$ after exactly $k$ iterations, i.e.*

$$(7.82) \qquad N_k(a; T_5) := |\{n : T_5^{(k)}(n) = a\}|.$$

*(2) Let $N_k^*(a; T_5)$ count the number of integers not divisible by 5 that forward iterate under the $5x + 1$ map $T_5(n)$ to $a$ after exactly $k$ iterations, i.e.*

$$(7.83) \qquad N_k^*(a; T_5) := |\{n : T_5^{(k)}(n) = a, \ n \not\equiv 0 \pmod 5)\}|.$$

The case $a = 1$ is of particular interest, since the quantities then count integers that iterate to 1, and in this case we let

$$N_{k,5} := N_k(1; T_5), \ N_{k,5}^* := N_k^*(1; T_5).$$

DEFINITION 7.6. *(1) For a given a the $5x + 1$ tree growth constant $\delta_5(a)$ for a is given by*

$$(7.84) \qquad \delta_5(a) := \limsup_{k \to \infty} \frac{1}{k} \left( \log N_k(a; T_5) \right).$$

*(2) The $5x + 1$ tree growth constant $\delta_5 = \delta_5(1)$.*

The constant $\delta_5(a)$ exists and is finite, as follows from the same upper bound as in (2.23).

The following result gives information on the sizes of depth $k$ pruned $5x + 1$ trees over all possible tree types.

THEOREM 7.5.   (Structure of Pruned $5x + 1$ Trees)
*(1) For $k \geq 1$ and $a \not\equiv 0( \bmod 5)$, the structure of the pruned level $k$ tree $\overline{\mathcal{T}}^*_k(a)$, and hence the number $N^*_k(a; T_5)$, is completely determined by $a$ (mod $5^{k+1}$).*

*(2) There are $4 \cdot 5^k$ residue classes $a$ (mod $5^{k+1}$) with $a \not\equiv 0$ (mod 5). For these*

$$(7.85) \qquad \sum_{\substack{a \ (\bmod \ 5^{k+1}) \\ a \not\equiv 0 \ (\bmod \ 5)}} N^*_k(a; T_5) = 4 \cdot 6^k.$$

It follows that if a residue class $a$ (mod $5^{k+1}$) with $a \not\equiv 0(\bmod 5)$ is picked with the uniform distribution, the expected number of leaves in the random tree $\overline{\mathcal{T}}^*_k(a)$ is exactly $\left( \frac{6}{5} \right)^k$.

PROOF. This result is shown by a method exactly similar to the $3x + 1$ tree case ([**23**, Theorem 3.1]). We omit details.                                    □

Theorem 7.5 suggests the following conjecture.

CONJECTURE 7.1. *For each $a \not\equiv 0$ (mod 5), the $5x + 1$ tree growth constant $\delta_5(a)$ is given by*

$$(7.86) \qquad \delta_5(a) = \log \left( \frac{6}{5} \right).$$

Compare this conjecture with the prediction of Theorem 8.7.

**7.6. $5x + 1$ Count Statistics: Total Inverse Iterate Counts.** In considering backwards iteration of the $5x + 1$ function from an integer $a$, the complete data is the set of integers that contain $a$ in their forward orbit. The following function describes this set.

DEFINITION 7.7.   *Given an integer $a$, the* inverse iterate counting function *$\pi_{a,5}(x)$ counts the number of integers $n$ with $|n| \leq x$ that contain $a$ in their forward orbit under the $3x + 1$ function. That is*

$$(7.87) \qquad \pi_{a,5}(x) := \#\{n : \ |n| \leq x \text{ such that some } T_5^{(k)}(n) = a, \ k \geq 0\}.$$

The inverse tree methods for the $3x + 1$ problem carry over to the $5x + 1$ problem, so that one can obtain a result qualitatively of the following type, by similar proofs.

THEOREM 7.6. (Inverse Iterate Lower Bound) *There is a positive constant $c_1$ such that the following holds. For each $a \not\equiv 0 \pmod 5$, there is some $x_0(a)$ such that for all $x \geq x_0(a)$,*

$$(7.88) \qquad\qquad \pi_{a,5}(x) \geq x^{c_1}.$$

The following statistics measure the size of the inverse iterate set in the sense of fractional dimension.

DEFINITION 7.8.  *Given an integer $a$, the upper and lower $5x + 1$ growth exponents for $a$ are given by*

$$\eta_5^+(a) := \limsup_{x \to \infty} \frac{\log \pi_{a,5}(x)}{\log x},$$

*and*

$$\eta_5^-(a) := \liminf_{x \to \infty} \frac{\log \pi_{a,5}(x)}{\log x}.$$

*If these quantities are equal, we define the $5x + 1$ growth exponent $\eta_5(a)$ to be $\eta_5(a) = \eta_5^+(a) = \eta_5^-(a)$.*

In parallel with conjectures for the $3x + 1$ map, we formulate the following conjecture.

CONJECTURE 7.2.  ($5x + 1$ Growth Exponent Conjecture) *For all integers $a \not\equiv 0 \pmod 5$, the $3x + 1$ growth exponent $\eta_5(a)$ exists, and takes a constant value $\eta_5$ independent of $a$. This value satisfies*

$$(7.89) \qquad\qquad \eta_5 < 1.$$

The stochastic models discussed in §8 suggest that the constant $\eta_5$ exists and takes a value strictly smaller than 1. There is some controversy concerning the conjectured value of the constant. In §8 we present a repeated random walk model and a branching random walk model that both suggest the value $\eta_5 \approx 0.649$. A different branching random walk model formulated by Volkov [**40**] suggests the value $\eta_5 \approx 0.678$. Lower bounds toward this conjecture can be rigorously established, cf. Theorem 7.6 above. We have not bothered to determine $c_1$ in (7.88), though we suspect it is well below either of the above predictions, and hence cannot distinguish between them.

## 8. $5x + 1$ Function: Stochastic Models and Results

We now discuss stochastic models for the $5x + 1$ problem paralleling those for the $3x + 1$ problem. These include random walk models for forward iteration of the $5x + 1$ map, analysis of the accelerated $5x + 1$ map, and branching random walks for the backwards iteration of the $5x + 1$ map.

### 8.1. $5x+1$ Forward Iteration: Multiplicative Random Product Model.
Concerning forward iteration, we may formulate a multiplicative random product model parallel to that in §3. Consider the random products

$$Y_k := X_1 X_2 \cdots X_k,$$

in which the $X_i$ are each independent identically distributed (i.i.d.) random variables $X_i$ having the discrete distribution

$$X_i = \begin{cases} \dfrac{5}{2} & \text{with probability } \tfrac{1}{2}, \\[2ex] \dfrac{1}{2} & \text{with probability } \tfrac{1}{2}. \end{cases}$$

We call this the $5x + 1$ *multiplicative random product* (MRP) model.

As before, this model does not include the choice of starting value of the iteration, which would correspond to $X_0$; the random variable $Y_k$ really models the *ratio* $\frac{T_5^{(k)}(X_0)}{X_0}$. We define for $\lambda^+ \geq 1$ the $\lambda^+$-*stopping time random variable*

$$(8.90) \qquad\qquad V_\lambda^+(\omega) := \inf\{k : Y_k \geq \lambda\}.$$

where $\omega = (X_1, X_2, X_3, \cdots)$ denotes a sequence of random variables as above. This random vector $\omega$ models the change in size of a random starting value $n = X_0$ that occurs on iteration of the $5x + 1$ map.

This stochastic model can be used to exactly account for the density of $\lambda^+$-stopping times, as follows.

THEOREM 8.1. ($\lambda^+$-Stopping Time Density Formula) *For the $5x+1$ map $T_5(n)$ and any fixed $\lambda > 1$, the natural density $\mathbb{D}(S_\lambda(k))$ for integers having $\lambda^+$-stopping time at most $k$ is given exactly by the formula*

$$(8.91) \qquad\qquad \mathbb{D}(S_\lambda^+(k)) = \mathbb{P}[V_\lambda^+(\omega) \leq k],$$

*in which $V_\lambda^+$ is the $\lambda^+$-stopping time random variable in the multiplicative random product (MRP) model.*

PROOF. This follows by a parallel argument to that in Borovkov and Pfeifer [**10**, Theorem 3] for the $3x + 1$ problem.                                                     $\square$

Theorem 8.1 is the stochastic model parallel of Theorem 7.2.

**8.2. $5x + 1$ Forward Iteration: Additive Random Walk Model.** We next formulate additive random walk models, obtained after logarithmic rescaling of the $5x + 1$ iteration. The $5x + 1$ iteration takes $x_0 = n$ and $x_k = T_5^{(k)}(n)$. Using a logarithmic rescaling with $y_k = \log x_k$ (natural logarithm) we have

$$y_k = \log x_k := \log T^{(k)}(n).$$

Then we have

$$(8.92) \qquad y_{k+1} = \begin{cases} y_k + \log \tfrac{5}{2} + e_k & \text{if } x \equiv 1 \ (\text{mod } 2), \\[2ex] y_k + \log \tfrac{1}{2} & \text{if } x \equiv 0 \ (\text{mod } 2), \end{cases}$$

with

$$(8.93) \qquad\qquad e_k := \log\left(1 + \frac{1}{5x_k}\right).$$

Here $e_k$ is small as long as $|x_k|$ is large.

We approximate the deterministic process above with the following random walk model with unequal size steps. We take random variables

$$W_k := -\log 2 + \delta_k \log 5,$$

in which $\delta_k$ are i.i.d. Bernoulli random variables. The random walk positions $\{Z_k : k \geq 0\}$, are then random variables having starting value $Z_0 = \log m$, for some fixed initial condition $m > 1$, and with

$$Z_k = Z_0 + W_1 + W_2 + \cdots + W_k.$$

The $Z_k$ define a biased random walk, whose expected drift $\mu$ is given by

$$\mu := E[W_k)] = -\log 2 + \frac{1}{2}\log 5 = \frac{1}{2}\log\left(\frac{5}{4}\right) \approx 0.11157.$$

The variance $\sigma$ of each step is given by

$$\sigma := \mathrm{Var}[W_k] = \frac{1}{2}\log 5 \approx 0.80472.$$

Call this random walk the $5x+1$ *Biased Random Walk Model* ($5x+1$ *BRW Model*).

Since the mean of this random walk is positive, this biased random walk has a *positive drift*. This positive drift implies that a random trajectory diverges with probability one.

THEOREM 8.2.  *For the $5x + 1$ BRW model, with probability one, a trajectory $\{Z_k : k \geq 0\}$ diverges to $+\infty$.*

PROOF.  This is an elementary fact about random walks with positive drift.  □

This result implies that a generic trajectory has total stopping time equal to $+\infty$. That is, starting from $Z_0 = \log n$, the probability $\mathbb{P}[E_n]$ of the event $E_n$ that for some $k \geq 1$, the total stopping time condition $Z_k \leq 0$ is satisfied, is strictly smaller than 1, i.e., $\mathbb{P}[E_n] < 1$. It is positive but decreases to 0 as $n$ increases to $+\infty$. (To not confuse this fact with Theorem 8.2, even if $Z_k$ dips below 0, it charges back up to infinity, almost surely.)

To obtain a result parallel to those of §3 on the average behavior of numbers $n$ having a finite total stopping time, one needs to condition on the set of $n$ that have a finite total stopping time. This appears an approachable problem, but requires a more complicated analysis than that given in [23] or Borovkov and Pfeifer [10].

**8.3. $5x + 1$ Forward Iteration: Repeated Random Walk Model.** Next, paralleling §4, we formulate a $5x + 1$ *Repeated Random Walk (RRW) model* as follows. A model trial is the countable set of random variables

(8.94)                    $$\omega := \{Z_{k,n} : k \geq 0, n \geq 1\},$$

having initial condition $Z_{0,n} = \log n$, with the individual random walks being $5x+1$ biased random walks, as above. In the following subsections we consider other predictions that RRW model makes for various statistics.

THEOREM 8.3.  *For the $5x + 1$ RRW model, with probability one, for every $n \geq 1$ the trajectory $\{Z_{k,n} : k \geq 0\}$ diverges to $+\infty$.*

PROOF.  This follows immediately from Theorem 8.2, since the complement of this event is a countable union of measure zero events.  □

One might misinterpret the above as suggesting that the $5x + 1$ RRW model predicts that *all* trajectories are unbounded. Of course this is an incorrect prediction. The $5x + 1$ iteration has some finite cycles, and furthermore there are infinite number of integers that eventually enter one of these cycles. The stochastic model above cannot account for such bounded trajectories! Instead we interpret the stochastic model prediction to be that a *density one* set of integers lie on unbounded trajectories.

This should make you *very worried* about relying on stochastic models to predict that $3x + 1$ trajectories decay! There could potentially be a set of measure zero escaping to infinity, which the model simply cannot see. Such a pathological trajectory is the *heart and soul* of the $3x+1$ problem, and root cause of its difficulty!

**8.4. $5x + 1$ RRW Model Prediction: Minimum Excursion Constant.**
The $5x + 1$ RRW model has the following analogues of minimal excursion values and of the minimum excursion constant.

DEFINITION 8.1. *For a realization $\omega = \{Z_{k,n} : k \geq 0, n \geq 1\}$ of the $5x + 1$ RRW model, the* minimal excursion value $t^-(n, \omega)$ *is given, for each $n \geq 1$, by*

$$(8.95) \qquad t^-(n, \omega) := \inf\{e^{Z_{k,n}} : k \geq 0\}.$$

Theorem 8.3 implies that with probability one the value $t^-(n, \omega)$ is well-defined and strictly positive.

DEFINITION 8.2. *For a realization $\omega$ of the $5x + 1$ RRW model, the* minimum excursion constant $\rho_5^-(\omega)$ *is given by*

$$(8.96) \qquad \rho_5^-(\omega) := \liminf_{n \to \infty} \frac{\log t^-(n, \omega)}{\log n}.$$

Now a large deviations analysis yields the following result.

THEOREM 8.4. *($5x + 1$ RRW Minimum Excursion Constant) For the $5x + 1$ RRW model, with probability one the quantities $t^-(n, \omega)$ are finite for every $n \geq 1$. In addition, with probability one the random quantity*

$$(8.97) \qquad \rho_{5,RRW}^-(\omega) := \liminf_{n \to \infty} \frac{\log t^-(n; \omega)}{\log n} = \liminf_{n \to \infty} \left( \inf_{k \geq 0} \frac{Z_{k,n}}{\log n} \right)$$

*equals the constant*

$$(8.98) \qquad \rho_{5,RRW}^- = 1 - \frac{1}{\theta^*} \approx -1.86466,$$

*in which $\theta^* \approx 0.3490813$ is the larger of the two real roots of the equation $M_{5,RRW}(\theta) = 1$, where $M_{5,RRW}(\theta) := \frac{1}{2}\left(2^\theta + \left(\frac{2}{5}\right)^\theta\right)$ is a moment generating function associated to the random walk.*

PROOF. This is proved by a large deviations argument similar to that used for the maximum excursion constant for the $3x + 1$ problem in Lagarias and Weiss [**23**, Theorem 2.3]. We sketch the main computation. We estimate the probability $P(r, H, x)$ on a single trial starting at $\log x$ of having

$$-Z_{r \log x, \log x} \geq H \log x.$$

We define $a$ by the condition $H = ar$ and find that the probability is given by Chernoff's bound as

$$P(r, H, x) = \exp\left(-g_{5,RRW}(a) r \log x (1 + o(1))\right),$$

in which

(8.99) $$g_{5,RRW}(a) := \sup_{\theta \in \mathbb{R}} \left(a\theta - \log M_{5,RRW}(\theta)\right)$$

is a large deviations rate function, which is the Legendre transform of the logarithm of the moment generating function $M_{5,RRW}(\theta) = \frac{1}{2}\left(2^\theta + \frac{5}{2}\right)^\theta$. The repeated random walk makes $x$ trials $1 \leq n \leq x$ so the probability of a success over these trials is $xP(r, H, x)$, and we want this to be at least $x^\epsilon$, so that a success occurs infinitely often as $x \to \infty$. (We also will let $\epsilon \to 0$, so we set it equal to zero in what follows.) We want therefore to maximize $H = ar$ subject to the constraint that $g_{5,RRW}(a)r \leq 1$. To maximize we may take $g(a)r = 1$, whence $r = \frac{1}{g(a)}$ can be used to eliminate the variable $r$. We now have the maximization problem to maximize $H := \frac{a}{g_{5,RRW}(a)}$ over $0 < a < \infty$. One finds an extremality condition for maximization which yields

$$H^* = \frac{1}{\theta(a^*)},$$

where $a^*$ achieves the maximum, and $\theta^*$ is the corresponding value in the Legendre transform. Uniqueness of the maximum follows from convexity properties of the function $\log M_{RRW}(\theta)$. Detailed error estimates are also needed to verify that this the maximum gives the dominant contribution. □

This constant $\rho_{5,RRW}^-$ found in Theorem 8.4 is *negative*, i.e. the minimum excursion in the model reaches a real number much smaller than 1! As a prediction for the $5x + 1$ problem, this disagrees with the exact answer for minimum excursion constant for the $5x + 1$ problem $\rho_5^- = 0$ given in Theorem 7.4.

We view this inaccurate prediction as stemming from the discrepancy that the $5x + 1$ function takes only values on the integer lattice, and that its additive correction term is not accounted for in this stochastic model. That is, the stochastic model will not necessarily make good predictions on behavior of an orbit once an orbit reaches a small value, e.g. $|x| < C$ for any fixed constant $C$. We may hope that the $5x + 1$ model still makes an accurate prediction concerns the question: how many integers reach some small value, for example reaching the interval $|x| < C$.

**8.5. $5x + 1$ RRW Model Prediction: Total Stopping Time Counts.** We can interpret the false prediction above for minimum excursions in a constructive way: as soon as a $5x + 1$ trajectory achieves a size $e^{Z_{k,n}} < 1$, it enters a periodic orbit. Therefore this condition can be treated as a "stopping time" condition that detects when a trajectory reaches the value 1.

THEOREM 8.5. ($5x + 1$ RRW Total Stopping Time Counts) *For the $5x + 1$ RRW model and for a given $\omega$, let*

$$S_\infty(\omega) := \{n \geq 1 : e^{Z_{k,n}} < 1 \text{ holds for some } k \geq 1\}.$$

*Collect those seeds $n$ whose trajectory according to $\omega$ "reaches 1". Let $\pi_5(\cdot; \omega)$ denote the corresponding counting function,*

$$\pi_5(x; \omega) := \#\{1 \leq n \leq x : n \in S_\infty(\omega)\}.$$

*Then*

$$\lim_{x\to\infty} \frac{\log \pi_5(x;\omega)}{\log x} = \eta_{5,RRW}, \qquad \textit{for almost every } \omega.$$

Here $\eta_{5,RRW} \approx 0.65049$ *is given by* $\eta_{5,RRW} = 1 - \theta_{5,RRW}$ *where* $\theta_{5,RRW} \approx 0.34951$ *is the unique positive solution to the equation*

$$(8.100) \qquad M_{5,RRW}(\theta) := \frac{1}{2}\left(2^\theta + \left(\frac{5}{2}\right)^\theta\right) = 1.$$

PROOF. This can be proved by a large deviations model similar in nature to those considered in Lagarias and Weiss [**23**, Theorem 2.4]. We sketch the main estimate. For $k = r \log x$, consider the probability $P(r,x)$ that for a single random walk $e^{Z_{k,\log x}} < 1$. Since we make $x$ draws for $1 \le n \le x$ in the repeated random walk, the expected number of such individuals satisfying this property will be $xP(r,x)$. This probability is estimated using Chernoff's bound to be

$$P(r,x) = \exp\left(-g_{5,RRW}(a) r \log x (1 + o(1))\right),$$

where $a = \frac{1}{r}$, and $g_{5,RRW}$ is the large deviations rate function (8.99) in Theorem 8.4. We now maximize this probability over $r$. To do this we eliminate $r$ using $r = \frac{1}{a}$, so we want to determine

$$\tau_{5,RRW} := \min_{0 \le a < \infty} \frac{g_{5,RRW}(a)}{a}.$$

Then we obtain $xP(r,x) \le x^{1-\tau_{5,RRW}+o(1)}$ for all $r$, with equality holding for $r = \frac{1}{a^*}$ where $a^*$ be the value that attains the maximum of $f(a) := \frac{g_{5,RRW}(a)}{a}$ taken on the positive half-line. The extremality conditions for the minimum leads to the condition $M_{RRW}(\theta(a^*)) = 1$, where $\theta$ is the Legendre transform variable, and also to the identity

$$\tau_{5,RRW} = \frac{g_{5,RRW}(a^*)}{a^*} = \theta(a^*) := \theta_{5,RRW}.$$

The strict convexity of the function $\log M_{RRW}(\theta)$ is used to get a unique minimum, with $\eta_{5,RRW} = 1 - \tau_{5,RRW}$. For a rigorous proof, one must control various error estimates to show the dominant contribution to the probability comes from a small region near $a^*$. □

REMARK. The value of $\theta_{5,RRW}$ in the minimization problem in the proof of Theorem 8.5 turns out to be identical to that in the maximization problem that is needed for proving Theorem 8.4.

**8.6. $5x+1$ Accelerated Forward Iteration: Brownian Motion.** Kontorovich and Sinai [**18**] extended the Structure Theorem (that is, Theorems 5.1 and 5.2) and the consequences on the Central Limit Theorem (Theorem 5.4) and geometric Brownian motion (Theorem 5.5) to a class of functions which they called $(d,g,h)$-maps. The case $d = 2$, $g = 5$, and $h = 1$ corresponds to the accelerated $5x+1$ function, $U_5(n)$.

The analogous distribution and Central Limit Theorems are proved in the same way, leading to the following.

THEOREM 8.6. *(Geometric Brownian Motion) The rescaled paths of the accelerated $5x+1$ map are those of a geometric Brownian motion with drift $\log(\frac{5}{4})$. By this we mean the following.*

*For an initial seed $x_0$ which is relatively prime to both 2 and 5, denote its iterates by $x_k := U_5^{(k)}(x_0)$, let $y_k := \log x_k$ and define the scaled variable*

$$\omega_k := \frac{y_k - y_0 - k\log(\frac{5}{4})}{\sqrt{2k}\log 2}.$$

*Partition the interval $[0,1]$ as $0 = t_0 < t_1 < \cdots < t_r = 1$, and set $k_j = \lfloor t_j k \rfloor$. Then for any $a_j < b_j$, $j = 1, \ldots, r$,*

$$\lim_{k \to \infty} \mathbb{P}\left[x_0 : a_j < \omega_{k_j} - \omega_{k_{j-1}} < b_j, \text{ for all } j = 1, 2, \ldots, r\right] = \prod_{j=1}^{r}\left(\Phi(b_j) - \Phi(a_j)\right),$$

*where $\Phi(a)$ is the cumulative distribution function for the standard normal distribution.*

PROOF. This is a consequence of Theorem 5 in Kontorovich-Sinai [**18**]. $\square$

REMARK. The accelerated drift, $\log(\frac{5}{4})$, is again double that of the Biased Random Walk model, which predicts a drift of $\frac{1}{2}\log(\frac{5}{4})$. A zero-mean, unit-variance Wiener process $W_t$ satisfies the "law of iterated logs" almost surely, that is:

$$\limsup_{t \to \infty} \frac{|W_t|}{\sqrt{2t \log \log t}} = 1,$$

with probability 1. Hence the drift being positive implies that almost every $5x+1$ trajectory escapes to infinity (yet we emphasize again that we do not know how to prove this for a single given trajectory!).

### 8.7. $5x+1$ Backwards Stochastic Models: Branching Random Walks.

We next formulate branching random walks to model the $5x+1$ iteration in exact analogy with the $3x+1$ models. We denote these models $\mathcal{B}[5^j]$ for $j \geq 0$.

$5x+1$ *Branching Random walk $\mathcal{B}[5^0]$.* There is one type of individual. With probability $\frac{4}{5}$ an individual has a single offspring located at a position shifted by $\log 2$ on the line from its progenitor, and with probability $\frac{1}{5}$ it has two offspring located at positions shifted $\log 2$ and $\log \frac{2}{5}$ on the line from their progenitor. If the progenitor is in generation $k-1$, the offspring are in generation $k$. The tree is grown from a single individual in generation 0, the root, with specified initial location $\log a$.

The more general models for $j \geq 1$ are given as follows.

$5x+1$ *Branching Random walk $\mathcal{B}[5^j], (j \geq 1)$.* There are $p = 4 \cdot 5^{j-1}$ types of individuals, indexed by residue classes $a \pmod{5^j}$ with $a \not\equiv 0 \pmod 5$. The distribution of offspring of an individual of type $a \pmod{5^j}$, at any given generation (or depth) $k$ in the branching, is determined as follows: Suppose $a \pmod{5^j}$ is the type of a node at depth $k-1$. Now regard it as being, with probability $\frac{1}{5}$ each, one of the five possible residue classes $\tilde{a} \pmod{5^{j+1}}$ consistent with its class $\pmod{5^j}$. A tree of depth 1 having $\tilde{a}$ as root node, then has either one or two progeny, at depth 1, given by $(T^*)^{-1}(\tilde{a})$, whose node labels are well-defined classes $\pmod{5^j}$, either $2\tilde{a}$ or, if it legally occurs, $\frac{2\tilde{a}-1}{3}\pmod{5^j}$. The branching random walk then produces an individual of type $2\tilde{a}$ at generation $k$ whose position is additively shifted by $\log 2$ from that of the generation $k-1$ progenitor node of type $\tilde{a}$ plus, if legal, another node of type $\frac{2\tilde{a}-1}{5} (\mathrm{mod} \ 5^j)$, which is shifted in position by $\log(\frac{2}{5})$ on the line from that of the generation $k-1$-node. The tree is grown from a single individual at depth 0, with specified type and location $\log a$.

Just as in the $3x+1$ branching random walk models, the behavior of the random walk part of the model can completely reconstructed from knowing the type of each node.

For the rest of this section, let $\omega$ denote a single realization of such a branching random walk $\mathcal{B}[5^j]$ which starts from a single individual $\omega_{0,1}$ of type 1 $(\mathrm{mod} \ 5^j)$ at depth 0, with initial position label $\log|a|$. Here $\omega$ describes a particular infinite tree. We let $N_k(\omega)$ denote the number of individuals at level $k$ of the tree. We let $S(\omega_k, j)$ denote the position of the $j$-th individual at level $k$ in the tree, for $1 \le j \le N_k(\omega)$.

These models are *supercritical branching processes* exactly as for the $3x+1$ case: In *every* random realization $\omega$, the number of nodes at level $d$ grows exponentially in $d$, and there are no extinction events.

In terms of growth of trees of inverse iterates, these models will accurately represent certain features of $5x+1$ trees, and not others. They might accurately describe tree sizes. However these branching random walks very likely do not accurately model positions of inverse iterates of the $5x+1$ in certain crucial ways. Namely, individuals whose branching walk position is negative (corresponding to a $5x+1$ iteration value $x$ falling in the interval $(0,1)$) are where the correction term $e_k$ in (8.93) in the $5x+1$ iteration becomes significant, breaking the size connection of the model iterates and the $5x+1$ iterates.

We now give some quantities of the trees associated to a realization $\omega$ of the branching random walk $\mathcal{B}[5^j]$. We let $N_k := N_k(\omega)$ denote the number of individuals in generation $k$, and let $\{\omega_{k,i} : 1 \le i \le N_k(\omega)\}$ denote the set of all individuals in generation $i$, ordered by their branching random walk locations on the line, denoted

$$L(\omega_{k,1}) \le L(\omega_{k,2}) \le \cdots \le L(\omega_{k,N_k}).$$

The *size* of the element $\omega_{k,i}$, viewed as analogues of the $5x+1$ iterates, is the exponentiated quantity

(8.101)                         $$Z_{k,i} := e^{L(\omega_{k,i})}.$$

The branching random walk has the property that the sizes of most individuals in a tree will tend to get larger. (This initially seems rather surprising, but note that if a forward orbit is unbounded, then necessarily all backward orbits leading to it must be unbounded as well!) We are interested in individuals whose size under the $5x+1$ iteration is around a given value $x$. The tree models will detect individuals whose size is larger than $x$.

In the following subsections we address for the $5x+1$ branching random walk models the following questions.

1. What is the exponential growth rate of the quantities $N_k(\omega)$, as a function of $k$?

2.  What is the maximum level $k$ that has some individual $Z_{k,i} \leq x$? This requires analyzing the size of the *first birth location* $L(\omega_{k,1})$.

3.  How does the total number of individuals $\pi_5(x; \omega)$ in the $5x+1$ tree having location $Z_{k,i} \leq x$ grow as a function of $x$?

**8.8. Backwards Iteration Prediction: $5x+1$ Tree Counts.** The size of $5x+1$ trees can be estimated for these models $\mathcal{B}[5^j]$ , as follows.

THEOREM 8.7.  ($5x+1$ Stochastic Tree Size) *For all $j \geq 0$ a realization $\omega$ of a tree grown in the $5x+1$ branching random walk model $\mathcal{B}[5^j]$ satisfies*

$$(8.102) \qquad \lim_{k \to \infty} \frac{1}{k} \left( \log N_k(\omega) \right) = \log \left( \frac{6}{5} \right), \quad \textit{almost surely.}$$

PROOF.  This is proved in exactly similar fashion to the $3x+1$ stochastic model case in Lagarias and Weiss [**23**, Corollary 3.1].    □

This result only uses the Galton-Watson process branching structure built into the branching random walk $\mathcal{B}[5^j]$. It does not depend on the sizes of the iterates.

The conclusion of Theorem 8.7, viewed as a prediction of the growth behavior of $5x+1$ trees, is consistent with the rigourous results on average tree size for pruned $5x+1$ trees given in Theorem 7.5.

**8.9. Backwards Iteration Prediction: Extremal Finite Total Stopping Times.**  As indicated above, most integers for the $5x+1$ map will not have a finite total stopping time. However it is of interest to analyze the small subset of integers that do have a total stopping time; these are exactly the integers in the tree of inverse iterates of $a = 1$. We analyze what is the maximum generation $k$ that contains an individual having size $e^{L(\omega_{k,i})} \leq x$.

Denote the location of this first birth individual in generation $k$ by $L_k^*(\omega) := L(\omega_{k,1})$, for a given realization $\omega$ of the random walk.

THEOREM 8.8.  (Asymptotic $5x+1$ First Birth Location) *There is a constant $\beta_{5,BP}$ such that, for all $j \geq 1$, the branching random walk model $\mathcal{B}[3^j]$ has asymptotic first birth (leftmost birth)*

$$(8.103) \qquad \lim_{k \to \infty} \frac{1}{k} L_k^*(\omega) = \beta_{5,BP} \quad a.\ s.$$

*This constant $\beta_{5,BP} \approx 0.01179816$ is determined uniquely by the properties that it is the unique constant with $\beta > 0$ that satisfies*

$$(8.104) \qquad \overline{g}_{5,BP}(\beta) = 0,$$

*where*

$$(8.105) \qquad \overline{g}_{5,BP}(a) \quad := \quad -\sup_{\theta \leq 0} \left( a\theta - \log \left( 2^\theta + \frac{1}{5} (\frac{2}{5})^\theta \right) \right).$$

PROOF.  This is proved by an argument analogous to the $3x+1$ case analyzed in Lagarias and Weiss [**23**, Theorem 3.4], cf. Theorem 6.3. Here we use a branching process (inverse) moment generating function

$$(8.106) \qquad M_{5,BP}(\theta) := 2^\theta + \frac{1}{5} \left( \frac{2}{5} \right)^\theta.$$

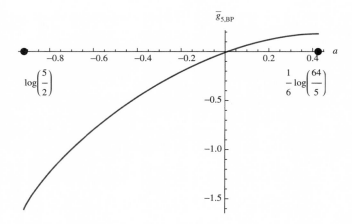

FIGURE 8.8. A plot of $a$ versus $\bar{g}_{5,BP}(a)$, in the range $\log(2/5) < a < \frac{1}{6}\log(64/5)$.

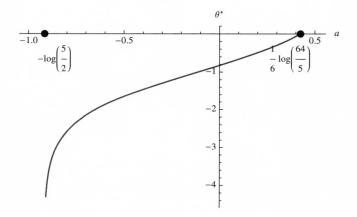

FIGURE 8.9. A plot of $a$ versus $\theta^*$, in the range $\log(2/5) < a < \frac{1}{6}\log(64/5)$.

in computing the rate function $\bar{g}_{5,BP}(a)$. We note that $\bar{g}_{5,BP}(a)$ is increasing for $\log\frac{2}{5} < a < \frac{1}{6}\log\frac{64}{5}$, (see Figure 8.8) and on this range the value $\theta^* := \theta(a)$ achieving the extremum in (8.105) is an increasing function of $a$, reaching the value $\theta = 0$ at the upper endpoint (see Figure 8.9). We have $\bar{g}_{5,BP}(a) = \log(\frac{6}{5})$ for $\frac{1}{6}\log\frac{64}{5} \le a < \infty$.                                                                    □

Now one defines a *branching random walk stopping limit*

$$\gamma_{5,BP}(\omega) := \limsup_{k\to\infty} \frac{k}{L_k^*(\omega)}.$$

Theorem 8.8 implies that this value is constant almost surely, equaling a value $\gamma_{5,BP}$ given by

(8.107)
$$\gamma_{5,BP} = \frac{1}{\beta_{5,BP}} \approx 84.76012.$$

One can show the constants $\gamma_{5,BP}$ and $\gamma_{5,RRW}$ agree, just as for the $3x + 1$ stochastic models.

THEOREM 8.9. *($5x + 1$ Random Walk-Branching Random Walk Duality) The $5x + 1$ repeated random walk (RRW) scaled stopping time limit $\gamma_{5,RRW}$ and the branching random walk stopping limit $\gamma_{5,BP}$ for the $5x + 1$ branching random walk (BP) model $\mathcal{B}[5^j]$ with $j = 0$, are related by*

$$(8.108) \qquad \gamma_{5,RRW} = \gamma_{5,BP}.$$

PROOF. This result is proved using a relation between moment generating functions

$$M_{5,BP}(\theta) = M_{5,RRW}(\theta + 1),$$

compare (8.100) and (8.105). It is identical in spirit to the proof in Lagarias and Weiss [**23**, Theorem 4.1]. $\qquad \square$

The analogue of this result applied to the $5x+1$ problem would be the following heuristic prediction: *For any constant $\gamma > \gamma_{5,BP}$ all but finitely many trajectories having total stopping time $\sigma_\infty(n) > \gamma \log n$ necessarily have $\sigma_\infty(n) = +\infty$.* We could take $\gamma = 85$, for example.

**8.10. Backwards Iteration Prediction: Total Preimage Counts.** The following result gives, for the simplest branching random walk model $\mathcal{B}[5^0]$, an almost sure asymptotic for the number of inverse iterates of size below a given bound.

THEOREM 8.10. *(Stochastic Inverse Iterate Counts) For a realization $\omega$ of the branching random walk $\mathcal{B}[1]$, let $I^*(t; \omega)$ count the number of progeny located at positions $Z(\omega_{k,j}) \leq x$, i.e.*

$$(8.109) \qquad I^*(x; \omega) := \#\{\omega_{k,j} : Z(\omega_{k,j}) \leq x, \text{for any } k \geq 1, \ 1 \leq j \leq N_k(\omega)\}.$$

*This quantity satsfies with probability one the asymptotic estimate*

$$(8.110) \qquad I^*(x; \omega) = x^{\eta_{5,BP} + o(1)} \ as \ x \to \infty,$$

*in which $\eta_{5,BP} \approx 0.650919$ is the maximum value of $f(a) := \frac{1}{a}\overline{g}_{5,BP}(a)$ taken over the interval $0 \leq a < \frac{1}{6}\log\frac{64}{5}$.*

PROOF. This is proved by a large deviations argument similar to that used in Lagarias and Weiss [**23**, Theorem 4.2]. One counts the number of progeny at level $k$ for each level $k$ satisfying the bound, by estimating the probability that a random leaf satisfies the appropriate large deviations bound. One shows that this number peaks for $k \approx \theta_{5,BP} \log x$, where $\theta_{5,BP} = \frac{1}{a^*} \approx 9.19963$, where $a^* \approx 0.1087$ is the value of $a$ achieving the maximum above. One shows that the right side is an upper bound for all levels $k$, and that the sum total of levels $k > 100 \log x$ contribute negligibly to the sum. $\qquad \square$

The model statistic $I^*(x; \omega)$ functions as a proxy for the $5x + 1$ count function $\pi_a^*(x)$, where $\log|a|$ gives the position of the root node of the branching random walk. This result is the stochastic analogue of Conjecture 2.1 about the $3x + 1$ growth exponent. The argument above also makes the prediction is that the levels $k$ at which the bulk of the members of $\pi_a(x)$ occur has $k \approx \frac{1}{a^*}\log x$.

REMARK. An entirely different set of branching random walk models has been developed by S. Volkov [40] to model the $5x+1$ problem. Volkov models counting all non-divergent trajectories of the $5x + 1$ problem, which are those which enter some finite cycle, and denotes the number of these below $x$ by $Q(x)$. Thus $\pi_5(x) \le Q(x)$, and conjecturally these should be of similar orders of growth. It is expected there are finitely many cycles, and each should absorb roughly the same number of integers below $x$, in the sense of the exponent in the power of $x$ involved.

Volkov's branching process stochastic models grow a complete binary tree, rather than a tree that may have either one or two branches from each node, as in the models above. He suggests that the $5x + 1$ problem can be modeled by such trees, using an unusual encoding of the iterates (some edges encode several iteration steps of the inverse Collatz function). In order to do this, his node weights are chosen differently than above. He arrives at a predicted exponent $\eta^*_{5,BP} \approx 0.678$, which differs from the prediction $\eta_{5,BP} \approx 0.650919$ made in Theorem 8.10 above. The empirical data Volkov presents seems insufficient to discriminate between these two predicted exponents. It would be interesting for this problem to be investigated further.

## 9. Benford's Law for $3x + 1$ and $5x + 1$ Maps

Another curious statistic satisfied by the $3x + 1$ function was discovered by Kontorovich and Miller [17]: *Benford's Law*.

In the late 1800s, Newcomb [29] noticed a surprising fact while perusing tables of logarithms: certain pages were significantly more worn than others. Numbers whose logarithm started with 1 were being referenced more frequently than other digits. Instead of observing one-ninth (about 11%) of entries having a leading digit of 1, as one would expect if the digits $1, 2, \ldots, 9$ were equally likely, over 30% of the entries had leading digit 1, and about 70% had leading digit less than 5. Since $\log_{10} 2 \approx 0.301$ and $\log_{10} 5 \approx 0.699$, Newcomb speculated that the probability of observing a digit less than $k$ was $\log_{10} k$. This logarithmic phenomenon became known as *Benford's Law* after Benford [6] collected and in 1938 popularized extensive empirical evidence of this distribution in diverse data sets.

Benford's law seems to hold for many sequences of numbers generated by dynamical systems having an "expanding" property, see Berger et al [7] and Miller and Takloo-Bighash [28, Chap. 9]. Benford behavior has been empirically observed for initial digits of the first iterates of the $3x + 1$ map or accelerated $3x + 1$ map for a randomly chosen initial number $n$. Here we survey some rigorous theorems quantifying this statement, for initial iterates. Similar Benford results can be proved for the $5x + 1$ function.

We emphasize that the Benford law behavior quantifed here concerns behavior on a fixed finite set of initial iterates of these maps. Indeed, the $3x + 1$ conjecture predicts that Benford behavior cannot hold for the full infinite set of forward iterates, since conjecturally they become periodic! However it remains possible that a strong form of Benford behavior could hold on (infinite) divergent orbits of the $5x + 1$ problem.

MODELS FOR THE $3x+1$ AND $5x+1$ PROBLEMS, AND RELATED PROBLEMS 181

**9.1. Benford's Law and Uniform Distribution of Logarithms.** To make Benford's law precise, we say that the *mantissa* function $\mathcal{M}(n) \in [1, 10)$ is the leading entry of $n$ in "scientific notation", that is, $n = \mathcal{M}(n) \cdot 10^{\lfloor \log_{10} n \rfloor}$. Benford's law concerns the distribution of leading digit of the mantissa, while one can also consider the distribution of the lower order digits of the mantissa.

DEFINITION 9.1. *An infinite sequence* $\{n_1, n_2, \dots, n_k, \dots\}$ *satisfies the* strong Benford's Law *(to base 10) if the logarithmic digit frequency holds for any order digits in the mantissa. That is, for any* $a \in [1, 10)$,

$$(9.111) \qquad \lim_{x \to \infty} \frac{\#\{k \le x : \mathcal{M}(n_k) < a\}}{x} = \log_{10}(a).$$

The strong version of Benford's law is well known to be equivalent to uniform distribution mod 1 of the base 10 logarithms of the numbers in the sequence, cf. Diaconis [**15**, Theorem 1].

THEOREM 9.1. *(Strong Benford Law Criterion) A sequence* $\{n_1, n_2, \dots\}$ *satisfies the strong Benford's Law (or "is strong Benford") to base 10 if and only if the sequence* $\{\log_{10} n_1, \log_{10} n_2, \dots\}$ *is equidistributed* (mod 1), *that is, for any* $a \in [0, 1)$,

$$(9.112) \qquad \lim_{x \to \infty} \frac{\#\{k \le x : \log_{10} n_k (\mathrm{mod}\ 1) < a\}}{x} = a.$$

The definition and theorem above extend to expansions in any integer base $B \ge 2$. This result suggests the following general definition of strong Benford's Law to any *real* base $B > 1$.

DEFINITION 9.2. *Let* $B > 1$ *be a real number. A sequence* $\{n_1, n_2, \dots, n_k, \dots\}$ *satisfies the* strong Benford's Law to base $B$ *if and only if the sequence* $\{\log_B(n_1), \log_B(n_2), \dots\}$ *is uniformly distributed modulo one.*

This definition is equivalent to the earlier one for integers expanded in a radix expansion to any base $B > 1$. One can similarly define the mantissa function to any real base $B > 1$, extending Definition 9.1.

Benford's Law is stated for infinite sequences. However one can obtain approximate results that apply to finite sequences $\{x_1, x_2, \dots, x_k\}$, by using the following discrepancy measure of approximation to uniform distribution of such sequences.

DEFINITION 9.3. *Given a finite set* $\mathcal{Y} = \{y_1, \dots, y_k\}$ *of size* $k$, *for each* $0 \le a < 1$, *set*

$$\mathcal{D}(\mathcal{Y}; a) := \frac{\#\{j \le k : y_j (\mathrm{mod}\ 1) < a\}}{k} - a.$$

*The* discrepancy $\mathcal{D}(\mathcal{Y})$ *is defined by*

$$\mathcal{D}(\mathcal{Y}) := \sup_{0 \le a < 1} \mathcal{D}(\mathcal{Y}; a) - \inf_{0 \le a < 1} \mathcal{D}(\mathcal{Y}; a).$$

One always has $\mathcal{D}(\mathcal{Y}) \le 1$. The smallest possible discrepancy of a finite set $\mathcal{Y}$ is $\mathcal{D}(\mathcal{Y}) = 1/k$, attained by equally spaced elements $y_j = \frac{j}{k}$, $1 \le j \le k$.

A small discrepancy indicates that the set $\mathcal{Y}$ is close to equidistributed modulo 1. In particular, for an infinite sequence $\mathcal{X} = \{x_j : j \ge 1\}$, if $\mathcal{X}_k = \{x_j : 1 \le j \le k\}$ then $\mathcal{X}$ is uniformly distributed ( mod 1) if and only if the discrepancies $\mathcal{D}(\mathcal{X}_k) \to 0$ as $k \to \infty$.

**9.2. Benford's Law for $3x + 1$ Function Iterates.** Kontorovich and Miller [**17**] considered iterates of the accelerated $3x+1$ function $U(n)$. Fix an odd integer $n = n_0$, and let $\{n_1, n_2, \dots\}$ be the sequence of iterates from the starting seed $n_0 \in \Pi$, where $\Pi$ consists of all positive integers relatively prime to 6. The main $3x + 1$ conjecture asserts that this sequence is eventually periodic, and hence it is impossible for (9.112) to hold!

The following was their interpretation of (weak) "Benford behavior" for the $3x + 1$ function:

THEOREM 9.2. *For $x_0 = n \in \Pi$, denote its accelerated $3x + 1$ iterates by $x_k := U^{(k)}(x_0)$. Now set $y_k := \log_{10} x_k$ and define the shifted variables*

$$\omega_k := y_k - y_0 - k\log_{10}\left(\frac{3}{4}\right).$$

*Then, for any $a \in [0, 1)$,*

$$\lim_{k \to \infty} \mathbb{D}_\Pi\left[x_0 : \ \omega_k(\text{mod } 1) < a\right] = a.$$

PROOF. This is established as Theorem 5.3 in Kontorovich and Miller [**17**].  □

Arguably, the normalization from $y_k$ to $\omega_k$ in Theorem 9.2 makes the above result only an approximation to "true" Benford behavior, which should be that $\mathbb{D}_\Pi[x_0 : \ y_k \ (\text{mod } 1) < a] \to a$ as $k \to \infty$.

Lagarias and Soundararajan [**22**] were able to use the non-accelerated $3x + 1$ function $T$ to show another approximation to Benford behavior, as follows.

THEOREM 9.3. (Approximate Strong Benford's Law for $3x+1$ Map) *Let $B > 1$ be any integer base. Then for a given $N \geq 1$ and each $X \geq 2^N$, most initial starting values $x_0$ in $1 \leq x_0 \leq X$ have first $N$ initial $3x + 1$ iterates $\{x_k : 1 \leq k \leq N\}$ that satisfy the discrepancy bound*

(9.113)           $$\mathcal{D}\left(\{\log_B x_k(\text{mod } 1) : 1 \leq k \leq N\}\right) \leq 2N^{-\frac{1}{36}}.$$

*The exceptional set $\mathcal{E}(X, B)$ of initial seeds $x_0$ in $1 \leq x_0 \leq X$ that do not satisfy the bound has cardinality*

(9.114)           $$|\mathcal{E}(X, B)| \leq c(B)N^{-\frac{1}{36}}$$

*where $c(B)$ is a positive constant depending only on the base $B$.*

PROOF. This is established as Theorem 2.1 in Lagarias and Soundararajan [**22**].  □

**9.3. Benford's Law for $5x + 1$ Function Iterates.** The $5x + 1$ map also exhibits similar "Benford" behavior for its iterates. The results of [**17**] apply to general $(d, g, h)$-Maps, in particular, to the $5x+1$ function, giving a direct analogue of Theorem 9.2.

The method of proof in [**22**] of Theorem 9.3 should also extend to give qualitatively similar results in the $5x + 1$ case. This proof relied on the Parity Sequence Theorem for the $3x + 1$ map which has an exact analogue for the $5x + 1$ map. The proof in [**22**] also used some Diophantine approximation results for the transcendental number $\alpha_3 := \log_2 3$, and qualitatively similar Diophantine approximation results are valid for $\alpha_5 := \log_2 5$ needed in the $5x + 1$ case.

These rigorous results concern only the initial iterates of $5x + 1$ trajectories. However since the $5x+1$ map conjecturally has divergent orbits, it seems a plausible guess that a strong form of Benford behavior might hold on all infinite divergent orbits of the $5x + 1$ map.

## 10. 2-Adic Extensions of $3x + 1$ and $5x + 1$ Maps

What happens if we put these probabilistic models in a more general context? We can obtain a perfect set of symbolic dynamics if we extend the domain of these maps to the 2-adic integers. Such extensions are possible for both the $3x + 1$ map $T_3(x)$ and the $5x + 1$ map $T_5(x)$.

THEOREM 10.1. *The $3x+1$ map $T_3$ and the $5x+1$ map $T_5$ extend continuously from maps on the integers to maps on the 2-adic integers $\mathbb{Z}_2$, viewing $\mathbb{Z}$ as a dense subset of $\mathbb{Z}_2$. Denoting the extensions by $\tilde{T}_3$ and $\tilde{T}_5$, respectively, these maps have the following properties.*

*(i) Both maps $\tilde{T}_3$ and $\tilde{T}_5$ are homeomorphisms of $\mathbb{Z}_2$ to itself.*

*(ii) Both maps $\tilde{T}_3$ and $\tilde{T}_5$ are measure-preserving maps on $\mathbb{Z}_2$ for the standard 2-adic measure $\mu_2$ on $\mathbb{Z}_2$.*

*(iii) Both maps $\tilde{T}_3$ and $\tilde{T}_5$ are strongly mixing with respect to the measure $\mu_2$, hence ergodic.*

PROOF. For the $3x + 1$ map, properties (i)-(iii) are stated in Lagarias [**21**, Theorem K]. The property of strong mixing is an ergodic-theoretic notion explained there. Akin [**1**] gives another proof of these facts for the $3x + 1$ map.

For the $5x + 1$ map, properties (i)-(iii) may be established by proofs similar to the $3x + 1$ map case. This is based on the fact that an analogue of Theorem 2.1 holds for the symbolic dynamics of iterating the $5x + 1$ map. It is also a corollary of results of Bernstein and Lagarias [**9**, Sect. 4], whose results imply that (i)-(iii) hold more generally for all $ax + b$-maps. Here the $ax + b$ map $T_{a,b}$ is

$$T_{a,b}(x) := \begin{cases} \dfrac{ax + b}{2} & \text{if } x \equiv 1 \ (\text{mod } 2), \\[2mm] \dfrac{x}{2} & \text{if } x \equiv 0 \ (\text{mod } 2), \end{cases}$$

where $a$ and $b$ are odd integers. $\qquad\square$

A much stronger ergodicity result is valid for the 2-adic extensions of these maps. Define the 2-*adic shift map* $S : \mathbb{Z}_2 \to \mathbb{Z}_2$ to be the 2-to-1 map given for $\alpha = \sum_{j=0}^{\infty} a_j 2^j = .a_0 a_1 a_2...$, with each $a_j = 0$ or 1, by

$$S(\alpha) = S(.a_0 a_1 a_2 \cdots) := .a_1 a_2 a_3 \cdots$$

That is,

(10.115) $$S(\alpha) = \begin{cases} \dfrac{\alpha - 1}{2} & \text{if } \alpha \equiv 1 \ (\text{mod } 2) \\[2mm] \dfrac{\alpha}{2} & \text{if } \alpha \equiv 0 \ (\text{mod } 2). \end{cases}$$

This map has the 2-adic measure as Haar measure, and is mixing in the strongest sense.

THEOREM 10.2.   *The 2-adic extensions $\tilde{T}_3$ of the $3x+1$ map and $\tilde{T}_5$ of the $5x+1$ map are each topologically conjugate to the 2-adic shift map, by a conjugacy map $\Phi_3$, resp. $\Phi_5$. That is, these maps are homeomorphisms of $\mathbb{Z}_2$ with $\Phi_3^{-1} \circ \tilde{T}_3 \circ \Phi_3 = S$ and $\Phi_5^{-1} \circ (\tilde{T})_5 \circ \Phi_5 = S$.*

*(1) The maps $\Phi_j$, $j = 3$ or $5$, are solenoidal, i.e. for each $n \geq 1$ they have the property*

$$x \equiv y \ (\mathrm{mod} \ \ 2^n) \longrightarrow \Phi_j(x) \equiv \Phi_j(y) \ (\mathrm{mod} \ \ 2^n).$$

*(2) The inverses of these conjugacy maps are explicitly given by*

$$\Phi_j^{-1}(\alpha) := \sum_{k=0}^{\infty} \left( T_j^{(k)}(\alpha) \ (\mathrm{mod} \ \ 2) \right) 2^k,$$

*for $j = 3$ or $5$, and the residue $(\mathrm{mod} \ 2)$ is taken to be $0$ or $1$.*

PROOF.  These results follow from Bernstein and Lagarias [**9**, Sect. 3, 4], where results are proved for a general class of mappings including both the $3x + 1$ map and $5x + 1$ map.                                                                    □

Theorem 10.2 immediately gives the following corollary.

COROLLARY 10.1.   *The 2-adic extensions $\tilde{T}_3$ of the $3x + 1$ map and $\tilde{T}_5$ of the $5x + 1$ map are topologically conjugate and metrically conjugate maps.*

The corollary shows that from the viewpoint of extensions to the 2-adic integers, the $3x + 1$ maps and the $5x + 1$ maps have identical ergodic theory properties, i.e. they are both conjugate to the shift map. That is, their symbolic dynamics is "the same" in the topological sense, and their dynamics is also identical in the measure-theoretic sense.

The original $3x + 1$ problem (resp. $5x + 1$ problem) concerns their behavior when restricted to the dense set $\mathbb{Z}$ inside $\mathbb{Z}_2$. This set $\mathbb{Z}$ is countable, so has 2-adic measure zero, so the general properties of ergodic theory allow no conclusion to be drawn about behavior of iteration on these maps on $\mathbb{Z}$. Indeed empirical data and the stochastic models above show that the dynamics of iteration of the $3x + 1$ map and $5x + 1$ map are "not the same" on $\mathbb{Z}$.

To conclude, we remark that the two accelerated functions $U_3$ and $U_5$ also make sense 2-adically, in a restricted domain. Let $\mathbb{Z}_2^{\times} = \{\alpha \in \mathbb{Z}_2 : \ \alpha \equiv 1 \ (\mathrm{mod} \ 2)\}$. We have $U_3 : \mathbb{Z}_2^{\times} \to \mathbb{Z}_2^{\times} \cup \{0\}$ (in the latter case we set $U(-\frac{1}{3}) = 0$.) and $U_5 : \mathbb{Z}_2^{\times} \to \mathbb{Z}_2^{\times} \cup \{0\}$ (in the latter case we set $U(-\frac{1}{5}) = 0$.) It might prove worthwhile to find invariant measures for these functions, and to study their ergodic-theoretic behavior.

## 11.  Concluding Remarks

We have presented results on stochastic models simulating aspects of the behavior of the $3x + 1$ function and $5x + 1$ problems. These models resulted in specific predictions about various statistics of the orbits of these functions under iteration, which can be tested empirically. The experimental tests done so far have generally been consistent with these predictions.

**11.1. Comparisons.** We compare and contrast the behavior of these two maps under iteration. The $3x+1$ map and $5x+1$ map are similar in the following dimensions.

(1) (*Symbolic dynamics*) The allowed symbolic dynamics of even and odd iterates is the same for the $3x+1$ and $5x+1$ maps. Every finite symbol sequence is legal.

(2) (*Periodic orbits on the integers*) Conjecturally, both the $3x+1$ map and $5x+1$ maps have a finite number of distinct periodic orbits on the domain $\mathbb{Z}$.

(3) (*Periodic orbits on rational numbers with odd denominator*) Every possible symbolic dynamics for a periodic orbit is the periodic orbit for some rational starting point, for both the $3x+1$ map and $5x+1$ map. That is, extensions of the maps $T_3$ and $T_5$ to rational numbers with odd denominator each have $2^p$ periodic orbits of period $p$, for each $p \geq 1$. Here the period $p$ may not be the minimal period of the orbit, so a period $k$ orbit is also counted as a period $p = kn$ orbit for each $k \geq 1$.

(4) (*Benford Law behavior*) Both the initial $3x+1$ function iterates of a random starting point, and the initial $5x+1$ iterates of a random starting point, with high probability exhibit strong Benford law behavior to any integer base $B \geq 2$.

(5) (*2-adic extensions*) The 2-adic extensions of the two maps are topologically and metrically conjugate. Therefore they have the same dynamics in the topological sense, and in the ergodic theory sense, on the domain $\mathbb{Z}_2$.

The main differences between the $3x+1$ maps and $5x+1$ maps concerns the change in size of their interates.

(1) (*Short-term behavior of iterates*) For the $3x+1$ map, the initial steps of most orbits shrink in size, while for the $5x+1$ map most orbits expand in size. This is rigorously quantified in §2 and §7.

(2) (*Long-term behavior of iterates*) The $3x+1$ and $5x+1$ conjecturally differ greatly in their long-term behavior of orbits on the integers. For the $3x+1$ map, conjecturally all orbits are bounded. For the $5x+1$ map, conjecturally a density one set of integers have unbounded orbits.

It is the long term behavior of iterates where all the difficulties connected with the $3x+1$ and $5x+1$ function lie.

**11.2. Insights.** Comparison of the results of these stochastic models, combined with deterministic results, deliver certain insights in understanding the $3x+1$ and $5x+1$ problem, and suggest topics for further work.

First, the 2-adic results indicate that the differences in of the dynamics of the $3x+1$ map $T_3$ and $5x+1$ map on the integers are invisible at the level of measure theory. Therefore these differences must depend in some way on number-theoretic features inside the integers $\mathbb{Z}$.

Second, the behavior of the iteration of these function of in $\mathbb{Z}$, viewed inside the 2-adic framework, must be encoded in the specific properties of the conjugacy maps $\Phi_3$ and $\Phi_5$ identifying these maps with the 2-adic shift map. Here we note that there is an explicit formula for the $3x+1$ conjugacy map, obtained by Bernstein [8], and there is an analogous formula for the $5x+1$ conjugacy map as well. These conjugacy maps have an intricate structure, detailed in [9], which might be worthy of further investigation.

Third, we observe that the ergodic behavior of the 2-adic extensions is exactly the behavior that served as a framework to formulate the random walk models presented in §3, §5, and §7. These random walk models yield information by combining these model iterations with estimates of the size of iterates in the standard absolute value on the real line $\mathbb{R}$. That is, they use information from an archimedean norm, rather than the non-archimedean norm on the 2-adic integers. Perhaps one needs to consider models that incorporate both norms at once, e.g. functions on $\mathbb{R} \times \mathbb{Z}_2$.

Fourth, a suitable maximal domain, larger than $\mathbb{Z}$, on which to understand the difference between the $3x + 1$ map $T_3$ dynamics and the $5x + 1$ map $T_5$ dynamics appears to be the domain

$$\mathbb{Q}_{(2)} := \mathbb{Q} \cap \mathbb{Z}_2,$$

i.e. the set of rational numbers that are 2-adic integers. The set $\mathbb{Q}_{(2)}$ is exactly the set of rational numbers having an odd denominator, and both $T_3$ and $T_5$ leave the set $\mathbb{Q}_{(2)}$ invariant. This set includes all periodic orbits of both $T_3$ and $T_5$, and from the viewpoint of existence of periodic orbits, these two maps are the same on $\mathbb{Q}_{(2)}$. The difference in the dynamics of these maps on $\mathbb{Z}$ seems to have something to do with the distribution of these periodic orbits. Viewing $\mathbb{Q}_{(2)}$ as having the topology induced from the 2-adic topology, one may conjecture that $T_3$ and $T_5$ are *not* topologically conjugate mappings on this domain.

Fifth, the $5x + 1$ map exhibits various "exceptional" behaviors. Although almost all of its integer orbits (conjecturally) diverge, nevertheless there exists an infinite exceptional set of integers that have eventually periodic orbits. The density (fractional dimension) of such integers is predicted (conjecturally) to be a constant $\delta_5 \approx 0.649$, solving a large deviations functional equation. This seems a hard problem to resolve rigorously. Now, for the $3x + 1$ map, a similar prediction is made by the models for the growth constant $g = 1$. It too is the solution of a large deviations functional equation. We currently know that $1 \geq g \geq 0.84$. This analogy suggests that rigorously proving that the growth constant $\delta_3 = 1$ may turn out to be a much harder problem than it seems at first glance.

Sixth, we note that there are extensions of the maps for backwards iteration to larger domains, to the invertible 3-adic integers $\mathbb{Z}_3^*$ for the $3x + 1$ map, and to the invertible 5-adic integers $\mathbb{Z}_5^*$ for the $5c+1$ map. In effect the branching random walk models may fruitfully be extended to allowing root node labels that are invertible 3-adic integers (resp. 5-adic integers), and this provides enough information to grow the entire infinite tree. Various interesting properties of the extended $3x + 1$ trees obtained this way have been obtained, cf. [4]. This is a topic worth further investigation.

## References

[1] Ethan Akin, Why is the $3x + 1$ Problem Hard?, In: *Chapel Hill Ergodic Theory Workshops* (I. Assani, Ed.), Contemp. Math. vol 356, Amer. Math. Soc. 2004, pp. 1–20.

[2] D. Applegate and J. C. Lagarias, *Density Bounds for the $3x + 1$ Problem I. Tree-Search Method,* Math. Comp., **64** (1995), 411–426.

[3] D. Applegate and J. C. Lagarias, *Density Bounds for the $3x + 1$ Problem II. Krasikov Inequalities,* Math. Comp., **64** (1995), 427–438.

[4] D. Applegate and J. C. Lagarias, *On the distribution of* $3x + 1$ *trees,* Experimental Mathematics **4** (1995), 101–117.

[5] D. Applegate and J. C. Lagarias, *Lower bounds for the for the total stopping time of* $3x + 1$ *iterates,* Math. Comp. **72** (2003), 1035–1049.

[6] F. Benford, *The law of anomalous numbers*, Proceedings of the American Philosophical Society **78** (1938), 551-572.

[7] A. Berger, L. Bunimovich and T. Hill, *One-dimensional dynamical systems and Benford's law,* Trans. Amer. Math. Soc. **357** (2005), 197–219.

[8] D. J. Bernstein, *A non-iterative 2-adic statement of the* $3x + 1$ *Conjecture,* Proc. Amer. Math. Soc. **121** (1994), 405–408.

[9] D. J. Bernstein and J. C. Lagarias, *The* $3x + 1$ *Conjugacy Map,* Canadian J. Math. **48** (1996), 1154–1169.

[10] K. Borovkov and D. Pfeifer, *Estimates for the Syracuse problem via a probabilistic model,* Theory of Probability and its Applications **45**, No. 2 (2000), 300–310.

[11] R. N. Buttsworth and K. R. Matthews, *On some Markov matrices arising from the generalized Collatz mapping,* Acta Arithmetica **55** (1990), 43–57.

[12] M. Chamberland, *An update on the* $3x + 1$ *problem, (Catalan)*, Butlettí Societat Catalana de Matemàtiques **18** (2003), No.1, 19–45.

[13] J. H. Conway, *Unpredictable Iterations,* Proc. 1972 Number Theory Conference (Univ. Colorado, Boulder, Colo., 1972 ), pp. 49–52. Univ. Colorado, Boulder, Colo. 1972.

[14] R. E. Crandall, *On the '3x+1' problem,* Math. Comp. **32** (1978), 1281–1292.

[15] P. Diaconis, *The distribution of leading digits and uniform distribution* (mod 1), Ann. Prob. **5** (1977), 72–81.

[16] C. J. Everett, *Iteration of the number theoretic function* $f(2n) = n, f(2n + 1) = 3n + 2,$ Advances in Math. **25** (1977), 42–45.

[17] A. V. Kontorovich and S. J. Miller, *Benford's law, values of L-functions, and the* $3x + 1$ *problem,* Acta Arithmetica **120** (2005), 269–297.

[18] A. V. Kontorovich and Ya. G. Sinai, *Structure Theorem for* $(d, g, h)$*-maps,* Bull. Braz. Math. Soc. (N.S.) **33** (2002), 213–224.

[19] I. Krasikov, *How many numbers satisfy the* $3x + 1$ *Conjecture?,* Internatl. J. Math. & Math. Sci. **12** (1989), 791–796.

[20] I. Krasikov and J. C. Lagarias, *Bounds for the* $3x + 1$ *problem using difference inequalities,* Acta Arithmetica **109** (2003), no. 3, 237–258.

[21] J. C. Lagarias, *The* $3x + 1$ *problem and its generalizations,* Amer. Math. Monthly **92** (1985), 3–23.

[22] J. C. Lagarias and K. Soundararajan, Benford's Law for the $3x + 1$ Function, J. London Math. Soc. **74** (2006), 289–303.

[23] J. C. Lagarias and A. Weiss, *The* $3x+1$ *Problem: Two Stochastic Models,* Annals of Applied Probability **2** (1992), 229–261.

[24] G. M. Leigh, A Markov process underlying the generalized Syracuse algorithm, Acta Arithmetica **46** (1986), 125–143.

[25] K. R. Matthews, *The generalized* $3x + 1$ *mapping: Markov chains and ergodic theory*, this volume.

[26] K. R. Matthews and A. M. Watts, *A generalization of Hasse's generalization of the Syracuse algorithm,* Acta. Arithmetica **43** (1984), 167–175.

[27] K. R. Matthews and A. M. Watts, *A Markov approach to the generalized Syracuse algorithm,* Acta Arithmetica **45** (1985), 29–42.

[28] S. J. Miller and R. Takloo-Bighash, *An Invitation to Modern Number Theory,* Princeton University Press: Princeton 2006.

[29] S. Newcomb, *Note on the frequency of use of the different digits in natural numbers*, Amer. J. Math. **4** (1881), 39-40.

[30] T. Oliveira e Silva, Maximum excursion and stopping time record-holders for the $3x + 1$ problem: Computational results, Math. Comp. **68** (1999) No. 1, 371-384.

[31] T. Oliveira e Silva, Empirical verification of the $3x + 1$ conjecture and related conjectures, in this volume.

[32] D. W. Rawsthorne, Imitation of an iteration, Math. Mag. **58** (1985), 172–176.

[33] E. Roosendaal, *On the 3x+1 problem,* website on distributed search for $3x + 1$ records, http://www.ericr.nl/wondrous

[34] A. Shwartz and A. Weiss, *Large deviations for performance analysis. Queues, communications and computing* With an appendix by Robert J. Vanderbei. Stochastic Modelling Series, Chapman & Hall: London 1995.

[35] Ya. G. Sinai, *Statistical $(3X + 1)$-Problem*, Dedicated to the memory of Jürgen K. Moser. Comm. Pure Appl. Math. **56** No. 7 (2003), 1016–1028.

[36] Ya. G. Sinai, *Uniform distribution in the $(3x + 1)$ problem*, Moscow Math. Journal **3** (2003), No. 4, 1429–1440. (S. P. Novikov 65-th birthday issue).

[37] Ya. G. Sinai, *A theorem about uniform distribution,* Commun. Math. Phys. **252** (2004), 581–588. (F. Dyson birthday issue)

[38] R. Terras, A stopping time problem on the positive integers, Acta Arith. **30** (1976), 241–252.

[39] R. Terras, On the existence of a density, Acta Arith. **35** (1979), 101–102.

[40] S. Volkov, A probabilistic model for the $5k + 1$ problem and related problems, Stochastic Processes and Applications **116** (2006), 662–674.

[41] S. Wagon, The Collatz Problem, Math. Intelligencer **7**, No. 1 (1985), 72–76.

[42] G. J. Wirsching, On the combinatorial structure of $3x + 1$ predecessor sets, Discrete Math. **148** (1996), No. 3, 265–286.

[43] G. J. Wirsching, *The dynamical system generated by the $3n + 1$ function*, Lecture Notes in Math. No. 1681, Springer-Verlag: Berlin 1998.

DEPARTMENT OF MATHEMATICS, BROWN UNIVERSITY, PROVIDENCE, RI
*E-mail address*: `alexk@math.brown.edu`

DEPARTMENT OF MATHEMATICS, UNIVERSITY OF MICHIGAN, ANN ARBOR, MI 48109-1109
*E-mail address*: `lagarias@umich.edu`

# Empirical verification of the $3x + 1$ and related conjectures

Tomás Oliveira e Silva

ABSTRACT. This paper describes the strategy used by the author to verify empirically the truth of the $3x + 1$ conjecture, also known as the Collatz conjecture, for all positive integers up to $20 \times 2^{58} \approx 5.7646 \times 10^{18}$. It also presents some empirical results concerning two generalized Collatz mappings which resemble in spirit the $3x+1$ mapping. For each of these two mappings the growth rate of the so-called maximum excursion record-holders is compared with the conjectured growth rate obtained from the study of a Markov process which was designed to mimic the behavior of the iterates of the generalized mapping.

## 1. Introduction

The $3x + 1$ conjecture, which, in one popular form, states that the iterates $n_0$, $T(n_0)$, $T\big(T(n_0)\big)$, ..., with $n_0 > 0$, of the function

$$(1) \qquad T(n) = \begin{cases} n/2, & \text{if } n \text{ is even} \\ (3n + 1)/2, & \text{if } n \text{ is odd} \end{cases}$$

eventually reach 1, has so far eluded proof. In the last decades the $3x+1$ conjecture has been verified by computer up to progressively larger limits (cf. Table 1). This was made possible not only by the use of faster and more numerous computing resources, but also by the use of better algorithms. All empirical evidence available so far suggests the truth of this conjecture.

The algorithm used in the verification effort reported in this paper is described in detail in section 2; on contemporary processor cores, this algorithm is faster than the one described in [12]. With it, and using considerable computing resources, we were able to prove the following theorem.

THEOREM 1. *The $3x + 1$ conjecture holds for all $x \leq 20 \times 2^{58}$.*

Let $T^{(0)}(n) = n$ and let, for $i > 0$, $T^{(i)}(n) = T\big(T^{(i-1)}(n)\big)$. The maximum excursion for a starting value of $n$, denoted by $t(n)$, is the maximum value attained by the iterates $T^{(i)}(n)$, $i = 0, 1, \ldots$, if it exists, and infinity otherwise. It turns out that the empirical observation made in [12], and predicted by a stochastic model of the iterates of the $3x+1$ mapping [7], that the maximum excursion of the iterates of the $3x+1$ mapping grows, for record-holders, like $n^2$ remains valid (a record-holder for a function $f$ is a value of $n$ for which $f(m) < f(n)$ for all $m < n$).

---

*Key words and phrases.* $3x+1$ problem, Kakutani $5x+1$ problem, Kakutani $7x+1$ problem, empirical verification, maximum excursion, generalized Collatz mappings, Markov chain model.

TABLE 1. Records of verification of the $3x+1$ conjecture. It seems
that there exist other (unpublished) records of verification before
the year 1992; Garner [**4**] mentions verifications up to $6 \times 10^7$ and
$2 \times 10^9$.

| limit | year | who |
|---|---|---|
| $22\,882\,247 \approx 2.3 \times 10^7$ | 1973 | R. Dunn [**2**] |
| $2^{40} \approx 1.1 \times 10^{12}$ | 1983 | N. Yoneda (unpublished, see ref. 2 of [**7**]) |
| $5.6 \times 10^{13}$ | 1992 | G. Leavens and M. Vermeulen [**8**] |
| $24 \times 2^{50} \approx 2.7 \times 10^{16}$ | 1999 | T. Oliveira e Silva [**12**] |
| $100 \times 2^{50} \approx 1.1 \times 10^{17}$ | 2000 | T. Oliveira e Silva (unpublished) |
| $309 \times 2^{50} \approx 3.5 \times 10^{17}$ | 2004 | E. Roosendaal (unpublished) |
| $20 \times 2^{58} \approx 5.8 \times 10^{18}$ | 2009 | T. Oliveira e Silva (this paper) |

The ideas used to test the $3x + 1$ conjecture can also be used to test other
similar conjectures. In this paper the behavior under iteration of two mappings of
the form

$$(2) \qquad\qquad C_{K,\mathcal{D}}(n) = \begin{cases} n/d_{\mathcal{D}}(n), & \text{if } d_{\mathcal{D}}(n) > 1 \\ Kn + 1, & \text{if } d_{\mathcal{D}}(n) = 1 \end{cases}$$

is also studied, where, in general, $K$ is a positive integer larger than 1, $\mathcal{D}$ is an
ordered set of distinct positive integers larger that 1, and $d_{\mathcal{D}}(n)$ is equal to 1 if $n$
is not divisible by any of the elements of $\mathcal{D}$, and is equal to the first such element
otherwise. Note that

$$(3) \qquad\qquad C_{3,\{2\}}(n) = C(n) = \begin{cases} n/2, & \text{if } n \text{ is even} \\ 3n + 1, & \text{if } n \text{ is odd} \end{cases}$$

is frequently used instead of $T(n)$ in the study of the $3x + 1$ problem. Remarkably,
for the mappings $C_{5,\{2,3\}}$ and $C_{7,\{2,3,5\}}$ empirical observations suggest that, starting from a positive integer, the iterates of each of these two mappings eventually
reach 1. (The study of the $C_{5,\{2,3\}}$ mapping was suggested to the author by Eric
Roosendaal.) Given the obvious similarity with what happens with the $T$ mapping,
one may thus introduce the Kakutani $5x + 1$ and $7x + 1$ conjectures, which state
the same as the $3x + 1$ conjecture, but with $T$ replaced by $C_{5,\{2,3\}}$ or $C_{7,\{2,3,5\}}$, respectively. Section 3 describes the algorithms used to test empirically the Kakutani
$5x + 1$ and $7x + 1$ conjectures. With them, and using relatively modest computing
resources, we were able to prove the following two theorems.

THEOREM 2. *The Kakutani $5x + 1$ conjecture, which states that for a positive
initial value the iterates of $C_{5,\{2,3\}}$ eventually reach one, holds for all $x \le 10^{16}$.*

THEOREM 3. *The Kakutani $7x + 1$ conjecture, which states that for a positive
initial value the iterates of $C_{7,\{2,3,5\}}$ eventually reach one, holds for all $x \le 10^{14}$.*

The mappings described by (2) are examples of so-called generalized Collatz
mappings [**10**]. In [**9**] it was conjectured that each mapping of this (more general)
type can be reasonably well modeled by a Markov process. The final section of this
paper, section 4, describes how to use the Markov process model to infer the rate
of growth of the maximum excursion record-holders for the $C_{5,\{2,3\}}$ and $C_{7,\{2,3,5\}}$
mappings. The main results of that section are summarized in the following two
conjectures, which are corroborated by the available empirical data.

CONJECTURE 1. *The maximum-excursion record-holders for the $C_{5,\{2,3\}}$ mapping grow like $n^{\rho_5}$, with $\rho_5 \approx 1.442\,762\,198\,279$.*

CONJECTURE 2. *The maximum-excursion record-holders for the $C_{7,\{2,3,5\}}$ mapping grow like $n^{\rho_7}$, with $\rho_7 \approx 2.215\,508\,001\,593$.*

## 2. Empirical verification of the $3x + 1$ conjecture

In this section the algorithm used to verify the $3x + 1$ conjecture up to $20 \times 2^{58}$ is described in detail. It represents a small improvement of the algorithm presented in [**12**]. While the old algorithm performed one iterate per step, the new one performs a variable number of them; increasing the average number of iterates performed per step makes the algorithm more efficient.

**2.1. Reducing the number of test cases.** In order to test the $3x + 1$ conjecture for all positive integers smaller than a given limit, it is only necessary to iterate $T$ for a given initial value $n_0 > 2$ as long as $T^{(i)}(n_0) > n_0$. This is so because once $T^{(i)}(n_0)$ falls below $n_0$ one has reached a case tested before; the first such $i$ is called the stopping time and is denoted by $\sigma(n)$. Also, if $T^{(i)}(n_0) = n_0$ for $n_0 > 2$ and $i > 0$ then there exists a cycle of length $i$ other that the cycle $1 \rightarrow 2 \rightarrow 1$, which would render the $3x + 1$ conjecture false. It is known that any such "non-trivial" cycle must have a length of at least $10\,439\,860\,591$ (this result can be inferred from [**3**], taking into consideration that the upper limit of verification of the $3x + 1$ conjecture is now above $1.08 \times 2^{60}$).

All recent empirical verification efforts of the $3x + 1$ conjecture are based on the fact that to determine the branch trajectory history up to the $i$-th iterate it is only necessary to know the value of $n \bmod 2^i$ (as in [**5**], $n \bmod m = n - m\lfloor n/m \rfloor$, where $\lfloor x \rfloor$ is the largest integer not smaller than $x$). In particular, the following property of the mapping $T$ is well known (see, for example, [**14**], [**1**], or [**12**]).

PROPOSITION 1. *Let $n_0 = 2^i n_i + m_i$, with $n_i = \lfloor n/2^i \rfloor$ and $m_i = n_0 \bmod 2^i$. Then,*

$$(4) \qquad T^{(i)}(n_0) = 3^{k(i;m_i)} n_i + T^{(i)}(m_i),$$

*where $k(i; m_i)$ is the number of "$n$ is odd" branches of $T$ that where taken in the computation of $T^{(i)}(m_i)$.*

A key observation used to speed-up the empirical verification of the $3x + 1$ conjecture is, perhaps, best explained by an example. Consider what happens to the iterates of all integers of the form $128n + 15$:

$$(5) \quad \begin{array}{ll} T^{(0)}(128n + 15) = 128n + 15 & T^{(4)}(128n + 15) = 648n + 80 \\ T^{(1)}(128n + 15) = 192n + 23 & T^{(5)}(128n + 15) = 324n + 40 \\ T^{(2)}(128n + 15) = 288n + 35 & T^{(6)}(128n + 15) = 162n + 20 \\ T^{(3)}(128n + 15) = 432n + 53 & T^{(7)}(128n + 15) = 81n + 10. \end{array}$$

Clearly, for any non-negative integer $n$, the first six iterates always stay above the initial value, and the seventh always falls below the initial value. Thus, it is not necessary to test numbers of this form. Careful analysis of this case reveals that if one rearranges the columns of the matrix

$$(6) \qquad \begin{bmatrix} 128 & 192 & 288 & 432 & 648 & 324 & 162 & 81 \\ 15 & 23 & 35 & 53 & 80 & 40 & 20 & 10 \end{bmatrix}$$

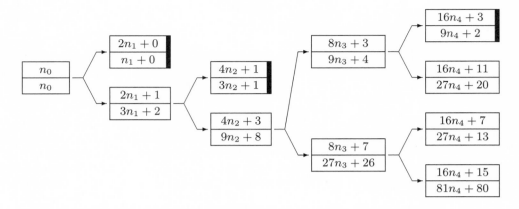

FIGURE 1. Prunned tree of depth 4. In each tree node, the top line presents the initial value and the bottom line presents the value after the first $i$ iterates have been performed, where $i$ is the depth of the node (from 0 on the left to 4 on the right). Subdividing a node is done by setting $n_i = 2n_{i+1} + 0$ for one of the branches, and $n_i = 2n_{i+1} + 1$ for the other, and then performing one more iterate.

in such a way that the first line becomes sorted in increasing order, then the second line also becomes sorted in increasing order. This remarkable property of the iterates has been verified empirically for all possible starting values modulo $2^{46}$, under the additional constraint than once it is known for certain that an iterate always falls below the initial value no further iterates are used. This verification was done using a pruned binary tree with a maximum depth of 46; figure 1 presents a very small part of that tree. Two initial values are exceptional (0 and 1), because they give rise to cycles when $n_i$ is zero; for these two cases the first and last elements of the unsorted second row are equal.

Table 2, which is an extension of table 1 of [**12**], presents the number $n_o(d)$ of surviving tree nodes at depth $d$. Each such node represents a congruence class modulo $2^d$ for which the first $d$ iterates always stay above the starting value. It may be tempting to use the last column of table 2 as an indicator of the speed-up factor one gains by using a pruned tree of a given depth. Doing so, however, would not be accurate, because it is also necessary to take into account the average number of iterates needed to test a typical representative of each surviving tree leaf (nodes of the tree at the prunning depth are called here leaves). To estimate this average, let $n_o(d, k)$ be the number of surviving tree nodes at depth $d$ for which $k$ "$n$ is odd" branches were taken in the first $d$ iterates (obviously, $0 \leq k \leq d$). If the so-called coefficient stopping time conjecture is true (it is for $d \leq 2\,592$ [**14**]), then $n_o(d, k) = 0$ when $3^k < 2^d$, and $n_o(d, k) = n_o(d - 1, k - 1) + n_o(d - 1, k)$ otherwise [**14, 12**]. Starting from $n_o(0, 0) = 1$, it is then possible to compute easily the (conjectured) number of surviving tree nodes at a given depth; starting from $n_o(d, k) = 1$ and $n_o(d, l) = 0$ for $l \neq k$, it is possible to do the same for a tree node at depth $d$ and corresponding to $k$ "$n$ is odd" branches.

Before presenting numerical values, it must be mentioned that the time spent generating the pruned tree becomes negligible if one tests many consecutive num-

TABLE 2. Number of surviving tree nodes, $n_o(d)$, at depth $d$; a tree node at depth $d$ is a survivor if and only if the first $d$ iterates stay above the initial value.

| $d$ | $n_o(d)$ | $2^d/n_o(d)$ | $d$ | $n_o(d)$ | $2^d/n_o(d)$ |
|---|---|---|---|---|---|
| 0 | 1 | 1.000 | 26 | 1 037 374 | 64.691 |
| 1 | 1 | 2.000 | 27 | 1 762 293 | 76.161 |
| 2 | 1 | 4.000 | 28 | 3 524 586 | 76.161 |
| 3 | 2 | 4.000 | 29 | 6 385 637 | 84.075 |
| 4 | 3 | 5.333 | 30 | 12 771 274 | 84.075 |
| 5 | 4 | 8.000 | 31 | 23 642 078 | 90.833 |
| 6 | 8 | 8.000 | 32 | 41 347 483 | 103.875 |
| 7 | 13 | 9.846 | 33 | 82 694 966 | 103.875 |
| 8 | 19 | 13.474 | 34 | 151 917 636 | 113.087 |
| 9 | 38 | 13.474 | 35 | 263 841 377 | 130.229 |
| 10 | 64 | 16.000 | 36 | 527 682 754 | 130.229 |
| 11 | 128 | 16.000 | 37 | 967 378 591 | 142.074 |
| 12 | 226 | 18.124 | 38 | 1 934 757 182 | 142.074 |
| 13 | 367 | 22.322 | 39 | 3 611 535 862 | 152.222 |
| 14 | 734 | 22.322 | 40 | 6 402 835 000 | 171.723 |
| 15 | 1 295 | 25.303 | 41 | 12 805 670 000 | 171.723 |
| 16 | 2 114 | 31.001 | 42 | 23 711 865 322 | 185.479 |
| 17 | 4 228 | 31.001 | 43 | 41 700 700 058 | 210.934 |
| 18 | 7 495 | 34.976 | 44 | 83 401 400 116 | 210.934 |
| 19 | 14 990 | 34.976 | 45 | 153 993 322 696 | 228.480 |
| 20 | 27 328 | 38.370 | 46 | 269 949 796 982 | 260.673 |
| 21 | 46 611 | 44.993 | 47 | 539 899 593 964 | 260.673 |
| 22 | 93 222 | 44.993 | 48 | 995 657 382 851 | 282.703 |
| 23 | 168 807 | 49.693 | 49 | 1 991 314 765 702 | 282.703 |
| 24 | 286 581 | 58.543 | 50 | 3 734 259 929 440 | 301.505 |
| 25 | 573 162 | 58.543 | | | |

bers belonging to a residue class associated with a surviving tree leaf. This is so because, according to proposition 1, only one addition is required to compute $T^{(d)}\big(2^d(n_d+1)+m_d\big)$ once $T^{(d)}(2^d n_d+m_d)$ is known. The first $d$ iterates can then be performed almost "for free"; they will not be accounted for in the average values reported below.

Table 3 presents the average number of iterates required to process a number belonging to a residue class of a random surviving leaf of a tree of depth $d$, as well as the corresponding speed-up factor (with respect to processing a random integer without using a pruned tree, i.e., depth 0). As expected, larger depths give larger speed-ups. In [12] a tree depth of 40 was chosen, because $3^{40}$ is the largest power of 3 which can be represented by a 64-bit integer, and because it only took about one hour to generate the entire pruned tree on the computers used at that time. The improved algorithm described in this paper uses a deeper tree ($d = 46$) to take advantage of a larger speed-up (44.866 versus 32.118, i.e., a 40% improvement).

Table 3 also presents the average number of iterates required to process a number belonging to a residue class associated with a surviving leaf with the smallest number of "$n$ is odd" branches taken. As explained later, to increase the efficiency of the inner loop of the improved algorithm this average should be large. According to the last column of this table, 37, 40, 43, 46, and 48 are reasonably good values for $d$. As mentioned above, the improved algorithm uses $d = 46$.

TABLE 3. Average number of iterates required to verify the $3x+1$ conjecture for many "consecutive" numbers belonging to the residue class associated with i) a random surviving leaf of a tree with depth $d$ ($d$-avg column), and ii) a surviving leaf at depth $d$ with the smallest number $k$ of "$n$ is odd" branches taken (($d,k$)-avg column); $k = \lceil d\frac{\log 2}{\log 3}\rceil$. The first $d$ iterates are not taken in consideration because, according to proposition 1, their computation takes constant time. The speed-up column presents the expected speed-up factor with respect to using a tree of zero depth.

| $d$ | $d$-avg | speed-up | $k$ | ($d,k$)-avg | | $d$ | $d$-avg | speed-up | $k$ | ($d,k$)-avg |
|---|---|---|---|---|---|---|---|---|---|---|
| 0 | 3.493 | 1.000 | 0 | 3.493 | | 26 | 16.131 | 14.006 | 17 | 5.853 |
| 1 | 4.985 | 1.401 | 1 | 4.985 | | 27 | 17.814 | 14.932 | 18 | 9.706 |
| 2 | 7.971 | 1.753 | 2 | 7.971 | | 28 | 16.814 | 15.820 | 18 | 4.943 |
| 3 | 6.971 | 2.004 | 2 | 3.634 | | 29 | 17.457 | 16.821 | 19 | 7.886 |
| 4 | 7.961 | 2.340 | 3 | 5.267 | | 30 | 16.457 | 17.843 | 19 | 3.601 |
| 5 | 10.441 | 2.676 | 4 | 8.534 | | 31 | 16.700 | 18.997 | 20 | 5.201 |
| 6 | 9.441 | 2.959 | 4 | 3.925 | | 32 | 17.954 | 20.207 | 21 | 8.402 |
| 7 | 10.389 | 3.310 | 5 | 5.851 | | 33 | 16.954 | 21.399 | 21 | 3.870 |
| 8 | 12.848 | 3.663 | 6 | 9.701 | | 34 | 17.369 | 22.740 | 22 | 5.740 |
| 9 | 11.848 | 3.972 | 6 | 4.941 | | 35 | 18.850 | 24.129 | 23 | 9.480 |
| 10 | 12.882 | 4.338 | 7 | 7.882 | | 36 | 17.850 | 25.481 | 23 | 4.827 |
| 11 | 11.882 | 4.703 | 7 | 3.599 | | 37 | 18.383 | 26.993 | 24 | 7.653 |
| 12 | 12.327 | 5.135 | 8 | 5.199 | | 38 | 17.383 | 28.546 | 24 | 3.495 |
| 13 | 13.950 | 5.588 | 9 | 8.398 | | 39 | 17.553 | 30.289 | 25 | 4.990 |
| 14 | 12.950 | 6.020 | 9 | 3.868 | | 40 | 18.674 | 32.118 | 26 | 7.981 |
| 15 | 13.547 | 6.524 | 10 | 5.736 | | 41 | 17.674 | 33.936 | 26 | 3.637 |
| 16 | 15.372 | 7.044 | 11 | 9.473 | | 42 | 18.009 | 35.971 | 27 | 5.274 |
| 17 | 14.372 | 7.534 | 11 | 4.824 | | 43 | 19.344 | 38.086 | 28 | 8.548 |
| 18 | 15.087 | 8.097 | 12 | 7.648 | | 44 | 18.344 | 40.162 | 28 | 3.930 |
| 19 | 14.087 | 8.672 | 12 | 3.493 | | 45 | 18.786 | 42.478 | 29 | 5.861 |
| 20 | 14.357 | 9.335 | 13 | 4.986 | | 46 | 20.292 | 44.866 | 30 | 9.721 |
| 21 | 15.662 | 10.034 | 14 | 7.973 | | 47 | 19.292 | 47.192 | 30 | 4.949 |
| 22 | 14.662 | 10.718 | 14 | 3.634 | | 48 | 19.838 | 49.771 | 31 | 7.899 |
| 23 | 15.089 | 11.502 | 15 | 5.269 | | 49 | 18.838 | 52.413 | 31 | 3.605 |
| 24 | 16.598 | 12.319 | 16 | 8.537 | | 50 | 19.025 | 55.352 | 32 | 5.210 |
| 25 | 15.598 | 13.109 | 16 | 3.926 | | | | | | |

Table 4 presents the number of surviving nodes at depths 40 and 46 with $k$ "$n$ is odd" branches taken, as well as the average number of iterates required to test each such node. Surviving nodes with a low value of $k$ are much more common than those with a moderate or high value of $k$; testing them uses, on average, a relatively small number of iterates. It is thus important to optimize the way a small number of iterates is handled by the algorithm.

Before describing how the iterates are performed in the new algorithm, two further optimizations that reduce the number of test cases deserve to be mentioned. The first, uses $T(2n + 1) = 3n + 2$ to state that it is not necessary to test initial values which are congruent to 2 mod 3, because testing $2n + 1$ automatically tests the larger $3n + 2$ [**1, 8**]. This observation alone accounts for a speed-up of 33.3%. Using also $T^{(3)}(8n+3) = 9n+4$ and $T^{(6)}(64n+7) = 81n+10$ gives rise to a speed-up of 45.7%, which is quite close to the best that can be achieved in this manner (about 46.7%). The second, communicated to the author by Eric Roosendaal, takes

TABLE 4. Number of surviving leaves at depths 40 and 46 with $k$
"$n$ is odd" branches taken, and their corresponding average number
of iterates.

| $k$ | $n_o(40, k)$ | $(40, k)$-avg |
|-----|--------------|---------------|
| 26 | 1 899 474 678 | 7.981 |
| 27 | 1 924 081 230 | 15.284 |
| 28 | 1 338 809 812 | 23.106 |
| 29 | 729 745 534 | 30.805 |
| 30 | 329 945 187 | 38.461 |
| 31 | 125 874 920 | 46.110 |
| 32 | 40 696 860 | 53.766 |
| 33 | 11 112 915 | 61.418 |
| 34 | 2 536 964 | 69.061 |
| 35 | 476 181 | 76.694 |
| 36 | 71 660 | 84.321 |
| 37 | 8 320 | 91.946 |
| 38 | 700 | 99.572 |
| 39 | 38 | 107.202 |
| 40 | 1 | 114.835 |

| $k$ | $n_o(46, k)$ | $(46, k)$-avg |
|-----|--------------|---------------|
| 30 | 84 141 805 077 | 9.721 |
| 31 | 80 085 991 810 | 17.219 |
| 32 | 53 866 144 103 | 24.786 |
| 33 | 29 645 444 727 | 32.451 |
| 34 | 13 863 626 575 | 40.107 |
| 35 | 5 593 869 346 | 47.770 |
| 36 | 1 958 889 113 | 55.428 |
| 37 | 595 070 080 | 63.071 |
| 38 | 156 077 950 | 70.699 |
| 39 | 35 039 652 | 78.321 |
| 40 | 6 646 079 | 85.944 |
| 41 | 1 045 517 | 93.569 |
| 42 | 132 855 | 101.199 |
| 43 | 13 110 | 108.832 |
| 44 | 943 | 116.469 |
| 45 | 44 | 124.107 |
| 46 | 1 | 131.747 |

advantage of the fact that the iterates of $n$ and of $n - 1$ have some tendency to coalesce [1]. For example, a starting number of the form $64n + 15$ does not need to be tested because $T^{(6)}(64n + 15) = T^{(6)}(64n + 14) = T^{(5)}(32n + 7) = 81n + 20$. By keeping track of the iterates of $n$ and of $n - 1$ while the tree is being expanded, and by discarding all nodes for which the two iterates become equal, about 17.5% of the leaves can be discarded at any given depth.

**2.2. Iterating $T$.** The main difference between the algorithm presented in [12] and the one presented in this paper has to do with how the iterates are performed. In [12] the iterates were performed one at a time; on the Alpha processors used at that time the idea, suggested in [8], of performing a fixed number of iterates in one step was tried and found to be slower, due to complications related to the detection of maximum excursions and to the application of the termination condition $T^{(i)}(n) < n$. To solve these problems, in the new algorithm a variable number of iterates is performed in one step. Doing so while retaining the ability to detect, for each $n$, the exact value of the maximum excursion, or to determine the exact $i$ for which $T^{(i)}(n) < n$, turns out to be inefficient. Fortunately, in the search for record-holders one is only interested in exceptionally large values of $t(n)$ or of $\sigma(n)$, which are very rare. It is therefore enough to be able to detect with certainty if $t(n)$ is above a given threshold, or if $\sigma(n)$ is above another given threshold; after finding each such exceptional value of $n$, a slower but exact algorithm can be used to determine $t(n)$ and $\sigma(n)$, which can then be tested against the appropriate current record-holder.

Let $i$ be the current iterate number, and let $\delta$ be the number of iterates to be performed in a single step, with $\delta \leq \Delta$. To ensure that a potential maximum excursion record-holder is not missed, and to ensure that the iterates are stopped when one iterate, skipped or not, falls below $n$, $\delta$ has to satisfy the following two conditions:

TABLE 5. Average number of iterates ($\delta$-avg) performed in one step for $\tau = 10$. $\Delta$ is the maximum number of iterates that can be performed in one step (the actual number of iterates may be smaller, as described in the main text); no skipped iterate can be above a non-skipped iterate by a factor larger than $\tau$.

| $\Delta$ | 6 | 7 | 8 | 9 | 10 | 11 | 12 | 13 | 14 | 15 |
|---|---|---|---|---|---|---|---|---|---|---|
| $\delta$-avg | 4.656 | 5.391 | 6.129 | 6.949 | 7.733 | 8.519 | 9.298 | 10.081 | 10.908 | 11.715 |

(1) $T^{(i+j)}(n) \leq \tau \max\big(T^{(i)}(n), T^{(i+\delta)}(n)\big)$ for $0 < j < \delta$, where $\tau \geq 1$ is a tolerance factor,

(2) if $T^{(i+j)}(n) < T^{(i)}(n)$ for some $j \leq \delta$ then, for $0 \leq j < \delta$, it must be true that $T^{(i+\delta)}(n) < T^{(i+j)}(n)$.

In words, the first condition makes sure that no skipped iterate can attain a value larger by a factor of $\tau$ than a non-skipped iterate, and the second condition makes sure that if one has to go down then one stops at a point lower than all the skipped ones. It is possible, due to proposition 1, to compute a table of values of $\delta$, of $3^{k(\delta;m_\delta)}$, and of $T^{(\delta)}(m_\delta)$, for each of the residue classes modulo $2^\Delta$ (for each residue class do the first $\Delta$ iterates and record all relevant information of the last iterate which satisfies the two conditions stated above). Such information, indexed by $T^{(i)}(n) \bmod 2^\Delta$, can be used to compute $T^{(i+\delta)}(n)$ quickly. This information is accessed randomly, so in practice $\Delta$ has to be small (all data should fit on the processor's level 1 data cache).

Table 5 presents the average value of $\delta$ for several values of $\Delta$ and for $\tau = 10$. It turns out that these averages are relatively insensitive to the value of $\tau$ as long as it is sufficiently larger than 1 (say, $\tau \geq 5$). After measuring the speed of the program for several values of $\Delta$ and on several different processors, $\Delta = 14$ was chosen (together with $\tau = 10$). The actual speed-up is smaller, by a factor of about three for 32-bit code, than what table 5 suggests, because the (assembly) code used to implement proposition 1 is considerably more complex than the code used to perform one iterate of $T$.

**2.3. Putting it all together.** In our algorithm the $3x + 1$ conjecture was tested on consecutive intervals of $2^{46+12} = 2^{58}$ integers. Before testing each interval, the maximum excursion threshold was set to one-thousandth of the largest maximum excursion available at the time (since $\tau = 10$, one-tenth would have been enough), and the stopping time threshold was set to the largest stopping time available at the time minus 200. Using hindsight, the thresholds were set to reasonable levels for the first interval; the missed record-holders were already known from earlier computations. Although these thresholds were somewhat conservative, they gave rise to a very small number of exceptional starting values (cf., for example, [**7**, theorem 2.4]). The surviving leaves of the depth 46 tree were then generated. For each such leaf $2^{12}$ initial values were tested. Those congruent to 2 mod 3, to 4 mod 9, or to 10 mod 81 were immediately discarded because, as explained above, testing them would be redundant. For each undiscarded initial value several iterates were performed in each computational step until i) the number of iterates crossed the stopping time threshold, or ii) the current iterate value was larger than the maximum excursion threshold, or iii) the current iterate value was smaller than the initial value. In the first two cases the initial value was exceptional, and required further processing: the exact values of $t(n)$ and $\sigma(n)$ were computed and were put, if necessary, in a sorted list of candidate record-holders. Since these exceptional values were very rare, testing them took an insignificant amount of time. After testing all surviving nodes, the record-holders, if any, were saved, and the entire process was done again for the next interval.

The entire work was parallelized in the following way. One computer, the "master," distributed work to a pool of "workers" and collected the results they produced. Each piece of work was distributed to two different workers, which processed it independently of each other. The results (record-holder candidates and a cyclic redundancy check sum which depended on the entire computation) were returned to the master, which made sure that they matched exactly; in the 81 CPU-years used, this redundancy test failed only once, which forced the incompatible results to be invalidaded and, thus, to be recomputed. The master generated the pieces of work by expanding the tree up to depth 20. Each depth-20 node received by a worker was expanded to depth 46, and each surviving leaf was processed as explained above (using assembly code in the inner loop to make it faster). It took between 5 minutes and 20 hours to process each piece of work on a typical worker computer (bought in 2004 or earlier), respectively for depth-20 nodes with the smallest or the highest $k$ (i.e., number of "$n$ is odd" branches taken). There were many more depth-20 nodes with a small $k$ than with a large $k$, so, by distributing the pieces of work corresponding to a high $k$ first, the disparity of execution times posed no problem because, as was our case, there were many more pieces of work to be done than there were workers to do them. Although it was possible to approximately equalize the execution times for all work units by using a tree expansion of variable depth on the master, we decided not to do so.

**2.4. Results.** The verification of the $3x+1$ conjecture using the algorithm described above started on the summer of 2004. Each interval of $2^{58}$ integers took about 4 CPU-years (between two and four months of real time) to be double-checked. Between 20 and about 100 different computers, all belonging to computing laboratories of the University of Aveiro, participated in the verification effort. On January 2009, when this distributed computation was terminated, 20 such intervals had been processed. For the record, our previous program was run up to $100 \times 2^{50}$, and was terminated in April 2000. Eric Roosendaal's program had reached $309 \times 2^{50}$ when our new program overtook his on December 2004; on June 2010 it had reached $864 \times 2^{50}$. (Hence, the record-holders reported below between $100 \times 2^{50}$ and $309 \times 2^{50}$ were discovered by Eric Roosendaal.)

As expected, no counter-example of the $3x+1$ conjecture with $n \leq 20 \times 2^{58}$ was found. Table 6 present all $t$ record-holders that are currently known, and table 7 does the same for $\sigma$ record-holders. As expected, the $t$ record-holders are relatively close to $n^2$. As can be seen in figure 2, it remains unclear whether $t(n)/n^2$ is bounded from above or not. For example, with the available empirical data it is not possible to discriminate between the following two possibilities: i) $t(n) = \mathcal{O}(n^2)$, or ii) $t(n) = \mathcal{O}(n^2 \log \log n)$.

## 3. Empirical verification of the Kakutani $5x+1$ and $7x+1$ conjectures

The computational verification of the Kakutani $5x+1$ or of the $7x+1$ conjectures can be done along lines similar to those used for the verification of the $3x+1$ conjecture. To avoid redundancy, only the main algorithmic differences between

## TABLE 6. List of the $t$ record-holders smaller than $20 \times 2^{58}$.

| $n$ | $t(n)$ | $t(n)/n^2$ |
|---:|---:|---:|
| 2 | 2 | 0.500 |
| 3 | 8 | 0.889 |
| 7 | 26 | 0.531 |
| 15 | 80 | 0.356 |
| 27 | 4 616 | 6.332 |
| 255 | 6 560 | 0.101 |
| 447 | 19 682 | 0.099 |
| 639 | 20 762 | 0.051 |
| 703 | 125 252 | 0.253 |
| 1 819 | 638 468 | 0.193 |
| 4 255 | 3 405 068 | 0.188 |
| 4 591 | 4 076 810 | 0.193 |
| 9 663 | 13 557 212 | 0.145 |
| 20 895 | 25 071 632 | 0.057 |
| 26 623 | 53 179 010 | 0.075 |

| $n$ | $t(n)$ | $t(n)/n^2$ |
|---:|---:|---:|
| 31 911 | 60 506 432 | 0.059 |
| 60 975 | 296 639 576 | 0.080 |
| 77 671 | 785 412 368 | 0.130 |
| 113 383 | 1 241 055 674 | 0.097 |
| 138 367 | 1 399 161 680 | 0.073 |
| 159 487 | 8 601 188 876 | 0.338 |
| 270 271 | 12 324 038 948 | 0.169 |
| 665 215 | 26 241 642 656 | 0.059 |
| 704 511 | 28 495 741 760 | 0.057 |
| 1 042 431 | 45 119 577 824 | 0.042 |
| 1 212 415 | 69 823 368 404 | 0.048 |
| 1 441 407 | 75 814 787 186 | 0.036 |
| 1 875 711 | 77 952 174 848 | 0.022 |
| 1 988 859 | 78 457 189 112 | 0.020 |
| 2 643 183 | 95 229 909 242 | 0.014 |

| $n$ | $t(n)$ | $t(n)/n^2$ |
|---:|---:|---:|
| 2 684 647 | 176 308 906 472 | 0.024 |
| 3 041 127 | 311 358 950 810 | 0.034 |
| 3 873 535 | 429 277 584 788 | 0.029 |
| 4 637 979 | 659 401 147 466 | 0.031 |
| 5 656 191 | 1 206 246 808 304 | 0.038 |
| 6 416 623 | 2 399 998 472 684 | 0.058 |
| 6 631 675 | 30 171 305 459 816 | 0.686 |
| 19 638 399 | 153 148 462 601 876 | 0.397 |
| 38 595 583 | 237 318 849 425 546 | 0.159 |
| 80 049 391 | 1 092 571 914 585 050 | 0.171 |
| 120 080 895 | 1 638 950 788 059 290 | 0.114 |
| 210 964 383 | 3 202 398 580 560 632 | 0.072 |
| 319 804 831 | 707 118 223 359 971 240 | 6.914 |
| 1 410 123 943 | 3 562 942 561 397 226 080 | 1.792 |
| 8 528 817 511 | 9 072 297 468 678 299 012 | 0.125 |
| 12 327 829 503 | 10 361 199 457 202 525 864 | 0.068 |
| 23 035 537 407 | 34 419 078 320 774 113 520 | 0.065 |
| 45 871 962 271 | 41 170 824 451 011 417 002 | 0.020 |
| 51 739 336 447 | 57 319 808 570 806 999 220 | 0.021 |
| 59 152 641 055 | 75 749 682 531 195 100 772 | 0.022 |
| 59 436 135 663 | 102 868 194 685 920 926 084 | 0.029 |
| 70 141 259 775 | 210 483 556 894 194 914 852 | 0.043 |
| 77 566 362 559 | 458 306 514 538 433 899 928 | 0.076 |
| 110 243 094 271 | 686 226 824 783 134 190 180 | 0.056 |
| 204 430 613 247 | 707 630 396 504 827 495 544 | 0.017 |
| 231 913 730 799 | 1 095 171 911 941 437 256 778 | 0.020 |
| 272 025 660 543 | 10 974 241 817 835 208 981 874 | 0.148 |
| 446 559 217 279 | 19 766 638 455 389 030 190 536 | 0.099 |
| 567 839 862 631 | 50 270 086 612 792 993 117 994 | 0.156 |
| 871 673 828 443 | 200 279 370 410 625 061 016 864 | 0.264 |
| 2 674 309 547 647 | 385 209 974 924 871 186 526 136 | 0.054 |
| 3 716 509 988 199 | 103 968 231 672 274 974 522 437 732 | 7.527 |
| 9 016 346 070 511 | 126 114 763 591 721 667 597 212 096 | 1.551 |
| 64 848 224 337 147 | 637 053 460 104 079 232 893 133 864 | 0.151 |
| 116 050 121 715 711 | 1 265 292 033 916 892 480 613 118 196 | 0.094 |
| 201 321 227 677 935 | 2 636 975 512 088 803 001 946 985 208 | 0.065 |
| 265 078 413 377 535 | 2 857 204 078 078 966 555 847 716 826 | 0.041 |
| 291 732 129 855 135 | 3 537 558 936 133 726 760 243 328 464 | 0.042 |
| 394 491 988 532 895 | 6 054 282 113 227 445 504 606 919 650 | 0.039 |
| 406 738 920 960 667 | 12 800 696 705 021 228 411 442 619 682 | 0.077 |
| 613 450 176 662 511 | 22 881 441 742 972 862 145 992 619 776 | 0.061 |
| 737 482 236 053 119 | 37 684 665 798 782 446 690 107 505 928 | 0.069 |
| 1 254 251 874 774 375 | 1 823 036 311 464 280 263 720 932 141 024 | 1.159 |
| 5 323 048 232 813 247 | 1 964 730 439 297 455 725 829 478 995 944 | 0.069 |
| 8 562 235 014 026 655 | 13 471 057 008 351 679 202 003 944 688 336 | 0.184 |
| 10 709 980 568 908 647 | 175 294 593 968 539 094 415 936 960 141 122 | 1.528 |
| 49 163 256 101 584 231 | 301 753 104 069 007 668 258 074 264 675 786 | 0.125 |
| 82 450 591 202 377 887 | 875 612 750 096 198 197 075 499 421 245 450 | 0.129 |
| 93 264 792 503 458 119 | 2 115 362 774 686 865 777 485 863 406 680 032 | 0.243 |
| 172 545 331 199 510 631 | 2 118 089 541 282 012 618 909 185 268 056 780 | 0.071 |
| 212 581 558 780 141 311 | 2 176 718 166 004 315 761 101 410 771 585 688 | 0.048 |
| 255 875 336 134 000 063 | 2 415 428 612 584 587 115 646 993 931 986 234 | 0.037 |
| 484 549 993 128 097 215 | 4 332 751 846 533 208 436 890 106 993 276 834 | 0.018 |
| 562 380 758 422 254 271 | 6 718 947 974 962 862 349 115 040 884 231 904 | 0.021 |
| 628 226 286 374 752 923 | 31 268 160 888 027 375 005 205 169 043 314 754 | 0.079 |
| 891 563 131 061 253 151 | 140 246 903 347 442 029 303 138 585 287 425 762 | 0.176 |
| 1 038 743 969 413 717 663 | 159 695 671 984 678 120 932 209 599 662 553 676 | 0.148 |
| 1 980 976 057 694 848 447 | 32 012 333 661 096 566 765 082 938 647 132 369 010 | 8.158 |

TABLE 7. List of the $\sigma$ record-holders smaller than $20 \times 2^{58}$.

| $n$ | $\sigma(n)$ | $n$ | $\sigma(n)$ |
|---:|---:|---:|---:|
| 2 | 1 | 63 728 127 | 376 |
| 3 | 4 | 217 740 015 | 395 |
| 7 | 7 | 1 200 991 791 | 398 |
| 27 | 59 | 1 827 397 567 | 433 |
| 703 | 81 | 2 788 008 987 | 447 |
| 10 087 | 105 | 12 235 060 455 | 547 |
| 35 655 | 135 | 898 696 369 947 | 550 |
| 270 271 | 164 | 2 081 751 768 559 | 606 |
| 362 343 | 165 | 13 179 928 405 231 | 688 |
| 381 727 | 173 | 31 835 572 457 967 | 712 |
| 626 331 | 176 | 70 665 924 117 439 | 722 |
| 1 027 431 | 183 | 739 448 869 367 967 | 728 |
| 1 126 015 | 224 | 1 008 932 249 296 231 | 886 |
| 8 088 063 | 246 | 118 303 688 851 791 519 | 902 |
| 13 421 671 | 287 | 180 352 746 940 718 527 | 966 |
| 20 638 335 | 292 | 1 236 472 189 813 512 351 | 990 |
| 26 716 671 | 298 | 2 602 714 556 700 227 743 | 1 005 |
| 56 924 955 | 308 | | |

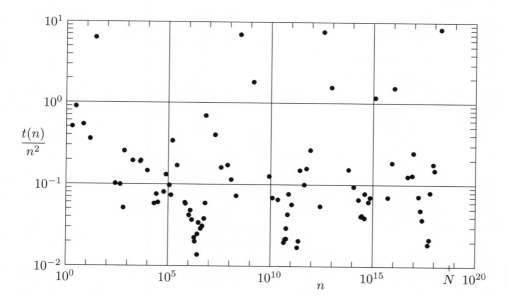

FIGURE 2. The values of $t(n)/n^2$ for the maximum excursion record-holders smaller than $N = 20 \times 2^{58}$.

the verification of the $3x + 1$ and the Kakutani $5x + 1$ conjectures are described in this section; the $7x + 1$ case is similar and its details are left for the reader to complete.

Proposition 1, which lies at the heart of the algorithm used to verify empirically the $3x+1$ conjecture, has a counterpart for the $C_{5,\{2,3\}}$ mapping. Given the nature of this mapping, it turns out that it is more convenient to put it in the following form [15]:

$$(7) \qquad C_{5,\{2,3\}}(n) = \begin{cases} n/2, & \text{if } n \bmod 2 = 0 \\ n/3, & \text{if } n \bmod 2 \neq 0 \text{ and } n \bmod 3 = 0 \\ 5n + 1, & \text{otherwise.} \end{cases}$$

PROPOSITION 2. *Let $k_2(i, m_i)$, $k_3(i, m_i)$, and $k_5(i, m_i)$ be respectively the number of $n/2$, $n/3$, and $5x + 1$ branches of $C_{5,\{2,3\}}(n)$ taken in the computation of $C^{(i)}_{5,\{2,3\}}(m_i)$. Let $r_i = 2^{k_2(i,m_i)}3^{k_3(i,m_i)}$, and let $n_0 = r_i n_i + m_i$, with $n_i = \lfloor n/r_i \rfloor$ and $m_i = n_0 \bmod r_i$. Then,*

$$(8) \qquad C^{(i)}_{5,\{2,3\}}(n_0) = 5^{k_5(i;m_i)}n_i + C^{(i)}_{5,\{2,3\}}(m_i).$$

The application of this proposition is not as straightforward as that of proposition 1, because the definition of $m_i$ is somewhat circular. Since the active branch of $C_{5,\{2,3\}}$ depends only on $n \bmod 6$, the values of $2^{k_2(i,m_i)}3^{k_3(i,m_i)}$, $5^{k_5(i,m_i)}$, and $C^{(i)}_{5,\{2,3\}}(m_i)$ can be extracted from a table indexed by $n_0 \bmod 6^i$. In practice, as described below, a shorter table is used.

As in the $3x + 1$ case, the iterates of $C_{5,\{2,3\}}$ can be sorted without knowing the value of $n_i$. For example, the first seven iterates of $48n + 17$ are

$$(9) \qquad \begin{aligned} T^{(0)}(48n + 17) &= 48n + 17 & T^{(4)}(48n + 17) &= 300n + 108 \\ T^{(1)}(48n + 17) &= 240n + 86 & T^{(5)}(48n + 17) &= 150n + 54 \\ T^{(2)}(48n + 17) &= 120n + 43 & T^{(6)}(48n + 17) &= 75n + 27 \\ T^{(3)}(48n + 17) &= 600n + 216 & T^{(7)}(48n + 17) &= 25n + 9, \end{aligned}$$

which implies that it is not necessary to test initial values of this form. Putting the polynomial coefficients in two rows of a matrix and rearranging its columns so that the first row becomes sorted in increasing order gives (the last row is the iterate number) gives

$$(10) \qquad \begin{bmatrix} 25 & 48 & 75 & 120 & 150 & 240 & 300 & 600 \\ 9 & 17 & 27 & 43 & 54 & 86 & 108 & 216 \\ 7 & 0 & 6 & 2 & 5 & 1 & 4 & 3 \end{bmatrix}.$$

Notice that the second row also becomes sorted. This property of the iterates, verified empirically for $k_2(i, m_i) \leq 16$ and for $k_3(i, m_i) \leq 8$, can be used to speed-up considerably the verification of the Kakutani $5x + 1$ conjecture, in a manner similar to that described in section 2.

Instead of generating the surviving leaves by pruning a tree at a large depth, in the $5x + 1$ case a simpler approach was used. The first iterates with $k_2(i, m_i) \leq 15$ and $k_3(i, m_i) \leq 6$ of all initial values modulo $2^{15}3^6$ were examined, and those for which one iterate fell below the initial value were discarded. Out of $23\,887\,872$ residues only $92\,242$ survived. The $k_2(i, m_i)$ and $k_3(i, m_i)$ limits given above were those that gave the smallest fraction of surviving residues under the constraint that the total number of them was below $10^5$ (cf. table 8). This constrain, and the one of the next paragraph, were imposed to limit the amount of memory used by the program; with this limit all frequently accessed data fits nicely in the processor's data caches. The verification of the $7x + 1$ case used the same approach; out of the $2^8 3^4 5^3 = 2\,592\,000$ residue classes, only $63\,507$ survived.

An in the $3x + 1$ case, in the $5x + 1$ case several iterates were performed in a single step. Table 9 presents the average number of iterates and the number

TABLE 8. Fraction of surviving residues and their number for $k_2(i, m_i) \leq K_2$ and for $k_3(i, m_i) \leq K_3$.

| $K2 \backslash K3$ | 3 | 4 | 5 | 6 | 7 |
|---|---|---|---|---|---|
| 12 | 0.015 263 | 0.009 745 | 0.007 697 | 0.006 725 | 0.006 440 |
|  | 1 688 | 3 233 | 7 661 | 20 080 | 57 687 |
| 13 | 0.015 006 | 0.008 919 | 0.006 370 | 0.005 436 | 0.004 991 |
|  | 3 319 | 5 918 | 12 681 | 32 464 | 89 418 |
| 14 | 0.014 289 | 0.008 121 | 0.005 502 | 0.004 319 | 0.003 900 |
|  | 6 321 | 10 778 | 21 904 | 51 580 | 139 738 |
| 15 | 0.014 196 | 0.007 731 | 0.005 105 | 0.003 861 | 0.003 308 |
|  | 12 560 | 20 520 | 40 651 | 92 242 | 237 059 |
| 16 | 0.014 078 | 0.007 589 | 0.004 730 | 0.003 529 | 0.002 929 |
|  | 24 911 | 40 284 | 75 329 | 168 579 | 419 811 |

TABLE 9. Average number of iterates performed in one step for $k_2(i, m_i) \leq K_2$ and for $k_3(i, m_i) \leq K_3$, and the number $2^{K_2} 3^{K_3}$ of residue classes for which data has to be recorded.

| $K2 \backslash K3$ | 2 | 3 | 4 | 5 | 6 |
|---|---|---|---|---|---|
| 4 | 3.500 000 | 4.319 444 | 4.788 580 | 5.020 062 | 5.136 488 |
|  | 144 | 432 | 1 296 | 3 888 | 11 664 |
| 5 | 4.034 722 | 5.020 833 | 5.677 469 | 6.048 354 | 6.246 999 |
|  | 288 | 864 | 2 592 | 7 776 | 23 328 |
| 6 | 4.451 389 | 5.719 907 | 6.543 210 | 7.054 913 | 7.342 635 |
|  | 576 | 1 728 | 5 184 | 15 552 | 46 656 |
| 7 | 4.692 708 | 6.260 417 | 7.324 267 | 8.000 514 | 8.406 143 |
|  | 1 152 | 3 456 | 10 368 | 31 104 | 93 312 |
| 8 | 4.818 576 | 6.630 787 | 7.966 821 | 8.859 986 | 9.423 547 |
|  | 2 304 | 6 912 | 20 736 | 62 208 | 186 624 |

of rows of the table required to store all relevant information for several maximal values of $k_2(i, m_i)$ and of $k_3(i, m_i)$. To conserve memory, the number of rows was constrained to be below 12 000, which gave $k_2(i, m_i) \leq 7$ and $k_3(i, m_i) \leq 4$ as the best choices. Given these limits, the maximum excursion threshold $\tau$ has to be larger than 45.212 ($\tau = 50$ was used in the program). Because the $7x + 1$ case was given less attention, the iterates of $C_{7, \{2,3,5\}}$ were performed one at a time.

**3.1. Results.** The verification of the Kakutani $5x + 1$ conjecture was performed on the spare time of an old 450MHz Pentium III computer (kindly lent by Miguel Oliveira e Silva) and was stopped after about 5 CPU-months when $10^{16}$ was reached. Table 10 presents the maximum-excursion record-holders of $C_{5, \{2,3\}}$ found during this verification. For this mapping, the maximum excursion for an initial value of $n$ was denoted by $t_5(n)$.

The verification of the Kakutani $7x + 1$ conjecture was also performed on the spare time of the same 450MHz Pentium III computer and was stopped after about 2 CPU-months when $10^{14}$ was reached. Table 11 presents the maximum-excursion record-holders of $C_{7, \{2,3,5\}}$ found during this verification. For this mapping, the maximum excursion for an initial value of $n$ was denoted by $t_7(n)$.

TABLE 10. List of all maximum excursion record-holders for $C_{5,\{2,3\}}$ up to $10^{16}$.

| $n$ | $t_5(n)$ | $n$ | $t_5(n)$ |
|---:|---:|---:|---:|
| 1 | 6 | 324 965 021 | 5 370 031 738 266 |
| 5 | 66 | 398 902 655 | 22 351 914 050 766 |
| 17 | 216 | 1 035 539 573 | 26 863 385 244 516 |
| 23 | 366 | 1 163 240 831 | 66 928 916 839 866 |
| 41 | 516 | 2 436 773 015 | 91 027 289 009 916 |
| 53 | 666 | 2 909 738 477 | 148 405 221 792 366 |
| 65 | 816 | 3 968 216 183 | 165 479 090 366 616 |
| 71 | 1 116 | 4 843 258 391 | 238 909 977 203 766 |
| 77 | 2 016 | 4 995 428 117 | 305 737 834 543 266 |
| 95 | 4 866 | 10 321 682 681 | 432 446 197 666 266 |
| 221 | 5 766 | 11 788 426 685 | 833 453 179 667 016 |
| 317 | 8 616 | 23 245 762 325 | 1 067 042 033 501 166 |
| 365 | 9 516 | 27 293 002 007 | 2 074 276 490 244 516 |
| 383 | 229 866 | 29 162 173 757 | 4 539 052 265 854 866 |
| 3 317 | 286 116 | 54 699 628 127 | 5 413 005 791 199 516 |
| 3 575 | 473 616 | 71 453 529 533 | 7 819 905 704 484 366 |
| 3 605 | 737 016 | 81 867 799 181 | 38 513 221 689 541 866 |
| 6 473 | 3 230 766 | 303 304 272 701 | 38 673 525 305 161 116 |
| 24 125 | 4 438 266 | 324 164 559 605 | 47 524 316 276 011 116 |
| 31 901 | 6 260 016 | 335 300 999 321 | 50 965 956 755 859 366 |
| 39 965 | 10 637 016 | 389 794 947 095 | 73 006 100 494 883 616 |
| 44 183 | 107 662 496 | 469 848 666 647 | 376 189 810 799 652 366 |
| 163 733 | 117 661 116 | 865 172 295 191 | 1 101 033 339 961 411 116 |
| 317 885 | 152 291 166 | 2 014 753 058 015 | 1 118 075 145 401 479 866 |
| 490 541 | 176 254 866 | 2 296 736 656 535 | 1 141 744 958 458 823 616 |
| 519 113 | 179 900 766 | 2 795 130 933 341 | 1 543 894 242 919 637 016 |
| 558 365 | 201 566 166 | 2 981 236 831 103 | 3 533 413 581 914 292 366 |
| 602 591 | 230 949 516 | 4 762 958 102 933 | 12 687 240 379 655 386 116 |
| 707 735 | 662 224 116 | 9 770 208 066 941 | 14 053 612 657 903 775 616 |
| 753 023 | 860 107 416 | 12 193 036 187 117 | 17 959 572 148 315 073 616 |
| 1 019 615 | 1 221 182 016 | 18 425 237 431 901 | 22 510 347 117 096 323 616 |
| 1 463 897 | 1 268 296 866 | 26 891 720 049 737 | 31 539 798 438 276 477 366 |
| 1 597 973 | 2 163 215 616 | 26 998 785 794 453 | 57 265 208 836 397 574 516 |
| 1 752 575 | 4 826 872 416 | 39 398 516 920 445 | 122 174 458 860 581 184 366 |
| 4 595 735 | 7 167 004 866 | 62 801 989 560 725 | 215 376 177 400 048 543 266 |
| 6 197 855 | 15 435 004 866 | 78 088 737 353 087 | 685 571 299 553 745 260 616 |
| 8 757 725 | 17 152 119 516 | 144 463 113 398 813 | 1 378 847 067 203 080 186 116 |
| 12 447 533 | 106 102 036 116 | 299 734 127 351 573 | 1 581 936 475 210 722 886 116 |
| 31 001 117 | 314 752 213 116 | 360 492 724 870 421 | 12 444 914 384 485 881 157 416 |
| 46 389 533 | 415 898 911 116 | 1 972 745 444 313 335 | 24 632 173 613 577 422 104 866 |
| 85 689 581 | 600 185 953 116 | 2 316 066 543 155 957 | 47 973 089 912 494 397 762 016 |
| 130 363 895 | 841 506 744 366 | 4 450 574 650 632 053 | 69 463 680 546 423 712 363 266 |
| 190 573 175 | 1 943 958 512 016 | 6 137 195 959 891 733 | 90 397 020 019 079 025 229 866 |
| 310 680 671 | 2 310 292 588 266 | 6 510 438 441 044 213 | 173 616 401 711 338 878 613 266 |
| 324 560 351 | 2 681 672 294 916 | | |

## 4. The rate of growth of the Kakutani $5x+1$ and $7x+1$ maximum excursion record-holders

Upon close examination, the record-holders of table 10 appear to satisfy $t_5(n) \approx n^{\rho_5}$, with $\rho_5 \approx 1.44$ (cf. figure 3), and whose of table 11 appear to satisfy $t_7(n) \approx n^{\rho_7}$, with $\rho_7 \approx 2.22$ (cf. figure 4). This section presents non-rigorous probabilistic arguments that support these empirical observations, and that yield conjectured values for $\rho_5$ and $\rho_7$ (conjectures 1 and 2). Only the $5x+1$ case will be described in detail; the same arguments can be used to study the maximum excursions of other generalized Collatz mappings [11, 9, 10].

Modulo 6, the iterates of $C_{5,\{2,3\}}$ for a large initial $n$ appear to be accurately modeled by a Markov process with six states, one for each of the residue classes modulo 6. Since the active branch of $C_{5,\{2,3\}}$ depends only on $n \bmod 6$, it turns

TABLE 11. List of all maximum excursion record-holders for $C_{7,\{2,3,5\}}$ up to $10^{14}$.

| $n$ | $t_7(n)$ |
|---:|---:|
| 1 | 8 |
| 7 | 50 |
| 11 | 162 |
| 19 | 470 |
| 31 | 9 584 |
| 43 | 28 400 |
| 163 | 91 890 |
| 283 | 193 040 |
| 403 | 265 070 |
| 1 111 | 291 824 |
| 1 123 | 337 100 |
| 1 243 | 388 830 |
| 1 303 | 524 070 |
| 1 549 | 3 231 824 |
| 1 963 | 18 052 070 |
| 4 123 | 30 949 850 |
| 9 643 | 31 021 880 |
| 10 003 | 65 552 768 |
| 11 539 | 774 659 600 |
| 21 431 | 10 276 607 888 |
| 76 963 | 38 128 783 428 |
| 97 031 | 190 067 750 300 |
| 468 109 | 4 835 458 627 140 |
| 1 351 963 | 25 515 567 812 750 |
| 4 553 323 | 53 371 819 453 590 |
| 4 778 471 | 175 961 993 364 704 |
| 5 163 139 | 188 057 534 682 690 |
| 6 563 551 | 2 887 456 873 364 610 |
| 7 618 843 | 33 593 011 618 349 100 |
| 45 214 123 | 42 910 318 316 713 730 |
| 65 704 243 | 68 432 765 698 835 210 |
| 161 738 803 | 168 168 088 356 840 650 |
| 202 903 723 | 215 493 326 348 709 620 |
| 208 854 751 | 238 207 407 183 611 090 |
| 240 605 389 | 3 148 201 366 707 916 940 |
| 315 329 203 | 3 170 755 559 888 262 200 |
| 539 397 283 | 4 611 059 471 298 024 368 |
| 598 768 243 | 4 743 389 727 551 888 120 |
| 1 201 681 111 | 39 208 014 304 697 980 190 |
| 1 797 386 923 | 111 445 879 791 914 102 660 |
| 2 379 632 923 | 546 700 615 046 844 620 510 |
| 4 577 697 439 | 1 530 529 988 512 829 992 850 |
| 8 118 341 803 | 20 217 336 401 527 498 400 420 |
| 21 681 207 511 | 133 286 378 943 790 685 629 430 |
| 49 996 533 811 | 1 772 586 503 063 406 674 881 070 |
| 120 645 183 403 | 9 093 063 348 479 883 964 868 720 |
| 247 292 934 871 | 11 775 623 360 776 250 795 887 460 |
| 251 172 653 443 | 14 908 129 093 778 286 535 042 490 |
| 588 222 381 043 | 40 790 644 396 515 322 811 208 918 |
| 713 540 981 143 | 7 157 403 932 644 319 420 401 386 080 |
| 11 408 899 638 811 | 9 192 294 603 131 852 785 127 719 788 |
| 14 192 648 902 631 | 1 310 933 600 826 889 791 266 912 256 770 |
| 16 877 362 155 871 | 3 241 740 218 240 695 200 803 510 633 870 |

out to be convenient to describe what happens to each of the six residue classes:

(11)
$$
\begin{aligned}
C_{5,\{2,3\}}(6n+0) &= 3n & C_{5,\{2,3\}}(6n+3) &= 2n+1 \\
C_{5,\{2,3\}}(6n+1) &= 30n+6 & C_{5,\{2,3\}}(6n+4) &= 3n+2 \\
C_{5,\{2,3\}}(6n+2) &= 3n+1 & C_{5,\{2,3\}}(6n+5) &= 30n+26.
\end{aligned}
$$

In this particular case the Markov chain description of the iterates modulo 6 can be easily constructed from this way of expressing the branches of $C_{5,\{2,3\}}$ [9]. For example, since $C_{5,\{2,3\}}(6n+0) = 3n$, it follows that if the chain is in state 0 (corresponding to the residue class 0 modulo 6) then the next state is either 0 or 3 with equal probability ($3n \bmod 6$ is equal to 0 when $n$ is even and is equal to 3 when $n$ is odd). Doing this analysis for all branches yields the irreducible state transition matrix

$$(12) \qquad \boldsymbol{P}_{5,\{2,3\}} = \frac{1}{6} \begin{bmatrix} 3 & 0 & 0 & 3 & 0 & 0 \\ 6 & 0 & 0 & 0 & 0 & 0 \\ 0 & 3 & 0 & 0 & 3 & 0 \\ 0 & 2 & 0 & 2 & 0 & 2 \\ 0 & 0 & 3 & 0 & 0 & 3 \\ 0 & 0 & 6 & 0 & 0 & 0 \end{bmatrix},$$

which has the stationary distribution

$$(13) \qquad \boldsymbol{\pi}_{5,\{2,3\}} = \frac{1}{27} \begin{bmatrix} 8 & 4 & 4 & 6 & 2 & 3 \end{bmatrix}.$$

(The empirical state transition matrix obtained by performing many consecutive iterates for many large initial values of $n$ agrees quite well with $\boldsymbol{P}_{5,\{2,3\}}$.) By ignoring the constant offset present in five of the six branches of (11), the value of $\log C_{5,\{2,3\}}^{(m)}(n)$ can be well approximated, for large $n$, by

$$(14) \qquad \log n \sum_{i=0}^{m-1} \log f_{5,\{2,3\}}(x_i), \qquad \text{with } x_i = C_{5,\{2,3\}}^{(i)}(n) \bmod 6,$$

where $f_{5,\{2,3\}}(x) = \lim_{n\to\infty} C_{5,\{2,3\}}(6n+x)/(6n)$ is the "expansion factor" function for the mapping $C_{5,\{2,3\}}$. It will be useful later on to place the values of $f_{5,\{2,3\}}(x)$ in a diagonal matrix; $f_{5,\{2,3\}}(0)$ on the first row, $f_{5,\{2,3\}}(1)$ on the second, and so on up to $f_{5,\{2,3\}}(5)$. Doing so gives

$$(15) \qquad \boldsymbol{F}_{5,\{2,3\}} = \operatorname{diag}(\boldsymbol{f}_{5,\{2,3\}}), \qquad \text{with } \boldsymbol{f}_{5,\{2,3\}} = \begin{bmatrix} \frac{1}{2} & 5 & \frac{1}{2} & \frac{1}{3} & \frac{1}{2} & 5 \end{bmatrix}.$$

Until further notice the subscript $_{5,\{2,3\}}$ will be omitted in vector, matrix, or function names; this simplifies the notation and makes the argument immediately applicable to other mappings. As mentioned above, the sequence $\{x_i\}$ can be mimicked by a realization of a Markov process $\{X_i\}$ with transition matrix $\boldsymbol{P}$, as long as the iterates do not enter a cycle. It is therefore natural to replace $\log n \sum_{i=0}^{m-1} \log f(x_i)$ by $\log n \sum_{i=1}^{m} \log f(X_i)$ in the study of record-breaking values of the former summation; one then becomes interested in the large deviation behavior of the latter summation. In particular, if

$$(16) \qquad \sum_{n=1}^{\infty} \operatorname{Prob}\left[\log n \sum_{i=1}^{m} \log f(X_i) \geq \rho \log n \text{ for some } m > 0\right]$$

diverges, then, by the Borel-Cantelli lemma [13], the value of $n \prod_{i=1}^{m} f(X_i)$ almost surely surpasses $n^\rho$ an infinite number of times (otherwise it does not); one may then hope that the same happens to the maximum excursions of the iterates of $C$. Of course, one is interested in the largest $\rho$ that makes (16) divergent.

Until The following non-rigorous argument is an adaptation of the proof of theorem 2.3 of [7], and has as objective the determination of the largest $\rho$ that makes (16) divergent. It assumes that the drift of the Markov chain, given by

$\boldsymbol{\pi f}^T$, is negative. Applying the Chernoff bound to $\sum_{i=1}^{m} \log f(X_i)$ yields

$$(17) \quad \text{Prob}\left[\sum_{i=1}^{m} \log f(X_i) \geq am\right] \leq \exp\left(-\sup_{\theta \geq 0}\left(am\theta - \log E[e^{\theta \sum_{i=1}^{m} \log f(X_i)}]\right)\right),$$

where $E[\cdot]$ denotes mathematical expectation; $E[e^{\theta \sum_{i=1}^{m} \log f(X_i)}]$ can be explicitly computed by summing the contribution of all possible paths of the chain starting from an arbitrary initial state, with initial state probability (row) vector $\boldsymbol{q}$, and ending in any of the states. Doing so yields

$$(18) \quad E[e^{\theta \sum_{i=1}^{m} \log f(X_i)}] = \boldsymbol{q} \underbrace{\boldsymbol{F}^\theta \boldsymbol{P} \cdots \boldsymbol{F}^\theta \boldsymbol{P}}_{m \text{ times}} \boldsymbol{1} = \boldsymbol{q}(\boldsymbol{F}^\theta \boldsymbol{P})^m \boldsymbol{1},$$

where $\boldsymbol{1}$ is a column vector of all ones. Assuming that $\boldsymbol{F}^\theta \boldsymbol{P}$ has only one eigenvalue of largest absolute value, denoted by $\lambda(\boldsymbol{F}^\theta \boldsymbol{P})$, the previous equation can be put, for large $m$, in the following form:

$$(19) \quad \log E[e^{\theta \sum_{i=1}^{m} \log f(X_i)}] = \left(m + o(m)\right) \log \lambda(\boldsymbol{F}^\theta \boldsymbol{P}).$$

Using this result in (17) gives

$$(20) \quad \text{Prob}\left[\sum_{i=1}^{m} \log f(X_i) \geq am\right] \leq \exp\left(-m \sup_{\theta \geq 0}\left(a\theta - (1 + o(1)) \log \lambda(\boldsymbol{F}^\theta \boldsymbol{P})\right)\right).$$

Empirical evidence suggests that this bound is quite tight. This leads to the conjecture that for large $m$

$$(21) \quad \text{Prob}\left[\sum_{i=1}^{m} \log f(X_i) \geq am\right] = \exp\left(-m(1 + o(1)) \sup_{\theta \geq 0}\left(a\theta - \log \lambda(\boldsymbol{F}^\theta \boldsymbol{P})\right)\right).$$

This formula is similar to the formula for $\text{Prob}[S_m > am]$ that appears in the proof of theorem 2.3 of [7]. The rest of the proof of that theorem can then be used with very few modifications. It turns out that the largest value of $\rho$ which is conjectured to make (16) divergent is given by

$$(22) \qquad\qquad \rho^* = 1 + 1/\theta^*,$$

$\theta^*$ being the only positive solution of the equation

$$(23) \qquad\qquad \lambda(\boldsymbol{F}^\theta \boldsymbol{P}) = 1.$$

It is interesting to observe that the perturbation theory of linear operators [6] gives

$$(24) \qquad\qquad \left.\frac{d\lambda(\boldsymbol{F}^\theta \boldsymbol{P})}{d\theta}\right|_{\theta=0} = \boldsymbol{\pi f}^T,$$

i.e., a negative drift appears to be necessary for (23) to have a positive solution (since $d^2\lambda(\boldsymbol{F}^\theta \boldsymbol{P})/d\theta^2$ appears to be positive for positive $\theta$).

For the $C_{5,\{2,3\}}$ mapping, numerical computations give for $\theta^*$ and $\rho^*$ the approximate values

$$(25) \qquad \theta^* \approx 2.258\,548\,728\,610 \qquad \text{and} \qquad \rho_5 = \rho^* \approx 1.442\,762\,198\,279.$$

Exactly the same values were obtained for the mapping $C_{5,\{3,2\}}$. That was to be expected because the mappings $C_{K,\mathcal{D}}$ and $C_{K,\mathcal{D}'}$, where $\mathcal{D}'$ is a permutation of $\mathcal{D}$, have exactly the same maximum excursions. For the $C_{7,\{2,3,5\}}$ mapping, and for

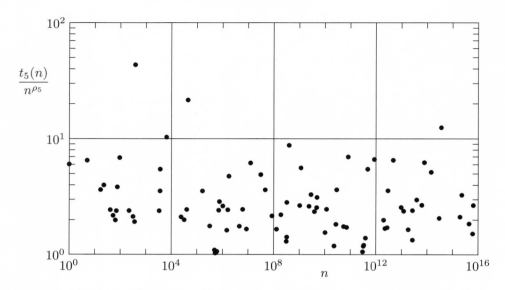

FIGURE 3. The values of $t_5(n)/n^{\rho_5}$ for the $C_{5,\{2,3\}}$ maximum excursion record-holders, with the value of $\rho_5$ given by conjecture 1.

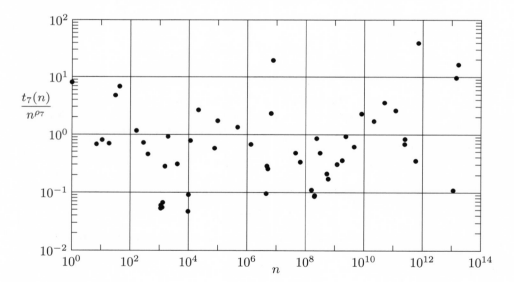

FIGURE 4. The values of $t_7(n)/n^{\rho_7}$ for the $C_{7,\{2,3,5\}}$ maximum excursion record-holders, with the value of $\rho_7$ given by conjecture 2.

its permuted variants, numerical computations give for $\theta^*$ and $\rho^*$ the approximate values

$$(26) \qquad \theta^* \approx 0.822\,701\,289\,247 \qquad \text{and} \qquad \rho_7 = \rho^* \approx 2.215\,508\,001\,593.$$

These conjectured rates of growth of the iterates of $C_{5,\{2,3\}}$ and of $C_{7,\{2,3,5\}}$ agree quite well with all available empirical data, as can be seen in figures 3 and 4.

# References

[1] Jacques Arsac. Algorithmes pour vérifier la conjecture de Syracuse. *Comptes Rendus de l'Académie de Sciences de Paris*, 303, Série I(4):155–159, 1986.

[2] R. Dunn. On Ulam's problem. Technical Report CU-CS-011-73, Department of Computer Science, University of Colorado, Boulder, 1973.

[3] Shalom Eliahou. The $3x + 1$ problem: New lower bounds on nontrivial cycle lengths. *Discrete Mathematics*, 118(1–3):45–56, 1993.

[4] Lynn E. Garner. On the collatz $3n + 1$ algorithm. *Proceedings of the American Mathematical Society*, 82(1):19–22, May 1981.

[5] Ronald L. Graham, Donald E. Knuth, and Oren Patashnik. *Concrete Mathematics*. Addison-Wesley, Reading, Massachusetts, second edition, 1994.

[6] Tosio Kato. *Perturbation Theory of Linear Operators*. Springer-Verlag, Berlin / New York, second edition, 1976.

[7] J. C. Lagarias and A. Weiss. The $3x + 1$ problem: Two stochastic models. *The Annals of Applied Probability*, 2(1):229–261, 1992.

[8] Gary T. Leavens and Mike Vermeulen. $3x + 1$ search programs. *Computers and Mathematics with Applications*, 24(11):79–99, 1992.

[9] G. M. Leigh. A Markov process underlying the generalized Syracuse algorithm. *Acta Arithmetica*, XLVI:125–143, 1986.

[10] K. R. Matthews. Generalized $3x + 1$ mappings: Markov chains and ergodic theory. This book.

[11] K. R. Matthews and A. M. Watts. A Markov approach to the generalized Syracuse algorithm. *Acta Arithmetica*, XLV:29–42, 1985.

[12] Tomás Oliveira e Silva. Maximum excursion and stopping time record-holders for the $3x + 1$ problem: Computational results. *Mathematics of Computation*, 68(225):371–384, January 1999.

[13] A. N. Shiryaev. *Probability*, volume 95 of *Graduate Texts in Mathematics*. Springer-Verlag, Berlin / New York, second edition, 1996.

[14] Riho Terras. A stopping time problem on the positive integers. *Acta Arithmetica*, XXX:241–252, 1976. See also Vol. XXXV, pp. 101-102, 1979.

[15] Cheng Yu Zhang. A generalization of Kakutani's conjecture. *Nature Magazine*, 13:267–269, 1990. In Chinese.

ELECTRONICS, TELECOMMUNICATIONS, AND INFORMATICS DEPT., UNIVERSITY OF AVEIRO, PORTUGAL

INSTITUTO DE ENGENHARIA ELECTRÓNICA E TELEMÁTICA DE AVEIRO
*E-mail address*: tos@ua.pt

# PART IV.

Reprinted Early Papers

# Cyclic Sequences and Frieze Patterns
## (The Fourth Felix Behrend Memorial Lecture)

## H. S. M. Coxeter

Although I never had the pleasure of meeting Felix Behrend, his friend Bernhard Neumann (who gave the first of these lectures) is also a friend of mine. Looking at one of the obituaries, I noticed that Behrend's published works include at least one paper on sequences, as well as a book *Ulysses' Father* (F.W. Cheshire. Melbourne, 1962). Mrs. Behrend kindly lent me her copy of this delightful book. In the form of a series of bedtime stories for their elder daughter, it elaborates the famous tale of Sisyphos and the indiscretion for which he was eternally condemned to roll a heavy rock up a mountain, never quite reaching the summit before the rock would slip from his grasp and fall crashing to the bottom. Behrend mentions also that Sisyphos was one of the unfortunates whose torments were eased for a moment by the music of Orpheus, visiting Hell in an attempt to bring back his wife, Eurydice.

Behrend was interested in analogies between Homer and the Old Testament. If he were with us today, I would ask him whether he would accept, as an instance, the analogy between Eurydice and Lot's wife. Both were leaving Hell, or the devastated city of Sodom which symbolizes Hell. But there are two differences: whereas Orpheus lost Eurydice by looking back too soon, in the Bible story it was the lady herself who looked back; and whereas the Greek Eurydice was drawn down again, Lot's wife was transformed into a pillar of salt.

In another part of the book, Behrend remarks that the best way to tell a tale is backwards, getting ever closer to the origin, which is the truth. So what was the background of the story of Lot and his wife? When I was as young as the daughter who listened to her father's stories, I used to enjoy, each evening, a chapter of the Bible read by my own father or mother. Of all these chapters (including those about Joseph and his brothers, that appealed so strongly to Thomas Mann), my own favourite was Genesis 18 (beginning at verse 23), wherein Abraham was pleading with God to spare Sodom if as many as fifty righteous people still lived there. Gradually the critical number was reduced from 50 to 46, 40, 30, 20, 10. The suspense was terrific, as each smaller number seemed to be the ultimate concession. I soon learnt to appreciate the poetic language, including the magnificent line

> Shall not the Judge of all the earth do right?

Monotony is subtly avoided by different turns of phrase each time a smaller number is proposed. At first I wondered why Abraham's final number was not reduced to 5, but the next Chapter explains why this would have made no difference. The actual number of good people in Sodom was exactly 4: Abraham's nephew Lot, Lot's wife, and their two daughters. Moreover, God arranged for these four to be evacuated in the nick of time, only imposing the condition that they should not look back. Of course Mrs. Lot did so, as any good lady would if her house were on fire, but the result was disastrous.

Abraham's 50, 45, 40, 30, 20, 10 was a steadily decreasing sequence of positive integers, and thus could not go on for ever. On the other hand, one increasing sequence that does go on for ever is the sequence of Fibonacci numbers $f_1, f_2, \ldots$ given by

$$f_1 = f_2 = 1, \quad f_{r+1} = f_r + f_{r-1}.$$

This can be transformed into a cyclic sequence by taking residues modulo $n$, for any integer $n > 1$. For instance, we may take $n = 4$ and obtain the "6-cycle"

$$1, 1, 2, 3, 1, 0, 1, 1, 2, 3, 1, \ldots.$$

Another cyclic sequence is given by

$$u_{r+1} = (1 + u_r)/u_{r-1},$$

where $u_1$ and $u_2$ can be any positive numbers. For instance, if $u_1 = u_2 = 1$, the sequence is

$$1, 1, 2, 3, 2, 1, 1, 2, 3, 2, 1, \ldots.$$

This 5-cycle seems to have been transmitted in the form of mathematical gossip for a long time. For instance, R. C. Lyness mentioned it in the Mathematical Gazette, *26* (1942), p. 62, *29* (1946), p. 231, and *45* (1961), pp. 207–209. I first heard of it in a lecture by H. F. Baker when I was a student in Cambridge. One mathematician, whom I met recently, ascribed it to Abel. The following neat proof is by my Canadian colleague Israel Halperin. From

$$u_1 u_3 = 1 + u_2, \quad u_2 u_4 = 1 + u_3, \quad u_3 u_5 = 1 + u_4, \quad u_4 u_6 = 1 + u_5$$

we deduce

$$u_1 u_3 u_4 = (1 + u_2)u_4 = u_4 + 1 + u_3 = u_3 + 1 + u_4 = u_3(1 + u_5) = u_3 u_4 u_6.$$

Since $u_3 u_4 \neq 0$, we conclude that $u_1 = u_6$.

This cyclic sequence may be exhibited as a *frieze pattern*:

$$
\begin{array}{ccccccc}
0 & 0 & 0 & 0 & 0 & 0 & \cdots \\
1 & 1 & 1 & 1 & 1 & 1 & \cdots \\
u_1 & u_3 & u_5 & u_2 & u_4 & u_1 & \cdots \\
u_2 & u_4 & u_1 & u_3 & u_5 & u_2 & \cdots \\
1 & 1 & 1 & 1 & 1 & 1 & \cdots \\
0 & 0 & 0 & 0 & 0 & 0 & \cdots
\end{array}
$$

with the rule that every four adjacent numbers

$$
\begin{array}{ccc}
 & b & \\
a & & d \\
 & c &
\end{array}
$$

satisfy the equation $ad - bc = 1$. Is such a pattern (with borders of 0's and 1's) necessarily periodic? Some children, nine or ten years old, find it a fascinating pastime to look for an answer experimentally. They should try first the narrow pattern

```
0     0     0     0     0     0     0
   1     1     1     1     1     1      . . .
   1     2     1     2     1     2
      1     1     1     1     1         . . .
      0     0     0     0     0
```

and gradually increase the width, beginning with a prescribed diagonal row of ones and twos on the left, either straight, as above, or bent as in the following more complicated example:

```
  0     0     0     0     0     0     0     0     0
1     1     1     1     1     1     1     1     1     1    . . .
   1     2     3     1     5     1     2     2     4
      1     5     2     4     4     1     3     7     3     . . .
         2     3     7     3     3     1    10     5
            1    10     5     2     2     3     7     3     . . .
               1     3     7     3     1     5     2     4
            1     2     2     4     1     2     3     1     5   . . .
         1     1     1     1     1     1     1     1     1
```

It can be explained to older children that a frieze pattern may be defined more precisely as an arrangement of numbers

$$
\begin{array}{cccccc}
(0,0) & (1,1) & (2,2) & (3,3) & (4,4) & (5,5) \\
(0,1) & (1,2) & (2,3) & (3,4) & (4,5) & \\
(0,2) & (1,3) & (2,4) & (3,5) & (4,6) &
\end{array}
$$

$$
\begin{array}{cccc}
\cdots & & \cdots & \\
(0,n-1) & (1,n) & (2,n+1) & \\
(0,n) & (1,n+1) & (2,n+2) &
\end{array}
$$

that $(s, s+n) = 0$ and $(s, s+1) = (s+1, s+n) = 1$ for all $s$, and

$$(s,t)(u,v) + (s,u)(v,t) + (s,v)(t,u) = 0$$

all $s, t, u, v$. It follows from this last relation that $(s, s) = 0$ and

$$(s,t) + (t,s) = 0.$$

I must confess that I wasted many hours before realizing that it is easy to establish the periodicity as follows. Setting $u = s + 1$ and $v = s + n$, we have

$$(s,t)(s+1, s+n) + (s, s+1)(s+n, t) + (s, s+n)(t, s+1) = 0.$$

Since $(s+1, s+n) = 1 = (s, s+1)$ and $(s, s+n) = 0$, it follows that

$$(s,t) + (s+n, t) = 0$$

and

$$(s,t) = (t, s+n) = (s+n, t+n).$$

This means that the pattern is of period $n$.

For further details, including applications to determinants, continued fractions, and geometry, see my forthcoming paper *Frieze patterns* in the Davenport memorial issue of Acta Arithmetica.

A more recent piece of mathematical gossip concerns the sequence of positive integers $x_1, x_2, \ldots$, where $x_1$ is given and

$$x_{r+1} = \frac{1}{2}x_r \quad \text{or} \quad 3x_r + 1$$

according as $x_r$ is even or odd. For instance, if $x_1 = 50$, as in the case of Abraham's sequence, the terms are

> 50, 25, 76, 38, 19, 58, 29, 88, 44, 22, 11, 34, 17, 52, 26, 13, 40,
> 20, 10, 5, 16, 8, 4, 2, 1, 4, 2, 1, 4, 2, ...

By trial one finds that the sequence will eventually settle down to 142, not only when $x_1 = 50$ but for *any* $x_1$, suggesting the following conjecture:

*For all choices of $x_1$, there exists an $n$ such that $x_n = 1$.*

This reminds me of another fascinating book which, like *Ulysses' Father*, is out of print. On page 165 of J. L. Synge, *Kandelman's Krim* (Jonathan Cape, London, 1957) we read the following description of a typical working mathematician:

> Even the moonlit track ahead of him faded from his consciousness; for into his head had come a theorem which might be true or might be false, and his mind darted hither and thither seeking proofs to establish its truth and counter-examples to show that it could not possibly be true.

I am tempted to follow the example of Paul Erdős, who offers prizes for the solution of certain problems. If the above conjecture is true, I will gladly award a prize of fifty dollars to the first person who sends me a proof that I can understand. If it is false, I offer a prize of a hundred dollars to the first one who can establish a counterexample.

I must warn you not to try this in your heads or on the back of an old envelope, because the result has been tested with an electronic computer for all $x_1 \leq 500,000$. This means that, if the conjecture is false, the prizewinner must either find a sequence of this kind which he can prove to be divergent, or else find a cyclic sequence of this kind whose terms are all greater than half a million.

Whenever you spend days or weeks struggling with a problem that comes to nothing, think of poor Sisyphos. As Felix Behrend puts it at the end of his book:

> Sisyphos and his stone are the symbol of man and his eternal striving, never ceasing, never fulfilled and yet always triumphant.
> What more can you ask?

*H. S. M. Coxeter*

## Editorial Commentary

(1) Coxeter's lecture refers to the life and work of Felix Adalbert Behrend (1911–1962), based on his reading biographical material compiled by B. H. Neumann. The following is the complete text of a biography written by B. H. Neumann[17] for the Journal of the London Mathematical Society. (A more detailed account of the life and work of F. A. Behrend appears in Cherry and Neumann [1].)

"Felix A Behrend was born at Berlin, Germany, on 23 April, 1911, the eldest of four children of Dr Felix W Behrend and his wife Maria, née Zoellner. Felix

Behrend senior was a mathematics and physics master at the Herderschule, a noted "Reform-Realgymnasium" in one of the western suburbs of Berlin (where he also taught this writer); he was a widely known educationalist, and later headmaster of an important school elsewhere in Berlin, until demoted and finally dismissed by the Nazis, partly because of some Jewish ancestry, partly because of his liberal political views. Felix Behrend junior also went to the Herderschule, and passed out of it in 1929, with high distinction, to study mathematics and theoretical physics at the Universities of Berlin and Hamburg. He soon showed himself a pure mathematician of originality and imagination; his first three papers in the theory of numbers were published in quick succession before he was 23 years old. After taking, in 1933, his Dr phil. at the University of Berlin, he emigrated from Nazi Germany in 1934, first to Cambridge, then to Zürich and Prague, where he worked as an actuary in a life insurance company and continued his work in pure mathematics. He took the degree of Doctor of Science at the Charles University of Prague in 1938, but Czechoslovakia became unsafe in 1939, and he returned to Zürich and then to England just before the outbreak of war.

During the brief scare in the summer of 1940 when Holland was overrun by the German armies, Felix Behrend, like most adult male "enemy aliens" in Britain, whether friends or foes of Hitler, was interned, and not long after transported to Australia. At the instance of G H Hardy, J H C Whitehead, and other prominent British mathematicians, his release from internment was authorised before the end of 1940, but he chose not to return to Britain - the journey to Australia on the Dunera had been a harrowing experience - and remained in Australian internment camps. The time spent there was not lost, but gave full play to his great pedagogical gifts; he gave courses of lectures in the "camp university", and awakened an abiding enthusiasm for mathematics in several younger fellow internees, among them Walter F Freiberger, F I Mautner, and J R M Radok. The students were prepared for examinations of the University of Melbourne, without textbooks because none were available: in spite of this, or perhaps because of this, the teaching was highly successful.

Release from internment came in 1942, through the efforts of T M Cherry, and Felix Behrend joined his department at the University of Melbourne as a tutor. There he remained, being successively promoted Lecturer (1943), Senior Lecturer (1948), and Associate Professor (1954). He would have been made a (personal) Professor, but the illness intervened that led to his premature death on 27 May, 1962, at the age of 51 years. He is survived by his widow Daisy, née Pirnitzer, whom he had married in 1945, and their two daughters.

Felix Behrend's sympathies within pure mathematics were wide, and his creativeness ranged over theory of numbers, algebraic equations, topology, and foundations of analysis. A problem that caught his fancy early and that still occupied him shortly before his death was that of finite models in Euclidean 3-space of the real projective plane. He remained productive for much of the two years of his final illness, and left many unfinished notes in which his work on foundations of analysis is continued. A more detailed evaluation of his mathematical work is to be included in a fuller obituary notice, by T M Cherry, to be published in the Journal of the Australian Mathematical Society.

Felix Behrend was, like his father, an outstanding teacher. He took meticulous care over the preparation of his lectures, which were always distinguished by great lucidity and a sense of beauty, and also by a sure feeling for the level of his audience. His students could always rely on his sympathetic understanding of their problems and on his imaginative but sound advice; he gave them unstintingly of his time.

Among Felix Behrend's recreations, classical music ranked high, but much the most important was creative writing. He had as a young man been deeply impressed by the novels of Thomas Mann, and he continued to admire him and to acknowledge him as his master; but in his later writings the influence of Thomas Mann becomes less noticeable, and his style more individual. Most of his essays and short stories were written in German, and circulated privately among his friends. But he also wrote in English, and his last work, completed only shortly before his illness took its final turn, is a book, "Ulysses' Father", published shortly after his death : it is a children's book of very great charm and will, like "Alice in Wonderland", surely captivate grown-ups, too.

I acknowledge with gratitude the help I have received from Mrs F A Behrend and from Dr Hilde Behrend, Felix Behrend's younger (and only surviving) sister, in collecting some of the factual material I have used."

(2) Frieze patterns share with the $3x+1$ problem the property of being generated by a kind of nonlinear recurrence. They have a very different structure however. Coxeter published four papers on frieze patterns, two jointly with John Conway. He also included this theory in his 1974 book on regular complex polytopes [3]. In the first paper in 1971 ([8]) Coxeter gives constructions of frieze patterns and proves that they have a certain involution symmetry. The first of two papers with John Conway ([11]) in 1973 poses 35 problems relating frieze patterns and triangulated polygons, and the second ([12]) outlines solutions to these problems. Finally, in a 1999 paper written jointly with A. V. Kharchenko [13], he relates frieze patterns to polytopes and honeycombs in 3-space and 4-space. He infers that the average number of 4-dimensional bubbles surrounding a central one, in a 4-dimensional froth, is 28.

(3) H. S. M. Coxeter (1907–2003) is famous as a geometer. Born in London and home-schooled, he was educated at Trinity Colleage at the University of Cambridge, having undergraduate supervisor J. E. Littlewood, where he came Senior Wrangler on the tripos. He had H. F. Baker as graduate advisor in geometry, and completed his PhD in at Cambridge in 1931 with a thesis on polytopes (higher dimensional polyhedra) having regular faces. The word "polytope" was coined by Alicia Boole Stott, a daughter of George Boole, about 1902 (see [10], [19].) He also began the study of groups generated by reflection symmetries ("kaledoscopes"). After a Trinity Fellowship and two years in Princeton, he moved to the University of Toronto in 1936, where he spent the rest of his career. He had an abiding interest in geometry, group structures, and patterns. His name appears in the terms: Coxeter groups, Coxeter graphs, Coxeter diagrams, and the Coxeter element in Lie group and Lie algebra theory. He wrote the books, in chronological order: *Non-Euclidean Geometry* [2], *Regular Polytopes* [3], *Generators and Relations for Discrete Groups* (with W. O. J. Moser)[14], *The Real Projective Plane* [4], *Introduction to Geometry* [6], *Projective Geometry* [7], and *Regular Complex Polytopes* [9].

Of significance here, G. H. Hardy suggested that Coxeter be the person to update the classic 1892 book of W. W. Rouse Ball [20] titled: *Mathematical recreations and problems of past and present times*. Rouse-Ball had been a tutor of Littlewood at Cambridge, so that Coxeter was his mathematical descendant. He did so, twice revising and updating the book, adding to it an extra chapter on Polyhedra and removing some weak chapters. His revision raised the mathematical level of the book, which appeared as W.W. Rouse-Ball and H. S. M. Coxeter, *Mathematical Recreations and Essays.* [**21**]). He had a continuing interest in mathematical games, exhibited for example in a 1953 paper on the golden ratio, which among other things surveyed the connection between Beatty sequences and winning positions in Wythoff's game [**5**].

Further biographical information on H. S. M. Coxeter can be found in Roberts [**18**], and Roberts and Weiss [**19**]. For two appreciations of Coxeter's work see Davis, Grúnbaum and Sherk [**16**] and Davis and Ellers [**15**].

## References

[1] T. M. Cherry and B. H. Neumann, Felix Adalbert Behrend, J. Australian Math. Soc.. **4** (1964), 264–270.

[2] H. S. M. Coxeter, *Non-Euclidean Geometry*, Sixth Edition, MAA Spectrum, Math. Assoc. America, Washington DC 1998. (First edition: Univ. of Toronto Press 1942)

[3] H. S. M. Coxeter, *Regular Polytopes*, Third edition. Dover, New York 1973. (First Edition: Methuen, London 1948)

[4] H. S. M. Coxeter, *The real projective plane*, Third edition. With an appendix by George Beck. Springer-Verlag, New York 1993. (First Edition 1949)

[5] H. S. M. Coxeter, The golden section, phyllotaxis and Wythoff's game, Scripta Math. **19** (1953), 135–143.

[6] H. S. M. Coxeter, *Introduction to Geometry*, Wiley Classics Library, John Wiley & Sons. Inc., New York 1989. (First edition 1961).

[7] H. S. M. Coxeter, *Projective Geometry*, Revised reprint of second (1974) edition. Springer-Verlag, New York 1994. (First edition: Blaisdell Publ Co. 1964)

[8] H. S. M. Coxeter, Frieze patterns, Acta Arithmetica **18** (1971), 297–310.

[9] H. S. M. Coxeter, *Regular Complex Polytopes*, Cambridge University Press, London-New York 1974.

[10] H. S. M. Coxeter, *Alicia Boole Stott (1860–1940)*. in: Women of mathematics, 220-224, Greenwood: Westport, CT 1987.

[11] H. S. M. Coxeter and J. H. Conway, Triangulated polygons and frieze patterns, Math. Gazette **57** (1973), No. 400, 87–94.

[12] H. S. M. Coxeter and J. H. Conway, Triangulated polygons and frieze patterns, Math. Gazette **57** (1973), No. 401, 175–183.

[13] H. S. M. Coxeter and A. V. Kharchenko, Frieze patterns for regular star polytopes and statistical honeycombs, Discrete geometry and rigidity (Budapest 1999), Period. Math. Hungar. **39** (1999), 51–63.

[14] H. S. M. Coxeter and W. O. J. Moser, *Generators and relations for discrete groups,* Fourth edition. Ergebnisse der Mathematik und ihre Grezgebiete, Neue Folge, Band 14, Springer-Verlag, Berline-Göttingen-Heidelberg, 1980. (First edition 1957)

[15] Chandler Davis and E. W. Ellers (Eds.), *The Coxeter Legacy*, American Math. Soc., Providence 2006.

[16] Chandler Davis, Branko Grünbaum and F. A. Scherk (Eds.), *The Geometric Vein. The Coxeter Festschrift*, Springer-Verlag, New York-Berlin 1981.

[17] B. H. Neumann, Felix Adalbert Behrend, Journal of the London Mathematical Society **38** (1963), 308-310.

[18] Siobhan Roberts, *King of Infinite Space: Donald Coxeter, the Man Who Saved Geometry,* 2006.

[19] S. Roberts and A. I. Weiss, Obituary: Harold Scott Macdonald Coxeter, FRS, 1907–2003, Bull. London Math. Soc. **41** (2009), no. 5, 943–960.

[20] W. W. Rouse Ball, *Mathematical recreations and problems of past and present times*, MacMillan, London 1892.

[21] W. W. Rouse Ball and H. S. M. Coxeter, *Mathematical Recreations and Essays*, Twelfth Edition. University of Toronto Press 1974.

# Unpredictable Iterations

## J.H. Conway

## Trinity College, Cambridge, England

Lothar Collatz has defined the function
$$g(n) = \begin{cases} n/2 & (n \text{ even}) \\ 3n+1 & (n \text{ odd}) \end{cases}$$
and conjectured that for any positive integer $n$, there exists $k$ such that
$$g^k(n) = 1.$$
The conjecture is still unproven, but has been verified for $n \leq 10^9$ by D.H. and Emma Lehmer and J.L.Selfridge. It prompts consideration of more general functions
$$g(n) = a_i n + b_i \quad (n \equiv i \mod P)$$
where $a_0, b_0, \ldots, a_{P-1}, b_{P-1}$ are rational numbers chosen so that $g(n)$ is always integral. What can be predicted about the iterates $g^k(n)$? Here we show that even when $b_i = 0$, the behavior is unpredictable, in general.

THEOREM. *If $f$ is any computable function, there is a function $g$ such that*

(1) *$g(n)/n$ is periodic (with rational values)*

(2) *$2^{f(n)} = g^k(2^n)$, where $k$ is minimal positive subject to $g^k(2^n)$ a power of 2.*

COROLLARY. *There is no algorithm, which, given a function $g$ with $g(n)/n$ periodic, and given a number $n$, determines whether or not there is $k$ with $g^k(n) = 1$. The word "computable" will mean "computable by a Minksy program", as defined below. This is equivalent to (partial) recursive.*

**Minsky machines.** These have *registers* $a, b, c, \ldots$ capable of holding arbitrary non-negative integers, and two types of *order*:

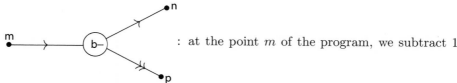

$m \longrightarrow (a+) \longrightarrow n$ : at the point $m$ in the program, we add 1 to register $a$, and proceed to $n$.

: at the point $m$ of the program, we subtract 1 from register $b$ and proceed to $n$, if $b > 0$, while if $b = 0$ we simply proceed to $p$.

A *Minsky program* consists of such orders:

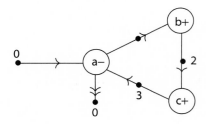

All entry and exit points are conventionally labelled 0. Other labels are distinct positive integers.

The example is a program to add the contents of register $a$ to each of registers $b$ and $c$, and clear $a$ to 0. It is easy to write Minsky programs to simulate the orders of more conventional computing machines (multiplication, etc.) so that functions computable by such machines are computable in our sense. Equally, all (partial) recursive functions are Minsky computable in the sense that there is a Minsky program which started with register contents $n, 0, 0, \ldots$ ends with register contents $f(n), 0, 0, \ldots$.

**Vector games.** Suppose wo are given a finite list of vectors with integer co-ordinates and the same dimension, e.g.:

$$(0, 0, 0 \mid 1, -1)$$
$$(-1, 0, 0 \mid -3, 1)$$
$$(0, 0, 0 \mid -3, 0)$$
$$(0, 0, 1 \mid -2, 3)$$
$$(0, 1, 0 \mid -1, 2)$$
$$(-1, 0, 0 \mid -0, 1)$$
$$(0, 0, 0 \mid -0, 0)$$

Then we can play the following game. Starting with a vector $v$ with *non-negative* integer coordinates, add to $v$ the first vector from the list which preserves this property. What happens when we repeat this indefinitely?

We show that for any computable $f$, there is a vector game, which when started at $(n, 0, 0, \ldots)$ reaches $(f(n), 0, 0, \ldots)$. In fact we use vector games to simulate Minsky programs, as follows.

We have one coordinate for each register, and two more coordinates (after the vertical bar). For an order $\overset{m}{\longrightarrow}\ \overset{}{\underset{}{a+}}\ \overset{n}{\longrightarrow}\bullet$ we have a vector

$(\overset{a}{1}, 0, \ldots \mid -m, n)$ and for an order $\overset{m}{\bullet\longrightarrow}\ \overset{}{\underset{}{b-}}$ we have a pair of

vectors $(0, \overset{b}{-1}, 0, \ldots \mid -m, n)$, $(0, \overset{b}{0}, 0, \ldots \mid -m, p)$ in that order. These vectors are listed in *decreasing* order of $m$, and preceded by the vector $(0, 0, \ldots \mid 1, -1)$. In this way, our example of a Minsky program yields the given vector-list.

It is easy to see how the vector game so defined simulates the corresponding Minsky program. Maybe some industrious reader will produce a short prime-generating vector game.

**Rational games.** Suppose we have a finite list of rational numbers $r_1, r_2, \ldots, r_s$. Then we can play another game. Starting with an integer $n$, replace $n$ by $g(n) = r_i \cdot n$ for the least $i$ for which this is an integer. What happens when we iterate?

Obviously, if we replace a vector $(a, b, c, \ldots)$ by the number $2^a 3^b 5^c \ldots$, we obtain this game from our previous one. So we have proved that for any computable function $f$, there is a rational game, which started at $2^n$, gets to $2^{f(n)}$ without finding any intermediate power of 2.

Our main theorem now follows from the observation that for this function $g(n)$, $g(n)/n$ is periodic, with period dividing the least common denominator of the $r_i$. Since it is undecidable whether or not a given partial recursive function is everywhere defined and identically zero, we obtain also the Corollary.

Of course, particular games of this type can still have predictable properties, so that (for instance) our theorem says nothing about the Collatz game. But it does prohibit any general solution to games of this type, and also shows that there exist special cases for which the prediction problem is unsolvable.

It is amusing to note that the Theorem contains the Kleene Normal Form Theorem for recursive functions, since the functions $g(n)$, $2^n$, etc., are obviously primitive recursive.

California Institute of Technology Jan.-Aug. 1972

University of Cambridge Aug. 1972 $\to \infty$

## Editorial Commentary

(1) This paper was one of the earliest mathematical papers on the $3x + 1$ problem. The results here grew out of J. H. Conway's interests in the theory of computation and games. Some of his early studies in computation appear in his 1971 book *Regular Algebra and Finite Machines* [2]. In this paper Conway challenges some industrious reader to produce a prime-generating vector game; this challenge was answered in 1983 by R. K. Guy [12]. The computational model underlying this paper using fractions was later formalized as FRACTRAN by Conway [4] in 1987. This name was clearly intended as a pun on FORTRAN. Just as FORTRAN has since become Fortran, FRACTRAN has since become Fractran.

(2) The *Minsky machines* defined in this paper are the counter machines described in the textbook of Marvin Minsky [15, Sect. 11.1]. Minsky originally introduced them to show the unsolvability of Post's problem of 'tag' in 1961 ([14]).

(3) The *Kleene Normal Form theorem* for partial recursive functions (Kleene [13, Sec. 63, Theorem XIX]) states, when specialized to the one-variable case, the following:

**Theorem.** *There are primitive recursive functions $U(y)$ and $T_1(y, e, x)$ such that for any one-variable partial recursive function $f(x)$ there is some positive integer $e$ ("Gödel number") such that*

$$f(x) = U(\mu y T_1(y, e, x) = 0)$$

*where $\mu y$ is a quantifier that enotes the least positive integer $y$ such that $T_1(y, e, x) = 0$, and is undefined otherwise ([**13**, Sec. 57]). Furthermore the primitive recursive function $U(y)$ has the additional property that*

$$f(x) = U(y), \text{ for any } y \text{ such that } T_1(y, e, x) = 0.$$

Kleene [**13**, Sec. 58, Theorem IX] gives a construction of suitable primitive recursive functions $U$ and $T_1(y, e, x)$.

(3) John Horton Conway (1937–) is well known for his work in geometry, group theory, algebra and various other subjects of his own invention. He was educated in Cambridge at Gonville and Caius College, where he received his BA in 1959. He completed a PhD at Cambridge with supervisor Harold Davenport in 1964 on Waring's problem for degree five, showing that each integer can be written as a sum of 37 fitth powers. This result was not published because it was independently proved and published first by Chen Jing-run. He is most well known for the construction of several of the sporadic finite simple groups, the *Conway groups*, which are obtained from automorphism groups of the Leech lattice. With Simon P. Norton he proposed conjectures giving connections between the Monster simple group and modular forms, termed "monstrous moonshine". These conjectures were eventually proved by his student Richard Borcherds, who was awarded a Fields medal in part for this work. Some of his work on groups and lattices appears in the book with N. J. A. Sloane on *Sphere Packings, Lattices and Groups* ([**10**]). In algebra he studied Quaternions and Icosians, both introduced by William Rowan Hamilton (see [**11**]). He is known for inventing a new system of numbers, the field of "surreal numbers", which include all ordinal numbers, and for an extended theory of combinatorial mathematical games (see [**3**], [**1**]). The set of all (equivalence classes of) games form a group containing the field of surreal numbers. This theory includes as a special case, the game of *Life*, which is a 2-dimensional cellular automaton, now proved to be universal. He has published on a wide range of subjects, on each of which he has his own viewpoint, full of stimulating ideas. These include books on algebra and computation ([**2**]), on numbers (with Richard Guy [**8**]) and on integral quadratic forms ([**7**]). Two of his more remarkable results involving recursive constructions are his study of "audioactive decay" sequences (Conway [**5**]) and constructions of lexicographic codes ([**9**], [**6**]).

## References

[1] E. Berlekamp, J. H. Conway and R. K. Guy, *Winning ways for your mathematical plays, Vols. 1- 4* Second edition. A. K. Peters Ltd., Natick Mass. 2001-2004. [First Edition: Harcourt-Brace-Jovanovich, London-New York 1982 (Two volumes).

[2] J. H. Conway, *Regular Algebra and Finite Machines,* Chapman and Hall: London 1971. (Zbl 0231.94041)

[3] J. H. Conway, *On Numbers and Games,* Second edition. A. K. Peters, Ltd. Natick, MA 2001. (First edition: London Math. Society Monographs, No. 6, Academic Press-London, New York 1976).

[4] J. H. Conway, FRACTRAN: A Simple Universal Computing Language for Arithmetic, In: *Open Problems in Communication and Computation* (T. M. Cover and B. Gopinath, Eds.), Springer-Verlag: New York 1987, pp. 3-27 [Reprinted in this volume]

[5] J. H. Conway, The weird and wonderful chemistry of audioactive decay, in: T. H. Cover and B. Gopinath, Eds., *Open Problems in Communication and Computation.* Springer-Verlag: New York 1987, pp. 173–188.

[6] J. H. Conway Integral lexicographic codes, Discrete Mathematics **83** (1990), no. 2-3, 219-235.

[7] J. H. Conway, *The sensual (quadratic) form,* With the assistance of Francis Y. C. Fung, Carus Mathematical Monographs No. 26, Math. Assoc. America: Washington DC. 1997.

[8] J. H. Conway and R. K. Guy, *The Book of Numbers,* Copernicus, New York 1996.

[9] J. H. Conway and N. J. A. Sloane, Lexicographic codes: error-correcting codes from game theory, IEEE Trans. Information Theory, **32** (1986), 337–348.

[10] J. H. Conway and N. J. A. Sloane, *Sphere Packings, Lattices and Groups,* Third edition. With additional contributions by E. Bannai, R. E. Borcherds, J. Leech, S. P. Norton, A. M. Odlyzko, R. A. Parker, L. Queen and B. B. Venkov. Grundlehren der Mathematischen Wissenschaften, Band 290. Springer-Verlag, New York 1999. (First Edition 1988)

[11] J. H. Conway and D. A. Smith, *On quaternions and octonions: their geometry, arithmetic, and symmetry,* A. K. Peters, Inc., Natick, MA 2003, xii+159pp.

[12] R. K. Guy, Conway's Prime Producing Machine, Mathematics Magazine **56** (1983), No. 1, 26–33.

[13] S. Kleene, *Introduction to Metamathematics,* Van Nostrand: Princeton 1952.

[14] M. Minsky, Recursive unsolvability of Post's problem of 'tag' and other topics in the theory of Turing machines, Annals of Math. **74** (1961). 437–455.

[15] M. Minsky, *Computation: Finite and Infinite Machines,* Prentice-Hall, Inc: Engelwood Cliffs, NJ 1967.

# ITERATION OF THE NUMBER-THEORETIC FUNCTION
$$f(2n) = n, \; f(2n + 1) = 3n + 2$$

C. J. EVERETT

It is a long-standing conjecture that, under iteration of the number-theoretic function,

$$f(2n) = n, \; f(2n + 1) = 3n + 2,$$

every integer $m$ has an iterate $f^k(m) = 1$. Since virtually nothing is known about the question, save that it seems to be true for $m$ up to the millions, it may be of interest to know that almost every $m$ has an iterate $f^k(m) < m$, a result proved in the present paper.

## I. Introduction

Under iteration of the function

(1)
$$f(m) = \begin{cases} m/2; & m \text{ even}, \\ (3m + 1)/2; & m \text{ odd}, \end{cases}$$

every integer $m \geq 0$ gives rise to an infinite sequence of integers

(2)
$$m \longrightarrow [m_0, m_1, m_2, ...],$$

where $m_n = f^n(m)$. Thus for $m = 7$, one finds that

$$7 \longrightarrow [7, 11, 17, 26, 13, 20, 10, 5, 8, 4, 2, 1, 2, 1, 2, ...]$$

with $f^{11}(7) = 1$. It has been conjectured that *every* $m \geq 1$ has an iterate $f^k(m) = 1$. It is shown here that, at any rate, almost every integer $m$ has an iterate $f^k(m) < m$.

## II. The Parity Sequence

The sequence (2) may be used to assign to every integer $m \geq 0$ a "parity sequence"

(3)
$$m \longrightarrow \{x_0, x_1, x_2, ...\},$$

where $x_n = 0$ if $m_n = f^n(m)$ is even, and $x_n = 1$ if it is odd. For example, one sees from above that

$$7 \longrightarrow \{1, 1, 1, 0, 1, 0, 0, 1, 0, 0, 0, 1, 0, 1, 0, ...\}$$

It is trivial that $m \to \{0, 0, 0, ...\}$ iff $m = 0$. Similarly $m \to \{1, 0, 1, 0, ...\}$ iff $m = 1$, and hence the parity sequence for any $m$ terminates in $\{x_k, x_{k+1}, ...\} = \{1, 0, 1, 0, ...\}$ iff $m_k = f^k(m) = 1$. Thus the above conjecture asserts that *every* parity sequence not the zero sequence terminates in $1, 0, 1, 0, ...$ If so, the list of parity sequences for $m = 1, 2, ...$ in (3) would be rather remarkable, in view of the following property, which it does indeed have; namely the $2^N$ parity sequences for the integers $m < 2^N$

have subsequences $\{x_0, ..., x_{N-1}\}$ ranging over the full set of $2^N$ $0, 1$ vectors. Thus for $N = 2$, one sees that

$$
\begin{array}{ccccc}
0 & \longrightarrow & [0, 0, ...] & \longrightarrow & \{0, 0, ...\} \\
1 & \longrightarrow & [1, 2, ...] & \longrightarrow & \{1, 0, ...\} \\
2 & \longrightarrow & [2, 1, ...] & \longrightarrow & \{0, 1, ...\} \\
3 & \longrightarrow & [3, 5, ...] & \longrightarrow & \{1, 1, ...\}.
\end{array}
$$

Moreover, all integers $m = a + 2^N Q$, $Q = 0, 1, 2, ...$ have identical parity sequences through component $x_{N-1}$. This is the substance of Theorem 1.

**Theorem 1.** *An arbitrary diadic sequence* $\{x_0, ..., x_{N-1}\}$ *arises via (3) from a unique integer* $m < 2^N$. *Specifically, the* $x_n$ *determine* $m$ *and* $m_N$ *to be of the forms*

$$
m = a_{N-1} + 2^N Q_N; \; 0 \le a_{N-1} < 2^N.
$$

(4)
$$
m_N = b_{N-1} + 3^X Q_N; \; 0 \le b_{N-1} < 3^X, \; X = \sum_0^{N-1} x_n.
$$

*Hence the correspondence (3) is one to one.*

Proof by induction on $N = 1, 2...$ For $N = 1$, $x_0 = 0$ implies $m = 2Q_1$, $M_1 = Q_1$, whereas $x_0 = 1$ implies $m = 1 + 2Q_1$, $m_1 = 2 + 3Q_1$. Assuming Eq. (4) for any $N \ge 1$, we must consider two cases, depending on the parity of the current $b_{N-1}$.

*Case I.* $b_{N-1}$ *even.*

(a) $x_N = 0$ implies $Q_N = 2Q_{N+1}$, $m = a_{N-1} + 2^{N+1}Q_{N+1}$, $m_N = b_{N-1} + 3^X \cdot 2Q_{N+1}$, $m_{N+1} = (b_{N-1}/2) + 3^X Q_{N+1}$.

(b) $x_N = 1$ implies $Q_N = 1 + 2Q_{N+1}$, $m = a_{N-1} + 2^N + 2^{N+1}Q_{N+1}$, $m_N = (b_{N-1} + 3^X) + 3^X \cdot 2Q_{N+1}$, $m_{N+1} = (\frac{1}{2})(3b_{N-1} + 3^{X+1} + 1) + 3^{X+1}Q_{N+1}$ where $\frac{1}{2}(3b_{N-1} + 3^{X+1} + 1) \le \frac{1}{2}[3(3^X - 1) + 3^{X+1} + 1] = 3^{X+1} - 1 < 3^{X+1}$.

*Case II.* $b_{N-1}$ *odd.*

(a) $x_N = 0$ implies $Q_N = 1 + 2Q_{N+1}$, $m = a_{N-1} + 2^N + 2^{N+1}Q_{N+1}$, $m_N = (b_{N-1} + 3^X) + 3^X \cdot 2Q_{N+1}$, $m_{N+1} = (\frac{1}{2})(b_{N-1} + 3^X) + 3^X Q_{N+1}$, where $(\frac{1}{2})(b_{N-1} + 3^X) < (\frac{1}{2})(2 \cdot 3^X) = 3^X$.

(b) $x_N = 1$ implies $Q_N = 2Q_{N+1}$, $m = a_{N-1} + 2^{N+1}Q_{N+1}$, $m_N = b_{N-1} + 3^X \cdot 2Q_{N+1}$, $m_{N+1} = (\frac{1}{2})(3b_{N-1} + 1) + 3^{X+1}Q_{N+1} <$ where $(\frac{1}{2})(3b_{N-1} + 1) < (\frac{1}{2})(3^{X+1} + 1) < 3^{X+1}$. Hence Eq. (4) is true at stage $N + 1$ in all cases.

**Corollary.** *The correspondence*

$$
m \longrightarrow [m_0, ..., m_{N-1}] \longrightarrow \{x_0, ..., x_{N-1}\}
$$

*induces a one to one mapping of all positive integers* $m \le 2^N$ *on the set of all* $2^N$ *diadic vectors* $\{x_0, ..., x_{N-1}\}$.

## III. A Density Theorem

Let $A(M)$ denote the number of positive integers $m \le M$ having some iterate $f^k(m) < m$. Our object is to show that the density $A(M)/M$ approaches 1 as $M \to \infty$.

Consider first the case $M = 2^N$, for which we have the correspondence of the corollary

$$m \longrightarrow [m_0, ..., m_{N-1}] \longrightarrow \{x_0, ..., x_{N-1}\}.$$

Roughly, the idea is that most diadic sequences have nearly the same number of 0's and 1's if $N$ is large. Moreover, $x_n = 0$ implies $m_{n+1}/m_n = \frac{1}{2}$, while $x_n = 1$ implies $m_{n+1}/m_n = (3Q + 2)/(2Q + 1) \leq 5/3$ if $m_n > 1$. Hence most integers $m \leq 2^N$ should have $m_N \simeq (\frac{1}{2})^{N/2}(\frac{5}{3})^{N/2}m_0 < m_0$. In fact, it follows from a well-known inequality of probability (Ref. 1) that the set $H_N$ of all diadic sequences $\{x_0, ..., x_{N-1}\}$ such that

(5) $$\frac{1}{2} - \epsilon < \frac{X}{N} < \frac{1}{2} + \epsilon,$$

where $X = \sum_0^{N-1} x_n$, $\epsilon \equiv L - (\frac{1}{2})$, $L \equiv \log 2/\log(10/3)$ satisfies the relation

(6) $$\#H_N/2^N \geq 1 - \frac{1}{4}\epsilon^2 N.$$

Since condition (5) implies the inequality

(7) $$X/N < L,$$

it is clear that the set $D_N$ of diadic sequences satisfying Eq. (7) contains the set $H_N$, and hence

(8) $$\#D_N \geq \#H_N.$$

Now except for the integer $m = 1$, the integers $m \leq 2^N$ whose parity sequences $\{x_0, x_1, ..., x_{N-1}\}$ satisfy (7) are of two kinds, those for which some $m_n \equiv f^n(m) = 1$ $(< m)$, $n \leq N - 1$, and the rest, for which no such $m_n = 1$. But for the latter, the equivalence of (7) with

$$(1/2)^{N-X}(5/3)^X < 1$$

shows that

$$m_N = m_0(m_1/m_0) \cdots (m_N/m_{N-1}) \leq (1/2)^{N-X}(5/3)^X m_0 < m_0 = m.$$

It follows that $A(2^N) \geq \#D_N - 1 \geq \#H_N - 1$, and hence by (6),

(9) $$\lim_{N\to\infty} A(2^N)/2^N = 1.$$

It remains to consider the case $2^N < M < 2^{N+1}$. If we set $A_N = A(2^N)$, and define

(10) $$n_N = 2^N + \{(2^{N+1} - 2^N) - (A_{N+1} - A_N)\} = A_N + (2^{N+1} - A_{N+1}) \geq 2^N,$$

then for $M = 2^N + 1, ..., n_N$, it is obvious that

(11) $$A(M)/M \geq A_N/n_N.$$

On the other hand, for $M = n_N + k$, $k = 1, 2, ..., 2^{N+1} - 1 - n_N$, one has

(12) $$A(M)/M \geq (A_N + k)/(n_N + k) \geq A_N/n_N,$$

since $n_N \geq 2^N \geq A_N$. Now by Eqs. (9) and (10)

$$A_N/n_N = \frac{A_N/2^N}{A_N/2^N + 2(1 - A_{N+1}/2^{N+1})} \longrightarrow 1,$$

and hence by (9), (11), and (12),

$$\lim_{M\to\infty} A(M)/M = 1.$$

In this way we are led to Theorem 2.

**Theorem 2.** *For the function f defined in Eq. (1), "almost every" integer m has some iterate $m_k = f^k(m) < m$, in the sense that the density $A(M)/M$ of such integers approaches unity.*

REFERENCES

[1] J. V. Uspensky, *Introduction to Mathematical Probability,* McGraw-Hill, New York (1937), p. 209.

LOS ALAMOS SCIENTIFIC LABORATORY, LOS ALAMOS, NEW MEXICO 89745

**Editorial Commentary.**

(1) This paper was first issued in 1976 as Los Alamos report LA-6449-MS. The main results of this paper were obtained independently, and published slightly earlier, by Riho Terras [14], [15] in 1976. The analysis of Terras proves a stronger result, namely that for each $m \geq 1$, there exists a natural density of the set of integers which have some iterate $f^k(n) < n$ with $1 \leq k \leq m$. This result is presented in Lagarias [8, Theorem A]. We have included Everett's paper here for its short and elegant proof.

(2) Cornelius Joseph Everett, Jr. (1914–1987) completed his Ph.D. at the University of Wisconsin in 1940 as a student of C. C. MacDuffee, with a thesis titled: "Rings as Groups with Operators." After the degree he joined the faculty at the Unversity of Wisconsin, where in the late 1940's he had as his sole doctoral student Herbert J. Ryser, jointly advising him with C. C. MacDuffee. Ryser became afterwards a distinguished combinatorialist.

Everett's career is intertwined with that of Stansiaw M. Ulam. He was a long time collaborator of Ulam, starting at the University of Wisconsin and continuing later at Los Alamos National Laboratories. Ulam was an instructor at the University of Wisconsin (1941-1943) before he moved to Los Alamos, and after the war he persuaded Everett to work at Los Alamos (Ulam [16, Chaps. 7 and 10]). About Everett's mathematical work habits, Ulam [16, p. 131] wrote:

> Everett exhibited a trait of mind whose aspects are, so to say, non-additive: persistence in thinking. Thinking continuously or almost continuously for an hour is at least for me-and I think for many mathematicians-more effective than doing it for two half-hour periods. It is like climbing a slippery slope. If one stops, one tends to slide back.

About Everett's personality, Ulam wrote ([16, p. 195])

> In Madison he was already a shy and retiring man, but as time passed he became more and more of a recluse. In the early days of his stay at Los Alamos, although he was reluctant to mingle with people, he could still be coaxed into coming to our house if one

made the solemn promise that no one else would be there at the
same time. Later he even refused to do that, and the only place
one can see him is in his little windowless cubicle of an office or in
a carrel in the excellent laboratory library."

In 1948 Everett and Ulam ([4], [5]) wrote several papers on branching processes
in $n$ dimensions. Paul Stein [12, p. 94] says:

I recently asked Everett how he and Stan had worked together on
these papers. Everett's reply was succinct. "Ulam told me what
to do and I did it."

In 1950 Everett and Ulam did extensive calculations to show that Edward Teller's
original design for the "super" [H-bomb] would ignite but then have the fusion
reaction fizzle out ([9, Chap. 14, p. 343]):

Group T-8 [at Los Alamos] consisted mainly of Ulam and his University of Wisconsin associate Dr. C. J. Everett. Ulam never
concealed his dependence on Everett: "I had some general, sometimes only vague, ideas. Everett supplied the rigor, the ingenuities
and the details of the proof, and final constructions." In these
Ulam-Everett calculations, sometimes miscalled the "debunking of
Teller's super," Everett "virtually wore out his slide rule." Ulam's
admirers said that at this stage he was working single-handedly;
his critics said the single hand was that of Everett.

This work eventually stimulated Ulam to have an idea that, improved by Teller,
led to a working H-bomb design, and to a joint patent of Teller and Ulam. For
Ulam's personal view of this history, see ([16, Chap. 11]) and for other views, see
Rhodes [10], Carlson [2] and Bernstein [1].

In the 1950's Ulam pioneered the investigation of iteration of nonlinear maps.
This included the Fermi, Pasta, Ulam problem ([7]), which involves non-linear
oscillations, and which, contrary to expectation, did not become completely random
but returned close to an initial configuration after not too long a time. Ulam's
investigations of non-linear problems continued into the 1960's jointly with Paul
Stein [11] and others. During this period Ulam continued work with Everett on
many problems, see Everett and Ulam [6] and Ulam [17].

From 1965 to 1975 Ulam left Los Alamos to became professor and chairman of
the mathematics department at the University of Colorado, Boulder, while Everett
remained at Los Alamos. Ulam continued to visit Los Alamos during summers.
Since the $3x + 1$ problem is a prototype "toy" non-linear iteration, Ulam took an
interest in it and circulated it. At some point he interested Everett in the problem,
resulting in his paper. Everett's interest in nonlinear problems continued for the
rest of his life, including his last paper, Stein and Everett [13].

## REFERENCES

[1] J. Bernstein, John von Neumann and Klaus Fuchs: an unlikely collaboration, Phys. Perspect.
**12** (2010), 36–50.

[2] B. Carlson, How Ulam set the stage. History has not given enough credit to the main man
behind the H-bomb, Bulletin of the Atomic Scientists **59** (2003), No. 4, 46–51.

[3] Necia Grant Cooper, (Ed), *From Cardinals to Chaos. Reflections on the Life and Legacy of
Stanislaw Ulam,* Cambridge University Press: Cambridge 1990.

[4] C. J. Everett and S. M. Ulam, Multiplicative systems I., Proc. Natl. Acad. Sci. U. S. A. **34** (1948), 403–405.

[5] C. J. Everett and S. M. Ulam, Multiplicative systems in several variables, I, II, III, Los Alamos Technical Reports LA-683, June 7, 1948, LA-690, June 11, 1948, and LA-707, Oct. 28, 1948. (Reprinted in Ulam [17, Chap. 3].)

[6] C. J. Everett and S. M. Ulam, The entropy of interacting populations, Los Alamos Technical Report LA-4256, 1969. (Reprinted in Ulam [17, Chap. 14].)

[7] E. Fermi, J. Pasta and S. Ulam, *Studies of Nonlinear Problems*, Document LA-1040 (May 1955). (Reprinted as paper 266 in: *Collected Papers of Enrico Fermi*, Vol 2, The University of Chicago Press: Chicago 1965, and in Ulam [17, Chap. 5].).

[8] J. C. Lagarias, The $3x + 1$ problem and its generalizations, Amer. Math. Monthly **92** (1985), 3–23.

[9] N. Macrae, *John von Neumann. The scientific genius who pioneered the modern computer, game theory, nuclear deterrence, and much more.* Pantheon Books: New York 1992. [Reprint: American Math. Society: Providence, RI 1999]

[10] R. Rhodes, *Dark sun: The Making of the Hydrogen Bomb,* Simon and Schuster: New York 1995.

[11] P. R. Stein and S. M. Ulam, Nonlinear transformation Studies on Electronic Computers, Los Alamos Technical Report LADC-5688, 1963. (Reprinted in Ulam [17], Chpater 11.)

[12] P. R. Stein, Iteration of maps, strange attractors, and number theory–An Ulamian potpourri, in: Cooper [3].

[13] P. R. Stein and C. J. Everett, On the iteration of a bijective transformation of integer $k$-tuples, Discrete Math. **63** (1987), No. 1, 67–79.

[14] R. Terras, A stopping time problem on the positive integers, Acta Arithmetica **30** (1976), No. 3, 241–252.

[15] R. Terras, On the existence of a density, Acta Arithmetica **35** (1979), no. 1, 101–102.

[16] S. Ulam, *Adventures of a Mathematician,* Charles Scribner's Sons: New York 1976. (Reprint: University of California Press: Berkeley 1991.)

[17] S. M. Ulam, *Analogies between analogies. The mathematical reports of S. M. Ulam and his Los Alamos collaborators.* Edited and with a foreword by A. R. Bednarek and Francoise Ulam. With a bibliography of Ulam by Barbara Hendry. Univ of California Press, Berkeley, CA 1990.

# Don't Try to Solve These Problems!

## Richard K. Guy

Such an exhortation will likely produce the opposite effect, but I'm serious, and will explain why. This article has been in mind for some time, but its eruption is triggered by a proposal from

Schmuel Schreiber, Department of Mathematics and Computer
Science, Bar-Ilan University, Ramat-Gan, Israel.

PROBLEM 0. *For an integer $a$ define the set $S_a$ inductively by*

(1) $a \in S_a$,   (2) if $k \in S_a$, then $2k + 2 \in S_a$,   (3) if $k \in S_a$, then $3k + 3 \in S_a$.

*Equivalently, define a function $s_a(n)$ on the integers by*

(1) $s_a(1) = a$,   (2) $s_a(2k) = 2s_a(k) + 2$,   (3) $s_a(2k + 1) = 3s_a(k) + 3$.

*For $a < -3$ or $a > 2$ is $s_a$ injective? Or does $S_a$ contain repeated elements?*

Some of you are already scribbling, in spite of the warning! More cautious readers may have been reminded of other problems, perhaps one or more of the following.

PROBLEM 1. *The diophantine equation $a^2 + b^2 + c^2 = 3abc$ has the* singular solutions $(1, 1, 1)$ *and* $(2, 1, 1)$. *Other solutions can be generated from these, because the equation is quadratic in each variable, for example, $b = 2$, $c = 1$ gives $a^2 - 6a + 5 = 0$, $a = 1$ or $5$ and $(5, 2, 1)$ is a solution. Each solution, apart from the singular ones, is a* neighbor *of just three others, and they form a binary tree. Is this a genuine tree, or can the same number be generated by two different routes through it?*

PROBLEM 2. *Consider the sequence $a_{n+1} = a_n/2$ ($a_n$ even), $a_{n+1} = 3a_n + 1$ ($a_n$ odd). For each positive integer $a_1$ is there a value of $n$ such that $a_n = 1$?*

PROBLEM 3. *Consider the mapping*

$$2m \to 3m, \quad 4m - 1 \to 3m - 1, \quad 4m + 1 \to 3m + 1.$$

*This generates the cycles $(1)$, $(2, 3)$, $(4, 6, 9, 7, 5)$ and $(44, 66, 99, 74, 111, 83, 62, 93, 70, 105, 79, 59)$. Are there others?*

Problem 0 can be visualized as a binary tree generated by the pair of *unary* functions $a \to 2a + 2$, $a \to 3a + 3$. For example, if $a = 1$, we have Figure 1.

The number 66 appears twice in Figure 1, by making three steps to the right, or by making one to the left, one to the right and two to the left. A right step multiplies by 3 (roughly); a left step multiplies by 2 (roughly); the coincidence is

232                           RICHARD K. GUY

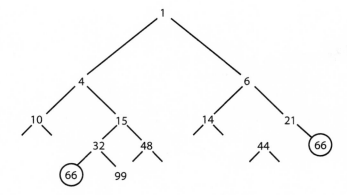

FIGURE 1. Binary tree generated by two unary functions.

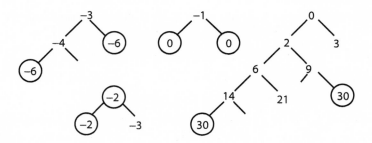

FIGURE 2. Small values of $a$ lead to coincidences.

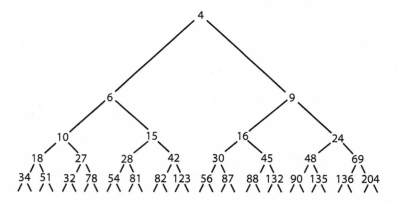

FIGURE 3. Binary tree generated by $a \to 2a - 2$ and $a \to 3a - 3$.

roughly explained by the approximation: two right ≈ three left; $3^2 \approx 2^3$. Is this another example of the strong law of small numbers [11]? If we try other small values for $a$, we find similar coincidences (Figure 2).

Let us look at $a = -4$. We've omitted the minus signs; alternatively, change the plus signs to minuses in each of conditions (2) and conditions (3) in Problem 0. The binary tree is now as in Figure 3.

The numbers that appear, when arranged in numerical order, are

4, 6, 9, 10, 15, 16, 18, 24, 27, 28, 30, 34, 42, 45, 46, 51, 52, 54,
58, 66, 69, 78, 81, 82, 87, 88, 90, 99, 100, 102, 106, 114, 123, 130,
132, 135, 136, 150, 153, 154, 159, 160, 162, 171, 172, 174, 178,
195, 196, 198, 202, 204, 210, . . . .

Does a number ever occur twice? I sent an earlier draft of this article to John
Leech, who disobeyed orders, started scribbling, and found that had I continued
Figure 3, I would have discovered 258 and 402 on the next row, with both of these
repeated two rows further down. In fact, reinstating the minus signs and using $L$
for a left step and $R$ for a right one, we can write these coincidences as

$$-4L^7 = -258 = -4RL^2R^2 \quad \text{and} \quad -4L^2RL^4 = -402 = -4R^2LR^2$$

and those starting from 1 and 0 in Figures 1 and 2 as

$$1LRL^2 = 66 = 1R^3 \quad \text{and} \quad 0L^4 = 30 = 0LR^2.$$

The last coincidence starts with a left step in either case, so can be regarded as
a coincidence starting with 2: $2L^3 = 30 = 2R^2$. Leech used his computer to find
coincidences starting with 3, 4 and 5:

$$3LRLRL^{10} = 177150 = 3R^7LR^2$$
$$4L^4R^8 = 626574 = 4RL^7RLRL^4$$
$$5L^7RL^2R^6 = 241662 = 5R^2L^{12}.$$

Since $3R = 12 = 5L$ and $3L^2 = 18 = 5R$, this last gives another coincidence starting
from 3. We have seen coincidences starting from $-1$, $-2$, $-3$ and $-4$; Leech also
found

$$-5L^3R^4 = -1986 = -5R^2L^6$$
$$-7LRL^2RL^3RL^4 = -143250 = -7R^2L^2R^6$$
$$-8LRL^8 = -9474 = -8R^4LR^2$$
$$-9L^{10} = -7170 = -9RL^2R^4$$
$$-10L^3RL^6 = -12354 = -10R^2LR^4$$
$$-11L^5R^2L^6 = -166146 = -11R^4LRL^2R^2$$

and conjectures that there will always be "duplicate fruits on the trees" no matter
where you start. If you want a coincidence starting with $-6$, note that $-6R = -10$,
hut it would be cleaner to find one of the form $-6L \cdots = -6R \ldots$, and I'm tempted
to strengthen Leech's conjecture to:

For each integer $n$, there are strings $L \ldots$ and $R \ldots$ such that $nL \ldots = nR \ldots$?

*Stop press!* In yet another letter Leech proves his conjecture, so don't bother
to solve this problem!

In problem 1 a binary tree is similarly generated by the pair of ternary functions

$$(a, b, c) \to (3ab - c, a, b), \quad (a, b, c) \to (3ac - b, a, c)$$

as in Figure 4.

We can exhibit more of the tree by simplifying it as in Figure 5. To recapture
the triples from this, choose any entry for $a$, and its immediate predecessor for $b$.
Then $c$ is found when travelling up the tree, just after the first step after the first
rightward step. For example, $a = 985$ has predecessor $b = 169$. When travelling

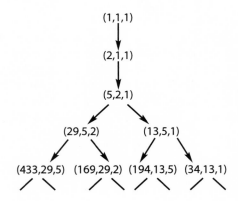

FIGURE 4. Binary tree of Markoff triples.

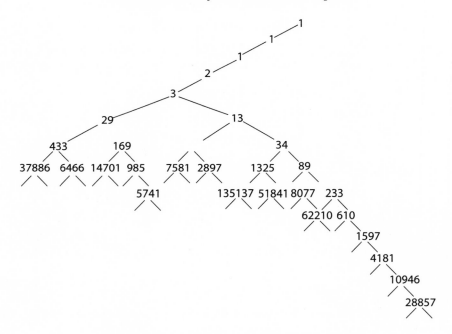

FIGURE 5. Simplified Markoff tree.

upwards from 985, the first rightward step is from 29 to 5. The next step is from 5 to 2, so $c = 2$.

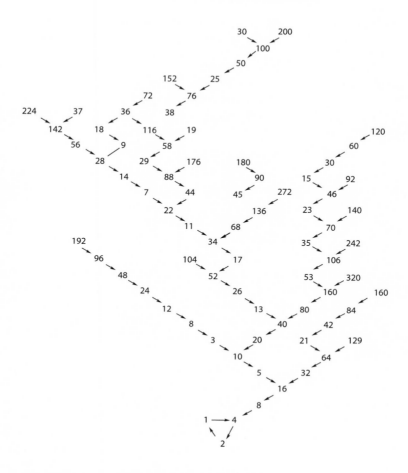

FIGURE 6. Does the Collatz algorithm give any more cycles?

Whether or not there are repetitions in the sequence of *Markoff numbers*

1, 2, 5, 13, 29, 34, 89, 169, 194, 233, 433, 610, 985, 1325, 1597, 2897, 4181, 5741, 6466, 7561, 9077, 10946, 14701, 28657, 37666, 51641, ...

has become a notorious problem. There are occasional claims to have proved uniqueness, but none seem to hold water [**19**]. Don Zagier [**23**] has some results on distribution, but none on distinctness. He can show that the problem is equivalent to the unsolvability of a certain system of diophantine equations, so we may be nearing the realm of Hilbert's tenth problem. Hence the title of this paper.

Problem 2 is associated with various names: Collatz, Hasse, Kakutani, Syracuse. It is just as notorious. Lothar Collatz told me that he thought of it when a student. One of its several waves of popularity started when he mentioned it to several people at the 1950 International Mathematical Congress in Cambridge (the wrong Cambridge). Presumably some mathematicians from Syracuse (the wrong Syracuse) became interested in it; the boys from Syracuse can perhaps fill in that bit of hi story.

Is the graph of the Collatz sequence unicyclic? Figure 6 includes all the numbers up to 26; the branch containing 27 is a much longer one, but still comes down to 1 after 111 steps.

After a long and inconclusive correspondence some years ago, a claimant to have a proof eventually admitted that "Erdős says that mathematics is not yet ripe enough for such questions." Hence the title of this paper.

Problem 3 is one of John Conway's *permutation sequences*. It is similar to the Collatz problem, but here the function has an inverse

$$3m \to 2m, \quad 3m - 1 \to 4m - 1, \quad 3m + 1 \to 4m + 1$$

(if the number's a multiple of 3, take a third off; otherwise add a third on) so the sequence can be pursued in either direction. Its graph consists of a number of disjoint cycles and doubly infinite chains. But it hasn't even been proved that an infinite chain exists! What is the status of the sequence containing the number 8?

$$\dots, 41, 31, 23, 17, 13, 10, 15, 11, 8, 12, 18, 27, 20, 30, 45, 34,$$
$$51, 38, 57, 43, 32, \dots.$$

What gives a cycle? Each term is either $3/2$ times the previous one, or approximately $3/4$ of it. Our best chance of getting back to an earlier value is to find a power of 3 which is close to a power of 2. The known cycles of lengths 1, 2, 5 and 12 correspond to the approximations of $3^1$, $3^2$, $3^5$ and $3^{12}$ by $2^2$, $2^3$, $2^8$ and $2^{19}$. The last is the ratio of $D$ sharp to $E$ flat! In fact in each of problems 0, 2 and 3, the convergents

$$\frac{1}{1} \, \frac{2}{1} \, \frac{3}{2} \, \frac{8}{5} \, \frac{19}{12} \, \frac{65}{41} \, \frac{84}{53} \, \frac{485}{306} \, \frac{1054}{665} \, \frac{24727}{15601} \, \frac{50508}{31867} \dots$$

to the continued fraction for $\log 3$ to the base 2 are of significance. Note that there arc cycles corresponding to the denominators 1, 2, 5 and 12. It has been shown that there are none of length 41, 53 or 306. Computers can push numerical results quite a long way, but it's not clear that they can be of any use with such problems.

In [**5**] Conway relates families of sequences similar to that in Problem 3 to the vector reachability problem and Minsky machines. Hence the title of this paper.

**Postscript.** After I'd written this, the following problem of David A. Klarner reached me from two different directions. There seem to be some significant similarities.

PROBLEM 4. *Let $S$ be the smallest set of numbers such that $1 \in S$ and if $x \in S$, then $2x$, $3x + 2$ and $6x + 3$ each belong to $S$, i.e.,*

$$S = \{1, 2, 4, 5, 8, 9, 10, 14, 15, 16, 17, 18, 20, 26, 27, 28, 29, 30, 32, 33, 34, \dots\}.$$

*Does $S$ have positive density? That is, as $n \to \infty$ is*

$$\liminf |S \cap \{1, 2, \dots, n\}|/n \to 0?$$

### References

[1] Michael Beeler, William Gosper, and Rich Schoeppel, Hakmera, Memo 239, Artificial Intelligence Laboratory, M.I.T., 1972, p. 64.
[2] Corrado Böhm and Giovanna Sontacchi, On the existence of cycles of given length in integer sequences like $x_{n+1} = x_n/2$ if $x_n$ even, and $x_{n+1} = 3x_n + 1$ otherwise, Atti Acead. Naz. Lincei Rend. Cl. Sci. Fiz. Mat. Natur. (8) 64 (1978) 260–264.
[3] J. W. S. Cassels, An Introduction to Diophantine Approximation, Cambridge, 1957, 27–44.

[4] H. Cohn, Approach to Markoffs minimal forms through modular functions, Ann. of Math., Princeton (2) 61 (1955) 1–12.

[5] J. H. Conway, Unpredictable iterations, Proc. Number Theory Conf., Boulder, 1972, 49–52.

[6] R. E. Crandall, On the "$3x+1$" problem, Math. Comput., 32 (197S) 1281–1292; MR58#494; Zbl. 395.10013.

[7] L. E. Dickson, Studies in the Theory of Numbers, Chicago Univ. Press, 1930, Chap. 7.

[8] C. J. Everett, Iteration of the number-theoretic function $f(2n) = n$, $f(2n+1) = 3n+2$, Adv. in Math., 25 (1977) 42–45; MR56#15552; Zbl. 352.10001.

[9] G. Frobenius, Über die Markoffschen Zahlen, S.-B. Preuss. Akad. Wlss. Berlin (1913) 458–487.

[10] Martin Gardner, Mathematical Games, A miscellany of transcendental problems, simple to state but not at all easy to solve, Scientific Amer., 226 #6 (Jun 1972) 114–118, esp. p. 115.

[11] Martin Gardner, Mathematical Games, Patterns in primes are a clue to the strong law of small numbers, Scientific Amer., 243 #6 (Dec 1980) 18–28.

[12] Richard K. Guy, Unsolved Problems in Number Theory, Springer, New York, 1981, Problems D12, E16, E17.

[13] E. Heppner, Eine Bernerkung zum Hasse-Syracuse-Algorithmus, Arch. Math. (Basel), 31 (1977/79) 317–320; MR80d:10007; Zbl. 377.10027.

[14] David C. Kay, Pi Mu Epsilon J., 5 (1972) 338.

[15] A. Markoff, Sur les formes quadratiques binaires indéfinies, Math. Ann., 15 (1879) 381–409.

[16] H. Möller, Über Hasses Verallgemeinerung der Syracuse-Algorithmus (Kakutani's Problem), Acta. Arith., 34 (1978) 219–226; MR57#16246; Zbl. 329.10008.

[17] R. Remak, Über indefinite binäre quadratische Minimalformen, Math. Ann., 92 (1924) 155–182.

[18] R. Remak, Über die geometrische Darstellung der indefinitiven binären quadratischen Minimalformen, Jber. Deutsch Math.-Verein, 33 (1925) 228–245.

[19] Gerhard Rosenberger, The uniqueness of the Markoff numbers, Math. Camp., 30 (1976) 361–365; but see MR53#280.

[20] Ray P. Steiner, A theorem on the Syracuse problem, Congressus Numerantium XX, Proc. 7th Conf. Numerical Math. Comput. Manitoba, 1977, 553–559; MR80g:10003.

[21] Riho Terras, A slopping time problem on the positive integers, Acta Arith., 30 (1976) 241–252; MR58#27879 (and see 35 (1979) 100–102; MR80h:10066).

[22] L. Ju. Vulah, The diophantine equation $p^2 + 2q^2 + 3r^2 = 6pqr$ and the Markoff Spectrum (Russian), Trudy Moskov. Inst. Radiotehn. Elektron. i Avtomat. Vyp. 67 Mat, (1973) 105–112, 152; MR58#21957.

[23] Don B. Zagier, Distribution of Markov numbers, Abstract 796–A37, Notices Amer. Math. Soc., 26 (1979) A-543.

[24] David A. Klarner, An algorithm to determine when certain sets have 0-density, J. Algorithms, 2 (1981) 31–43; Zbl. 464.10046.

DEPARTMENT OF MATHEMATICS AND STATISTICS, THE UNIVERSITY OF CALGARY, CALGARY, ALBERTA, CANADA T2N 1N4

## Editorial Commentary

(1) The present paper is one of many problem collections assembled by R. K. Guy. Here is the current status of these problems. As the paper notes, Problem 0 is solved (in the negative) by John Leech (unpublished). Problem 1, the uniqueness of Markoff numbers, remains open. Problems 2 is the $3x + 1$ problem, which remains open. Problem 3 has a history that predates John Conway. This problem has been attributed to Collatz, who states he thought of it in 1932, see Lagarias [24, p. 3], where it is termed the *original Collatz problem*. Independent of this, it was posed in 1963 by Murray Klamkin [18] as a SIAM Review problem, with 1965 commentary supplied by Dan Shanks [25] giving the heuristic above, and further commentary by A. O. L. Atkin [1] giving a method to determine all cycles of a given length, and showing there are no cycles of length less than 200 other than the known ones.

This problem remains open. Problem 4 was posed in 1982 by David Klarner [**20**, pp. 147–148]. In the early 1970's he studied, in joint work with Richard Rado [**23**], the structure of sets of integers closed under a finite set of affine operations. Under some conditions these consist of a finite number of arithmetic progressions, under other conditions they are complicated sets, which may or may not have a positive density. Other work of Klarner on this topic (some with others) includes [**16**], [**17**], [**19**], [**21**], [**22**]. In [**20**] he reports that Erdös posed a prize problem of deciding whether the smallest set of integers that includes 1 and which is closed under iteration by the affine maps $f_1(x) = 2x+1$, $f_2(x) = 3x+1$ and $f_3(x) = 6x+1$ has a positive density. He states that it was settled by G. J. Crampin and A. J. W. Hilton (unpublished) who showed it has density 0, using the fact that the semigroup generated by $\{f_i(x) : 1 \leq i \leq 3\}$ under composition has a nontrivial relation. Klarner [**20**] establishes that the semigroup generated by $g_1(x) = 2x$, $g_2(x) = 3x+2$ and $g_3(x) = 6x + 3$ is a free semigroup, which leaves unresolved the question of the lower density of the set of integers it generates. Problem 4 is still open. These four problems all appear in Guy's 2004 book *Unsolved Problems in Number Theory* (Third Edition) [**11**] as Problems D12, E16, E17 and E36, respectively.

(2) Richard K. Guy (1916–) was born in Warwickshire. He received a B. A. and M. A. at the University of Cambridge, the latter in 1941. After the war he studied number theory with Theodor Estermann at the University College, London. He held positions at the University of Singapore 1951-1963, then was Professor and Head of Department of Math and Humanities, Indian Institute of Technology, New Delhi, 1962–1965, culminating in his 1965 move to the University of Calgary, where he is now a Faculty Professor.

He is well known for his work in the theory of combinatorial games. The latter started with his first paper in 1956 (with C. A. B. Smith) [**9**]. He interested John H. Conway in this subject, later joined by Elwyn Berlekamp, and this culminated in their four volume treatise on combinatorial games [**2**]. About this subject he says in 1991 ([**13**]):

> "Perhaps I've never done any serious mathematics. Perhaps there isn't any serious mathematics. The subject of combinatorics is only slowly acquiring respectability and combinatorial games will clearly take longer than the rest of combinatorics."

He is also well known for many collections of unsolved problems, with commentary, of which a sampling are Guy [**10**], Erdős and Guy [**8**], Guy [**11**], and Croft, Falconer and Guy [**6**]. About these, the first edition (1981) of *Unsolved problems in number theory* [**11**] says:

> "If the book is to serve its purpose it will start becoming out of date the moment it appears."

It and its second and third editions all include the $3x + 1$ problem, of course [Problem E16: Collatz's sequence]. In connection with the $3x+1$ problem, his 1983 paper [**12**] presents an implementation of the computing scheme of Conway [**3**] in the $3x + 1$ problem framework to compute the successive prime numbers. This can be formalized as a FRACTRAN program, see Conway [**4**].

He is known for composing and solving chess endgames. He notes that chess problems were described by G. H. Hardy [**15**, p. 88] as "genuine mathematics, but in some sense 'trivial' mathematics." Nevertheless an interesting combinatorial game: "Dawson's Chess," arose from such considerations.

## References

[1] A. O. L. Atkin, Comment on problem 63 − 13*, SIAM Review **8** (1966), 234–236.

[2] E. Berlekamp, J. H. Conway and R. K. Guy, *Winning ways for your mathematical plays.* Volumes 1–4. Second edition. A. K. Peters Ltd., Natick Mass. 2001-2004. [First Edition: Harcourt-Brace-Jovanovich, London-New York 1982 (Two volumes)].

[3] J. H. Conway, Unpredictable Iterations, Proc. of the Number Theory Confeence (Univ. Colorado, Boulder, Colorado 1972), pp. 49–52, Univ. Colorado, Boulder, CO. 1972.

[4] J. H. Conway, FRACTRAN- A Simple Universal Computing Language for Arithmetic, in: *Open Problems in Communication and Computation* (T. M. Cover and B. Gopinath, Eds.), Springer-Verlag: New York 1987, pp. 3–27.

[5] J. H. Conway and R. K. Guy, *The Book of Numbers,* Copernicus, New York 1996.

[6] H. T. Croft, K. J. Falconer, and R. K. Guy, *Unsolved problems in geometry. Corrected reprint of the 1991 original,* Unsolved Problems in Intuitive Mathematics II, Springer-Verlag: New York 1994. (198pp)

[7] P. Erdős and R. L. Graham, *Old and New Problems and Results in Combinatorial Number Theory,* Mongraphie Enseign. Math. No. 28 , Kundig: Geneva 1980

[8] P. Erdős and R. K. Guy, Crossing number problems, Amer. Math. Monthly **80** (1973), 52–58.

[9] R. K. Guy and C. A. B. Smith, The G-values of various gmes, Proc. Camb. Phil. Soc. **52** (1956), 514–526.

[10] R. K. Guy, Unsolved combinatorial problems, 1971 Combinatorial Mathematics and its Applications (Proc. Conf. , Oxford 1969), pp. 121–127, Academic Press: London.

[11] R. K. Guy, *Unsolved problems in number theory.* Third Edition. Unsolved Problems in Intuitive Mathematics, I, Springer-Verlag 2004, 437pp. (First Edition: 1981, Second Edition: 1994).

[12] R. K. Guy, Conway's Prime Producing Machine, Mathematics Magazine **56** (1983), No. 1, 26–33.

[13] Mathematics from fun & fun from mathematics: an informal autobiographical history of combinatorial games, pp. 287–295 in: *Paul Halmos: Celebrating 50 Years of Mathematics* (J. H. Ewing and F. W. Gehring, Eds.), Springer-Verlag: New York 1991.

[14] R. K. Guy, Aviezri Fraenkel and combinatorial games, Electronic Journal of Combinatorics **8** (2001), no. 2, 6pp.

[15] Godfrey Harold Hardy, *A Mathematician's Apology,* reprinted with a Foreword by C. P. Snow, Cambridge University Press: Cambridge 1967.

[16] D. G. Hoffman and D. A. Klarner, Sets of integers closed under affine operators-the closure of finite sets, Pacific J. Math. **78** (1978), No. 2, 337–344.

[17] D. G. Hoffman and D. A. Klarner, Sets of integers closed under affine operators-the finite basis theorems, Pacific J. Math. **83** (1979), No. 1, 135–144.

[18] M. S. Klamkin, Problem 63 − 13*, SIAM Review **5** (1963), 275–276.

[19] D. A. Klarner, An algorithm to determine when certain sets have 0-density, J. Algorithms **2** (1981), 31–43.

[20] D. A. Klarner, A sufficient condition for certain semigroups to be free, J. Algebra **74** (1982), 140–148.

[21] D. A. Klarner, m-recognizability of sets closed under certain affine functions, Discrete Applied Math. **21** (1988), 207–214.

[22] D. A. Klarner and K. Post, Some fascinating integer sequences, Discrete Math. **106/107** (1992), 303–309.

[23] D. Klarner and R. Rado, Arithmetic properties of certain recursively defined sets, Pacific J. Math. **53** (1974), No. 2, 445–463.

[24] J. C. Lagarias, The $3x + 1$ problem and its generalizations, Amer. Math. Monthly **92** (1985), 3–23.

[25] D. Shanks, Comment on Problem 63 − 13*, SIAM Review **7** (1965), 284–286.

# On the Motivation and Origin of the $(3n + 1)$-Problem

## Lothar Collatz

## 1. Graph Theory and Number Theoretic Functions

There are numerous connections between elementary number theory and elementary graph theory. One such connection can be made using the fact that one can picture a number theoretic function $f(n)$ with a directed graph. One takes the whole numbers $n = 1, 2, 3, \ldots$ as the vertices of the graph and for each draws an arrow from $n$ to $f(n)$.

Through the lectures of Edmund Landau, Oscar Perron, Issai Schur and others during my time as a student from 1928 through 1933, I became interested in problems in the areas of number theoretic functions and graph theory. For the graphs of number theoretic functions, I have undertaken the following classification:

### I. Single-Valued Functions

(1) **Line.**

The graph is an unbounded line. Example: $f(n) = n + 1$.

(2) **Tree.**

Example: $f(n) = n - g(n)$ where $g(n)$ is the largest proper divisor of $n$. For example, $f(21) = 21 - 7 = 14$. In the case of a prime number $p$, $f(p) = p - 1$. A part of the tree appears in Figure 1, for the numbers 1 to 30.

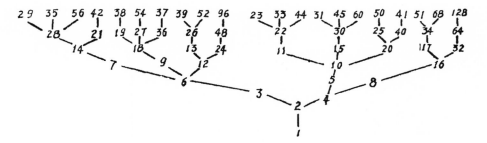

FIGURE 1

(3) **Forest.**

The graph consists of $k$ unconnected trees ($2 \leq k \leq \infty$).
Example: $f(n) = n + n^2$.

(4) **Fixed Point.**

The graph contains at least one fixed point, but no cycles. If $f(q) = q$ then we call $q$ a fixed point. A graph with 2 or more fixed points is not connected.

Example: $f(n) = \frac{1}{2}(n^2 - 9n) + 9$ for whole numbers $n \geq 1$. The graph in Figure 2 contains 2 fixed points, namely $n = 2$ and $n = 9$.

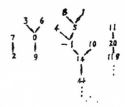

FIGURE 2

(5) **Cycles.**

The graph contains at least one cycle (and possibly also fixed points). The number $s$ of points on the cycle is called the length of the cycle. It is assumed that $s$ is finite and $\geq 2$.

Example:

$$f(n) = \begin{cases} 3n & \text{for } n \text{ prime} \\ \text{the sum of all proper divisors of } n & \text{otherwise.} \end{cases}$$

Thus $f(1) = 3$ and $f(21) = 7 + 3 = 10$. A small piece of the graph is shown in Figure 3.

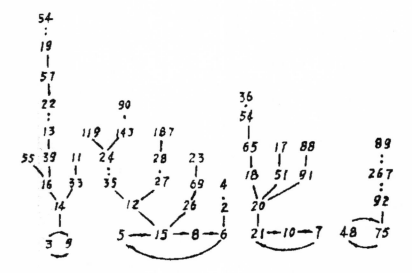

FIGURE 3

## II. Multi-Valued Functions

Here several cases are possible. It is therefore more reasonable to employ the idea of a hypergraph than to use graphs. (See pp. 16-20 in L. Collatz, Verzweigungsdiagramme und Hypergraphen [1]). We consider here two examples.

(1) In the case of an $n$ which is representable as $a \cdot b$, $1 < a < b$, we define $f(n) = b^2 - a^2$. This definition gives a multi-valued function. Thus for $n = 36 = 2 \times 18 = 3 \times 12 = 4 \times 9$, we take $f(n)$ to be $324 - 4 = 320$, $144 - 9 = 135$, and $81 - 16 = 65$. One piece of the complex graph is shown in Figure 4. Above each edge are the factors $b$ and $a$ (sometimes just $b$) which generate it. The cycle $(231, 320, 144)$ is marked with arrows in the figure.

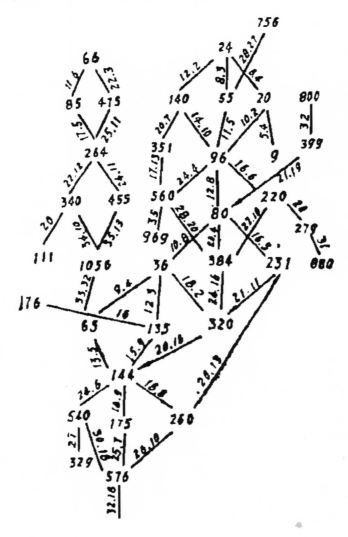

FIGURE 4

(2) A further example is the hypergraph which arises from the factorization of the whole numbers. Each whole number $n$ may be decomposed as $n = u \cdot v$. One considers all such pairs. (See L. Collatz, Different types of hypergraphs and several applications, in: *Graphs and other combinatorial topics (Prague 1982)*, Tuebner-Texte Math: Teubner: Leipzig 1983, pp. 30–49.)

## 2. The $(3n + 1)$-Problem

As stated, I have found many graphs with cycles. Through factorization one can easily find cycles. I tried, however, to get cycles for the functions above using only elementary calculations. These examples for this purpose should be as simple as possible, but also representative. If a graph contains a cycle, then $f(n) < n$ for some $n$, but $f(n) > n$ for other $n$.

We define $f(n) = \frac{n}{2}$ for even $n$, so $f(n) < n$, and $\hat{f}(n) = n + 1$ for odd $n$, so $\hat{f}(n) > n$. This admits only the trivial cycle $(1, 2)$. If instead we define $\hat{f}(n) = 2n + 1$, then there are no cycles to find, since the odd integers consistently increase. Now we define the "$3n + 1$" function:

$$f(n) = \begin{cases} 3n + 1 & \text{for odd } n \\ \frac{n}{2} & \text{for even } n. \end{cases}$$

In the graph, which is shown in Figure 5, I have only found the trivial cycle $(4, 2, 1)$.

I have experimented with many numbers. Before 1950 computers were not as well-developed as they are today. Therefore I could not try any large numbers. Using the iterative procedure

$$f_k(n) = f(f_{k-1}(n)), \qquad k = 1, 2, \ldots,$$
$$f_0(n) = f(n)$$

the following conjecture can be stated:

**The $(3n + 1)$-Conjecture**
   **Form a)** For each $n$ there is an index $k(n)$ so that $f_k(n) = 1$
or, graph-theoretically,
   **Form b)** The graph of the given function is connected.

If one can prove this conjecture, or give examples of cycles other than $(4, 2, 1)$, then this problem will be solved.

Since I could not solve the problem, I have not published the conjecture. I have mentioned the problem at many meetings and in many talks. In 1952, when I came to Hamburg, I explained the problem to my colleague Prof. Dr. Helmut Hasse. He was also interested in it. He has explained it in talks in other cities.

There are already numerous publications on the $(3n + 1)$-Problem. Only a few are listed here ([**2**], [**3**], [**4**]).

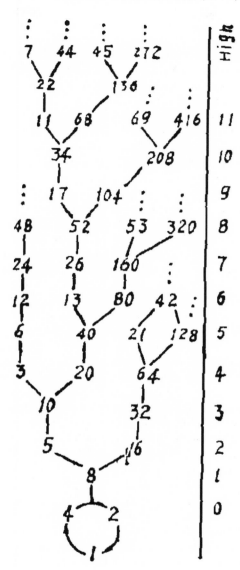

FIGURE 5

## References

[1] L. Collatz, Verzweigungsdiagramme und Hypergraphen, In: *Numerik und Anwendungen von Eigenwertaufgaben und Verzweigungsproblemen (Tagung, Math. Forschungsinst. Oberwolfach, 1976)*, pp. 9–42, Internat. Schrifteneihe Numer. Math., 38, Birkhäuser: Basel 1977.
[2] Lynn E. Garner, On heights in the Collatz $3n+1$ problem, Discrete Math., **55** (1985), 57–64.
[3] Jeffrey C. Lagarias, The $3x + 1$ problem and its generalizations, Amer. Math. Monthly **92** (1985), 3–23.
[4] Stan Wagon, The Collatz problem, Math. Intelligencer **7** (1985), No. 1, 72–76.

INSTITUT FÜR ANGEWANDTE MATHEMATIK DER UNIVERSITÄT HAMBURG

### Editorial Commentary

(1) Lothar Collatz is widely acknowledged as the originator of the $3x + 1$ problem. This paper, published in Chinese in *J. Qufu Normal University, Natural Science Edition* **12** (1986), no. 3, 9–11, is the only paper that Collatz wrote addressing the problem. The Chinese text states it is a translation by Zhi-Ping Ren, presumably of a German text. This English translation, made by Mark Conger, is based on a German retranslation of the Chinese paper, with checks of the Chinese text made by Chris Xiu. The Chinese title translates literally as: "The cause of the beginning of the (3n+1) problem." The original article has figures drawn by Collatz which are reproduced here.

(2) Lothar Collatz (1910-1990) was a well-known German applied mathematician. He was born in Arnsberg, Westphalia. Between 1928 and 1933 he studied at Göttingen, Berlin, Greifswald and Munich. In Berlin he attended lectures of Issai Schur, Alfred and Richard Brauer, Walter Ledermann and Helmut Wielandt. At Göttingen he studied with Richard von Mises, Richard Courant and Edmund Landau. He was attracted to problems in numerical analysis and mathematical physics, and was influenced by work of Carl Runge. He started work as a student of Richard von Mises, and took his oral exam hurriedly in November 1933, when von Mises chose to leave Germany for Turkey to become professor at the University of Istanbul. (Segal [**13**, p. 318]). He was awarded the doctorate in 1935 at Berlin, officially with Alfred Klose, an astronomer, as advisor. He completed his Habilitation in 1937 at Technische Hochschule Karlsruhe. He worked on difference schemes for numerical solving ordinary and partial differential equations. In 1938 he published work on spline interpolation of Fourier coefficients (with W. Quade [**12**]). He did early work on numerical computation of eigenvalues, publishing two papers in 1939 ([**3**]) and a book on them in 1945 ([**5**]). He remained at Karlsruhe till 1943, when he accepted a mathematics chair at Technische Hochschule Hannover. There he worked with engineers and wrote papers on mechanical and mathematical physics. In 1951 he published a standard textbook on numerical methods to solve differential equations [**6**], containing also theoretical error analysis. In 1952 he accepted a chair at the University of Hamburg, and remained there until his retirement in 1978. He early on recognized the importance of electronic computers. He did fundamental work in numerical linear algebra, see Elsner and Hadeler [**2**]. In this area in 1942 he proved the Collatz inclusion theorem on spectra of symmetric matrices [**4**]. In 1952 he introduced the notion of problems of monotone type (monotone matrices) [**7**], useful for obtaining error estimates, and applied it to some non-linear boundary value problems. These are finite difference analogues of elliptic operators, cf. Varga [**14**]. He did important work in approximation theory, surveyed in Meinardus and Nürnberger, Reissinger and Walz [**11**]. He had a long term interest in graph theory, starting from when he was a student, as indicated in this paper. He wrote one of the earliest papers on the spectra of finite graphs (with U. Sinogowitz) [**9**]. (See Cvetković, Doob and Sachs [**10**] for further development of this subject.) His formulation of the $3x + 1$ problem seems to have arisen from study of graphical representation of iterative processes. He wrote over 150 papers. Further information on Collatz's life and work is given in the memorial volume [**1**].

# References

[1] L. Brendendiek, H. Burchad, U. Grothkopf, H. J. Oberle, G. Opfer and B. Werner, Eds. *Lothar Collatz 1910–1990*, Inst. für Angewande Math. der Universität Hamburg, Bericht 16, 1991, 118pp.

[2] C. Elsner and K. P. Hadeler, Lothar Collatz– On the occation of his 75-th Birthday, Lin. Alg. Appl. **68** (1985), 1–8.

[3] L. Collatz, Gen äherte Berechnung von Eigenwerten, Z. Angew. Math. Mech. **19** (1939), 224–249, 297–318.

[4] L. Collatz, Einschlie$\beta$ungssatz für chakterische Zahlen von Matizen, Math. Z. **48** (1942), 221–226.

[5] L. Collatz, *Eigenwert probleme und ihre numersched Behandlung,* Akademisched Verlagsgeschellshaft, Leipzig 1945, 338pp.

[6] L. Collatz, *Numerische Behandlung von Differentialgleichungen*, Die Grundlehren der mathematischen Wissenscheften, Band 60, Springer-Verlag: Berlin, Göttingen, Heidelberg 1951. (English Translation: *The Numerical Treatment of Differential Equations,* Third Edition, Springer-Verlag: Berlin-Göttingen -Heidelberg 1960)

[7] L. Collatz, Aufgaben monotoner Art, Arch. Math. **3** (1952), 366–376.

[8] L. Collatz, *Funktionalanalysis und numerische Mathematik* Die Grundlehren der mathematischen Wissenschaften, Band 120. Springer-Verlag: Berlin 1964. (English Translation: *Functional analysis and numerical mathematics*, Translation from the German by Hasjörg Oser, Academic Press, New York-London 1966)

[9] L. Collatz and U. Sinogowitz, Spektren endlicher Grafen, Abh. Math. Sem. Univ. Hamburg **21** (1957), 63–77.

[10] D. Cvetković, M. Doob, H. Sachs, *Spectra of Graphs. Theory and applications. Third edition*, Johann Ambrosius Barth: Heidelberg 1995.

[11] G. Meinardus, G. Nürnberger, Th. Riessinger, and G. Walz, In memoriam: the work of Lothar Collatz in approximation theory, J. Approx. Th. **67** (1991). 119–128.

[12] W. Quade and L. Collatz, Zur Interpolationstheorie der reellen periodischen Funktioned, Proc. Preuss. Akad. Wiss. (Math.-Phys. Kl.) **30** (1938), 383–429.

[13] Sanford L. Segal, *Mathematicians under the Nazis,* Princeton University Press: Princeton, NJ 2003.

[14] R. Varga, On a discrete maximum principle, SIAM J. Numerical Analysis **3** (1966), No. 2, 355–359.

# FRACTRAN: A Simple Universal Programming Language for Arithmetic

## J.H. Conway

### 1. Your Free Samples of FRACTRAN

To play the *fraction game* corresponding to a given list

$$f_1, f_2, \ldots, f_k$$

of fractions and starting integer $N$, you repeatedly multiply the integer you have at any stage (initially $N$) by the earliest $f_i$ in the list for which the answer is integral. Whenever there is no such $f_i$, the game *stops*.

(Formally, we define the sequence $\{N_n\}$ by $N_0 = N$, $N_{n+1} = f_i N_n$, where $i$ $(1 \le i \le k)$ is the least $i$ for which $f_i N_n$ is integral, as long as such an $i$ exists.)

THEOREM 1. *When PRIMEGAME:*

$$\frac{17}{91} \; \frac{78}{85} \; \frac{19}{51} \; \frac{23}{38} \; \frac{29}{33} \; \frac{77}{29} \; \frac{95}{23} \; \frac{77}{19} \; \frac{1}{17} \; \frac{11}{13} \; \frac{13}{11} \; \frac{15}{2} \; \frac{1}{7} \; \frac{55}{1}$$

*is started at 2, the other powers of 2 that appear, namely,*

$$2, 2^3, 2^5, 2^7, 2^{11}, 2^{13}, 2^{17}, 2^{19}, 2^{23}, 2^{29}, \ldots,$$

*are precisely those whose indices are the prime numbers, in order of magnitude.*

THEOREM 2. *When PIGAME:*

$$\frac{365}{46} \; \frac{29}{161} \; \frac{79}{575} \; \frac{679}{451} \; \frac{3159}{413} \; \frac{83}{407} \; \frac{473}{371} \; \frac{638}{355} \; \frac{434}{335} \; \frac{89}{235} \; \frac{17}{209} \; \frac{79}{122}$$

$$\frac{31}{183} \; \frac{41}{115} \; \frac{517}{89} \; \frac{111}{83} \; \frac{305}{79} \; \frac{23}{73} \; \frac{73}{71} \; \frac{61}{67} \; \frac{37}{61} \; \frac{19}{59} \; \frac{89}{57} \; \frac{41}{53} \; \frac{833}{47} \; \frac{53}{43}$$

$$\frac{86}{41} \; \frac{13}{38} \; \frac{23}{37} \; \frac{67}{31} \; \frac{71}{29} \; \frac{83}{19} \; \frac{475}{17} \; \frac{59}{13} \; \frac{41}{291} \; \frac{1}{7} \; \frac{1}{11} \; \frac{1}{1024} \; \frac{1}{97} \; \frac{89}{1}$$

*is started at $2^n$, the next power of 2 to appear is $2^{\pi(n)}$, where for*

| $n$ = | 0 | 1 | 2 | 3 | 4 | 5 | 6 | 7 | 8 | 9 | 10 | 11 | 12 | 13 | 14 | 15 | 16 | 17 | 18 | 19 | 20 | ... |
|---|---|---|---|---|---|---|---|---|---|---|---|---|---|---|---|---|---|---|---|---|---|---|
| $\pi(n)$ = | 3 | 1 | 4 | 1 | 5 | 9 | 2 | 6 | 5 | 3 | 5 | 8 | 9 | 7 | 9 | 3 | 2 | 3 | 8 | 4 | 6 | ... |

For an arbitrary natural number $n$, $\pi(n)$ is the $n$th digit after the point in the decimal expansion of the number $\pi$.

THEOREM 3. *Define* $f_c(n) = m$ *if POLYGAME:*

$$\frac{583}{559}\ \frac{629}{551}\ \frac{437}{527}\ \frac{82}{517}\ \frac{615}{329}\ \frac{371}{129}\ \frac{1}{115}\ \frac{53}{86}\ \frac{43}{53}\ \frac{23}{47}\ \frac{341}{46}$$

$$\frac{41}{43}\ \frac{47}{41}\ \frac{29}{37}\ \frac{37}{31}\ \frac{37}{31}\ \frac{299}{29}\ \frac{47}{23}\ \frac{161}{15}\ \frac{527}{19}\ \frac{159}{7}\ \frac{1}{17}\ \frac{1}{13}\ \frac{1}{3}$$

*when started at* $c2^{2^n}$, *stops at* $2^{2^m}$, *and otherwise leave* $f_c(n)$ *undefined. Then every computable function appears among* $f_0, f_1, f_2, \ldots$.

## 2. The Catalogue

We remark that the "catalogue numbers" $c$ are easily computed for some quite interesting functions. Table 1 and its notes give $f_c$ for any $c$ whose largest odd divisor is less than $2^{10} = 1024$.

TABLE 1. The Catalogue

| $c$ | All defined values of $f_c$ |
|---|---|
| 0 | none |
| 1 | $n \to n$ |
| 2 | $0 \to 1$ |
| 4 | $0 \to 2$ |
| 8 | $1 \to 2$ |
| 16 | $2 \to 3$ |
| 64 | $1 \to 3$ |
| 77 | $n \to 0$ |
| 128 | $0 \to 3$ |
| 133 | $0 \to 0$ |
| 255 | $n + 1 \to n + 1$ |
| 256 | $3 \to 4$ |
| 847 | $n \to 1$ |
| 37485 | $0 \to 0,\ n + 1 \to n$ |
| 2268945 | $n \to n + 1$ |
| $2^k$ | $a \to b$ if $2^b - 2^a = k$ |
| $7 \cdot 11^{2^k}$ | $n \to k$ |
| $\frac{15}{7} \cdot 1029^{2^{k-1}}$ | $n \to n + k$ |
| $c_\pi$ | $n \to \pi(n)$ |

In this Table, $n$ denotes an arbitrary non-negative integer.

We also have

$$f_{2^k A} = f_0;$$
$$f_{2^k B} = f_{2^k};\quad f_{2^k B'} = f_{2^{k+1}};$$
$$f_{2^k C} = f_{77};\quad f_{2^k C'} = f_{847};$$
$$f_{2^k D} = f_{133}\ (k = 0)\quad \text{or}\quad f_0\ (k > 0);$$
$$f_{2^k E} = f_{255}\ (k = 0)\quad \text{or}\quad f_{2^k}\ (k > 0);$$

where

| | | |
|---|---|---|
| $A$ | is | any odd number $< 1024$ not visible below: |
| $B$ | is | 1, 3, 9, 13, 17, 27, 39, 45, 51, 81, 105, 115, 117, 135, 145, 153, 155, 161, 169, 185, 195, 203, 205, 217, 221, 235, 243, 259, 287, 289, 315, 329, 345, 351, 405, 435, 459, 465, 483, 507, 555, 585, 609, 615, 651, 663, 705, 729, 777, 861, 945, 975, 987, 1017, ... |
| $B'$ | is | 165, 495, ... |
| $C$ | is | 77, 91, 231, 273, 385, 455, 539, 1015, ... |
| $C'$ | is | 847, 1001, ... |
| $D$ | is | 133, 285, 399, 665, 855, ... |
| $E$ | is | 255, .... |

Figure 1 gives a $c$ for which $f_c(n)$ is the above function $\pi(n)$.

$$3 \left( \begin{matrix} 2^{100!} + 2^{\frac{365}{46}}101 \cdot 100! + 2^{\frac{29}{161}}101^2 100! + 2^{\frac{79}{575}}101^3 100! + 2^{\frac{7}{451}}101^4 100! \\ + 2^{\frac{3159}{413}}101^5 100! + 2^{\frac{83}{497}}101^6 100! + 2^{\frac{473}{371}}101^7 100! + 2^{\frac{638}{355}}101^8 100! \\ + 2^{\frac{434}{335}}101^9 100! + 2^{\frac{89}{235}}101^{10} 100! + 2^{\frac{17}{209}}101^{11} 100! + 2^{\frac{79}{122}}101^{12} 100! \\ + 2^{\frac{31}{183}}101^{13} 100! + 2^{\frac{41}{115}}101^{14} 100! + 2^{\frac{517}{89}}101^{15} 100! + 2^{\frac{111}{83}}101^{16} 100! \\ + 2^{\frac{305}{79}}101^{17} 100! + 2^{\frac{23}{73}}101^{18} 100! + 2^{\frac{73}{71}}101^{19} 100! + 2^{\frac{61}{67}}101^{20} 100! \\ + 2^{\frac{37}{61}}101^{21} 100! + 2^{\frac{19}{59}}101^{22} 100! + 2^{\frac{89}{57}}101^{23} 100! + 2^{\frac{41}{53}}101^{24} 100! \\ + 2^{\frac{883}{47}}101^{25} 100! + 2^{\frac{53}{43}}101^{26} 100! + 2^{\frac{86}{41}}101^{27} 100! + 2^{\frac{13}{38}}101^{28} 100! \\ + 2^{\frac{23}{37}}101^{29} 100! + 2^{\frac{67}{31}}101^{30} 100! + 2^{\frac{71}{29}}101^{31} 100! + 2^{\frac{83}{19}}101^{32} 100! \\ + 2^{\frac{475}{17}}101^{33} 100! + 2^{\frac{59}{13}}101^{34} 100! + 2^{\frac{41}{3}}101^{35} 100! + 2^{\frac{1}{7}}101^{36} 100! \\ + 2^{\frac{1}{11}}101^{37} 100! + 2^{\frac{1}{1024}}101^{38} 100! + 2^{101^{39} 100!} \end{matrix} \right) \times 5^{2^{89 \cdot 101!} + 2^{90 \cdot 101!}} \times 17^{101!-1} \times 23$$

FIGURE 1. The constant $c_\pi$.

## 3. Avoid Brand X

Works that develop the theory of effective computation are often written by authors whose interests are more logical than computational, and so they seldom give elegant treatments of the essentially computational parts of this theory. Any effective enumeration of the computable functions is probably complicated enough to spread over a chapter, and we might read that "of course the explicit computation of the index number for any function of interest is totally impracticable." Many of these defects stem from a bad choice of the underlying computational model.

Here we take the view that it is precisely because the particular computational model has no great logical interest that it should be carefully chosen. The logical points will be all the more clear when they don't have to be disentangled by the reader from a clumsy program written in an awkward language, and we can then "sell" the theory to a wider audience by giving simple and striking examples explicitly. (It is for associated reasons that we use the easily comprehended term "computable function" as a synonym for the usual "partial recursive function.")

## 4. Only FRACTRAN Has These Star Qualities

FRACTRAN is a simple theoretical programming language for arithmetic that has none of the defects described above.

- *Makes workday really easy!*

FRACTRAN needs no complicated programming manual—its entire syntax can be learned in 10 seconds, and programs for quite complicated and interesting functions can be written almost at once.

- *Gets those functions really clean!*

The entire configuration of a FRACTRAN machine at any instant is held as a single integer—there are no messy "tapes" or other foreign concepts to be understood by the fledgling programmer.

- *Matches any machine on the market!*

Your old machines (Turing, etc.) can quite easily be made to simulate arbitrary FRACTRAN programs, and it is usually even easier to write a FRACTRAN program to simulate other machines.

- *Astoundingly simple universal program!*

By making a FRACTRAN program that simulates an arbitrary other FRACTRAN program, we have obtained the simple universal FRACTRAN program described in Theorem 3.

## 5. Your PRIMEGAME Guarantee!

In some ways, it is a pity to remove some of the mystery from our programs such as PRIMEGAME. However, it is well said [2] that "A mathematician is a conjurer who gives away his secrets," so we'll now prove Theorem 1.

To help in Figure 2, we have labeled the fractions:

| $A$ | $B$ | $C$ | $D$ | $E$ | $F$ | $G$ | $H$ | $I$ | $J$ | $K$ | $L$ | $M$ | $N$ |
|---|---|---|---|---|---|---|---|---|---|---|---|---|---|
| $\frac{17}{91}$ | $\frac{78}{85}$ | $\frac{19}{51}$ | $\frac{23}{38}$ | $\frac{29}{33}$ | $\frac{77}{29}$ | $\frac{95}{23}$ | $\frac{77}{19}$ | $\frac{1}{17}$ | $\frac{11}{13}$ | $\frac{13}{11}$ | $\frac{15}{2}$ | $\frac{1}{7}$ | $\frac{55}{1}$ |

and we note that $AB = \frac{2 \times 3}{5 \times 7}$, $EF = \frac{7}{3}$, $DG = \frac{5}{2}$.

We let $n$ and $d$ be numbers with $0 < d < n$ and write $n = qd + r$ $(0 \le r < d)$. Figure 2 illustrates the action of PRIMEGAME on the number $5^n 7^d 13$. We see that this leads to $5^n 7^{d-1} 13$ or $5^{n+1} 7^n 13$ according as $d$ does or does not divide $n$. Moreover, the only case when a power of 2 arises is as the number $2^n 7^{d-1}$ when $d = 1$.

It follows that when the game is started at $5^n 7^{n-1} 13$, it tests all numbers from $n-1$ down to 1 until it first finds a divisor of $n$, and then continues with $n$ increased by 1. In the process, it passes through a power of $2^n$ of 2 only when the largest divisor of $n$ that is less than $n$ is $d = 1$, or in other words, only when $n$ is prime.

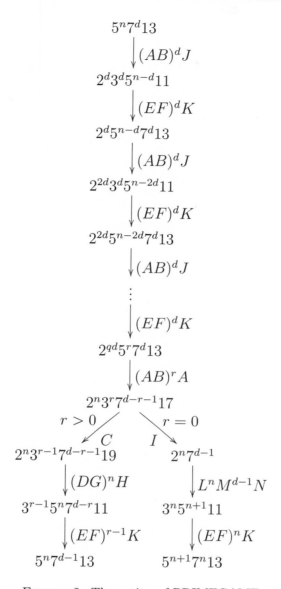

$$5^n 7^d 13$$

$$\downarrow (AB)^d J$$

$$2^d 3^d 5^{n-d} 11$$

$$\downarrow (EF)^d K$$

$$2^d 5^{n-d} 7^d 13$$

$$\downarrow (AB)^d J$$

$$2^{2d} 3^d 5^{n-2d} 11$$

$$\downarrow (EF)^d K$$

$$2^{2d} 5^{n-2d} 7^d 13$$

$$\downarrow (AB)^d J$$

$$\vdots$$

$$\downarrow (EF)^d K$$

$$2^{qd} 5^r 7^d 13$$

$$\downarrow (AB)^r A$$

$$2^n 3^r 7^{d-r-1} 17$$

$r > 0$  $\qquad$  $r = 0$

$C \qquad I$

$$2^n 3^{r-1} 7^{d-r-1} 19 \qquad\qquad 2^n 7^{d-1}$$

$$\downarrow (DG)^n H \qquad\qquad\qquad \downarrow L^n M^{d-1} N$$

$$3^{r-1} 5^n 7^{d-r} 11 \qquad\qquad 3^n 5^{n+1} 11$$

$$\downarrow (EF)^{r-1} K \qquad\qquad \downarrow (EF)^n K$$

$$5^n 7^{d-1} 13 \qquad\qquad 5^{n+1} 7^n 13$$

FIGURE 2. The action of PRIMEGAME.

## 6. FRACTRAN—Your Free Introductory Offer

TRAN program may have any number of lines, and a typical line mig       rm

$$\text{line 13}: \quad \frac{2}{3} \to 7, \quad \frac{4}{5} \to 14.$$

At this lin         machine replaces the current working integer $N$ by $\frac{2}{3}N$, if this is again an integer,    nd    oes to line 7. If $\frac{2}{3}N$ is not an integer, but $\frac{4}{5}N$ is, we should instead replace $N$ by $\frac{4}{5}N$, and go to line 14. If neither $\frac{2}{3}N$ nor $\frac{4}{5}N$ is integral, we should *stop* at line 13.

More generally, a FRACTRAN program line has the form

$$\text{line } n: \quad \frac{p_1}{q_1} \to n_1, \quad \frac{p_2}{q_2} \to n_2, \quad \ldots, \quad \frac{p_k}{q_k} \to n_k.$$

The action of the machine at this line is to replace $N$ by $\frac{p_i}{q_i}N$ for the least $i$ $(1 \leq i \leq k)$ for which this is integral, and then go to line $n_1$; or, if no $\frac{p_i}{q_i}N$ is integral, to *stop* at line $n$. (A line with $k = 0$ is permitted and serves as an unconditional stop order.)

A FRACTRAN program that has just $n$ lines is called a FRACTRAN-$n$ program. We introduce the convention that a line that cannot be jumped to counts as a $\frac{1}{2}$-line. (Sensible programs will contain at most one $\frac{1}{2}$-line, the initial line.)

We write

$$\left[\frac{p_1}{q_1}\frac{p_2}{q_2}\ldots\frac{p_k}{q_k}\right]$$

for the FRACTRAN-1 program

$$\text{line } 1: \quad \frac{p_1}{q_1} \to 1, \quad \frac{p_2}{q_2} \to 1, \quad \ldots, \quad \frac{p_k}{q_k} \to 1.$$

We shall see that every FRACTRAN program can be simulated by a FRACTRAN-1 program which starts at a suitable multiple of the original starting number. With a FRACTRAN-$1\frac{1}{2}$ program, we can make this multiple be 1.

The FRACTRAN-$1\frac{1}{2}$ program

$$\text{line } 0: \quad \frac{P_1}{Q_1} \to 1, \quad \frac{P_2}{Q_2} \to 1, \quad \ldots, \quad \frac{P_j}{Q_j} \to 1$$

$$\text{line } 1: \quad \frac{p_1}{q_1} \to 1, \quad \frac{p_2}{q_2} \to 1, \quad \ldots, \quad \frac{p_k}{q_k} \to 1$$

is symbolized by

$$\frac{P_1}{Q_1}\frac{P_2}{Q_2}\ldots\frac{P_j}{Q_j}\left[\frac{p_1}{q_1}\frac{p_2}{q_2}\ldots\frac{p_k}{q_k}\right]$$

Note that the FRACTRAN-$1\frac{1}{2}$ program

$$m[f_1 f_2 \ldots f_k]$$

started at $N$, simulates the FRACTRAN-1 program

$$[f_1 f_2 \ldots f_k]$$

started at $mN$.

We shall usually suppose tacitly that our FRACTRAN programs are only applied to working numbers $N$ whose prime divisors appear among the factors of the numerators and denominators of the fractions mentioned.

## 7. Beginners' Guide to FRACTRAN Programming

It's good practice to write FRACTRAN programs as flowcharts, with a node for each program line and arrows between these nodes marked with the appropriate fractions. We use the different styles of arrowhead

$$\xrightarrow{f} \quad \xrightarrow{f} \quad \xrightarrow{f} \quad \xrightarrow{f}$$

for the options with decreasing priorities from a given node, and if several options with fractions $f, g, h$ at a node have adjacent priorities, we often amalgamate them into a single arrow:

$$\xrightarrow{\quad f,g,h \quad}$$

The different primes that arise in the numerators and denominators of the various fractions may be regarded as storage registers, and in a state in which the current working integer is

$$N = 2^a 3^b 5^c 7^d \ldots,$$

we say that

register 2 holds $a$, or $r_2 = a$
register 3 holds $b$, or $r_3 = b$
register 5 holds $c$, or $r_5 = c$
register 7 holds $d$, or $r_7 = d$
etc.

FRACTRAN program lines are then regarded as instructions to change the contents of these registers by various small amounts, subject to the overriding requirement that no register may ever contain a negative number. Thus the line

$$\text{line 13}: \quad \frac{2}{3} \to 7, \quad \frac{4}{5} \to 14$$

| | | |
|---|---|---|
| either | *replaces $r_2$ by $r_2 + 1$, $r_3$ by $r_3 - 1$* | (if $r_3 > 0$) |
| or | *replaces $r_2$ by $r_2 + 2$, $r_5$ by $r_5 - 1$* | (if $r_5 > 0$) |
| or | *stops* | (if $r_3 = r_5 = 0$). |

In our figures, unmarked arrows are used when the associated fractions are 1. A tiny incoming arrow to a node indicates that that node will be used as a starting node; a tiny outgoing arrow marks a node that may be used as a stopping node. A few simple examples should convince the reader the FRACTRAN really does have universal computing power. (Readers familiar with Minsky's register machines will see that FRACTRAN can trivially simulate them.)

The program

is a destructive adder: when started with $r_2 = a$, $r_3 = b$, it stops with $r_2 = a + b$, $r_3 = 0$. We can make it less destructive by using register 5 as working space: the program

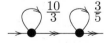

when started with $r_2 = a$, $r_3 = b$, $r_5 = 0$, stops with $r_2 = a + b$, $r_3 = b$, $r_5 = 0$.

By repeated addition, we can perform multiplication: the program

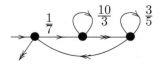

started with $r_2 = a$, $r_3 = b$, $r_5 = 0$, $r_7 = c$, stops with $r_2 = a + bc$, $r_3 = b$, $r_5 = r_7 = 0$. We add an order $\frac{1}{3}$ ("clear 3") at the starting/finishing node and formulate the result as an official FRACTRAN program:

$$\text{line 1}: \quad \frac{1}{7} \to 2, \quad \frac{1}{3} \to 1$$

$$\text{line 2}: \quad \frac{10}{3} \to 2, \quad \frac{1}{1} \to 3$$

$$\text{line 3}: \quad \frac{3}{5} \to 3, \quad \frac{1}{1} \to 1$$

When started at line 1 with $N = 3^b 7^c$, it stops at line 1, with $N = 2^{bc}$. The program obtained by preceding this one by a new

$$\text{line 0}: \quad \frac{21}{2} \to 0, \quad \frac{1}{1} \to 1,$$

when started at line 0 with $N = 2^n$, stops at line 1 with $N = 2^{n^2}$.

## 8. How to Use the FRACTRAN-1 Model

You can use a FRACTRAN-1 machine to simulate arbitrary FRACTRAN programs. You must first clear the given program of loops, in a way we explain later, and then label its lines (nodes) with prime numbers $P, Q, R, \ldots$ larger than any of the primes appearing in the numerators and denominators of any of its fractions. The FRACTRAN-1 program simulates

$$\text{line } P: \quad \frac{a}{b} \to Q, \quad \frac{c}{d} \to R, \quad \frac{e}{f} \to S, \quad \ldots$$

by the fractions

$$\frac{aQ}{bP} \frac{cR}{dP} \frac{eS}{fP}$$

in that order. If the FRACTRAN-0 program when started with $N$ in state $P$ stops with $M$ at line $Q$, the simulating FRACTRAN-1 program when started a $PN$ stops at $QM$.

*Manufacturer's note. Our guarantee is invalid if you use your FRACTRAN-1 machine in this way to simulate a FRACTRAN program that has loops at several nodes. Such loops may be eliminated by splitting nodes into two.*

The third of our examples

becomes

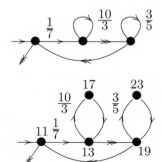

when each of the two nodes with a loop is split in this way, and the new nodes are labeled with the primes 11, 13, 17, 19, 23. Accordingly, it is simulated by the FRACTRAN-1 program

$$\left[\frac{13}{77} \quad \frac{170}{39} \quad \frac{19}{13} \quad \frac{13}{17} \quad \frac{69}{95} \quad \frac{11}{19}\right]$$

If started with $N = 2^a 3^b 7^c 11$, this program stops with $N = 2^{a+bc} 3^b 11$. (The factors of 11 here correspond to the starting and stopping states of the simulated machine.)

We note that it is permissible to label one of the states with the number 1, rather than a large prime number. The fractions corresponding to transitions from this state should be placed (in their proper order) at the *end* of the FRACTRAN-1 program. If this is done, loops, provided they have lower priority than any other transition, are permitted at node 1. Thus the FRACTRAN-1 program

$$\left[\frac{170}{39} \quad \frac{19}{13} \quad \frac{13}{17} \quad \frac{69}{95} \quad \frac{1}{19} \quad \frac{13}{7} \quad \frac{1}{3}\right]$$

simulates the previous program with a loop order $\frac{1}{3}$ adjoined at the starting/stopping node, which has been relabelled 1. This program, started at $3^b 7^c$, stops at $2^{bc}$.

A given FRACTRAN program can always be cleared of loops and adjusted so that 1 is its only stopping node. It follows that we can simulate it by a FRACTRAN-1 program that starts at $PN$ and stops at $M$ when the original program started at N and stopped at $M$. As we remarked in Section 6, we can simulate this by a FRACTRAN-$1\frac{1}{2}$ program

$$P[\ldots]$$

which starts at $N$ and stops at $M$.

## 9. Your PIGAME Guarantee

We now prove Theorem 2, which is equivalent to the assertion that the program

$$\left[\frac{365}{46} \quad \frac{29}{161} \quad \cdots \quad \frac{1}{11} \quad \frac{1}{1024}\right]$$

(obtained by ignoring factors of 97 and dropping the final fraction $\frac{89}{1}$ of PIGAME), when started at $2^n \cdot 89$, stops at $2^{\pi(n)}$. This FRACTRAN-1 program has been obtained from the FRACTRAN program of Figure 3 by the method outlined in the last section. The pairs of nodes 13 & 59, 29 & 71, 23 & 73, 31 & 67, and 43 & 53 were originally single nodes with loops.

We shall only sketch the action of this program, which we separate into three phases. The first phase ends when the program first reaches node 37, the second phase when it first reaches node 41, and the third phase when it finally stops, at node 1.

The first phase, started at 89 with register contents

$$r_2 = n, \quad r_3 = r_5 = r_7 = r_{11} = 0,$$

reaches 37 with contents

$$r_2 = 0, \quad r_3 = 1, \quad r_5 = E, \quad r_7 = 2 \cdot 10^n, \quad r_{11} = 0,$$

where $E$ is a very large even number. To see this, ignore the 5 and 11 registers for a moment, and see that it initially sets $r_7 = 2$. Then each pass around the triangular

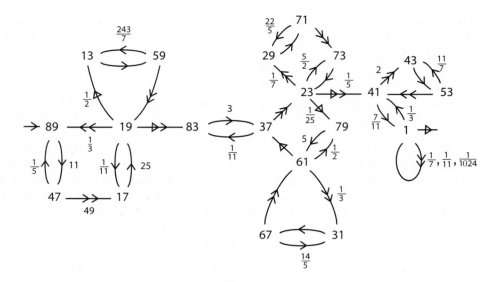

FIGURE 3. A FRACTRAN program for digits of $\pi$.

region multiplies $r_7$ by 5 and puts it into $r_3$ and is followed by passes around the square region which double $r_3$ and put it back into $r_7$. This is done $n$ times, so that at the end of this phase we have $r_7 = 2 \cdot 10^n$, as desired.

The first pass around the square ends with 4 in $r_5$, and each subsequent pass at least doubles this number, while keeping it even. At the last stage we pass around this region $10^n$ times and finish with an even number $E \geq 4 \times 2^{10^n}$ in $r_5$. It's easy to check that registers 2, 3, and 11 end with the indicated values.

At the end of the second phase, we shall have

$$r_2 = r_5 = r_7 = 0,$$

$$r_3 = 2 \times 10^n \times E(E-2)(E-2)(E-4)(E-4)(E-6)\cdots 4 \cdot 4 \cdot 2 \cdot 2 \triangleq N,$$

$$r_n = 1 \times (E-1)(E-1)(E-3)(E-3)(E-5)(E-5)\cdots 5 \cdot 3 \cdot 3 \cdot 1 \triangleq D.$$

This is fairly easy to check, the essential point being that each sojourn in the upper region multiplies $r_7$ by $r_5$ and puts it into $r_{11}$ (preserving the value of $r_5$ but clearing $r_7$), while in the lower region, we multiply $r_3$ by $r_5$ into $r_7$ in a similar way, and then (at the left) transfer $r_{11}$ back to $r_3$. Register 5 is decreased by 1 as we pass from the upper to the lower region; but when $r_5 = 1$ we instead clear it and pass to node 41, entering the third phase.

Now *Wallis' product* is

$$\frac{\pi}{2} = \frac{2}{1}\,\frac{2}{3}\,\frac{4}{3}\,\frac{4}{5}\,\frac{6}{5}\,\frac{6}{7}\,\frac{8}{7}\,\frac{8}{9}\,\frac{10}{9}\,\frac{10}{11}\cdots,$$

in which the successive fractions are obtained by alternately increasing the denominator and numerator. If we truncate it so as only to include all factors whose numerator and denominator are at most $K$, we obtain an approximation $\pi_K$ for $\pi$ which is within at most $\frac{\pi}{K}$ of $\pi$. So our $\frac{N}{D} = 10^n \cdot \pi_E$, where $\pi_E$ is a very good approximation indeed to $\pi$. It is in fact so good that the $n$th decimal digit of $\pi_E$ is the same as that of $\pi$. This digit can be obtained by reducing the integer part of $\frac{N}{D}$ modulo 10, and it is easy to check that the third phase of our program does just this, putting the answer in register 2 and clearing all other registers.

The assertion about the $n$th decimal digit of $\pi_E$ is not trivial. For $n = 0$, our approximation $\pi E$ is $\pi_4 = \frac{32}{9}$. For $n = 1$ or 2, we have $|\pi_E - \pi| < \frac{\pi}{4 \times 2^{10}}$ which is less than $\frac{1}{1000}$, and since $\pi = 3.141\ldots$ the $n$th digits ($n = 1$ and 2) after the decimal point in $\pi_E$ must both be correct.

For $n \geq 3$, the error in $\pi_E$ is at most

$$\frac{\pi}{4 \times 2^{10^n}} < \frac{1}{(1000)^{10^{n-1}}} = 10^{-3 \times 10^{n-1}} < 10^{-42n}.$$

The desired assertion now follows from Mahler's [4] famous irrationality measure for $\pi$: if $\frac{p}{q}$ (in least terms) is any nonintegral rational number, then

$$\left| \pi - \frac{p}{q} \right| > \frac{1}{q^{42}}.$$

## 10. How to Use Our Universal Program

In this section, we prove Theorem 3, using an ingenious lemma due to John Rickard. We shall call a FRACTRAN-1 program $[f_1, f_2, \ldots, f_k]$ *monotone* if $f_1 < f_2 < f_3 < \cdots < f_k$.

LEMMA. *Any FRACTRAN-1 program can be simulated by a monotone one that starts and stops with the same numbers.*

PROOF. Choose a new prime $P$ that is bigger than the ratio between any two of the $f_i$ and bigger than the inverse of any $f_i$. Then $\left[\frac{1}{P}, Pf_1, P^2 f_2, P^3 f_3, \ldots, P^k f_k\right]$ simulates $[f_1, f_2, f_3, \ldots, f_k]$ and is monotone. The new program behaves exactly like the old one, except that at each step a power of $P$ is introduced, only to be immediately cleared away before we copy the next step.                          □

We shall call a FRACTRAN-$1\frac{1}{2}$ program

$$f_1^* f_2^* \ldots, f_j^* [f_1, f_2, \ldots, f_k]$$

monotone if

$$f_1^* < f_2^* < \cdots < f_j^* \quad \text{and} \quad f_1 < f_2 < \cdots < f_k.$$

Then our universal program simulates monotone FRACTRAN-$1\frac{1}{2}$ programs. It codes such a program by three numbers, $M^*$, $M$, and $d$, defined as follows.

We take $d$ to be any common denominator of all the fractions mentioned and suppose the given FRACTRAN-$1\frac{1}{2}$ program is

$$\frac{m_1^*}{d} \frac{m_2^*}{d} \cdots \frac{m_j^*}{d} \left[ \frac{m_1}{d} \frac{m_2}{d} \cdots \frac{m_k}{d} \right].$$

We then adjoin dummy numbers $m_{j+1}^*$ and $m_{k+1}$, which are both multiples of $d$ and which satisfy

$$m_1^* < m_2^* < \cdots < m_j^* < m_{j+1}^*, \quad m_1 < m_2 < \cdots < m_k < m_{k+1},$$

$$\text{and} \quad \left[ \frac{1}{2} M^* \right] \leq M$$

where

$$M^* = 2^{m_1^*} + 2^{m_2^*} + \cdots + 2^{m_{j+1}^*}$$
$$M = 2^{m_1} + 2^{m_2} + \cdots + 2^{m_{k+1}}.$$

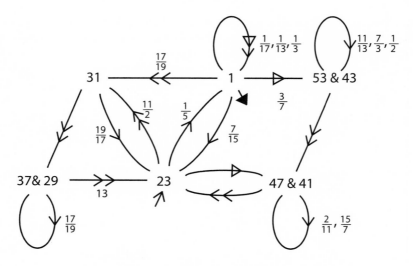

FIGURE 4. A flowchart for POLYGAME.

The universal program POLYGAME, started at

$$2^N 3^M 5^{M^*} 17^{d-1} 23$$

will simulate the given FRACTRAN-$1\frac{1}{2}$ program, started at $N$. This universal FRACTRAN-1 program was obtained from the FRACTRAN program shown in Figure 4, and accordingly, we consider starting the latter with $r_2 = N$, $r_3 = M$, $r_5 = M^*$, $r_{17} = d - 1$, at the node 23.

This works roughly as follows. After a new $N$ has been found, the program computes successive multiples $N, 2N, 3N, \ldots, mN$, and simultaneously repeatedly halves $M$ to get $[M/2], [M/4], \ldots, [M/2^m]$. If $[M/2^m]$ is odd, so that $m$ is one of the $m_i$, it sees whether $Nm$ is a multiple of $d$, and if so resets $M$ and takes a new $N = mN/d$, *unless* $m$ was $m_{k+1}$ (i.e., $[M/2^m] = 1$), when it arranges to stop at node 1 with register 2 containing $N$ and all other registers empty. For the first $p$ it uses $M^*$ in place of $M$.

Registers 13, 17, 19 function as a counter, whose count is stored in form from which we can see at once if it is a multiple of $d$. If

$$r_{13} = q, \quad r_{19} = r, \quad r_{17} = d - 1 - r \quad \text{with } 0 \le r < d,$$

then *the count* is the number $qd + r$. If the machine arrives at node 31 ("enters the counter") with these values, then when it next arrives at node 23 ("leaves the counter"), we shall have

$$r_{13} = q, \quad r_{19} = r + 1, \quad r_{17} = d - 1 - (r + 1), \quad \text{if } r < d - 1,$$
$$r_{13} = q + 1, \quad r_{19} = 0, \quad r_{17} = d - 1, \quad \text{if } r = d - 1.$$

In other words, the value of the count will have increased by 1.

So if the machine is started at 23, with $r_5 = r_{11} = 0$ and $r_2 = N$, it will increase the count by $N$ while transferring $N$ from register 2 to register 11, and then go to node 47 (where its first action will be to retransfer $N$ from register 11 back to register 2).

## TABLE 2. The Action of POLYGAME

| node | 2 | 3 | 5 | 7 | 11 | 13 | 17 | 19 | action |
|---|---|---|---|---|---|---|---|---|---|
| | | | | | | | | Contents of registers: | |
| 23 | $N$ | $M$ | $M_m$ | 0 | 0 | $q_m$ | $d{-}1{-}r_m$ | $r_m$ | |
| 1 | $N$ | $M{-}M_{m+1}$ | 0 | $M_{m+1}$ | 0 | $q_m$ | $d{-}1{-}r_m$ | $r_m$ | |
| 23 | $N$ | $M{-}M_{m+1}$ | 0 | $M_{m+1}$ | 0 | $q_m$ | $d{-}1{-}r_m$ | $r_m$ | |
| 47 & 41 | 0 | $M{-}M_{m+1}$ | 0 | $M_{m+1}$ | $N$ | $q_{m+1}$ | $d{-}1{-}r_{m+1}$ | $r_{m+1}$ | |
| 23 | $N$ | $M$ | $M_{m+1}$ | 0 | 0 | $q_{m+1}$ | $d{-}1{-}r_{m+1}$ | $r_{m+1}$ | |
| 47 & 41 | 0 | 0 | 0 | $M$ | $\frac{mN}{d}$ | 0 | $d{-}1$ | 0 | |
| 23 | $\frac{mN}{d}$ | $M$ | $M$ | 0 | 0 | 0 | $d{-}1$ | 0 | |
| 1 | $N$ | 0 | 0 | 0 | 0 | 0 | 0 | 0 | |

(In the action column: $M_m$ odd / $M_m$ even; $r_m \neq 0$ / $r_m = 0$; $M_{m+1} \neq 0$ / $M_{m+1} = 0$.)

$$mN = q_m \cdot d + r_m \ (0 \le r_m < d) \qquad M_m = [M_0/2^m]$$

After these remarks, the reader should have little difficulty in verifying the transitions between particular configurations shown in Table 2.

We suppose that for particular positive numbers $d$, $N$, $M$, and $M_0$ with $\left[\frac{1}{2}M_0\right] \le M$ we define for varying values of $m$ the numbers $M_m, q_m, r_m$ by

$$M_m = [M_0/2^m]$$
$$mN = q_m d + r_m \quad (0 \le r_m < d).$$

Then Table 2 shows that unless $M_m$ is odd and $r_m = 0$, the special type of configuration in the first line of the table leads to a similar one (in the fifth line) with $m$ increased by 1. In the excepted case, if $M_{m+1} \neq 0$, we obtain another such special configuration (in the seventh line), but with $m$ (and the count) reset to 0, the new initial value $M_0 = M$ for $M_m$, and $\frac{mN}{d}$ as the new $N$. If instead $M_{m+1}$ was 0, we arrive at the last line of a the table, and *stop* at node 1, with $N$ in register 2 and all other registers empty. The cases with $M_m$ odd and $r_m = 0$ are called *resets*.

Now suppose we start the machine in the special configuration in the top line of the table, with $m = 0$, and the initial value $M_0$ of $M_m$ set to the number

$$2^{m_0} + 2^{m_1} + \cdots + 2^{m_{k+1}},$$

where

$$m_0 < m_1 < \cdots < m_{k+1}$$

and $m_{k+1}$ is divisible by $d$. Then before the next reset, we have the equivalences

$$M_m \text{ odd} \iff m \text{ is one of the } m_i$$
$$r_m = 0 \iff mN/d \text{ is an integer}$$
$$M_{m+1} = 0 \iff m = m_k.$$

So the next reset will be at the first of the $m_i$ for which $m_i N/d$ is integral, and will *either*

replace $N$ by $m_i N/d$, and reset $m$ to 0 and $M_m$ to $M$ (if $i < k$),

or stop at node 1, with $N$ in register 2 and the rest empty ($i = k$).

This completes the required verifications. Initially, we set $m = 0$ and $M_0 = M^*$, but all subsequent resets will put $M_0 = M$, in accordance with the rules for FRACTRAN-$1\frac{1}{2}$ programs.

A FRACTRAN-1 program is a FRACTRAN-$1\frac{1}{2}$ program with $M = M^*$. For this we can use the alternate catalogue number $7^M 17^{d-1} 41$.

## 11. Applications, Improvements, Acknowledgments

For the function

$$g(N) = \begin{cases} \frac{1}{2}N & (N \text{ even}) \\ 3N + 1 & (N \text{ odd}), \end{cases}$$

the *Collatz problem* asks whether for every positive integer $N$ there exists a $k$ for which $g^k(N) = 1$. See [3] for a survey of this problem.

We can ask similar questions for more general *Collatz functions*

$$g(N) = a_N N + b_N,$$

where $a_N$ and $b_N$ are rational numbers that only depend on the value of $N$ modulo some fixed number $D$. We proved in [1] that there is no algorithm for solving arbitrary Collatz problems. Indeed, for any computable function $f(n)$, there is a FRACTRAN-1 program $[f_1 f_2 \ldots f_k]$ with the property that when we start it at $2^n$, the first strictly later power of 2 will be $2^{f(n)}$. In other words, we can define $f$ by

$$2^{f(n)} = g^k(2^n),$$

where $k$ is the smallest positive integer for which $g^k(2^n)$ is a power of 2, and the function $g(N)$, which has the above form, is just $f_i N$ for the least $i$ which makes this an integer. This result is an explicit version of *Kleene's Normal Form Theorem*.

We note that $g(N)/N$ is a periodic function with rational values, so that $g(N)$ is a Collatz function for which $b_N$ is always 0. So even for Collatz functions of this special type there can be no decision procedure. By applying the argument to a universal fraction game, we can get a *particular* Collatz-type problem with no decision procedure.

(We remark that of course Collatz problems with arbitrary $b_N$ are harder to solve, rather than easier. We might, for instance, define one that simulates a program written in 10 segments, each segment using only the numbers ending in a given decimal digit, and in which control is transferred between the segments only at certain crucial—and recursively unpredictable—times.)

John Rickard tells me that he has found a seven fraction universal program of type $2^{2^n} \cdot c \to 2^{2^{f(n)}}$ and a nine fraction one of type $2^n \cdot c \to 2^{f(n)}$. However, it seems that his fractions are much too complicated ever to be written down. I used one of Rickard's ideas in Section 10. Mike Guy gave valuable help in computing the catalogue numbers in Section 2. Of course, the responsibility for any errors in these numbers rests entirely with him.

## References

[1] J.H. Conway, "Unpredictable Iterations," in Proceedings of the Number Theory Conference, Boulder, Colorado, pp. 49–52 (1972).
[2] J.H. Conway, "FRACTRAN—A Simple Universal Programming Language for Arithmetic," *Open Problems Commun. Comput.*, pp. 4–26 (1986).
[3] J.C. Lagarias, "The $3x+1$ Problem and Its Generalizations," *Am. Math. Monthly*, 92, No. 1, pp. 3–25 (1985).
[4] K. Mahler, "On the Approximation of $\pi$," *Indagationes Math.*, 15, pp. 30–42 (1953).

DEPARTMENT OF MATHEMATICS, PRINCETON UNIVERSITY, PRINCETON, NJ 08544

## Editorial Commentary

(1) The FRACTRAN programming model extends the formalism of Conway's 1972 paper on the $3x+1$ problem ([1]); see the editorial commentary for that paper. FRACTRAN is intended as a name parallel to FORTRAN. Just as FORTRAN has now become Fortran, FRACTRAN is now called Fractran. The prime-producing algorithm given in Theorem 1 was presented earlier in Guy [5]. It is also discussed in the book of Conway and Guy [2]. In 2002 Kenneth G. Monks [8] encoded the $3x+1$ iteration as a Fractran program and used it to derive some properties of cycles of the iteration.

(2) To interpret Figure 1, the integer constant $c_\pi$ is a label for the modified PIGAME program given at the beginning of Section 9, encoded using the method of Section 10 in the universal program. It first converts this PIGAME program to a different FRACTRAN-1 program given by the Lemma in Section 10, using the new prime $P = 101$. The first ten lines of Figure 1 represent the exponent of 3; parentheses are added to the figure to clarify this point. The number $c_\pi$ is extracted from the number $2^N 3^M 5^{M^*} 17^{d-1} 23$ given in Section 10 by omitting the factor $2^N$ giving input value $N = 2^n$ to the program, as follows: it takes $d = 101!$, $M^* = 2^{89 \cdot 101!} + 2^{90 \cdot 101!}$, with $j = 1$ and $m_1^* = 89 \cdot 101!$, $m_2^* = 90 \cdot 101!$, and $M = \sum_{j=1}^{k+1} 2^{m_j}$, in which $m_1 = \frac{1}{P} 101!$ and $m_j = f_{j-1} 101!$ for $2 \le j \le k+1$.

(3) The Fractran computational model is related to Marvin Minsky's computational model of register machines (also called counter machines). These machines may store an arbitrarily large nonnegative integer in a single register, as described in Minsky ([6], [7, Sect. 11.1]). In the Fractran model the exponents of certain given primes function as registers, as explained here and in Conway [1]. Recently it has been noted that register machine models are relevant to making computations using chemical reaction networks, and in biological computation, as discussed in Soloveichik et al [9] and Cook et al [3]. Concerning the computational power of the Fractran model, J. Endrullis, C. Grabmeyer and D. Hendricks [4] showed that the problem of deciding if a Fractran program halts on all positive integer inputs is a $\Pi_2^0$-complete problem.

## References

[1] J. H. Conway, Unpredictable Iterations, In: Proc. 1972 Number Theory Conference, University of Colorado, Boulder, CO. 1972, pp. 49–52.
[2] J. H. Conway and R. K. Guy, *The Book of Numbers,* Copernicus, New York 1996.
[3] M. Cook, D. Soloveichik, E. Winfree and J. Bruck, Programmability of Chemical Reaction Networks, to appear in Festscrift for Grzegorz Rozenberg, Springer-Verlag.

[4] J. Endrullis, C. Grabmeyer and D. Hendriks, Complexity of Fractran and Productivity, in: R. A. Schmidt (Ed.)., *Automated Deduction- CADE 22*, Lecture Notes in Computer Science, 2009, Volume 5663, Springer-Verlag, 2009, pp. 371–387.

[5] R. K. Guy, Conway's Prime Producing Machine, Mathematics Magazine **56** (1983), No. 1, 26–33.

[6] M. Minsky, Recursive unsolvability of Post's problem of 'tag' and other topics in the theory of Turing machines, Annals of Math. **74** (1961). 437–455.

[7] M. Minsky, *Computation: Finite and Infinite Machines*, Prentice-Hall, Inc: Engelwood Cliffs, NJ 1967.

[8] K. G. Monks, $3x+1$ Minus the $+$, Discrete Math. and Theoretical Computer Science **5** (2002), 47–54.

[9] D. Soloveichik, M. Cook E. Winfree and J. Bruck, Computation with finite stochastic chemical reaction networks, Natural Computing **7** (2008), 615–633.

# PART V.

Annotated Bibliography

# The $3x + 1$ Problem: An Annotated Bibliography (1963-1999)

Jeffrey C. Lagarias

## 1. Introduction

The $3x + 1$ problem is most simply stated in terms of the *Collatz function $C(x)$* defined on integers as "multiply by three and add one" for odd integers and "divide by two" for even integers. That is,

$$C(x) = \begin{cases} 3x + 1 & \text{if } x \equiv 1 \ (\text{mod } 2) \,, \\ \frac{x}{2} & \text{if } x \equiv 0 \ (\text{mod } 2) \,, \end{cases}$$

The $3x + 1$ *problem*, or *Collatz problem*, is to prove that starting from any positive integer, some iterate of this function takes the value 1.

Much work on the problem is stated in terms of the $3x + 1$ *function*

$$T(x) = \begin{cases} \frac{3x+1}{2} & \text{if } x \equiv 1 \ (\text{mod } 2), \\ \frac{x}{2} & \text{if } x \equiv 0 \ (\text{mod } 2) \,. \end{cases}$$

The $3x + 1$ *conjecture* states that every $m \geq 1$ has some iterate $T^{(k)}(m) = 1$.

The $3x + 1$ problem is generally attributed to Lothar Collatz and has reportedly circulated since at least the early 1950's. It also goes under the names *Syracuse Problem, Hasse's Algorithm, Kakutani's Problem* and *Ulam's Problem*. The first published reference to a $3x + 1$-like function that I am aware of is Klamkin (1963), which concern's Collatz's original function, which is a permutation of the integers. Collatz has stated that he studied this function in the 1930's, see Lagarias (1985). The $3x + 1$ problem itself was reportedly described in an informal lecture of Collatz in 1950 at the International Math. Congress in Cambridge, Massachusetts (see Shanks (1965) and Trigg et al (1976)). The first journal appearance of the $3x + 1$ problem itself seems to be the text of a 1970 lecture of H. S. M. Coxeter, which appeared in Coxeter (1971). This was followed by Beeler et al (1972), Conway (1972), Gardner (1972), Kay (1972) and Ogilvy (1972). See Bryan Thwaites (1985) for his assertion to have (independently) formulated the problem in 1952. See Collatz (1986) for his assertions on formulating the $3x + 1$ problem prior to 1952.

The $3x + 1$ problem can also be rephrased as a problem concerning sets of integers generated using affine maps. Let $T$ be the smallest set of integers including 1 and closed under iteration of the affine maps $x \mapsto 2x$ and $3x + 2 \mapsto 2x + 1$. Here the latter map is the affine map $y \to \frac{2y-1}{3}$, with input restricted to integers $y \equiv 2 \ (\text{mod } 3)$, so that the output is an integer. The $3x + 1$ conjecture asserts that $T$ is the set of all positive integers. Therefore this bibliography includes work on sets of integers generated by iteration of affine maps, tracing back to Isard and Zwicky (1970) and Klarner and Rado (1974), which includes a problem of Erdős, described in Klarner (1982).

As of 1999 the $3x+1$ conjecture was verified up to $2 \times 10^{16}$ by a computation of Oliveira e Silva (1999). It has since been verified to at least $2 \times 10^{18}$ in independent computations by Oliveira e Silva and Roosendaal. At present the $3x+1$ conjecture remains unsolved.

This annotated bibliography mainly covers research articles and survey articles on the $3x + 1$ problem and related problems. It includes the surveys: Lagarias (1985), Müller (1991) and the first chapter of Wirsching (1998a). For more recent information see the surveys in this volume. It provides additional references to earlier history, much of it appearing as problems, with the first of these appearing in 1963. It also includes a few influential technical reports that were never published. The papers are grouped in 10 year blocks: 1960-1969, 1970-1979, 1980-1989, 1990-1999. The list includes 8, 34, 52 and 103 papers in these decades. References are ordered by author's surname within 10 year blocks; Chinese authors have surnames written first.

## 2. Terminology

We use the following definitions. The *trajectory* or *forward orbit* of an integer $m$ is the set

$$\{m,\ T(m)\ ,\ T^{(2)}(m), \ldots\}\ .$$

The *stopping time* $\sigma(m)$ of $m$ is the least $k$ such that $T^{(k)}(m) < m$, and is $\infty$ if no such $k$ exists. The *total stopping time* $\sigma_\infty(m)$ is the least $k$ such that $m$ iterates to 1 under $k$ applications of the function $T$ i.e.

$$\sigma_\infty(m) := \inf\ \{k\ :\ T^{(k)}(m) = 1\}.$$

The *scaled total stopping time* or *gamma value* $\gamma(m)$ is

$$\gamma(m) := \frac{\sigma_\infty(m)}{log\ m}$$

The *height* $h(m)$ the least $k$ for which the Collatz function $C(x)$ has $C^{(k)}(m) = 1$. It is easy to show that

$$h(m) = \sigma_\infty(m) + d(m),$$

where $d(m)$ counts the number of iterates $T^{(k)}(m)$ in $0 \le k < \sigma_\infty(m)$ that are odd integers. Finally, the function $\pi_a(x)$ counts the number of $n$ with $|n| \le x$ whose forward orbit under $T$ includes $a$.

## 3. Bibliography 1960–1969

All published results in the 1960's were "prehistory". They concern related problems, without stating the $3x+1$ problem itself, with the first of these appearing in 1963. We include work on a Collatz permutation, proposed as an unsolved problem by Klamkin (1963), and commented on by Atkin (1965) and Shanks (1965). We also include work on $(3x+1)$-like function which arose in connection with poetic rhyming patterns, in a problem raised by Queneau (1963), a founding member of the French mathematical-literary group Oulipo. The work of Mahler (1968) on $Z$-numbers involves yet another $(3x + 1)$-like function. During this period the $3x + 1$

problem itself circulated by word of mouth, and computations were reportedly done testing it up to $n = 500,000$.

**(60-1)**   Arthur O. L. Atkin (1966), *Comment on Problem $63 - 13^*$*, SIAM Review **8** (1966), 234–236.

This comment gives more information of the problem of Klamkin (1963) concerning iteration of the original Collatz function, which is a permutation of the positive integers. Adding to the comment of Shanks (1965), he notes there is a method which in principle can determine all the cycles of a given period $p$ of this map. This method determines upper and lower bounds on the integers that can appear in such a cycle. By computer calculation he shows that aside from the known cycles of periods 1, 2, 5 and 12 on the nonnegative integers, there are no other cycles of period less than 200.

He gives an example casting some doubt on the heuristic of Shanks (1965) concerning the possible lengths of periods. He shows that for the related permutation $f(3n) = 4n+3$, $f(3n+1) = 2n$, $f(3n+2) = 4n+1$, which should obey a similar heuristic, that it has a cycle of period 94 (least term $n = 140$), and 94 is not a denominator of the continued fraction convergent to $\log_2 3$.

Atkin presents a heuristic argument asserting that the Collatz permutation should only have a finite number of cycles, since the iterates grow "on average" at an exponential rate.

**(60-2)**   Monique Bringer (1969), *Sur un problème de R. Queneau*, Mathématiques et Sciences Humaines [Mathematics and Social Science], **27** Autumn 1969, 13–20.

This paper considers a problem proposed by Queneau (1963) in connection with rhyming patterns in poetry. It concerns, for a fixed $n \geq 2$, iteration of the map on the integers

$$\delta_n(x) := \begin{cases} \frac{x}{2} & \text{if } x \text{ is even,} \\ \frac{2n+1-x}{2} & \text{if } x \text{ is odd.} \end{cases}$$

This map acts as a permutation on the integers $\{1, 2, \ldots, n\}$ and it also has the fixed point $\delta_n(0) = 0$. It is called by the author a spiral permutation of $\{1, 2, \ldots, n\}$. The paper studies for which $n$ this spiral is a cyclic permutation, and calls such numbers *admissible*.

The motivation for this problem was that this permutation for $n = 6$ represents a poetic stanza pattern, the sestina, used in poems by an 11th century Troubadour, Arnaut Daniel. This pattern for general $n$ was studied by Raymond Queneau (1963), who determined small values of $n$ giving a cyclic permutation. His colleague Jacques Roubaud (1969) termed these rhyme schemes $n$−ines or *quenines*. Later he called these numbers *Queneau numbers*, cf. Robaud (1993).

In this paper the author, a student of Roubaud, shows that a necessary condition for a number $n$ to be admissible is that $p = 2n + 1$ be prime. She shows that a sufficient condition to be admissible is that 2 be a primitive root (mod $p$). She deduces that if $n$ and $2n+1$ are both primes then $n$ is admissible, and that the numbers $n = 2^k$ and $n = 2^k - 1$ are never admissible. Finally she shows that all $p \equiv 1 \pmod{8}$ are not admissible.

*Note.* The function $\delta_n(x)$ is defined on the integers, and is of $3x + 1$ type (i.e. it is a periodically linear function). Its long term behavior under iteration is analyzable because the function $\delta_n(x)$ decreases absolute value on each iteration for any $x$ with $|x| > 2n+1$. Thus all orbits eventually enter $-2n-1 \le x \le 2n+1$ and become eventually periodic. One can further show that all orbits eventually enter the region $\{0, 1, \ldots, n\}$ on which $\delta_n(x)$ is a permutation.

**(60-3)** Murray S. Klamkin (1963), *Problem $63 - 13^*$*, SIAM Review **5** (1963), 275–276.

The problem is: "Consider the infinite permutation

$$P \equiv \begin{pmatrix} 1 & 2 & 3 & 4 & 5 & 6 & \cdots \\ 1 & 3 & 2 & 5 & 7 & 4 & ,,, \end{pmatrix}$$

taking $n \mapsto f(n)$ where $f : \mathbb{N}^+ \to \mathbb{N}^+$ is given by $f(3n) = 2n, f(3n - 1) = 4n - 1, f(3n - 2) = 4n - 3$. We now write $P$ as a product of cycles

$$P \equiv (1)\ (2, 3)\ (4, 5, 7, 9, 6)\ (8, 11, 15, \ldots$$

It is conjectured that the cycle $(8, 11, 15, \ldots)$ is infinite. Other problems concerning $P$ are:

(a) Does the permutation $P$ consist of finitely many cycles?

(b) Are there any more finite cycles than those indicated? "

This function was the original function proposed by L. Collatz in his private notes in 1932. See Shanks (1965) and Atkin (1966) for comments on this problem. This problem remains unsolved concerning the orbit of $n = 8$ and part (a). Concerning (b) one more cycle was found, of period 12 with smallest element $n = 144$. Atkin (1966) presents a heuristic argument suggesting there are finitely many cycles.

*Note.* In 1963 M. Klamkin proposed another problem, jointly with A. L. Tritter, concerning the orbit structure of a different infinite permutation of the integers [*Problem 5109*, Amer. Math. Monthly **70** (1963), 572–573]. For this integer permutation all orbits are cycles. A solution to this problem was given by G. Bergman, Amer. Math. Monthly **71** (1964), 569–570.

**(60-4)** Kurt Mahler (1968), *An unsolved problem on the powers of 3/2*, J. Australian Math. Soc. **8** (1968), 313–321. (MR 37 #2694, Zbl. 155 #95.).

A $Z$-number is a real number $\xi$ such that $0 \le \{\{(\frac{3}{2})^k \xi\}\} \le \frac{1}{2}$ holds for all $k \ge 1$, where $\{\{x\}\}$ denotes the fractional part of $x$. Do $Z$-numbers exist? The $Z$-number problem was originally proposed by Prof. Saburo Uchiyama (Tsukuba Univ.) according to S. Ando (personal communication), and was motivated by a connection with the function $g(k)$ in Waring's problem, for which see G. H. Hardy and E. M. Wright, *An Introduction to the Theory of Numbers* (4-th edition), Oxford Univ. Press 1960, Theorem 393 ff. and Stemmler (1964).

Mahler shows that existence of $Z$-numbers relates to a question concerning the iteration of the $3x + 1$-like function

$$g(x) = \begin{cases} \frac{3x+1}{2} & \text{if } x \equiv 1 \pmod 2, \\ \frac{3x}{2} & \text{if } x \equiv 0 \pmod 2 \, . \end{cases}$$

Mahler showed that a $Z$-number exists in the interval $[n, n+1)$ if and only if no iterate $g^{(k)}(n) \equiv 3 \pmod 4$. He uses this relation to prove that the number of $Z$-numbers below $x$ is at most $O(x^{0.7})$. He conjecures that no $Z$-numbers exist, a problem which is still unsolved.

*Note.* Leopold Flatto (1992) subsequently improved Mahler's upper bound on $Z$-numbers below $x$ to $O(x^\theta)$, with $\theta = \log_2 \frac{3}{2} \approx 0.59$.]

**(60-5)**  Raymond Queneau (1963), *Note complémentaire sur la Sextine*, Subsidia Pataphysica, No.1 (1963), 79–80.

This short note is a comment on a preceding article in Subsidia Pataphysica by A. Taverna, *Arnaut Daniel et la Spirale*, Subsidia Pataphysica, No. 1 (1963), 73–78. Arnaut Daniel was a 12-th century troubadour who composed poems in Occitan having a particular rhyming pattern, called a sestina. Dante admired him and honored him in several works, including the *Divine Comedy*, where Daniel is depicted as doing penance in Purgatory in *Purgatorio*. The rhyming pattern of the sestina had six sextets with rhyme pattern involving a cyclic permutation of order 6, followed by a triplet. The work of Taverna observes that this cyclic permutation can be represented using a spiral pattern.

Queneau considers the "spiral permutation" on numbers $\{1, 2, \dots, n\}$ which which takes $2p$ to $p$ and $2p + 1$ to $n - p$. He raises the question: For which $n$ is a similar spiral permutations in the symmetric group $S_n$ a cyclic permutation? Call these allowable $n$. The example of the sestina is the case $n = 6$; this pattern he terms the sextine. He says it is easy to show that numbers of the form $n = 2xy + x + y$, with $x, y \geq 1$, are not allowable numbers; this excludes $n = 4, 7, 10$ etc. He states that 31 integers $n \leq 100$ are allowable, namely $n = 1, 2, 3, 5, 6, 9, 11, 14, 18, 23, 26, 29, 30, 33, 35, 39, 41, 50, 51, 53, 65, 69, 74, 81, 83, 86, 89, 90, 95, 98, 99$.

Queneau's question was later formulated as the behavior under iteration of a $(3x + 1)$-like function $\delta_n(x)$ on the range $x \in \{1, 2, \dots, n\}$, as observed in Roubaud (1969) and Bringer (1969). See also the paper Queneau (1972) which includes a $3x + 1$-like function.

*Notes.* (i) Raymond Queneau (1903-1976) was a French poet and novelist, and a founding member in 1960 of the French mathematical-literary group Oulipo (Ouvroir de littérature potentielle). He was a member of the mathematical society of France starting in 1948, and published in 1972 a mathematical paper in additive number theory. His final essay in 1976 was titled: "Les fondaments de la littérature d'après David Hilbert" (La Bibliothèque Oulipienne, No. 3); it set out an axiomatic foundation of literature in imitiation of Hilbert's *Foundations of Geometry*, replacing "points", "straight line", and "plane" with "word" , "sentence" and "paragraph", respectively, in some of Hilbert's axioms. [English translation: The Foundations of Literature (after David Hilbert), in: R. Queneau, I. Calvino, P. Fournel, J. Jouet, C. Berge and H. Mathews, *Oulipo Laboratory, Texts from the Bibliothèque Oulipienne*, Atlas Press: Bath 1995.] He deduces from his axioms: "THEOREM 7. *Between two words of a sentence there exists an infinity of other words.*" To explain this result, he posits the existence of "imaginary words."

(ii) This note is in the Oulipo spirit, considering literature obtainable when mathematical restrictions are placed on its allowable form. Queneau also discussed the topic of "Sextines" in his 1965 essay on the aims of the group Oulipo, "Littérature potentielle", published in *Bâtons, chiffres et lettres*, 2nd Edition, Gallimard: Paris 1965. [English translation: Potential Literature, pp. 181–196 in R. Queneau, *Letters, Numbers, Forms: Essays 1928–1970* (Jordan Stump, Translator), Univ. of Illinois Press: Urbana and Chicago 2007.]

**(60-6)**   Jacques Roubaud (1969) , *Un problème combinatoire posè par la poésie lyrique des troubadours,* Mathématiques et Sciences Humaines [Mathematics and Social Science], **27** Autumn 1969, 5–12.

This paper addresses the question of suitable rhyming schemes for poems, suggested by the schemes used by medieval troubadours, and raised in Queneau (1963). This leads to combinatorial questions concerning the permutation structure of such rhyming schemes. The author classifies the movement of rhymes by permutation patterns. In section 5 he formulates three questions suggested by the rhyme pattern of the sestina of Arnaut Daniel. One of these (problem $(Pa)$) is the question of Queneau (1963), which he states as iteration of the $3x + 1$-like function

$$\delta_n(x) := \begin{cases} \frac{x}{2} & \text{if } x \text{ is even} \\ \frac{2n+1-x}{2} & \text{if } x \text{ is odd} \end{cases}$$

restricted to the domain $\{1, 2, \ldots, n\}$. This paper goes together with the analysis of this function done by his student Monique Bringer (1969). In Robaud (1993) he proposes further generalizations of these problems.

*Note.* Jacques Roubaud is a mathematician and a member of the literary group Oulipo. In 1986 he wrote a spoof, an obituary for N. Bourbaki (see pages 73 and 115 in: M. Mashaal, *Bourbaki* (A. Pierrehumbert, Trans.), Amer. Math. Soc., Providence 2006). He discussed the mathematical work of Raymond Queneau in J. Roubaud, La mathematique dans la methode de Raymond Queneau, Critique: revue générale des publications fracaises et étrangères **33** (1977), no. 359, 392–413. [English Translation: Mathematics in the Method of Raymond Queneau, pp. 79–96 in: Warren F. Motte, Jr. (Editor and Translator), *Oulipo, A Primer of Potential Literature*, Univ. of Nebraska Press: Lincoln, NB 1986.]

**(60-7)**   Daniel Shanks (1965), *Comments on Problem 63 − 13\**, SIAM Review **7** (1965), 284–286.

This note gives comments on Problem $63-13^*$, proposed by Klamkin (1963). This problem concerns the iteration of Collatz's original function, which is a permutation of the integers. He states that these problems date back at least to 1950, when L. Collatz mentioned them in personal conversations at the International Math. Congress held at Harvard University. He gives the results of a computer search of the orbit of 8 for Collatz's original function, observing that it reaches numbers larger than $10^{10}$. He observes there are known cycles of length 1, 2, 5 and 12, the last having smallest element $n = 144$. He observes that this seems related to the fact that the continued fraction expansion of $\log_2 3$ has initial convergents having denominators 1, 2, 5, 12, 41, ... However he does not know of any cycle of period 41. He notes that it is not known whether the only cycle lengths that can occur must be denominators of such partial quotients. Later Atkin (1966) proved there exists no cycle of period 41.

**(60-8)**   Rosemarie M. Stemmler (1964), *The ideal Waring problem for exponents 401 − 200,000,* Math. Comp. **18** (1964), 144–146. (MR 28 #3019)

The ideal Waring theorem states that for a given $k \geq 2$ each positive integer is the sum of $2^k + \lfloor (\frac{3}{2})^k \rfloor$ non-negative $k$-th powers, provided that the fractional part $\{\{x\}\} := x - \lfloor x \rfloor$ of $(\frac{3}{2})^k$ satisfies

$$0 \leq \{\{(\frac{3}{2})^k\}\} < 1 - (\frac{3}{4})^k.$$

Study of the distribution of the fractional parts of powers of $(\frac{3}{2})^k$ motivated the work of Mahler (1968) on $Z$-numbers. This paper checks that the inequality above holds for $401 \leq k \leq 200000$.

## 4. Bibliography 1970–1979

The first statement in print of the $3x + 1$ problem appears to be in a 1970 talk of Coxeter, documented in Coxeter (1971). It then was stated in Beeler et al (1972), Conway (1972), Gardner (1972), Kay (1972) and Ogilvy (1972).

**(70-1)**   Jean-Paul Allouche (1979), *Sur la conjecture de "Syracuse-Kakutani-Collatz",* Séminaire de Théorie des Nombres 1978–1979, Expose No. 9, 15pp., CNRS Talence (France), 1979. (MR 81g:10014).

This paper studies generalized $3x + 1$ functions of the form proposed by Hasse. These have the form

$$T(n) = T_{m,d,R}(n) := \begin{cases} \frac{mn+r_j}{d} & \text{if } n \equiv j \pmod{d},\ 1 \leq j \leq d-1 \\ \frac{x}{d} & \text{if } n \equiv 0 \pmod{d}. \end{cases}$$

in which the parameters $(d, m)$ satisfy $d \geq 2$, $gcd(m, d) = 1$, and the set $R = \{r_j : 1 \leq j \leq d - 1\}$ has each $r_j \equiv -mj \pmod{d}$. The author notes that it is easy to show that all maps in Hasse's class with $1 \leq m < d$ have a finite number of cycles, and for these maps all orbits eventually enter one of these cycles. Thus we may assume that $m > d$.

Allouche's paper proves two theorems. Theorem 1 improves on the results of Heppner (1978). Let $T(\cdot)$ be a function in Hasse's class with parameters $d, m$. Let $a > 1$ be fixed and set $k = \lfloor \frac{\log x}{a \log m} \rfloor$. Let $A, B$ be two rationals with $A < B$ that are not of the form $\frac{m^i}{d^j}$ for integers $i \geq 0, j \geq 1$, and consider the counting function

$$F_{a,A,B}(x) := \sum_{n=1}^{x} \chi_{(A,B)} \left( \frac{T^{(k)}(n)}{n} \right),$$

in which $\chi_{(A,B)}(u) = 1$ if $A < u < B$ and 0 otherwise. Let $C$ be the maximum of the denominators of $A$ and $B$. Then:

(i) If $A < B < 0$ then

$$F_{a,A,B}(x) = O\left(Cx^{\frac{1}{a}}\right),$$

where the constant implied by the $O$-symbol is absolute.

(ii) If $B > 0$ and there exists $\epsilon > 0$ such that

$$\frac{d-1}{d} \geq \frac{\log d}{\log m} + \frac{\log B}{k \log m} + \epsilon,$$

then for $k > k_0(\epsilon)$, there holds

$$F_{a,A,B}(x) = O\left(Cx^{\frac{1}{a}} + x^{1-\frac{|\log \eta|}{a \log m}}\right)$$

for some $\eta$ with $0 < \eta < 1$ which depends on $\epsilon$. This is true in particular when $m > d^{\frac{d}{d-1}}$ with $B$ fixed and $x \to \infty$.

(iii) If $B > A > 0$ and if there exists $\epsilon$ such that

$$\frac{d-1}{d} \leq \frac{\log d}{\log m} + \frac{\log A}{k \log m} - \epsilon,$$

then for $k > k_0(\epsilon)$, there holds

$$F_{a,A,B}(x) = O\left(Cx^{\frac{1}{a}} + x^{1-\frac{|\log \eta|}{a \log m}}\right)$$

for some $\eta$ with $0 < \eta < 1$ which depends on $\epsilon$. This is true in particular when $m < d^{\frac{d}{d-1}}$, with $A$ fixed and $x \to \infty$.

Theorem 2 constructs for given values $d, m$ with $\gcd(m, d) = 1$ with $m > d \geq 2$ two functions $F(\cdot)$ and $G(\cdot)$ which fall slightly outside Hasse's class and have the following properties:

(i) The function $F(\cdot)$ has a finite number of periodic orbits, and every $n$ when iterated under $F(\cdot)$ eventually enters one of these orbits.

(ii) Each orbit of the function $G(\cdot)$ is divergent, i.e. $|G^{(k)}(n)| \to \infty$ as $k \to \infty$, for all but a finite number of initial values $n$.

To define the first function $F(\cdot)$ pick an integer $u$ such that $d < m < d^u$, and set

$$F(x) = \begin{cases} \frac{mx+r_j}{d} & \text{if } n \equiv j(\text{mod } d^u), \ \gcd(j,d) = 1, \\ \frac{x}{d} + s_j & \text{if } x \equiv j \ (\text{mod } d^u), \ j \equiv 0(\text{mod } d). \end{cases}$$

in which $0 < r_j < d^u$ is determined by $mj + r_j \equiv 0(\text{mod } d^u)$, and $0 \leq s_j < d^{u-1}$ is determined by $\frac{j}{d} + r_j \equiv 0(\text{mod } d^{u-1})$. The second function $G(x)$ is defined by $G(x) = F(x) + 1$.

These functions fall outside Hasse's class because each is linear on residue classes $n(\text{mod } d^u)$ for some $u \geq 2$, rather than linear on residue classes (mod $d$). However both these functions exhibit behavior qualitatively like functions in Hasse's class: There is a constant $C$ such that

$$\left|F(n) - \frac{n}{d}\right| \leq C \ \text{if } \ n \equiv 0 \ (\text{mod } d).$$

$$\left|F(n) - \frac{mn}{d}\right| \leq C \ \text{if } \ n \not\equiv 0 \ (\text{mod } d).$$

and similarly for $G(n)$, taking $C = d^{u-1} + 1$. The important difference is that functions in Hasse's class are mixing on residue classes (mod $d^k$) for all powers of $k$, while the functions $F(\cdot)$ and $G(\cdot)$ are not mixing in this fashion. The nature of the non-mixing behaviors of these functions underlies the proofs of properties (i), resp. (ii) for $F(\cdot)$, resp. $G(\cdot)$.

**(70-2)** Michael Beeler, William Gosper and Richard Schroeppel (1972), *HAK-MEM*, Memo 239, Artificial Intelligence Laboratory, MIT, 1972. (ONR Contract N00014-70-A-0362-0002).

The influential memorandum, never published in a journal, is a collection of problems and results. The list contains solved and unsolved problems, computer programs to write, programming hacks, computer hardware to design. There are 191 items in all. Example:

"**Problem 95:**    Solve *chess*. There are about $10^{40}$ possible positions; in most of them, one side is hopelessly lost."

It contains one of the earliest statements of the $3x+1$ problem, which appears as item 133. It was contributed by A. I. Lab members Richard Schroeppel, William Gosper, William Henneman and Roger Banks. It asks if there are any other cycles on the integers other than the five known ones. It asks if any orbit diverges.

**(70-3)**  Corrado Böhm and Giovanna Sontacchi (1978), *On the existence of cycles of given length in integer sequences like $x_{n+1} = x_n/2$ if $x_n$ even, and $x_{n+1} = 3x_n + 1$ otherwise*, Atti Accad. Naz. Lincei Rend. Cl. Sci. Fis. Mat. Natur. **64** (1978), 260–264. (MR 83h:10030)

The authors are primarily concerned with cycles of a generalization of the $3x + 1$ function. They consider the recursion in which $x_{n+1} = ax_n + b$, if a given recursive predicate $P(x_n)$ is true, and $x_{n+1} = cx_n + d$ if the predicate $P(x_n)$ false, where $a, b, c, d$ and $x_n$ are rational numbers. They observe that as a consequence of linearity alone there are at most $2^k$ possible cycles of period $k$, corresponding to all possible sequences of "true" and "false" of length $n$. Furthermore one can effectively determine the $2^n$ rationals that are the solutions to each of these equations and check if they give cycles. Thus in principle one can determine all cycles below any given finite bound. They observe that a rational number $x$ in a cycle of the $3x + 1$-function $T(\cdot)$ of period $n$ necessarily has the form

$$x = \frac{\sum_{k=0}^{n-1} 3^{m-k-1} 2^{v_k}}{2^n - 3^m}$$

with $0 \le v_0 < v_1 < \cdots < v_m = n$. They deduce that every integer $x$ in a cycle of length $n$ necessarily has $|x| < 3^n$.

*Note.* Further study of rational cycles of the $3x + 1$ function appears in Lagarias (1990).

**(70-4)**  Vasik Chvatal, David A. Klarner and Donald E. Knuth (1972) *Selected combinatorial research problems,* Stanford Computer Science Dept., Technical Report STAN-CS-72-292, June 1972, 31 pages.

This report contains 37 research problems, the first 16 of which are due to Klarner, the next 9 to Chvatal, and the remaining 11 to Knuth. This list contains two problems about iterating affine maps. Problem 1 asks whether the set of all positive integers reachable from 1 using the maps $x \mapsto 2x + 1$ and $x \mapsto 3x + 1$ can be partitioned into a disjoint union of infinite arithmetic progressions. Problem 14 considers for nonnegative integers $(m_1, \ldots, m_r)$ the set $S = \langle m_1 x_1 + \cdots + m_r x_r : 1 \rangle$ which is the smallest set of natural numbers containing 1 and which is closed under the operation of adjoining $m_1 x_1 + \cdots + m_r x_r$ whenever $x_i$ are in the set. It states that Klarner has shown that $S$ is a finite union of arithmetic progressions provided that (*i*) $r \ge 2$, (*ii*) the greatest common divisor $(m_1, \ldots, m)r) = 1$, and, (*iii*) the greatest common divisor $(m_1 + m_2 + \cdots + m_r, \prod_i m_i) = 1$. It asks if the same conclusion holds if hypothesis (*iii*) is dropped.

*Note.* Problem 1 was solved in the affirmative in Coppersmith (1975). Problem 14 relates to the theory developed in Klarner and Rado (1974), and was solved affirmatively in Hoffman and Klarner (1978), (1979).

**(70-5)** John H. Conway (1972), *Unpredictable Iterations*, In: Proc. 1972 Number Theory Conference, University of Colorado, Boulder, CO. 1972, pp. 49–52. (MR 52 #13717).

This paper states the $3x+1$ problem, and shows that a more general function iteration problem similar in form to the $3x + 1$ problem is computationally undecidable. It considers functions $f : \mathbb{Z} \to \mathbb{Z}$ for which there exists a finite modulus $N$ and rational numbers $\{a_j : 0 \le j \le N - 1\}$ such that

$$g(n) = a_j n \text{ if } n \equiv j \pmod{N}.$$

In order that the map take integers to integers it is necessary that the denominator of $a_j$ divide $gcd(j, N)$. The computationally undecidable question becomes: Given an $f(\cdot)$ in this class and an input value $n = 2^k$ decide whether or not some iterate of $n$ is a power of 2. More precisely, he shows that for any recursive function $f(n)$ there exists a choice of $g(\cdot)$ such that for each $n$ there holds $2^{f(n)} = g^k(2^n)$ for some $k \ge 1$ and this is the smallest value of $k$ for which the iterate is a power of 2. It follows that there is no decision procedure to recognize if the iteration of such a function, starting from input $n = 2^j$, will ever encounter another power of 2, particularly whether it will encounter the value 1.

The proof uses an encoding of computations in the exponents $e_p$ of a multiplicative factorization $n = 2^{e_2} \cdot 3^{e_3} \cdots$, in which only a fixed finite number of exponents $(e_2, e_3, \ldots, e_{p_r})$ control the computation, corresponding to the primes dividing the numerators and denominators of all $a_j$. The computation is based on a machine model with a finite number registers storing integers of arbitrary size, (the exponents $(e_2, e_3, \ldots, e_{p_r})$) and there is a finite state controller. These are called *Minsky machines* in the literature, and are described in Chapter 11 of M. Minsky, *Computation: Finite and Infinite Machines*, Prentice-Hall: Englewood Cliffs, NJ 1967 (especially Sec. 11.1).

*Note.* Conway (1987) later formalized this computational model as FRACTRAN, and also constructed a universal function $f(\cdot)$. See Burckel (1994) for other undecidability results.

**(70-6)** Don Coppersmith (1975), *The complement of certain recursively defined sets*, J. Combinatorial Theory, Series A **18** (1975), No. 3, 243–251. (MR 51 #5477).

This paper studies sets of nonnegative integers generated using a family of affine functions. $x \mapsto a_i x + b_i$, starting from a given set of seeds $A := \{c_1, \ldots, c_s\}$. The author assumes all $a_i \ge 2, b_i \ge 0, c_i \ge 1$ are integers. He calls any such set an *RD-set*. One may denote them $S = \langle a_1 x + b_1, \ldots, a_k x + b_k : c_1, \ldots, c_s \rangle$, as in Klarner and Rado (1974). Such a set is called *good* if its complement can be expressed as a disjoint union of infinite arithmetic progressions, otherwise it is *bad*. He proves a lemma stating that a set is good if its complement is a union of (not necessarily disjoint) infinite arithmetic progressions. He uses this lemma to reduce the problem to the case of a single seed (i.e. $s = 1$). A (possibly negative) integer is called a *feedback element* if it is a fixed point of some sequence of iterates of the maps above; it is determined by the affine functions, and such elements need not be part any particular $RD$-set. Theorem 1 says that for a fixed set of maps, if either there are no feedback elements, or if there are but no image of any feedback element under repeated iteration of the maps becomes positive, then for any seed $c_1 \geq 1$ the associated $RD$-set is good. Theorem 2 then gives a complicated sufficient condition on a set of maps for there to exist at least one seed giving a bad $RD$-set. In particular, Corollary 2b says that if $a \geq 2, m \geq 0$ and $b \geq 1$ then the set $S(c) := \langle ax + m(a-1), ax + m(a-1) + b : c \rangle$ is bad for some $c \geq 1$. Theorem 3 gives a stronger sufficient condition, more complicated to state, for a set of operators to have some seed giving a bad $RD$-set. The author asserts that this sufficient condition is "almost necessary."

*Note.* In Chvatal, Klarner and Knuth (1972), Problem 1 of Klarner asked whether it was true that $S := \langle 2x + 1, 3x + 1 : 1 \rangle$ is a good set. That it is a good set follows using Theorem 1, since it is easy to show by induction that all operators obtained by composition necessarily have the form $ax + b$ with $a > b \geq 1$, so such operators have a fixed point $x$ satisfying $-1 < x < 0$. Thus this set of operators has no feedback elements. Theorem 1 imples that $S(c) := \langle 2x + 1, 3x + 1 : c \rangle$ is good for all $c \geq 1$.

**(70-7)**   Harold Scott MacDonald Coxeter (1971), *Cyclic Sequences and Frieze Patterns, (The Fourth Felix Behrend Memorial Lecture)*, Vinculum **8** (1971), 4–7.

This lecture was given at the University of Melbourne in 1970. Coxeter discusses various integer sequences, including the Lyness iteration $u_{n+1} = \frac{1+u_n}{u_{n-1}}$, which has the orbit $(1, 1, 2, 3, 2)$ as one solution. He observes that a general solution to the Lyness iteration can be produced by a frieze pattern with some indeterminates. He then introduces the $3x + 1$ iteration as "a more recent piece of mathematical gossip." He states the $3x + 1$ conjecture and then says: "I am tempted to follow the example of Paul Erdős, who offers prizes for the solutions of certain problems. It the above conjecture is true, I will gladly offer a prize of fifty dollars to the first person who send me a proof that I can understand. If it is false, I offer a prize of a hundred dollars to the first one who can establish a counterexample. I must warn you not to try this in your heads or on the back of an old envelope, because the result has been tested with an electronic computer for all $x_1 \leq 500,000$."

*Note.* Based on knowledge of this talk, Ogilvy (1972) mistakenly credits Coxeter with proposing the $3x + 1$ problem. For more on Frieze patterns, see H. S. M. Coxeter, *Frieze patterns*, Acta Arithmetica **18** (1971), 297–310, and J. H. Conway and H. S. M. Coxeter, *Triangulated polygons and Frieze patterns,* Math. Gazette **57** (1973) no. 400, 87-94; no 401, 175–183. Vinculum is the journal of the Mathematical Association of Victoria (Melbourne, Australia).

**(70-8)**   Richard E. Crandall (1978), *On the "$3x + 1$" problem*, Math. Comp. **32** (1978), 1281–1292. (MR 58 #494).

This paper studies iteration of the "3x+1" map and more generally the "$qx + r$" map

$$T_{q,r}(x) = \begin{cases} \frac{qx+r}{2} & \text{if } x \equiv 1 \pmod 2 \,. \\ \frac{x}{2} & \text{if } x \equiv 0 \pmod 2 \,. \end{cases}$$

in which $q > 1$ and $r \geq 1$ are both odd integers. He actually considers iteration of the map $C_{q,r}(\cdot)$ acting on the domain of positive odd integers, given by

$$C_{q,r}(x) = \frac{qx + r}{2^{e_2(qx+r)}},$$

where $e_2(x)$ denotes the highest power of 2 dividing $x$.

Most results of the paper concern the map $C_{3,1}(\cdot)$ corresponding to the $3x+1$ map. He first presents a heuristic probabilistic argument why iterates of $C_{3,1}(\cdot)$ should decrease at an exponential rate, based on this he formulates a conjecture that the number of steps $H(x)$ starting from $x$ needed to reach 1 under iteration of $C_{3,1}(\cdot)$ should be approximately $H(x) \approx \frac{\log x}{\log \frac{16}{9}}$ for most integers. He proves that the number of odd integers $n$ taking exactly $h$ steps to reach 1 is at least $\frac{1}{h!}(\log_2 x)^h$. He deduces that the function $\pi_1(x)$ which counts the number of odd integers below $x$ that eventually reach 1 under iteration of $C_{3,1}(\cdot)$ has $\pi_1(x) > x^c$ for a positive constant $c$. (He does not compute its value, but his proof seems to give $c = 0.05$.) He shows there are no cycles of length less than 17985 aside from the trivial cycle, using approximations to $\log_2 3$.

Concerning the "$qx + r$" problem, he formulates the conjecture that, aside from $(q, r) = (3, 1)$, every map $C_{q,r}(\cdot)$ has at least one orbit that never visits 1. He proves that this conjecture is true whenever $r \geq 3$, and in the remaining case $r = 1$ he proves it for $q = 5$, $q = 181$ and $q = 1093$. For the first two cases he exhibits a periodic orbit not containing 1, while for $q = 1093$ he uses the fact that there are no numbers of height 2 above 1, based on the congruence $2^{q-1} \equiv 1 \pmod{q^2}$. (This last argument would apply as well to $q = 3511$.) He argues the conjecture is true in the remaining cases because a heuristic probabilistic argument suggests that for each $q \geq 5$ the "$qx + 1$" problem should have a divergent trajectory.

(70-9)  J. Leslie Davison (1977), *Some Comments on an Iteration Problem*, Proc. 6-th Manitoba Conf. On Numerical Mathematics, and Computing (Univ. of Manitoba-Winnipeg 1976), Congressus Numerantium XVIII, Utilitas Math.: Winnipeg, Manitoba 1977, pp. 55–59. (MR 58 #31773).

The author considers iteration of the map $f : \mathbb{Z}^+ \to \mathbb{Z}^+$ given by

$$f(n) = \begin{cases} \frac{3n+1}{2} & \text{if } n \equiv 1 \pmod 2, n > 1 \\ \frac{n}{2} & \text{if } n \equiv 0 \pmod 2 \\ 1 & \text{if } n = 1, \end{cases}$$

This is essentially the $3x + 1$ function, except for $n = 1$. He calls a sequence of iterates of this map a *circuit* if it starts with an odd number $n$, produces a sequence of odd numbers followed by a sequence of even numbers, ending at an odd number $n^*$. A circuit is a *cycle* if $n = n^*$.

Based on computer evidence, he conjectures that the number of circuits required during the iteration of a number $n$ to 1 is at most $K \log n$, for some absolute constant $K$. He presents a probabilistic heuristic argument in support of this conjecture.

He asks whether a circuit can ever be a cycle. He shows that this question can be formulated as the exponential Diophantine equation: There exists a circuit that is a cycle if and only if there exist positive integers $(k, l, h)$ satisfying

$$(2^{k+l} - 3^k)h = 2^l - 1.$$

(Here there are $k$ odd numbers and $l$ even numbers in the cycle, and $h = \frac{n+1}{2^k}$ where $n$ is the smallest odd number in the cycle.) The trivial cycle $\{1, 2\}$ of the $3x+1$ map corresponds to the solution $(k, l, h) = (1, 1, 1)$, Davison states he has been unable to find any other solutions. He notes that the $5x + 1$ problem has a circuit that is a cycle.

Steiner (1978) subsequently showed that $(1, 1, 1)$ is the only positive solution to the exponential Diophantine equation above.

**(70-10)**   Richard Dunn (1973), *On Ulam's Problem,* Department of Computer Science, University of Colorado, Boulder, Technical Report CU-CS-011-73, 15pp.

This report gives early computer experiments on the $3x + 1$ problem. The computation numerically verifies the $3x+1$ conjecture on a CDC 6400 computer up to 22,882,247. Dunn also calculates the densities $F(k)$ defined in equation (2.16) of Lagarias (1985) for $k \leq 21$.

**(70-11)**   Paul Erdős and R. L. Graham (1979), *Old and new problems and results in combinatorial number theory: van der Waerden's theorem and related topics,* Enseign. Math. **25** (1979), no. 3-4, 325–344. (MR 81f:10005).

This problem list includes problems raised in Klarner and Rado (1974), Hoffman (1976), and Hoffman and Klarner (1978), (1979) on the smallest set of nonnegative integers obtained from a given set $A$ under iteration of a finite set $R$ of functions $\rho(x_1, \ldots, x_r) = m_0 + m_1 x_1 + \cdots + m_r x_r$, with nonnegative integer $m_i$ for $i \geq 1$. Denoting this set $< R : A >$, one can ask for the size and structure of this set. In the case of one variable functions $R = \{a_1 x + b_1, \ldots, a_r x + b_r\}$ Erdős showed (see Klarner and Rado (1974)) that if $\sum \frac{1}{a_i} < 1$, then the set has density 0. The case when $\sum \frac{1}{a_i} = 1$ is pointed to as a source of unresolved problems. Erdős had proposed as a prize problem: For $R = \{2x + 1, 3x + 1, 6x + 1\}$ and $A = \{1\}$ is the set $< R : A >$ of positive density? This was answered in the negative by D. J. Crampin and A. J. W. Hilton (unpublished), as summarized in Klarner (1982) and Klarner (1988).

**(70-12)**   C. J. Everett (1977), *Iteration of the number theoretic function $f(2n) = n$, $f(2n + 1) = 3n + 2$,* Advances in Math. **25** (1977), 42–45. (MR 56#15552).

This is one of the first research papers specifically on the $3x + 1$ function. Here $f(\cdot)$ is the $3x + 1$-function $T(\cdot)$. The author shows that the set of positive integers $n$ having some iterate $T^{(k)}(n) < n$ has natural density one. The result was obtained independently and contemporaneously by Terras (1976).

*Note.* Cornelius J. Everett, Jr. was one of Stanisław M. Ulam's long-time collaborators at Los Alamos. Everett did joint work with Ulam on many topics, including several papers on branching processes in 1948. In 1950 he carried out detailed calculations for Ulam on questions concerning behavior of nuclear fission and fusion reactions and design questions on the feasibility of a hydrogen bomb. Ulam later circulated the $3x + 1$ problem and it is sometimes called "Ulam's problem." This paper was first issued in 1976 as Los Alamos Technical Report LA-6449-MS.

**(70-13)**   Michael Lawrence Fredman (1972) *Growth properties of a class of recursively defined functions*, Ph. D. Thesis, Stanford University, June 1972, 81 pages (D. E. Knuth, advisor).

Let $g(n)$ be a given nonnegative function defined on $n \geq 0$. This thesis discusses solutions of the general recurrence $M(0) = g(0)$,

$$M(n+1) := g(n+1) + \min_{0 \leq k \leq n} \left( \alpha M(k) + \beta M(n-k) \right),$$

in which $\alpha, \beta > 0$. It has three chapters and a conclusion. The introduction states that D. E. Knuth did earlier (unpublished) work with R. W. Floyd in 1964 on the case $\alpha = 1, \beta = 2, g(n) = n$, which arose in analysis of an algorithm for merging two sorted sequences of numbers into a single sorted sequence. Chapter 2 concern analysis in the special case where $g(n) = n$. Chapter 3 considers more general cases. The author how the quantity $M(n)$ has an interpretation in terms of minimum total weight of weighted binary trees having $n$ nodes. For the analysis in the $h(n) = n$ case. Theorem 2.1 sets $D(n) = M(n) - M(n-1)$ and sets $h(x) = \sum_{\{j:D(j)) \leq x\}} 1$, and shows that $h(x)$ satisfies $h(x) = 1$ for $0 \leq x < 1$ and the functional equation

$$h(x) = 1 + h(\frac{x-1}{\alpha}) + h(\frac{x-1}{\beta}).$$

It determines the growth rates of $h(x)$ and $M(x)$ in many circumstances. We describe here thesis results mainly for the case $\min(\alpha, \beta) > 1$ (Section 2.3) Let $\gamma$ be the unique positive solution to $\alpha^{-\gamma} + \beta^{-\gamma} = 1$. It is shown that the function $h(x)$ has order of magnitude $x^\gamma$ while $M(x)$ has order of magnitude $x^{1+\frac{1}{\gamma}}$. Theorem 2.3.2 states that $\lim_{x \to \infty} h(x) x^{-\gamma}$ exists if and only if $\lim_{x \to \infty} M(x) x^{-1-\frac{1}{\gamma}}$ exists. It is shown the limits always exist if $\frac{\log \alpha}{\log \beta}$ is irrational, but in general do not exist when $\frac{\log \alpha}{\log \beta}$ is rational. Chapter 3 obtains less precise growth rate information for a wide class of driving functions $g(n)$. Some of the proofs use complex analysis and Tauberian theorems for Dirichlet series. The conclusion of the thesis states applications and open problems. One given application is to answer a question raised by Klarner, see Klarner and Rado (1974). Fredman shows that the set $S$ of integers obtained starting from 1 and iterating the affine maps $x \mapsto 2x + 1$, $x \mapsto 3x + 1$ has density 0, and in fact the number of such integers below $x$ is at most $O(x^\gamma)$ where $2^{-\gamma} + 3^{-\gamma} = 1$. (Here $\gamma \approx 0.78788$.) This result follows from an upper bound on growth of $h(x)$ above when $\alpha = 2, \beta = 3, g(n) = n$. It uses the fact that the function $h(x)$ has an interpretation as counting the number of elements $\leq x$ in the multiset $\tilde{S} := \cup_{j=0}^\infty S_j$ generated by initial element $S_0 = \{1\}$ and inductively letting $S_{j+1}$ being the image of $S_j$ under iteration of the two affine maps $x \mapsto \alpha x + 1$, $x \mapsto \beta x + 1$ (counting elements in $\tilde{S}$ with the multiplicity they occur).

*Note.* This thesis also appeared as Stanford Computer Science Technical Report STAN-CS-72-296. Some results of this thesis were subsequently published in Fredman and Knuth (1974).

**(70-14)**   Michael Lawrence Fredman and Donald E. Knuth (1974) *Recurrence relations based on minimization*, J. Math. Anal. Appl. **48** (1974), 534–559 (MR 57#12364).

This paper studies the asymptotics of solutions of the general recurrence $M(0) = g(0)$,

$$M(n + 1) := g(n + 1) + \min_{0 \leq k \leq n} (\alpha M(k) + \beta M(n - k)),$$

for various choices of $\alpha, \beta, g(n)$. They denote this $M_{g\alpha\beta}(n)$ In §2-§4 they treat the case $g(n) = n$, where they develop an interpretation of this quantity in terms of weighted binary trees.; $M_{g\alpha\beta}$ is the minimum total weight of any rooted binary tree with $n$ nodes. The weight of a nodes in a finite binary tree $T$ is given by assigning the root node $\sigma = \epsilon$ the weight $w(\emptyset) = 1$, and then inductively defining $w(L\sigma) = 1 + \alpha w(\sigma), w(R\sigma) = 1 + \beta w(\sigma)$; for example $w(LRR) = 1 + \alpha + \alpha\beta + \alpha\beta^2$. The total weight function $\mathcal{M}(T)$ of a tree is the sum of the weights of all nodes in it. They set $M(n) := \min_{T:|T|=n} \mathcal{M}(T)$, and show this quantity is $M_{g\alpha\beta}(n)$ for $g(n) = n$. In §3 they analyze the asymptotic behavior of $M(n)$ in the case $\min(\alpha, \beta) = 1$, in §4 in the case $\min(\alpha, \beta) > 1$. They let $H(x) = h(x) + 1$ where $h(x)$ counts the number of node weights $w(\sigma) \leq x$ and observe that it satisfies $H(x) = 1$ for $0 \leq x < 1$ and the functional difference equation

$$H(x) = H(\frac{x - 1}{\alpha}) + H(\frac{x - 1}{\beta}).$$

They note that $H(x)$ and $M(x)$ are related using the fact that a depth $n$ node has weight $w(\sigma_n) \leq x$ if and only if $H(x) > n$. Now let $\gamma$ be the unique positive solution to $\alpha^{-\gamma} + \beta^{-\gamma} = 1$. For the case $\min(\alpha, \beta) \leq 1$ Theorems 1 and 3 show that $M_{g\alpha\beta}$ grows no faster than a polynomial in $n$. For the case $\min(\alpha, \beta) > 1$, Lemma 4.1 shows that $H(x)$ is on the order of $x^\gamma$, and $M(x)$ is on the order of $x^{1+\frac{1}{\gamma}}$, so that one can write $H(x) = c(x)x^\gamma$, $M(x) = C(x)x^{1+\frac{1}{\gamma}}$, where $c(x)$ and $C(x)$ are positive bounded functions. The asymptotic behaviors of $H(x)$ and $M(x)$ now depends on properties of the positive real number $\frac{\log \alpha}{\log \beta}$. Theorem 4.1 shows that when $\frac{\log \alpha}{\log \beta}$ is rational, the function $C(x)$ is usually an oscillatory function having no limiting value at $\infty$. Theorem 4.3 shows that when $\frac{\log \alpha}{\log \beta}$ is irrational, $C(x)$ has a positive limiting value as $x \to \infty$ so that $M(x) \sim Cx^{1+\frac{1}{\gamma}}$. (Similar results hold for $c(x)$; this was explicitly shown in Fredman (1972).) The remainder of the paper treats other $g(n)$. In §5 the case $g(n) = 1$, in §6 the case $g(n) = \delta_{n0}$, and in §7 the case $g(n) = n^2$. Some proofs directly use recurrence relations, others use complex analysis and Tauberian theorems, discussed in an Appendix. See also Pippenger (1993) for proofs of some results by more elementary methods.

*Note.* Many of these results appear in the PhD thesis of M. Fredman (1972). Some cases are new. The analysis of §4 can be applied to upper bound the number of elements below $x$ of the minimal set of integers $S$ containing 1 and closed under the affine maps $x \to 2x + 1$, $x \mapsto 3x + 1$. Here $\frac{\log 2}{\log 3}$ is irrational, and the upper bound is asymptotic to $Cx^\gamma$, for $\gamma \approx 0.78788$. The size and structure of the set $S$ was a problem raised in Chvatal, Klarner and Knuth (1972), and in Klarner and Rado (1974).

**(70-15)**   Martin Gardner (1972), *Mathematical Games*, Scientific American **226**, No. 6, (June 1972), 114–118.

   This article is one of the first places the $3x + 1$ problem is stated in print. Gardner attributes the problem to a technical report issued by M. Beeler, R. Gosper, and R. Schroeppel, HAKMEM, Memo 239, Artificial Intelligence Laboratory, M.I.T., 1972, p. 64. The $3x + 1$ problem certainly predates this memo of Beeler et al., which is a collection of problems.

**(70-16)**   Helmut Hasse (1975), *Unsolved Problems in Elementary Number Theory*, Lectures at University of Maine (Orono), Spring 1975. Mimeographed notes.

   Hasse discusses the $3x + 1$ problem on pp. 23–33. He calls it the Syracuse (or Kakutani) algorithm. He asserts that A. Fraenkel checked it for $n < 10^{50}$, which is not the case (private communication with A. Fraenkel). He states that Thompson has proved the Finite Cycles Conjecture, but this seems not to be the case, as no subsequent publication has appeared.
   He suggests a generalization (mod $m$) for $m > d \geq 2$ and in which the map is

$$T_d(x) := \frac{mx + f(r)}{d} \text{ if } x \equiv r \pmod{d}$$

in which $f(r) \equiv -mr \pmod{d}$. He gives a probabiliistic argument suggesting that all orbits are eventually periodic, when $m = d + 1$.
   Hasse's circulation of the problem motivated some of the first publications on it. He proposed a class of generalized $3x + 1$ maps studied in Möller (1978) and Heppner (1978).

**(70-17)**   Ernst Heppner (1978), *Eine Bemerkung zum Hasse-Syracuse Algorithmus*, Archiv. Math. **31** (1978), 317–320. (MR 80d:10007)

   This paper studies iteration of generalized $3x+1$ maps that belong to a class formulated by H. Hasse. This class consists of maps depending on parameters of the form

$$T(n) = T_{m,d,R}(n) := \begin{cases} \frac{mn+r_j}{d} & \text{if } n \equiv j \pmod{d}, \ 1 \leq j \leq d - 1 \\ \frac{x}{d} & \text{if } n \equiv 0 \pmod{d} . \end{cases}$$

in which the parameters $(d, m)$ satisfy $d \geq 2$, $gcd(m, d) = 1$, and the set $R = \{r_j : 1 \leq j \leq d - 1\}$ has each $r_j \equiv -mj \pmod{d}$. The qualitative behavior of iterates of these maps are shown to depend on the relative sizes of $m$ and $d$.
   Heppner proves that if $m < d^{d/d-1}$ then almost all iterates get smaller, in the following quantitative sense: There exist positive real numbers $\delta_1, \delta_2$ such that for any $x > d$, and $N = \lfloor \frac{\log x}{\log d} \rfloor$, there holds

$$\#\{n \leq x : T^{(N)}(n) \geq nx^{-\delta_1}\} = O(x^{1-\delta_2}).$$

He also proves that if $m > d^{d/d-1}$ then almost all iterates get larger, in the following quantitative sense: There exist positive real numbers $\delta_3, \delta_4$ such that for any $x > d$, and $N = \lfloor \frac{\log x}{\log d} \rfloor$, there holds

$$\#\{n \leq x : T^{(N)}(n) \leq nx^{\delta_3}\} = O(x^{1-\delta_4}).$$

In these results the constants $\delta_j$ depend on $d$ and $m$ only while the implied constant in the $O$-symbols depends on $m, d$ and $R$. This results improve on those of Möller (1978).

**(70-18)**  Dean G. Hoffman and David A. Klarner (1978), *Sets of integers closed under affine operators- the closure of finite sets,* Pacific Journal of Mathematics **78** (1978), No. 2, 337–344. (MR 80i:10075)

This work extends work of Klarner and Rado (1974), concerning sets of integers generated by iteration of a single multi-variable affine function. It considers iteration of the affine map $f(x_1, \ldots, x_m) = m_1 x_1 + \cdots + m_r x_r + c$ assuming that : (i) $r \geq 2$, (ii) each $m_i \neq 0$ and the greatest common divisor $(m_1, m_2, \ldots, m_r) = 1$. The author supposes that $T$ is a set of (not necessarily positive) integers that is closed under iteration of $f$ in the sense that if $t_1, \ldots, t_r \in T$ then $f(t_1, \ldots, t_r) \in T$. The main result, Theorem 12, states that if in addition $f(t, t, \ldots, t) > t$ holds for all $t \in T$, then the following two statements are equivalent: (1) $T$ is a finite union of infinite arithmetic progressions, and (2) T is generated by some finite set $A$ under iteration of the map $f$, using $A$ as a "seed."

**(70-19)**  Dean G. Hoffman and David A. Klarner (1979), *Sets of integers closed under affine operators- the finite basis theorems,* Pacific Journal of Mathematics **83** (1979), No. 1, 135–144. (MR 83e:10080)

This paper strengthens the results in Hoffman and Klarner (1978) concerning sets of integers generated by iteration of a single multi-variable affine function. The same hypotheses (i), (ii) are imposed on the function $f$ as in that paper. Without any further hypothesis, the authors now conclude that if $T$ is any set closed under iteration by $f$, then $T$ is generated by a finite set of "seeds" $A$, so we may write $T = \langle f : A \rangle$ in the notation of Klarner and Rado (1974). The second main result (Theorem 13) is the conclusion that either $T$ is a one-element set, or a finite union of one-sided infinite arithmetic progessions, bounded below, or a finite union of two-sided infinite arithmetic progressions.

**(70-20)**  Stephen D. Isard and Harold M. Zwicky (1970), *Three open questions in the theory of one-symbol Smullyan systems,* SIGACT News, Issue No. 7, 1970, 11-19.

Smullyan systems are described in R. Smullyan, *First Order Logic,* [Ergebnisse der Math. Vol. 43, Springer-Verlag, NY 1968 (Corrected Reprint: Dover, NY 1995)] In this paper open question 2 concerns the set of integers that are generated by certain Smullyan systems with one symbol $x$. If we let $n$ label a string of $n$ $x$'s, then the system allows the two string rewriting operations $f(n) = 3n$ and $g(4m + 3) = 2m + 1$. The problem asks whether it is true that, starting from $n_0 = 1$, we can reach every positive number $n \equiv 1 \pmod 3$ by a sequence of such string rewritings. The authors note that one can reach every such number except possibly those with $n \equiv 80 \pmod{81}$, and that the first few numbers in this congruence class are reachable. In the reverse direction we are allowed instead to apply either of the rules $F(3n) = n$ or $G(2m + 1) = 4m + 3$, and we then wish to get from an arbitrary $n \equiv 1 \pmod 3$ to 1. An undecidability result given at the end of the paper, using Minsky machines, has some features in common with Conway (1972).

*Note.* This symbol rewriting problem involves many-valued functions in both the forward and backward directions, which are linear on congruence classes to some modulus. Trigg et. al (1976) cited this problem in a discussion of the

$3x+1$ problem, as part of its prehistory. The $3x+1$ problem has similar feature (cf. Everett (1977)) with the difference that the $3x+1$ function is single-valued in the forwards direction and many-valued only in the backwards direction. Michel (1993) considers other rewriting rules which are functions in one direction.

**(70-21)**    David Kay (1972), *An algorithm for reducing the size of an Integer*, Pi Mu Epsilon Journal **4** (1972), 338.

     This short note proposes the $3x + 1$ problem as a possible undergraduate research project. The Collatz map is presented, being denoted $k(n)$, and it is noted that it is an unsolved problem to show that all iterates on the positive integers go to 1. It states this has been verified for all integers up to a fairly large bound.

     The proposed research project asks if one can find integers $p, q, r$ with $p, q \geq 2$, having the property that a result analogous to the $3x + 1$ Conjecture can be rigorously proved for the function

$$k(n) := \begin{cases} \frac{n}{p} & \text{if } n \equiv 0 \ (\text{mod } p) . \\ qn + r & \text{if } x \not\equiv 0 \ (\text{mod } p) . \end{cases}$$

     *Note.* This project was listed as proposed by the Editor of Pi Mu Epsilon; the Editor was David Kay.

**(70-22)**    David A. Klarner and Richard Rado (1973), *Linear combinations of sets of consecutive integers*, American Math. Monthly **80** (1973), No. 9, 985–989. (MR 48 #8378).

     This paper arose from questions concerning the iteration of integer-valued affine maps, detailed in Klarner and Rado (1974). This paper proves results on additive unions of sets of consecutive positive integers, containing certain larger sets of consecutive integers. It proves a vector-valued form of such a result. The results allow improvement of certain results in Klarner and Rado (1974), implying for example that when $m, n$ are positive integers having greatest common divisor $(m, n) = 1$ the set $S = \langle mx + ny + 1 : 0 \rangle$ of integers generated starting from 0 by iteration of the function $\rho(x, y) = mx + ny + 1$ contains all but finitely many positive integers in each arithmetic progressions (mod $mn$) that it can allowably reach.

**(70-23)**    David A. Klarner and Richard Rado (1974), *Arithmetic properties of certain recursively defined sets*, Pacific J. Math. **53** (1974), No. 2, 445–463. (MR 50 #9784).

     This paper studies the smallest set of nonnegative integers obtained from a given set $A$ under iteration of a finite set $R$ of affine maps

$$\rho(x_1, \ldots, x_r) = m_0 + m_1 x_1 + \cdots + m_r x_r,$$

in which all $m_i$ are integers, and $m_i \geq 0$ for $i \geq 1$. They denote this set $\langle R : A \rangle$. One can now ask questions concerning the size and structure of this set. This paper gives some sufficient conditions for such a set to be a finite union of arithmetic progressions, which are called *per-sets*; such sets necessarily have positive density. Theorem 4 shows that if $A$ is already a per-set, then closure under a map having greatest common divisor $(m_1, \ldots, m_r) = 1$ will give a per-set. Theorem 5 gives a general condition for a set $< R : A >$ to be closed under multiplication. Conjecture 1 states that if $R$ includes a function having greatest common divisor $(m_1, \ldots, m_r) = 1$, and $A = \{1\}$ then $< R : A >$ is a per-set. (The paper also announces a subsequent proof by Klarner of this conjecture; this was carried out in Hoffman and Klarner (1978), (1979).) Conjecture 1 is proved here in some cases, showing (Theorem 11) that $\langle 2x + ny : 1 \rangle$ is a per-set for all odd integers $n \geq 1$. Conjecture 2 asserts that for all $m, n \geq 1$ the set $< mx + ny : 1 >$ contains a non-empty per-set. In the case of one variable functions $R = \{m_1 x + b_1, \ldots, m_r x + b_r\}$ it includes a theorem attributed to Erdős (Theorem 8) showing that if $\sum \frac{1}{m_i} < 1$, then the set has density 0. The authors mention a numerical study of the set $S = \langle 2x + 1, 3x + 1 : 1 \rangle$, which has density 0 by Theorem 8. For this set Klarner (1972) had conjectured that the complement $\mathbb{N} \setminus S$ could be written as a disjoint union of infinite arithmetic progressions; this was proved by Coppersmith (1975).

*Note.* This work originally appeared as a series of Stanford Computer Science Dept. Technical Reports in 1972, numbered: STAN-CS-72-269. Related sequels by Klarner are STAN-CS-72-275, STAN-CS-73-338. The results in these reports are superseded by Hoffman and Klarner (1978), (1979).

**(70-24)**  Herbert Möller (1978), *Uber Hasses Verallgemeinerung des Syracuse-Algorithmus (Kakutanis Problem)*, Acta Arith. **34** (1978), No. 3, 219–226. (MR 57 #16246).

This paper studies for parameters $d, m$ with $d \geq 2, m \geq 1$ and $(m, d) = 1$ the class of maps $H : \mathbb{Z}^+ \setminus \mathbb{Z}^+ \to \mathbb{Z}^+ \setminus d\mathbb{Z}^+$ having the the form

$$H(x) = \frac{1}{d^{a(x)}}(mx - r_j) \text{ when } rx \equiv j \pmod{m}, \ 1 \leq j \leq d-1,$$

in which $R(d) := \{r_j : 1 \leq j \leq d-1\}$ is any set of integers satisfying $r_j \equiv mj \pmod{d}$ for $1 \leq j \leq d-1$, and $d^{a(x)}$ is the maximum power of $d$ dividing $mx - r_j$. The $3x + 1$ problem corresponds to $d = 2, m = 3$ and $R(d) = \{-1\}$. The paper shows that if

$$m \leq d^{d/(d-1)}$$

then the set of positive integers $n$ which have some iterate $H^{(k)}(n) < n$, has full natural density $1 - \frac{1}{d}$ in the set $\mathbb{Z}^+ \setminus d\mathbb{Z}^+$. He conjectures that when $m \leq d^{d/(d-1)}$ the exceptional set of positive integers which don't satisfy the conditon is finite.

The author's result generalizes that of Terras(1976) and Everett (1977). In a note added in proof the author asserts that the proofs of Terras (1976) are faulty. Terras's proofs seem essentially correct to me, and in response Terras (1979) provided further details.

**(70-25)**  Jürg Nivergelt (1975), *Computers and Mathematics Education,* Computers & Mathematics, with Applications, **1** (1975), 121–132.

The paper argues there is a strong connection between mathematics education and computers. It points out that mathematics is an experimental science. In an appendix the $3x + 1$ problem is formulated as an example for experimentation, and a number of its properties are derived. The author formulates the $3x + 1$ conjecture, says he does not know the answer, but that it can be numerically studied. For the Collatz function he gives a heuristic that predicts that the number of elements that take exactly $s$ steps to iterate to 1 should grow like $\alpha^s$, with $\alpha = \frac{1}{2}\left(1 + \sqrt{\frac{7}{3}}\right) \approx 1.26376$.

*Note.* A similar discussion of the $3x + 1$ problem also appears in pages 211–217 of the book: Jürg Nivergelt, J. Craig Farrar and Edward M. Reingold (1974), *Computer Approaches to Mathematical Problems*, Prentice-Hall, Inc.: Englewood Cliffs, NJ 1974.

**(70-26)**   C. Stanley Ogilvy (1972), *Tomorrow's Math: unsolved problems for the amateur*, Second Edition, Oxford University Press: New York 1972.

The $3x + 1$ problem is discussed on pages 103-104. He states: "H. S. M. Coxeter, who proposed it in 1970, stated then that it had been checked for all $N \le 500,000$. However if the conjecture is true, which seems likely, a proof will have nothing to do with computers." (In fact Coxeter (1971) does not claim to propose the problem but instead reports it "as a piece of mathematical gossip.") He notes that the analogous conjecture for the $5x + 1$ problem is false, since there is a cycle not reaching 1.

**(70-27)**   Raymond Queneau (1972), *Sur les suites s-additives,* J. of Combinatorial Theory, Series A, **12** (1972), 31-71. (MR 46 # 1741)

This is the detailed paper following the announcement in Comptes Rendus Acad. Sci. Paris (A-B) **266** (1968), A957–A958. This paper studies sequences of integers constructed by a "greedy" algorithm, where the first $2s$ integers $0 < u_1 < u_2 < \ldots, < u_{2s}$ are arbitrary, and thereafter each integer $a_t$ is the smallest integer that can be written in exactly $s$ distinct ways as $S_{ij} = u_i + u_{i+1} + \cdots + u_j$, for some $1 \le i < j < t$. He shows that in order for the series not to terminate the initial set must be a union of two arithmetic progressions of length $s$, having the same common difference, one being $\{u, 2u, \ldots, su\}$ and the other $\{v, v + u, \ldots, v + (s - 1)u\}$. Letting $S(s, u, v)$ denoting the sequence generated this way, he shows that for $s \ge 2$ and $u \ge 3$, then the sequence is infinite, consisting of $\{v + nu : n \ge s\}$, together with the single term $2v + (2s - 1)u$. He then considers cases with $s \ge 2$ and $u = 1$ or $u = 2$. Some sequences of the above form are finite, and some are infinite; he presents some results and conjectures. For $s = 0, 1$ all $s$-sequences are infinite, and may have a complicated structure.

On page 63 there appears a $3x + 1$-like function $\sigma(2p) = p - 1, \sigma(2p + 1) = p$ of the form considered in Queneau (1963).

**(70-28)**   Lee Ratzan (1973), *Some work on an Unsolved Palindromic Algorithm,* Pi Mu Epsilon Journal **5** (Fall 1973), 463–466.

The $3x + 1$ problem and a generalization were proposed as an Undergraduate Research Project by David Kay (1972). The author gives computer code to test the conjecture, and used it to verify the $3x + 1$ Conjecture up to 31,910. He notices some patterns in two consecutive numbers having the same total stopping time and makes conjectures when they occur.

**(70-29)**   Daniel Shanks (1975), *Problem #5*, Western Number Theory Conference 1975, Problem List, (R. K. Guy, Ed.).

The problem concerns iteration of the Collatz function $C(n)$. Let $l(n)$ count the number of distinct integers appearing in the sequence of iterates $C^{(k)}(n)$,

$k \geq 1$, assuming it eventually enters a cycle. Thus $l(1) = 3, l(2) = 3, l(3) = 8$, for example. Set $S(N) = \sum_{n=1}^{N} l(n)$. The problem asks whether it is true that

$$S(N) = AN \log N + BN + o(N) \text{ as } N \to \infty,$$

where

$$A = \frac{3}{2} \log \frac{4}{3} \approx 5.21409 \text{ and } B = A(1 - \log 2) \approx 1.59996.$$

In order for $S(N)$ to remain finite, there must be no divergent trajectories. This problem formalizes the result of a heuristic probabilistic calculation based on assuming the $3x + 1$ Conjecture to be true.

**(70-30)**  Ray P. Steiner (1978), *A Theorem on the Syracuse Problem*, Proc. 7-th Manitoba Conference on Numerical Mathematics and Computing (Univ. Manitoba-Winnipeg 1977), Congressus Numerantium XX, Utilitas Math.: Winnipeg, Manitoba 1978, pp. 553–559. (MR 80g:10003).

This paper studies periodic orbits of the $3x + 1$ map, and a problem raised by Davidson (1976). A sequence of iterates $\{n_1, n_2, \ldots, n_p, n_{p+1}\}$ with $T(n_j) = n_{j+1}$ is called by Davidson (1977) a *circuit* if it consists of a sequence of odd integers $\{n_1, n_2, \ldots, n_j\}$ followed by a sequence of even integers $\{n_{j+1}, n_{j+2}, \ldots, n_p\}$, with $n_{p+1} = T(n_p)$ an odd integer. A circuit is a *cycle* if $n_{p+1} = n_1$.

This paper shows that the only circuit on the positive integers that is a cycle is $\{1, 2\}$. It uses the observation of Davison (1977) that these corresponds to positive solutions $(k, l, h)$ to the exponential Diophantine equation

$$(2^{k+l} - 3^k)h = 2^l - 1.$$

The paper shows that the only solution to this equation in positive integers is $(k, l, h) = (1, 1, 1)$. The proof uses results from transcendence theory, Baker's method of linear forms in logarithms (see A. Baker, *Transcendental Number Theory*, Cambridge Univ. Press 1975, p. 45.)

*Note.* Solutions of this exponential Diophantine equation allowing negative integers correspond to periodic orbits of the $3x + 1$ function of this type on the negative integers. One can prove by similar methods that there are exactly three circuits that are cycles on the non-positive integers, namely $\{-1\}$, and $\{-5, -7, -10\}$. These correspond to the solutions to the exponential Diophantine equation $(k, l, h) = (k, 0, 0)$ for any $k \geq 1$; and $(2, 1, -1)$, respectively. A further solution $(0, 2, 1)$ corresponds to the cycle $\{0\}$ which is not a circuit by the definition above.

**(70-31)**  Riho Terras (1976), *A stopping time problem on the positive integers*, Acta Arithmetica **30** (1976), 241–252. (MR 58 #27879).

This is the first significant research paper to appear that deals directly with the $3x + 1$ function. The $3x + 1$ function was however the motivation for the paper Conway (1972). The main result of this paper was obtained independently and contemporaneously by Everett (1977).

A positive integer $n$ is said to have *stopping time* $k$ if the $k$-th iterate $T^{(k)}(n) < n$, and $T^{(j)}(n) \geq n$ for $1 \leq j < k$. The author shows that the set of integers having stopping time $k$ forms a set of congruence classes (mod $2^k$), minus a finite number of elements. He shows that the set of integers having a finite stopping time has natural density one. Some further details of this proof were supplied later in Terras (1979).

This paper introduces the notion of the *coefficient stopping time* $\kappa(n)$ of an integer $n > 1$. Write $T^{(k)}(n) = \alpha(n)n + \beta(n)$ with $\alpha(n) = \frac{3^{a(n)}}{2^n}$, where $a(n)$ is the number of iterates $T^{(j)}(n) \equiv 1 (\mathrm{mod}\ 2)$ with $0 \le j < k$. Then $\kappa(n)$ is defined to be the least $k \ge 1$ such that $T^{(k)}(n) = \alpha(n)n + \beta(n)$ has $\alpha(n) < 1$, and $\kappa(n) = \infty$ if no such value exists. It is clear that $\kappa(n) \le \sigma(n)$, where $\sigma(n)$ is the stopping time of $n$. Terras formulates the *Coefficient Stopping Time Conjecture*, which asserts that $\kappa(n) = \sigma(n)$ for all $n \ge 2$. He proves this conjecture for all values $\kappa(n) \le 2593$. This can be done for $\kappa(n)$ below a fixed bound by upper bounding $\beta(n)$ and showing bounding $\kappa(n) < 1 - \delta$ for suitable $\delta$ and determining the maximal value $\frac{3^l}{2^j} < 1$ possible with $j$ below the given bound. The convergents of the continued fraction expansion of $\log_2 3$ play a role in determining the values of $j$ that must be checked.

**(70-32)**  Riho Terras (1979), *On the existence of a density*, Acta Arithmetica **35** (1979), 101–102. (MR 80h:10066).

This paper supplies additional details concerning the proof in Terras (1976) that the set of integers having an infinite stopping time has asymptotic density zero. The proof in Terras (1976) had been criticized by Möller (1978).

**(70-33)**  Robert Tijdeman (1972), *Note on Mahler's 3/2-problem*, Det Kongelige Norske Videnskabers Selskab Skrifter No. 16, 1972, 4 pages. (Zbl. 227: 10025.)

This paper concerns the Z-number problem of Mahler (1968), which asks whether there exists any nonzero real number $\eta$ such that the fractional parts $0 \le \{\{\eta(\frac{3}{2})^n\}\} \le \frac{1}{2}$ for all $n \ge 0$. By an elementary argument Tijdeman shows that analogues of $Z$-numbers exist in a related problem. Namely, for every $k \ge 2$ amd $m \ge 1$ there exists a real number $\eta \in [m, m+1)$ such that $0 \le \{\{\eta(\frac{2k+1}{2})^n\}\} \le \frac{1}{2k+1}$ holds for all $n \ge 0$.

**(70-34)**  Charles W. Trigg, Clayton W. Dodge and Leroy F. Meyers (1976), *Comments on Problem 133*, Eureka (now Crux Mathematicorum) **2**, No. 7 (August-Sept.) (1976), 144–150.

Problem 133 is the $3x + 1$ problem. It was proposed by K. S. Williams (Concordia Univ.), who said that he was shown it by one of his students. C. W. Trigg gives some earlier history of the problem. He remarks that Richard K. Guy wrote to him stating that Lothar Collatz had given a lecture on the problem at Harvard in 1950 (informally at the International Math. Congress). He reported that in 1970 H. S. M. Coxeter offered a prize of $50 for proving the $3x + 1$ Conjecture and $100 for finding a counterexample, in his talk: "Cyclic Sequences and Frieze Patterns" (The Fourth Felix Behrend Memorial Lecture in Mathematics), The University of Melbourne, 1970, see Coxeter (1971). He also referenced a discussion of the problem in several issues of *Popular Computing* No. 1 (April 1973) 1–2; No. 4 (July 1973) 6–7; No. 13 (April 1974) 12–13; No. 25 (April 1975), 4–5. Dodge references the work of Isard and Zwicky (1970).

*Note.* Lothar Collatz was present at the 1950 ICM as part of the DMV delegation, but did not give an official lecture at the ICM.

## 5. Bibliography 1980–1989

This decade includes the statements of Collatz (1986) and Thwaites (1985) to have invented the $3x + 1$ problem. It also includes the treatment of Conway (1987) on FRACTRAN.

**(80-1)**  Sergio Albeverio, Danilo Merlini and Remiglio Tartini (1989), *Una breve introduzione a diffusioni su insiemi frattali e ad alcuni essempi di sistemi dinamici semplici,* Note di matematica e fisica, Edizioni Cerfim Locarno **3** (1989), 1–39.

This paper discusses dimensions of some simple fractals, starting with the Sierpinski gasket and the Koch snowflake. These arise as a fixed set from combining several linear iterations. It then considers an "arithmetical fractal" the Collatz function iteration as analogous, as it is given by set of two linear transformations. It looks at the tree of inverse iterates ("chalice") and estimates emprically the number of leaves at depth at most $k$ as growing like $c^k$ for $c \approx 1.265$. It discusses various cascades of points that arrive at the cycle $\{1, 4, 2\}$.

It then looks at a "geometric fractal" : a probalistic iteration schemes of type: start with vertices of equilateral triangle, and a new point $x_0$ not on the triangle. Then form $f(x_0) = \frac{x_0 + x_k}{\xi}$ where $\xi > 0$ is a parameter. then iterate with $k$ varying in any order through $1, 2, 3$. Depending on the value of $\xi$ for $\xi \geq 2$ get an orbit supported on a set like the Sierpinski gasket, for $1 \leq \xi < 2$ get an orbit that diffuses through the plane, for $\xi = 1$ get an orbit on a lattice, and for $\xi < 1$ get a divergent orbit. Finally it discusses an iterative physical model, connects with statistical mechanics, critical exponents, the Ising model.

**(80-2)**  S. Anderson (1987), *Struggling with the $3x + 1$ problem,* Math. Gazette **71** (1987), 271–274.

This paper studies simple analogues of the $3x + 1$ function such as

$$g(x) = \begin{cases} x + k & \text{if } x \equiv 1 \ (\text{mod } 2) , \\ \frac{x}{2} & \text{if } x \equiv 0 \ (\text{mod } 2) . \end{cases}$$

For $k = 1$ when iterated this map gives the binary expansion of $x$. The paper also reformulates the $3x + 1$ Conjecture using the function:

$$f(x) = \begin{cases} \frac{x}{3} & \text{if } x \equiv 0 \ (\text{mod } 3) , \\ \frac{x}{2} & \text{if } x \equiv 2 \text{ or } 4 \ (\text{mod } 6) , \\ 3x + 1 & \text{if } x \equiv 1 \text{ or } \ (\text{mod } 6) . \end{cases}$$

**(80-3)**  Jacques Arsac (1986), *Algorithmes pour vérifier la conjecture de Syracuse,* C. R. Acad. Sci. Paris **303**,Serie I, no. 4, (1986), 155–159. [Also: RAIRO, Inf. Théor. Appl. **21** (1987), 3–9.] (MR 87m:11128).

This paper studies the computational complexity of algorithms to compute stopping times of $3x + 1$ function on all integers below a given bound $x$.

**(80-4)**  Jacek Błażewitz and Alberto Pettorossi (1983), *Some properties of binary sequences useful for proving Collatz's conjecture,* J. Found. Control Engr. **8** (1983), 53–63. (MR 85e:11010, Zbl. 547.10000).

This paper studies the $3x + 1$ Problem interpreted as a strong termination property of a term rewriting system. They view the problem as transforming binary strings into new binary strings and look in partcular at its action on the patterns $1^n$, $0^n$ and $(10)^n$ occurring inside strings. The $3x + 1$ map exhibits regular behavior relating these patterns.

**(80-5)**  David W. Boyd (1985), *Which rationals are ratios of Pisot sequences?*, Canad. Math. Bull. **28** (1985), 343–349. (MR 86j:11078).

The Pisot sequence $E(a_0, a_1)$ is defined by $a_{n+2} = \left[ \frac{a_{n+1}^2}{a_n} + \frac{1}{2} \right]$, where $a_0, a_1$ are integer starting values. If $0 < a_0 < a_1$ then $\frac{a_n}{a_{n+1}}$ converges to a limit $\theta$ as $n \to \infty$. The paper asks: which rationals $\frac{p}{q}$ can occur as a limit? If $\frac{p}{q} > \frac{q}{2}$ then $\frac{p}{q}$ must be an integer. If $\frac{p}{q} < \frac{q}{2}$ the question is related to a stopping time problem resembling the $3x + 1$ problem.

**(80-6)**  Charles C. Cadogan (1984), *A note on the $3x + 1$ problem*, Caribbean J. Math. **3** No. 2 (1984), 69–72. (MR 87a:11013).

Using an observation of S. Znam, this paper shows that to prove the $3x + 1$ Conjecture it suffices to check it for all $n \equiv 1 \pmod 4$. This result complements the obvious fact that to prove the $3x + 1$ Conjecture it suffices to check it for all $n \equiv 3 \pmod 4$. Korec and Znam (1987) obtained other results in this spirit, for odd moduli.

**(80-7)**  Thomas Cloney, Eric C. Goles and Gérard Y. Vichniac (1987), *The $3x + 1$ Problem: a Quasi-Cellular Automaton*, Complex Systems **1**(1987), 349–360. (MR 88d:68080).

The paper presents computer graphics pictures of binary expansions of $\{T^{(i)}(m) : i = 1, 2, \ldots\}$ for "random" large $m$, using black and white pixels to represent 1 resp. 0 (mod 2). It discusses patterns seen in these pictures. There are no theorems.

**(80-8)**  Lothar Collatz (1986), *On the Motivation and Origin of the $(3n + 1)$-Problem*, J. of Qufu Normal University, Natural Science Edition [Qufu shi fan da xue xue bao. Zi ran ke xue ban] **12** (1986) No. 3, 9–11 (Chinese, translated by Zhi-Ping Ren).

Lothar Collatz describes his interest since 1928 in iteration problems represented using associated graphs and hypergraphs. He describes the structure of such graphs for several different problems. He states that he invented the $3x + 1$ problem and publicized it in many talks. He says: "Because I couldn't solve it I never published anything. In 1952 when I came to Hamburg I told it to my colleague Prof. Dr. Helmut Hasse. He was very interested in it. He circulated the problem in seminars and in other countries."

*Note.* Lothar Collatz was part of the DMV (Deutsche Mathematiker Vereinigung) delegation to the 1950 International Congress of Mathematicians in Cambridge, Massachusetts. There H. S. M. Coxeter, S. Kakutani and S. M. Ulam were invited speakers, who had papers in the conference proceedings. Collatz reportedly described the problem at this time to Kakutani and others in private conversations, cf. Trigg et al (1976).

**(80-9)**   John H. Conway (1987), *FRACTRAN- A Simple Universal Computing Language for Arithmetic*, in: *Open Problems in Communication and Computation* (T. M. Cover and B. Gopinath, Eds.), Springer-Verlag: New York 1987, pp. 3–27. (MR 89c:94003).

FRACTRAN is a method of universal computation based on Conway's (1972) earlier analysis in "Unpredictable Iterations." Successive computations are done by multiplying the current value of the computation, a positive integer, by one of a finite list of fractions, according to a definite rule which guarantees that the resulting value is still an integer. A FRACTRAN program iterates a function $g(.)$ of the form

$$g(m) := \frac{p_r}{q_r} m \text{ if } m \equiv r \pmod{N},\ 0 \le r \le N - 1,$$

and each fraction $\frac{p_r}{q_r}$ is positive with denominator $q_r$ dividing $gcd(N, r)$, so that the function takes positive integers as positive integers. If the input integer is of the form $m = 2^n$ then the FRACTRAN program is said to *halt* at the first value encountered which is again a power of 2, call it $2^{f(n)}$. The output $f(n) = *$ is undefined if the program never halts. A FRACTRAN program is regarded as computing the partial recursive function $\{f(n)\ :\ n \in \mathbb{Z}_{>0}\}$. FRACTRAN programs can compute any partial recursive function. The paper gives a number of examples of FRACTRAN programs, e.g. for computing the decimal digits of $\pi$, and for computing the successive primes. The prime producing algorithm was described earlier in Guy (1983b).

Section 11 of the paper discusses generalizations of the 3x+1 problem encoded as FRACTRAN programs, including a fixed such function for which the halting problem is undecidable.

**(80-10)**   James M. Dolan, Albert F. Gilman and Shan Manickam (1987), *A generalization of Everett's result on the Collatz $3x + 1$ problem*, Adv. Appl. Math. **8** (1987), 405–409. (MR 89a:11018).

This paper shows that for any $k \ge 1$, the set of $m \in \mathbb{Z}+$ having $k$ distinct iterates $T^{(i)}(m) < m$ has density one.

**(80-11)**   Peter D. T. A. Elliott (1985), *Arithmetic Functions and Integer Products*, Springer-Verlag, New York 1985. (MR 86j:11095)

An *additive function* is a function with domain $\mathbb{Z}^+$, which satisfies $f(ab) = f(a) + f(b)$ if $(a, b) = 1$. In Chapters 1–3 Elliott studies additive functions having the property that $|f(an + b) - f(An + B)| \le c_0$ for all $n \ge n_0$, for fixed positive integers $a, b, A, B$ with $\det \begin{bmatrix} a\ b \\ A\ B \end{bmatrix} \ne 0$, and deduces that $|f(n)| \le c_1 (\log n)^3$. For the special case $\begin{bmatrix} a\ b \\ A\ B \end{bmatrix} = \begin{bmatrix} 1\ 1 \\ 1\ 0 \end{bmatrix}$ an earlier argument of Wirsching yields a bound $|f(n)| \le c_2 (\log n)$. On page 19 Elliott indicates that the analogue of Wirsching's argument for $\begin{bmatrix} a\ b \\ A\ B \end{bmatrix} = \begin{bmatrix} 3\ 1 \\ 1\ 0 \end{bmatrix}$ leads to the $3x + 1$ function, and implies that $|f(n)| \le c_3 \sigma_\infty(n)$. A strong form of the $3x + 1$ Conjecture claims that $\sigma_\infty(n) \le c_4 \log n$, see Lagarias and Weiss (1992). Elliott proves elsewhere by other arguments that in fact $|f(n)| \le c_4 \log n$ holds. [P. D. T. A. Elliott, J. Number Theory **16** (1983), 285–310.]

292     JEFFREY C. LAGARIAS

**(80-12)**  Paul Erdős and R. L. Graham (1980), *Old and new problems and results in combinatorial number theory* Monographie No. 28 de L'Enseignement Mathématique, Kundig: Geneva 1980.

This book includes Erdős and Graham (1979) as one chapter. Thus it includes problems raised in Klarner and Rado (1974), Hoffman (1976), and Hoffman and Klarner (1978), (1979) on the smallest set of nonnegative integers obtained from a given set $A$ under iteration of a finite set $R$ of functions $\rho(x_1, \ldots, x_r) = m_0 + m_1 x_1 + \cdots + m_r x_r$, with nonnegative integer $m_i$ for $i \geq 1$. Denoting this set $< R : A >$, one can ask for the size and structure of this set. In the case of one variable functions $R = \{a_1 x + b_1, \ldots, a_r x + b_r\}$ Erdős showed (see Klarner and Rado (1974)) that if $\sum \frac{1}{a_i} < 1$, then the set has density 0. The case when $\sum \frac{1}{a_i} = 1$ is pointed to as a source of unresolved problems. Erdős had proposed as a prize problem: For $R = \{2x + 1, 3x + 1, 6x + 1\}$ and $A = \{1\}$ is the set $< R : A >$ of positive density? This was answered in the negative by D. J. Crampin and A. J. W. Hilton (unpublished), as summarized in Klarner (1982) and Klarner (1988).

**(80-13)**  Lynn E. Garner (1981), *On the Collatz $3n + 1$ algorithm*, Proc. Amer. Math. Soc. **82** (1981), 19–22. (MR 82j:10090).

The *coefficient stopping time* $\kappa(n)$ introduced by Terras (1976) is the least iterate $k$ such that $T^{(k)}(n) = \alpha(n)n + \beta(n)$, with $\alpha(n) < 1$. Here $\alpha(n) = \frac{3^{a(n)}}{2^n}$ where $a(n)$ is the number of iterates $T^{(j)}(n) \equiv 1 (\bmod\ 2)$ with $0 \leq j < k$. One has $\kappa(n) \leq \sigma(n)$, where $\sigma(n)$ is the stopping time of $n$, and the Coefficient Stopping Time Conjecture of Terras (1976) asserts that $\kappa(n) = \sigma(n)$ for all $n \geq 2$. This paper proves that $\kappa(n) < 105,000$ implies that $\kappa(n) \leq \sigma(n)$. The proof methods used are those of Terras (1976), who proved the conjecture holds for $\kappa(n) < 2593$. They invove the use of the continued fraction expansion of $\log_2 3$ and the truth of the $3x + 1$ Conjecture for $n < 2.0 \times 10^9$.

**(80-14)**  Lynn E. Garner (1985), *On heights in the Collatz $3n + 1$ problem,* Discrete Math. **55** (1985), 57–64. (MR 86j:11005).

This paper shows that infinitely many pairs of consecutive integers have equal (finite) heights and equal total stopping times. To do this he studies how trajectories of consecutive integers can coalesce. Given two consecutive integers $m, m + 1$ having $T^{(i)}(m) \neq T^{(i)}(m + 1)$ for $i < k$ and $T^{(k)}(m) = T^{(k)}(m + 1)$, associate to them the pair $(\mathbf{v}, \mathbf{v}')$ of $0 - 1$ vectors of length $k$ encoding the parity of $T^{(i)}(m)$ (resp. $T^{(i)}(m + 1)$) for $0 \leq i \leq k - 1$. Call the set $\mathcal{A}$ of pairs $(\mathbf{v}, \mathbf{v}')$ obtained this way *admissible pairs*. Garner exhibits collections $\mathcal{B}$ and $\mathcal{S}$ of pairs of equal-length $0 - 1$ vectors $(\mathbf{b}, \mathbf{b}')$ and $(\mathbf{s}, \mathbf{s}')$ called *blocks* and *strings*, respectively, which have the properties: If $(\mathbf{v}, \mathbf{v}') \in \mathcal{A}$ and $(\mathbf{b}, \mathbf{b}') \in \mathcal{B}$ then the concatenated pair $(\mathbf{bv}, \mathbf{b'v'}) \in \mathcal{A}$, and if $(\mathbf{s}, \mathbf{s}') \in \mathcal{S}$ then $(\mathbf{sv'}, \mathbf{s'v}) \in \mathcal{A}$. Since $(001, 100) \in \mathcal{A}$, $(10, 01) \in \mathcal{B}$ and $(000011, 101000) \in \mathcal{S}$, the set $\mathcal{A}$ is infinite. He conjectures that: (1) a majority of all positive integers have the same height as an adjacent integer (2) arbitrarily long runs of integers of the same height occur.

**(80-15)**  Wolfgang Gaschütz (1982), *Linear abgeschlossene Zahlenmengen I.* [Linearly closed number sets I.] J. Reine Angew. Math. **330** (1982), 143–158. (MR 83m:10095).

This paper studies subsets $S$ of $\mathbb{Z}$ or $\mathbb{N}$ closed under iteration of an affine map $f(x_1, \cdots, x_r) = w_0 + w_1 x_1 + w_2 x_2 + \cdots + w_r x_r$ with integer $w_0, w_1, \ldots, w_r$. He shows for a general function one can reduce analysis to the case $w_0 = 1$ and initial seed value 0. A polynomial $f(x_1, \ldots, x_r) \in \mathbb{N}[x_1, \ldots, x_r]$ is *controlled* ("gebremst") if $f(x_1, \ldots, x_r) \preccurlyeq (x_1 + \cdots + x_r)f(x_1, \ldots, x_r) + f(0, 0, \ldots, 0)$, where $f \preccurlyeq g$ means each coefficient of $f$ is no larger than the corresponding coefficient of $g$. It is *m-controlled* if $f(0, 0, \cdots, 0) \le m$. Let $F_{r,m}$ denote the set of $m$-controlled polynomials in $r$ variables, and let $F_{r,m}(w_1, \ldots, w_r) = \{f(w_1, \ldots, w_r) : f \in F_{r,m}\}$. Theorem 4.1 asserts that for $f := 1 + w_1 x_1 + \cdots + w_r x_r$ the smallest set containing 0 and closed under iteration of $f$ is $F_{r,1}(w_1, w_2, \ldots, w_r)$. He also shows that if the greatest common divisor of $(w_1, \ldots, w_r) = 1$ then (i) if $w_i$ are nonnegative then $F_{r,m}(w_1, .., w_r) = \mathbb{N}$ for all large enough $m$; (ii) if some $w_i$ is negative and some other $w_i$ nonzero then for $F_{r,m}(w_1, .., w_r) = \mathbb{Z}$ for large enough $m$. He applies these results to obtain a characterization of those functions of two variables $f(x_1, x_2) = w_0 + w_1 x_1 + w_2 x_2$ such that the smallest set of integers containing 0 and closed under its action consists of all nonnegative integers $\mathbb{N}$. They are exactly those functions with $w_0 = 1$, $w_1, w_2 \ge 0$, with greatest common divisor $(w_1, w_2) = 1$, having the extra property that iteration (mod $w_1 w_2$) visits all residue classes (mod $w_1 w_2$). He also obtains a criterion for the smallest set to be $\mathbb{Z}$. He notes that similar results were obtained earlier by Hoffman and Klarner (1978), (1979).

*Note.* This study was motivated by the author's earlier work on single-word criteria for subgroups of abelian groups: W. Gaschütz, *Untergruppenkriterien für abelsche Gruppen*, Math. Z. **146** (1976), 89–99.

**(80-16)**   H. Glaser and Hans-Georg Weigand (1989), *Das-ULAM Problem- Computergestütze Entdeckungen,* DdM (Didaktik der Mathematik) **17**, No. 2 (1989), 114–134.

This paper views the $3x+1$ problem as iterating the Collatz function, and views it as an algorithmic problem between mathematics and computer science. It views study of this problem as useful as training in exploration of mathematical ideas. It formulates exploration as a series of questions to ask about it, and answers some of them. Some of these concern properties of the trees of inverse iterates of the Collatz function starting from a given number. It proves branching properties of the trees via congruence properties modulo powers of 3. This paper also discusses programming the $3x+1$ iteration in the programming languages Pascal and LOGO.

**(80-17)**   Richard K. Guy (1981), *Unsolved Problems in Number Theory*, Springer-Verlag, New York 1981. [Second edition: 1994. Third Edition: 2004.]

Problem E16 discusses the $3x+1$ Problem. Problem E17 discusses permutation sequences, includes Collatz's original permutation, see Klamkin (1963). Problem E18 discusses Mahler's Z-numbers, see Mathler (1968). Problem E36 (in second edition) discusses Klarner-Rado sequences, see Klarner and Rado (1974).

*Note.* Richard Guy (private conversation) informs me that he first heard of the problem in the early 1960's from his son Michael Guy, who was a student at Cambridge University and friends with John Conway. John Conway (private conversation) confirms that he heard of the problem and worked on it as a Cambridge undergraduate (BA 1959).

**(80-18)**   Richard K. Guy (1983a) *Don't try to solve these problems!*, Amer. Math. Monthly **90** (1983), 35–41.

The article gives six unsolved questions on iterations of arithmetic maps of $3x + 1$ type, one of which is the $3x + 1$ problem itself. It mentions at second hand a statement of P. Erdős regarding the $3x + 1$ problem: "Mathematics is not yet ripe enough for such questions." The $3x + 1$ problem appears as Problem 2. Problem 3 concerns cycles in the original Collatz problem, for which see Klamkin (1963). Problem 4 asks the question : Let S be the smallest set of positive integers containing 1 which is closed under $x \mapsto 2x$, $x \mapsto 3x + 2$, $x \mapsto 6x + 3$. Does this set have a positive lower density?

**(80-19)**   Richard K. Guy (1983b), *Conway's prime producing machine*, Math. Magazine **56** (1983), 26–33. (MR 84j:10008).

This paper gives a $3x + 1$-like function $g(\cdot)$ of the type in Conway (1972) having the following property. If $p_j$ denotes the $j$-th prime, given in increasing order, then starting from the value $n = 2^{p_j}$ and iterating under $g(\cdot)$, the first power of 2 that is encountered in the iteration is $2^{p_{j+1}}$.

He shows that the associated register machine uses only four registers. See also the paper Conway (1987) on FRACTRAN.

**(80-20)**   Richard K. Guy (1986), *John Isbell's Game of Beanstalk and John Conway's Game of Beans Don't Talk,* Math. Magazine **59** (1986), 259–269. (MR 88c:90163).

John Isbell's game of Beanstalk has two players alternately make moves using the rule

$$n_{i+1} = \begin{cases} \frac{n_1}{2} & \text{if } n_i \equiv 0 \ (\text{mod } 2) , \\ 3n_i \pm 1 & \text{if } n_i \equiv 1 \ (\text{mod } 2) , \end{cases}$$

where they have a choice if $n_i$ is odd. The winner is the player who moves to 1. In Conway's game the second rule becomes $\frac{3n\pm1}{2^*}$, where $2^*$ is the highest power of 2 that divides the numerator. It is unknown whether or not there are positions from which neither player can force a win. If there are then the $3x + 1$ problem must have a nontrivial cycle or a divergent trajectory.

**(80-21)**   Brian Hayes (1984), *Computer recreations: The ups and downs of hailstone numbers*, Scientific American **250**, No. 1 (January 1984), 10–16.

The author introduces the $3x + 1$ problem to a general audience under yet another name — hailstone numbers.

**(80-22)**   Hong, Bo Yang (1986), *About 3X + 1 problem* (Chinese), J. of Hubei Normal University, Natural Science Edition, [Hubei shi fan xue yuan xue bao. Zi ran ke xue ban] (1986), No. 1, 1–5.

Theorem 1 shows that if there is a positive odd number such that the $3x + 1$ conjecture fails for it, then there are infinitely many such odd numbers, and the smallest such number must belong to one of the congruence classes $7, 15, 27, 31 \ (\text{mod } 32)$. Theorem 2 shows that if there is a positive odd number such that the $3x + 1$ conjecture fails for it, then there is such an odd number that is 53 (mod 64).

**(80-23)**  Frazer Jarvis (1989), *13, 31 and the $3x + 1$ problem*, Eureka **49** (1989), 22–25.

This paper studies the function $g(n) = h(n + 1) - h(n)$, where $h(n)$ is the height of $n$, and observes empirically that $g(n)$ appears unusually often to be representable as $13x + 31y$ with small values of $x$ and $y$. It offers a heuristic explanation of this observation in terms of Diophantine approximations to $\log_6 2$. Several open problems are proposed, mostly concerning the height function $h(n)$.

**(80-24)**  David A. Klarner (1981), *An algorithm to determine when certain sets have 0 density*, Journal of Algorithms **2** (1981), 31–43. (MR 84h:10076).

This paper studies sets of integers that are closed under the iteration of certain one variable affine maps. It considers the case when all maps $f_i(x) = mx + a_i$ are expanding maps with the same ratio $m \geq 2$. The problem is reduced to the study of sets $S(c) = \langle f_1, \ldots, f_k : c \rangle$ which denotes the closure under iteration of of these maps starting from a single seed element $c$. The *density* $\delta(S)$ of a sequence of nonnegative integers is the lower asymptotic density

$$\delta(S) := \liminf_{n \to \infty} \frac{1}{n+1} |S \cap \{0, 1, \ldots, n\}|.$$

If $k < m$ then each sequence $S(c)$ automatically has zero density. If $k = m$ and the semigroup of affine maps having generators $A := \{f_i : 1 \leq i \leq m\}$ has a nontrivial relation, then all sequences $S(c)$ have zero density, while if this semigroup is free on $k$ generators, then they have positive density. The paper gives an algorithm to determine whether a sequence $S(c)$ has zero density that works for all $k$. It is based on the fact that the number $n_t := |A^t / \sim|$ of distinct affine functions obtained by composition exactly $t$ times satisfies a linear homogeneous difference equation with constant coefficients. It follows that the generating function $F(z) := \sum_{t=0}^{\infty} n_t z^t$ is a rational function $F(z) = \frac{P(z)}{Q(z)}$ for relatively prime polynomials $P(z)$, $Q(z)$. The problem is reduced to testing whether $Q(\frac{1}{m}) = 0$ holds.

**(80-25)**  David A. Klarner (1982), *A sufficient condition for certain semigroups to be free*, Journal of Algebra **74** (1982), 140–148. (MR 83e:10081).

This paper studies sets of integers that are closed under the iteration of certain one variable affine maps $f_i(x) = m_i x + b_i$, where $m_i, b_i$ are integers with $m_i \geq 2$. It determines sufficient conditions for the semigroup generated by the maps $f_i$ under composition to be a free semigroup. As motivation, it mentions the Erdös problem asking whether the set of integers generated starting from seed $S = \{1\}$ by the maps $f_1(x) = 2x + 1$, $f_2(x) = 3x + 1$ and $f_3(x) = 6x + 1$ is of positive density. This problem was solved by D. J. Crampin and A. J. W. Hilton (unpublished) who observed the density must be zero using the fact that the semigroup generated by these functions is not free. The paper Klarner (1981) gives an effective algorithm for the case when all $m_i$ are equal to an integer $m \geq 2$. This paper first observes that a necessary condition for freeness is that $\frac{1}{m_1} + \frac{1}{m_2} + \cdots + \frac{1}{m_k} \leq 1$. Next renumber the functions so that the quantities $p_j = \frac{b_j}{m_j - 1}$ satisfy $p_1 \leq p_2 \leq \cdots \leq p_k$. Strict inequalities now required here for freeness because an equality $p_j = p_{j+1}$ implies that the generators $f_j$ and $f_{j+1}$ commute, giving a nontrivial relation. A sufficient condition for freeness is that on each collection $L(i_1, \ldots, i_k)$ of compositions using exactly

$i_j$ generators of type $j$ the linear ordering of all these functions evaluated at 0 coincides with the lexicographic ordering induced on functional composition, ordering the generators as $f_1 < f_2 < \cdots < f_k$. Theorem 2.1 characterizes the latter condition, showing that these two orderings coincide if and only if, the $p_j$ satisfy the two conditions (i) $p_1 < p_2 < \cdots < p_k$ and, (ii) $\frac{p_k+b_j}{m_j} \le \frac{p_1+b_{j+1}}{m_{j+1}}$ holds for $1 \le j \le k-1$. It then deduces some examples where the semigroups are free. These include $f_1(x) = 2x + b_1$, $f_2(x) = 3x + b_2$, $f_3(x) = 6x + b_3$) for the six cases $(b_1, b_2, b_3) = (0, 3, 10)$, $(0, 2, 3)$, $(2, 0, 15)$, $(1, 0, 2)$, $(2, 1, 0)$, and $(1, 6, 0)$. The second of these cases gives the functions $2x, 3x + 2, 6x + 3$ posed as unsolved Problem 4 in Guy (1983a).

*Note.* Anthony J. W. Hilton (private communication) reports that Erdős and Klarner were both at University of Reading about 1971, and from their discussions Erdős formulated the problem above. He could not immediately solve it, and offered £10 for a solution. A. J. W. Hilton thought of an approach for finding a proof of density 0 by finding a relation in the semigroup, which was completed with calculations by Dorothy Joan Crampin. They each received a cheque. He reports: "I was so broke at the time that I immediately cashed mine after xeroxing it." He also reports that Klarner's original interest in iterating affine functions was stimulated by their work on another combinatorial problem. D. J. Crampin and A. J. W. Hilton used iteration of (vector-valued) affine functions to show the existence of $n \times n$ Latin squares orthogonal to their transpose of all sufficiently large $n$. Roughly speaking they used constructions that took Latin squares of one size (with extra features) and used them to build Latin rectangles of larger sizes, whose sizes were affine functions of the size of the original squares They could thus produce suitable Latin squares of sizes generated by various vector-valued affine functions, and needed to show the resulting sizes included every sufficiently large integer. A computer calculation was required, and is described in: D. J. Crampin and A. J. W. Hilton, *Remarks on Sade's disproof of the Euler conjecture with an application to Latin squares orthogonal to their transpose*, J. Comb. Theory, Series A, **18** (1975), 47–59.

**(80-26)** David A. Klarner (1988), *m-recognizability of sets closed under certain affine functions*, Discrete Applied Mathematics **21** (1988), no. 3, 207–214. (MR 90m:68075)

This paper studies sets of integers $T$ that are closed under the iteration of certain one variable affine maps. It assumes that all maps have the special form $f_i(x) = m^{e_i}x + a_i$, all with the same $m \ge 2$, with all $e_i \ge 1, a_i \ge 0$. The author writes $T = \langle A : S \rangle$, where $A$ denotes the finite set of maps and $S$ a finite set of "seeds". The main idea of the paper is to study the base $m$-representations of the integers and to show these are described by languages accepted by a finite automaton; such sets are called here *m-recognizable*. This is the content of Theorem 1. This is exhibited on the example $f_1(x) = 3x$, $f_2(x) = 3x + 1$, $f_3(x) = 3x + 4$. Theorem 2 asserts that if $S$ is an $m$-recognizable set, then so is $T = \langle A : S \rangle$. Theorem 3 asserts that if $T$ is $m$-recognizable so is the translated set $T + h$.

At the end of the paper more general cases are discussed. He remarks that for $f_1(x) = 2x + 1$, $f_2(x) = 3x + 1$, $f_3(x) = 6x + 1$ with $S = \{1\}$, suggested by Erdős, the set $T = \langle A : S \rangle$ has no discernible structure. He mentions that

Crampin and A. W. S. Hilton (unpublished) showed that this sequence has (lower asymptotic) density zero, answering a question of Erdős. They used the existence of the nontrivial relation under composition

$$f_1 \circ f_1 \circ f_2 = f_3 \circ f_1 = 12x + 7,$$

of these affine functions, so that $A$ is not a free semigroup under composition in this case. Klarner also notes that for the functions considered in this paper there is an effective algorithm to test whether the resulting set has density zero, generalizing results in Klarner (1981).

*Note.* This paper shows these sets are *m-automatic sequences*, in the terminology of J.-P. Allouche and J. O. Shallit, *Automatic Sequences*, theory, applications, generalizatins. Cambridge University Press, Cambridge 2003.

**(80-27)** Ivan Korec and Stefan Znam (1987), *A Note on the $3x + 1$ Problem,* Amer. Math. Monthly **94** (1987), 771–772. (MR 90g:11023).

This paper shows that to prove the $3x + 1$ Conjecture it suffices to verify it for the set of all numbers $m \equiv a \pmod{p^n}$, for any fixed $n \geq 1$, provided that 2 is a primitive root $\pmod{p}$ and $(a, p) = 1$. This set has density $p^{-n}$.

**(80-28)** Ilia Krasikov (1989), *How many numbers satisfy the $3x + 1$ Conjecture?,* Internatl. J. Math. & Math. Sci. **12** (1989), 791–796. (MR 91c:11013).

This paper shows that the number of integers $\leq x$ for which the $3x + 1$ function has an iterate that is 1 is at least $x^{3/7}$. More generally, if $\theta_a(x) = \{n \leq x : \text{ some } T^{(k)}(n) = a\}$, then he shows that, for $a \not\equiv 0 \pmod{3}$, $\theta_a(x)$ contains at least $x^{3/7}$ elements, for large enough $x$. For each fixed $k \geq 2$ this paper derives a system of difference inequalities based on information $\pmod{3^k}$. The bound $x^{3/7}$ was obtained using $k = 2$, and by using larger values of $k$ better exponents can be obtained. This was done in Applegate and Lagarias (1995b) and Krasikov and Lagarias (2003).

**(80-29)** Jeffrey C. Lagarias (1985), *The $3x + 1$ problem and its generalizations,* Amer. Math. Monthly **92** (1985), 3–23. (MR 86i:11043).

This paper is a survey with extensive bibliography of known results on $3x+1$ problem and related problems up to 1984. It also contains improvements of previous results and some new results, including in particular Theorems D, E, F, L and M.

Theorem O has several misprints. The method of Conway (1972) gives, for any partial recursive function $f$, a periodic piecewise linear function $g$, with the property: iterating $g$ with the starting value $2^n$ will never be a power of 2 if $f(n)$ is undefined, and will eventually reach a power of 2 if $f(n)$ is defined, and the first such power of 2 will be $2^{f(n)}$. Thus, in parts (ii) and (iii) of Theorem O, occurrences of $n$ must be replaced by $2^n$. On page 15, the thirteenth partial quotient of $\log_2(3)$ should be $q_{13} = 190537$.

**(80-30)** George M. Leigh (1986), *A Markov process underlying the generalized Syracuse algorithm,* Acta Arithmetica **46** (1986), 125–143. (MR 87i:11099)

This paper considers mappings $T(x) = \frac{m_i x - r_i}{d}$ if $x \equiv i \pmod{d}$, where all $r_i \equiv i m_i \pmod{d}$. This work is motivated by earlier work of Matthews and Watts (1984), (1985), and introduces significant new ideas.

Given such a mapping $T$ and an auxiliary modulus $m$ it introduces two Markov chains, denoted $\{X_n\}$ and $\{Y_n\}$, whose behavior encodes information on the iterates of $T$ (modulo $md^k$) for all $k \geq 1$. In general, both Markov chains have a countable number of states; however in many interesting cases both these chains are finite state Markov chains. The states of each chain are labelled by certain congruence classes $B(x_j, M_j) := \{x : x \equiv x_j(\text{mod } M_j)\}$ with $M_j|md^k$ for some $k \geq 1$. (In particular $B(x_j, M_j) \subset B(x_k, M_k)$ may occur.) Any finite path in such a chain can be realized by a sequence of iterates $x_j = T^j(x_0)$ for some $x_0 \in \mathbb{Z}$ satisfying the congruences specified by the states. The author shows the two chains contain equivalent information; the chain $\{X_n\}$ has in general fewer states than the chain $\{Y_n\}$ but the latter is more suitable for theoretical analysis, in particular one chain is finite if and only if the other one is. If such a chain has a positive ergodic class of states, then the limiting frequency distribution of state occupation of a path in the chain in this class exists, and from this the corresponding frequencies of iterates in a given congruence class (mod $m$) can be calculated (Theorem 1- Theorem 4).

For applications to the map $T$ on the integers, the author 's guiding conjecture is: *any condition that occurs with probability zero in the Markov chain model does not occur in divergent trajectories on $T$.* Thus the author conjectures that in cases where the Markov chain is finite, there are limiting densities that divergent trajectories for $T$ (should they exist) spend in each residue class $j$ (mod $m$), obtained from frequency distribution in ergodic states of the chain.

The author gets a complete analysis whenever the associated Markov chains are finite. Furthermore Theorem 7 shows that any map $T$ satisfying the condition $gcd(m_i, d^2) = gcd(m_i, d)$ for $0 \leq i \leq d-1$, has, for every modulus $m$, both Markov chains $\{X_n\}$ and $\{Y_n\}$ being finite chains. All maps studied in the earlier work of Matthews and Watts (1984, 1985) satisfy this condition, and the author recovers nearly all of their results in these cases.

In Sect. 5 the author suggests that when infinite Markov chains occur, that one approximate them using a series of larger and larger finite Markov chains obtained by truncation. Conjecture 2 predicts limiting frequencies of iterates (mod $m$) for the map $T$, even in these cases. The paper concludes with several worked examples.

*Note.* This Markov chain approach was extended further by Venturini (1992).

**(80-31)** Heinrich Lunkenheimer (1988), *Eine kleine Untersuchung zu einem zahlentheoretischen Problem*, PM (Praxis der Mathematik in der Schule) **30** (1988), 4–9.

The paper considers the Collatz problem, giving no prior history or references. It observes the coalescences of some orbits. It lists some geometric series of integers which iterate to 1.

**(80-32)** Keith R. Matthews and George M. Leigh (1987), *A generalization of the Syracuse algorithm to $F_q[x]$*, J. Number Theory **25** (1987), 274–278. (MR 88f:11116).

This paper defines mappings analogous to the $3x+1$ function on polynomials over finite fields, e.g. $T(f) = \frac{f}{x}$ if $f \equiv 0$ (mod $x$) and $\frac{(x+1)^3 f + 1}{x}$ if $f \equiv 1$ (mod

$x$), over GF(2). It proves that divergent trajectories exist for certain such maps. These divergent trajectories have a regular behavior.

**(80-33)**   Keith R. Matthews and Anthony M. Watts (1984), *A generalization of Hasse's generalization of the Syracuse algorithm*, Acta. Arithmetica **43** (1984), 167–175. (MR 85i:11068).

This paper studies functions $T(x) = \frac{m_i x - r_i}{d}$ for $x \equiv i$ ( mod  $d$), where all $m_i$ are positive integers and $r_i \equiv i m_i$ (mod  $d$). It is shown that if $\{T^{(k)}(m) : k \geq 0\}$ is unbounded and uniformly distributed (mod  $d$) then $m_1 m_2 \cdots m_d > d^d$ and $\lim_{k \to \infty} |T^{(k)}(m)|^{1/k} = \frac{1}{d}(m_1 \cdots m_d)^{1/d}$. The function $T$ is extended to a mapping on the $d$-adic integers and is shown to be strongly mixing, hence ergodic, on $\mathbb{Z}_d$. The trajectories $\{T^{(k)}(\omega : k \geq 0\}$ for almost all $\omega \in \mathbb{Z}_d$ are equidistributed (mod  $d^k$) for all $k \geq 1$.

**(80-34)**   Keith R. Matthews and Anthony M. Watts (1985), *A Markov approach to the generalized Syracuse algorithm,* Acta Arithmetica **45** (1985), 29–42. (MR 87c:11071).

This paper studies the functions $T(x) = \frac{m_i x - r_i}{d}$ for $x \equiv i$ (mod  $d$), where all $m_i$ are positive integers and $r_i \equiv i m_i$ (mod  $d$), which were considered in Matthews and Watts (1984). Given a modulus $m$ one associates to $T$ a row-stochastic matrix $Q = [q_{jk}]$ in which $j, k$ index residue classes ( mod  $m$) and $q_{jk}$ equals $\frac{1}{md}$ times the number of residue classes (mod  $md$) which are $\equiv k$(mod  $m$) and whose image under $T$ is $\equiv j$(mod  $m$). It gives sufficient conditions for the entries of $Q^l$ to be the analogous probabilities associated to the iterated mapping $T^{(l)}$. Matthews and Watts conjecture that if $\mathcal{S}$ is an ergodic set of residues (mod  $m$) and $(\alpha_i; \ i \in \mathcal{S})$ is the corresponding stationary vector on $\mathcal{S}$, and if $A = \Pi_{i \in \mathcal{S}} \left(\frac{m_i}{d}\right)^{\alpha_i} < 1$ then all trajectories of $T$ starting in $\mathcal{S}$ will eventually be periodic, while if $A > 1$ almost all trajectories starting in $\mathcal{S}$ will diverge. Some numerical examples are given.

**(80-35)**   Władysław Narkiewicz (1980), *A note on a paper of H. Gupta concerning powers of 2 and 3*, Univ. Beograd. Publ. Elecktrotech. Fak. Ser. Mat. Fiz. No **678-715** (1980), 173-174.

P. Erdős raised the question: "Does there exist an integer $m \neq 0, 2, 8$ such that $2^m$ is a sum of distinct integral powers of 3?" This was motivated by work of H. Gupta, *Powers of 2 and sums of distinct powers of 3*, Univ. Beograd. Publ. Elecktrotech. Fak. Ser. Mat. Fiz. Nos. **602–633** (1980), 151–158. Gupta checked numerically that the only solutions were $m = 0, 2$ and 8, for $m \leq 4734$. This paper shows that if $N(T)$ denotes the number of such $m \leq T$ then $N(T) \leq 1.62 T^\theta$ where $\theta = \frac{\log 2}{\log 3}$. Diophantine properties of powers of 2 and 3 play a role in analyzing periodic orbits of the $3x + 1$ problem and similar iterations, cf. Steiner (1978). However the Erdős question arose in another context.

**(80-36)**   Clifford A. Pickover (1989), *Hailstone $3n + 1$ Number graphs*, J. Recreational Math. **21** (1989), 120–123.

This paper gives two-dimensional graphical plots of $3x + 1$ function iterates revealing "several patterns and a diffuse background of chaotically-positioned dots."

**(80-37)**  Susana Puddu (1986), *The Syracuse problem (Spanish)*, 5th Latin American Colloq. on Algebra – Santiago 1985, Notas Soc. Math. Chile **5** (1986), 199–200. (MR88c:11010).

This note considers iterates of the Collatz function $C(x)$. It shows every positive $m$ has some iterate $C^k(m) \equiv 1$ (mod4). If $m \equiv 3$ (mod4) the smallest such $k$ must have $C^k(m) \equiv 5$ (mod 12).

**(80-38)**  Daniel A. Rawsthorne (1985), *Imitation of an iteration,* Math. Magazine **58** (1985), 172–176. (MR 86i:40001).

This paper proposes a multiplicative random walk that models imitating the "average" behavior of the $3x+1$ function and similar functions. It compares the mean and standard deviation that this model predicts with empirical $3x + 1$ function data, and with data for several similar mappings, and finds good agreement between them.

**(80-39)**  H. J. J. te Riele (1983a), *Problem 669*, Nieuw Archief voor Wiskunde, Series IV, **1** (1983), p. 80.

This problem asks about the iteration of the function

$$f(x) = \begin{cases} \frac{n}{3} & \text{if } n \equiv 0 \ (\text{mod } 3) \\ \lfloor n\sqrt{3} \rfloor & \text{if } n \not\equiv 0 \ (\text{mod } 3) \ . \end{cases}$$

on the positive integers. The value $n = 1$ is a fixed point , and all powers $n = 3^k$ eventually reach this fixed point. The problem asks to show that if two consecutive iterates are not congruent to 0 (mod 3) then the trajectory of this orbit thereafter grows monotonically, so diverges to $+\infty$; otherwise the orbit reaches 1.

*Note.* Functions similar in form to $f(x)$, but with a condition (mod 2), were studied by Mignosi (1995) and Brocco (1995).

**(80-40)**  H. J. J. te Riele (1983b), *Iteration of Number-theoretic functions*, Nieuw Archief voor Wiskunde, Series IV, **1** (1983), 345–360. [MR 85e:11003].

The author surveys work on a wide variety of number-theoretic functions which take positive integers to positive integers, whose behavior under iteration is not understood. This includes the $3x + 1$ function (III.1) and the $qx + 1$ function (III.2). The author presents as Example 1 the function in te Riele (1983a),

$$f(x) = \begin{cases} \frac{n}{3} & \text{if } n \equiv 0 \ (\text{mod } 3) \\ \lfloor n\sqrt{3} \rfloor & \text{if } n \not\equiv 0 \ (\text{mod } 3) \ . \end{cases}$$

He showed that this function has divergent trajectories, and that all non-divergent orbits converge to the fixed point 1. He conjectures that almost all orbits on positive integers tend to $+\infty$. Experimentally only 459 values of $n < 10^5$ have orbits converging to 1.

**(80-41)**  Benedict G. Seifert (1988), *On the arithmetic of cycles for the Collatz-Hasse ('Syracuse') conjectures,* Discrete Math. **68** (1988), 293–298. (MR 89a: 11031).

This paper gives criteria for cycles of $3x + 1$ function to exist, and bounds the smallest number in the cycle in terms of the length of the cycle. Shows that if the $3x + 1$ Conjecture is true, then the only positive integral solution of $2^l - 3^r = 1$ is $l = 2, r = 1$.

*Note.* All integer solutions of $2^l - 3^r = 1$ have been found unconditionally. This can be done by a method of C. Størmer, Nyt. Tidsskr. Math. B **19** (1908) 1–7. It was done as a special case of various more general results by S. Pillai, Bull. Calcutta Math. Soc. **37** (1945), 15–20 (MR# 7, 145i); W. LeVeque, Amer. J. Math. **74** (1952), 325–331 (MR 13, 822f); R. Hampel, Ann. Polon. Math. **3** (1956), 1–4 (MR 18, 561c); etc.]

**(80-42)** Ray P. Steiner (1981a), *On the "$Qx + 1$" Problem, Q odd,* Fibonacci Quarterly **19** (1981), 285–288. (MR 84m:10007a)

This paper studies the $Qx + 1$-map

$$h(n) = \begin{cases} \frac{Qn+1}{2} & \text{if } n \equiv 1 \ (\text{mod } 2), \ n > 1 \\ \frac{n}{2} & \text{if } n \equiv 0 \ (\text{mod } 2) \\ 1 & \text{if } n = 1, \end{cases}$$

when $Q$ is odd and $Q > 3$. It proves that the only circuit which is a cycle when $Q = 5$ is $\{13, 208\}$, that there is no circuit which is a cycle for $Q = 7$. Baker's method is again used, as in Steiner (1978).

**(80-43)** Ray P. Steiner (1981b), *On the "$Qx + 1$" Problem, Q odd II,* Fibonacci Quarterly **19** (1981), 293–296. (MR 84m:10007b).

This paper continues to study the $Qx + 1$-map

$$h(n) = \begin{cases} \frac{Qn+1}{2} & \text{if } n \equiv 1 \ (\text{mod } 2), n > 1 \\ \frac{n}{2} & \text{if } n \equiv 0 \ (\text{mod } 2) \\ 1 & \text{if } n = 1, \end{cases}$$

when $Q$ is odd and $Q > 3$. It makes general remarks on the case $Q > 7$, and presents data on from the computation of $\log_2 \frac{5}{2}$ and $\log_2 \frac{7}{2}$ used in the proofs in part I.

**(80-44)** Bryan Thwaites (1985), *My conjecture,* Bull. Inst. Math. Appl. **21** (1985), 35–41. (MR86j:11022).

The author states that he invented the $3x + 1$ problem in 1952. He derives basic results about iterates, and makes conjectures on the "average" behavior of trajectories.

*Note.* The mathematical community generally credits L. Collatz as being the first to propose the $(3x + 1)$-problem, see the comment of Shanks (1965) and the paper of Collatz (1986). This would be an independent discovery of the problem.

**(80-45)** Giovanni Venturini (1982), *Sul Comportamento delle Iterazioni di Alcune Funzioni Numeriche,* Rend. Sci. Math. Institute Lombardo **A 116** (1982), 115–130. (MR 87i:11015; Zbl. 583.10009).

The author studies functions $g(n) = a_r n + b_r$ for $n \equiv r \ (\text{mod } p)$ where $a_r$ $(0 \le r \le p)$ are positive rationals with denominator $p$. He mainly treats the case that the $a_r$ take two distinct values. If $\tau = (a_0 a_1 \cdots a_{p-1})^{1/p}$ has $\tau < 1$ then for almost all $n$ there is some $k$ with $g^{(k)}(n) < n$, while if $\tau > 1$ then the iterates tend to increase.

*Note.* The Zentralblatt reviewer says that proofs are incomplete but contain an interesting idea. Rigorous versions of these results have since been established, see Lagarias (1985), Sect. 3.2.

**(80-46)**  Giovanni Venturini (1989), *On the* $3x+1$ *Problem*, Adv. Appl. Math **10** (1989), 344–347. (MR 90i:11020).

This paper shows that for any fixed $\rho$ with $0 < \rho < 1$ the set of $m \in \mathbb{Z}^+$ which either have some $T^{(i)}(m) = 1$ or some $T^{(i)}(m) < \rho m$ has density one. This result improves on Dolan, Gilman and Manickam (1987).

**(80-47)**  Carlo Viola (1983), *Un Problema di Aritmetica (A problem of arithmetic)* (Italian), Archimede **35** (1983), 37–39. (MR 85j:11024).

The author states the $3x+1$ problem, and gives a very brief survey of known results on the problem, with pointers to the literature.

**(80-48)**  Stanley Wagon (1985), *The Collatz problem,* Math. Intelligencer **7**, No. 1, (1985), 72-76. (MR 86d:11103, Zbl. 566.10008).

This article studies a random walk imitation of the "average" behavior of the $3x + 1$ function, computes its expected value and compares it to data on $3x + 1$ iterates.

**(80-49)**  Wang, Shi Tie (1988), *Some researches for transformation over recursive programs*, (Chinese, English summary) J. Xiamen University, Natural Science Ed. [Xiamen da xue xue bao. Zi ran ke xue ban] **27**, No. 1 (1988), 8–12. [MR 89g:68057, Zbl. 0689.68010]

English Abstract: "In this paper the transformations over recursive programs with the fixpoint theory is reported. The termination condition of the duple-recursive programs to compute the "91" function and to prove the $3x + 1$ problem is discussed."

*Note.* The author gives a general iteration scheme for computing certain recursively defined functions. The 91 function $F(x)$ is defined on positive integers by the condition, if $x > 100$ then $F(x) = x - 10$, otherwise $F(x) = F(F(x+11))$.

**(80-50)**  Blanton C. Wiggin (1988), *Wondrous Numbers – Conjecture about the $3n + 1$ family,* J. Recreational Math. **20**, No. 2 (1988), 52–56.

This paper calls Collatz function iterates "Wondrous Numbers" and attributes this name to D. Hofstadter, *Gödel, Escher, Bach.* He proposes studying iterates of the class of "MU" functions

$$F_D(x) = \begin{cases} \frac{x}{D} & \text{if } x \equiv 0 \ (\text{mod } D) \,, \\ (D+1)x - j & \text{if } x \equiv j \ (\text{mod } D), \ 1 \le j \le D - 2 \,, \\ (D+1)x + 1 & \text{if } x \equiv -1 \ (\text{mod } D) \,, \end{cases}$$

$F_D$ is the Collatz function for $D = 2$. Wiggin's analogue of the $3x+1$ conjecture for a given $D \ge 2$ is that all iterates of $F_D(n)$ for $n \ge 1$ reach some number smaller than $D$. Somewhat surprisingly, no $D$ is known for which this is false. It could be shown false for a given $D$ by exhibiting a cycle with all members $> D$; no such cycles exist for $x < 3 \times 10^4$ for $2 \le D \le 12$.

**(80-51)** Masaji Yamada (1980), *A convergence proof about an integral sequence*, Fibonacci Quarterly **18** (1980), 231–242. (MR 82d:10026)

This paper claims a proof of the $3x + 1$ Conjecture. However the proof is faulty, with specific mistakes pointed out in Math. Reviews. In particular, Lemma 7 (iii) and Lemma 8 are false.

**(80-52)** Yang, Zhi and Zhang, Zhongfu (1988), *Kakutani conjecture and graph theory representation of the black hole problem*, Nature Magazine [Zi ran za zhi (Shanghai)] **11** (1988), No. 6, 453-456.

This paper considers the Collatz function (denoted $J(n)$) viewing iteration of the function as a directed graph, with edges from $n \to J(n)$. The set $N_J$ of Kakutani numbers is defined to be the set of all numbers that iterate to 1; these form a connected graph. It phrases the $3x + 1$ conjecture as Kakutani's conjecture, asserting that all positive integers are Kakutani numbers.

It also discusses the "black hole" problem, which concerns iteration on $n$-digit integers (in base 10), taking $K(n) = n^+ - n^-$ where $n^+$ is the $n$-digit integer obtained from $n$, arranging digits in decreasing order, and $n^-$ arranging digits in increasing order. This iteration can be arranged in a directed graph with edges from $n$ to $T(n)$. It is known that for 3-digit numbers, $T(\cdot)$ iterates to the fixed point $n = 495$, and for 4-digit numbers it iterates to the fixed point $n = 6174$ (the "*Kaprekar constant*"). For more than 4-digits, iterations may enter cycles of period exceeding one, called here "black holes". For $n = 6$ there is a cycle of length 7 with $n = 840852$, and for $n = 9$ there is a cycle of length 14 starting with $n = 864197532$. The paper is descriptive, with examples but no proofs.

*Note.* Study of the "*Kaprekar constant*" to various bases has a long history. The starting point was: D. R. Kaprekar, *An interesting property of the number 6174*, Scripa Math. **21** (1955) 244-245. Relevant papers include H. Hasse and G. D. Prichett, *The determination of all four-digit Kaprekar constants*, J. reine Angew. Math. **299/300** (1978), 113–124; G. Prichett, A. Ludington and J. F. Lapenta, *The determination of all decadic Kaprekar constants*, Fibonacci Quarterly **19** (1981), 45–52. The latter paper proves that "black holes" exist for every $n \geq 5$.

## 6. Bibliography 1990–1999

This decade saw extensive work on the problem, including the Springer Lecture Notes volume of Wirsching (1998a).

**(90-1)** Amal S. Amleh, Edward A. Grove, Candace M. Kent, and Gerasimos Ladas (1998), *On some difference equations with eventually periodic solutions*, J. Math. Anal. Appl. **223** (1998), 196–215. (MR 99f:39002)

The authors consider the boundedness and periodicity of solutions of the set of difference equations

$$x_{n+1} = \begin{cases} \frac{1}{2}(\alpha x_n + \beta x_{n-1}) & \text{if } x \equiv 0 \ (\text{mod } 2) \, . \\ \gamma x_n + \delta x_{n-1} & \text{if } x \equiv 1 \ (\text{mod } 2) \, , \end{cases}$$

where the parameters $\alpha, \beta, \gamma, \delta \in \{-1, 1\}$, and the initial conditions $(x_0, x_1)$ are integers. There are 16 possible such iterations. Earlier Clark and Lewis (1995) considered the case $(\alpha, \beta, \gamma, \delta) = (1, 1, 1, -1)$, and showed that all orbits with initial conditions integers $(x_0, x_1)$ with $gcd(x_0, x_1) = 1$ converge to one of three periodic orbits . Here the authors consider all 16 cases, showing first a duality between solutions of $(\alpha, \beta, \gamma, \delta)$ and $(-\alpha, \beta, -\gamma, \delta)$, taking a solution $\{x_n\}$ of one to $\{(-1)^{n+1}x_n\}$ of the other. This reduces to considering the eight cases with $\alpha = +1$. They resolve six of these cases, as follows, leavng open the cases $(\alpha, \beta, \gamma, \delta) = (1, -1, 1, 1)$ and $(1, -1, -1, -1)$.

For the parameters $(1, 1, 1, 1)$ they show all orbits are eventually constant or unbounded, and that unbounded orbits occur.

For the parameters $(1, 1, 1, -1)$, the work of Clark and Lewis (1995) showed all orbits are eventually periodic.

For the parameters $(1, 1, -1, 1)$ all solutions are eventually periodic, and there are five relatively prime cycles, the fixed points $(1), (-1)$, the 4-cycles $(2, -1, 3, 1)$, $(-2, 1, -3, 1)$ and the 6-cycle $(1, 0, 1, -1, 0, -1)$.

For the parameters $(1, 1, -1, -1)$ all solutions are eventually periodic, and there are four relatively prime cycles, the fixpoints $(1), (-1)$ and the 3-cycles $(-1, 0, 1)$, $(1, 0, -1)$.

**(90-2)** Stefan Andrei and Cristian Masalagiu (1998), *About the Collatz Conjecture,* Acta Informatica **35** (1998), 167–179. (MR 99d:68097).

This paper describes two recursive algorithms for computing $3x + 1$-trees, starting from a given base node. A $3x + 1$-tree is a tree of inverse iterates of the function $T(.)$. The second algorithm shows a speedup of a factor of about three over the "naive' first algorithm.

**(90-3)** David Applegate and Jeffrey C. Lagarias (1995a), *Density Bounds for the $3x + 1$ Problem I. Tree-Search Method,* Math. Comp., **64** (1995), 411–426. (MR 95c:11024)

Let $n_k(a)$ count the number of integers $n$ having $T^{(k)}(n) = a$. Then for any $a \not\equiv 0 \pmod{3}$ and sufficiently large $k$, $(1.299)^k \leq n_k(a) \leq (1.361)^k$. Let $\pi_k(a)$ count the number of $|n| \leq x$ which eventually reach $a$ under iteration by $T$. If $a \not\equiv 0 \pmod{3}$ then $\pi_a(x) > x^{.643}$ for all sufficiently large $x$. The extremal distribution of number of leaves in $3x + 1$ trees with root $a$ and depth $k$ (under iteration of $T^{-1}$) as $a$ varies are computed for $k \leq 30$. The proofs are computer-intensive.

**(90-4)** David Applegate and Jeffrey C. Lagarias (1995b), *Density Bounds for the $3x + 1$ Problem II. Krasikov Inequalities,* Math. Comp., **64** (1995). 427–438. (MR 95c:11025)

Let $\pi_a(x)$ count the number of $|n| \leq x$ which eventually reach $a$ under iteration by $T$. If $a \not\equiv 0 \pmod{3}$, then $\pi_a(x) > x^{.809}$ for all sufficiently large $x$. It is shown that the inequalities of Krasikov (1989) can be used to construct nonlinear programming problems which yield lower bounds for the exponent $\gamma$ in $\pi_a(x) > x^\gamma$. The exponent above was derived by computer for such a nonlinear program having about 20000 variables.

**(90-5)** David Applegate and Jeffrey C. Lagarias (1995c), *On the distribution of $3x + 1$ trees,* Experimental Mathematics **4** (1995), 101–117. (MR 97e:11033).

The extremal distribution of the number of leaves in $3x + 1$ trees with root $a$ and depth $k$ (under iteration of $T^{-1}$) as $a$ varies were computed for $k \leq 30$ in Applegate and Lagarias (1995a). These data appear to have a much narrower spread around the mean value $(\frac{4}{3})^k$ of leaves in a $3x + 1$ tree of depth $k$ than is predicted by (repeated draws from) the branching process models of Lagarias and Weiss (1992). Rigorous asymptotic results are given for the branching process models.      The paper also derives formulas for the expected number of leaves in a $3x + 1$ tree of depth $k$ whose root node is $a$ (mod $3^{\ell}$). A 3-adic limit is proved to exist almost everywhere as $k \to \infty$, the expected number of leaves being $W_{\infty}(a) \left(\frac{4}{3}\right)^k$ where the function $W_{\infty} : \mathbb{Z}_3^{\times} \to \mathbb{R}$ almost everywhere satisies the 3-adic functional equation

$$W_{\infty}(\alpha) = \frac{3}{4}\left(W_{\infty}(2\alpha) + \psi(\alpha \bmod 9)W_{\infty}\left(\frac{2\alpha - 1}{3}\right)\right), \qquad (*)$$

in which $\psi(\alpha) = 1$ if $\alpha \equiv 2$ or $8 \pmod 9$ and is 0 otherwise. (Here $\mathbb{Z}_3^* = \{\alpha \in \mathbb{Z}_3 : \alpha \not\equiv 0 \pmod 3)\}$). It is conjectured that $W_{\infty}$ is continuous and everywhere nonzero. It is an open problem to characterize solutions of the functional equation $(*)$.

**(90-6)**   Charles Ashbacher (1992), *Further Investigations of the Wondrous Numbers*, J. Recreational Math. **24** (1992), 1–15.

This paper numerically studies the "MU" functions $F_D(x)$ of Wiggin (1988) on $x \in \mathbb{Z}^+$ for $2 \le D \le 12$. It finds no exceptions for Wiggin's conjecture that all cycles of $F_D$ on $\mathbb{Z}^+$ contain an integer smaller than $D$, for $x < 1.4 \times 10^7$. It tabulates integers in this range that have a large stopping time, and observes various patterns. These are easily explained by observing that, for $n \not\equiv 0 \pmod D$, $F_D^{(2)}(n) = \frac{n(D+1)-R}{D}$ if $n \equiv R \pmod D$, $-1 \le R \le D - 2$, hence, for most $n$, $F_D^{(2)}(n) > n$, although $F_D$ decreases iterates on the average.

**(90-7)**   Michael R. Avidon (1997), *On primitive 3-smooth partitions of $n$*, Electronic J. Combinatorics **4** (1997), no.1 , 10pp. (MR 98a:11136).

The author studies the number $r(n)$ of representations of $n$ as sums of numbers of the form $2^a 3^b$ which are primitive (no summand divides another). Iterates of $3x+1$ function applied to $n$ that get to 1 produces a representation of $n$ of this kind. The author proves results about the maximal and average order of this function. See also Blecksmith, McCallum and Selfredge (1998) for more information on the topic of 3-smooth representations of $n$.

**(90-8)**   Claudio Baiocchi (1998), *3N+1, UTM e Tag-Systems* (Italian), Dipartimento di Matematica dell'Università "La Sapienza" di Roma, Report **98/38** (1998).

This technical report constructs small state Turing machines that simulate the $3x + 1$ problem. Let $T(k, l)$ denote the class of one-tape Turing machines with $k$ state, with $l$-symbols, with one read head, and the tape is infinite in two directions. The author constructs Turing machines for simulating the $3x + 1$ iteration in the classes $T(10, 2), T(5, 3), T(4, 4), T(3, 5)$ and $T(2, 8)$, working on unary inputs. It follows that no method is curently known to decide the reachability problem for such machines. The author then produces a universal Turing machine in the class $T(22, 2)$.

*Note.* This work was motivated by a conference paper of M. Margenstern (1998), whose journal version is: M. Margenstern, Theor. Comp. Sci. **231** (2000), 217-251.

**(90-9)**  Ranan B. Banerji (1996), *Some Properties of the* $3n + 1$ *Function,* Cybernetics and Systems **27** (1996), 473–486.

The paper derives elementary results on forward iterates of the $3x + 1$ function viewed as binary integers, and on backward iterates of the map $g$ taking odd integers to odd integers, given by

$$g(n) := \frac{3n + 1}{2^k} \ , \ \text{where } 2^k \| 3n + 1 \ .$$

Integers $n \equiv 0 (\text{mod } 3)$ have no preimages under $g$. If $n \not\equiv 0 (\text{mod } 3)$ define $g^{-1}(n)$ to be the unique integer $t$ such that $g(t) = n$ and $t \not\equiv 5 (\text{mod } 8)$. Note that each odd $n$ there are infinitely many $\tilde{t}$ with $g(\tilde{t}) = n$. If $d(n) = 4n + 1$, these preimages are just $\{d^{(j)}g^{-1}(n) : j \geq 1\}$. The ternary expansion of $g^{-1}(n)$ is asserted to be computable from the ternary expansion of $n$ by a finite automaton. The author conjectures that given any odd integer $n$, there is some finite $k$ such that $(g^{-1})^{(k)}(n) \equiv 0 (\text{mod } 3)$. Here we are iterating the partially defined map

$$g^{-1}(6n + 1) = 8n + 1 \ ,$$
$$g^{-1}(6n + 5) = 4n + 3 \ ,$$

and asking if some iterate is $0 (\text{mod } 3)$. The problem resembles Mahler's $Z$-number iteration [J. Australian Math. Soc. **8** (1968), 313–321].

**(90-10)**  Enzo Barone (1999), *A heuristic probabilistic argument for the Collatz sequence.* (Italian), Ital. J. Pure Appl. Math. **4** (1999), 151–153. (MR 2000d:11033).

This paper presents a heuristic probabilistic argument which argues that iterates of the $3x + 1$-function should decrease on average by a multiplicative factor $(\frac{3}{4})^{1/2}$ at each step. Similar arguments appear earlier in Lagarias (1985) and many other places, and trace back to the original work of Terras (1976), Everett (1977) and Crandall (1978).

**(90-11)**  Edward Belaga (1998), *Reflecting on the $3x + 1$ Mystery: Outline of a Scenario- Improbable or Realistic?* U. Strasbourg report 1998-49, 10 pages. (http://hal.archives-ouvertes.fr/IRMA-ACF, file hal-0012576)

This is an expository paper, discussing the possibility that the $3x + 1$ conjecture is an undecidable problem. Various known results, pro and con, are presented.

**(90-12)**  Edward Belaga and Maurice Mignotte (1999), *Embedding the $3x + 1$ Conjecture in a $3x + d$ Context,* Experimental Math. **7**, No. 2 (1999), 145–151. (MR 200d:11034).

The paper studies iteration on the positive integers of the $3x + d$ function

$$T(x) = \begin{cases} \frac{3x+d}{2} & \text{if } x \equiv 1 \ (\text{mod } 2) \ , \\ \frac{x}{2} & \text{if } x \equiv 0 \ (\text{mod } 2) \ , \end{cases}$$

where $d \geq -1$ and $d \equiv \pm 1 \ (\text{mod } 6)$. It proves that there is an absolute constant $c$ such that there are at most $dk^c$ periodic orbits which contain at most $k$ odd integers. Furthermore $c$ is effectively computable. This follows using a transcendence result of A. Baker and G. Wüstholz [J. reine Angew. **442** (1993), 19–62.]

**(90-13)** Stefano Beltraminelli, Danilo Merlini and Luca Rusconi (1994), *Orbite inverse nel problema del 3n + 1*, Note di matematica e fisica, Edizioni Cerfim Locarno **7** (1994), 325–357.

This paper discusses the tree of inverse iterates of the Collatz function $C(n)$, which it terms the "chalice." It states the Collatz conjecture, and notes that it fails for the map

$$C_{3,5}(x) := \begin{cases} 3x + 5 & \text{if } x \equiv 1 \ (\text{mod } 2) \ , \\ \frac{x}{2} & \text{if } x \equiv 0 \ (\text{mod } 2) \ , \end{cases}$$

where there are at least two periodic orbits $\{1, 8, 4, 2\}$ and $\{5, 20, 10, 5\}$; here orbits with initial term $x \equiv 0 \ (\text{mod } 5)$ retain this property throughout the iteration.

It studies patterns of inverse iterates in the tree with numbers written in binary. The occurrence of patterns of odd iterates is described by congruence classes ( mod $2 \cdot 3^k$), for the $k$-th odd iterate backwards. The authors conjecture that if $l(n)$ is the Collatz function total stopping time, then $l(n) \leq n$ for all sufficiently large $n$. They note that empirical data can be fitted by $l(n) \leq 5.19 + 31.92 \log n$ up to $10^{10}$.

**(90-14)** Lothar Berg and Günter Meinardus (1994), *Functional equations connected with the Collatz problem*, Results in Math. **25** (1994), 1–12. (MR 95d:11025).

The $3x + 1$ Conjecture is stated as Conjecture 1. The paper proves its equivalence to each of Conjectures 2 and 3 below, which involve generating functions encoding iterations of the $3x + 1$ function $T(x)$. For $m, n \geq 0$ define $f_m(z) = \sum_{n=0}^{\infty} T^{(m)}(n) z^n$ and $g_n(w) = \sum_{n=0}^{\infty} T^{(m)}(n) w^m$. The paper shows that each $f_m(z)$ is a rational function of form

$$f_m(z) = \frac{p_m(z)}{(1 - z^{2^m})^2},$$

where $p_m(z)$ is a polynomial of degree $2^{m+1} - 1$ with integer coefficients. Conjecture 2 asserts that each $g_n(w)$ is a rational function of the form

$$g_n(w) = \frac{q_n(w)}{1 - w^2},$$

where $q_n(w)$ is a polynomial with integer coefficients, with no bound assumed on its degree. Concerning functional equations, the authors show first that the $f_m(z)$ satisfy the recursions

$$f_{m+1}(z^3) = f_m(z^6) + \frac{1}{3z} \sum_{j=0}^{2} \omega^j f_m(\omega^j z^2),$$

in which $\omega := \exp\left(\frac{2\pi i}{3}\right)$ is a nontrivial cube root of unity. They also consider the bivariate generating function $F(z, w) = \sum_{m=0}^{\infty} \sum_{n=0}^{\infty} T^{(m)}(n) z^n w^m$, which converges for $|z| < 1$ and $|y| < \frac{2}{3}$ to an analytic function of two complex variables. The authors show that it satisfies the functional equation

$$F(z^3, w) = \frac{z^3}{(1-z^3)^2} + wF(z^6, w) + \frac{w}{3z} \sum_{j=0}^{2} \omega^j F(\omega^j z^2, w) \ .$$

They prove that this functional equation determines $F(z, w)$ uniquely, i.e. there is only one analytic function of two variables in a neighborhood of $(z, w) = (0, 0)$ satisfying it. Next they consider a one-variable functional equation obtained from this one by formally setting $w = 1$ (note this falls outside the known region of analyticity of the function), and dropping the non-homogenous term. This functional equation is

$$h(z^3) = h(z^6) + \frac{1}{3z} \sum_{j=0}^{2} \omega^j h(\omega^j z^2).$$

Conjecture 3 asserts that the only solutions $h(z)$ of this functional equation that are analytic in the unit disk $|z| < 1$ are $h(z) = c_0 + c_1(\frac{z}{1-z})$ for complex constants $c_0, c_1$.

**(90-15)** Lothar Berg and Günter Meinardus (1995), *The 3n+1 Collatz Problem and Functional Equations,* Rostock Math. Kolloq. **48** (1995), 11-18. (MR 97e:11034).

This paper reviews the results of Berg and Meinardus (1994) and adds some new results. The first new result considers the functional equation

$$h(z^3) = h(z^6) + \frac{1}{3z} \sum_{j=0}^{2} \omega^j h(\omega^j z^2).$$

with $\omega := \exp\left(\frac{2\pi i}{3}\right)$, and shows that the only solutions $h(z)$ that are entire functions are constants. The authors next transform this functional equation to an equivalent system of two functional equations:

$$h(z) + h(-z) = 2h(z^2)$$
$$h(z^3) - h(-z^3) = \frac{2}{3z} \sum_{j=0}^{2} \omega^j h(\omega^j z^2).$$

They observe that analytic solutions on the open unit disk to these two functional equations can be studied separately. The second one is the most interesting. Set $\Phi(z) := \int_0^z h(z)dz$ for $|z| < 1$. Making the change of variable $z = e^{\frac{2\pi i}{3}\xi}$, the unit disk $|z| < 1$ is mapped to the upper half plane $Im(\xi) > 0$. Letting $\phi(\xi) := \Phi(e^{\frac{2\pi i}{3}\xi})$ the second functional equation above becomes

$$\phi(3\xi) + \phi(3\xi + \frac{3}{2}) = \phi(2\xi) + \phi(2\xi + 1) + \phi(2\xi + 2).$$

Here we also require $\phi(\xi) = \phi(\xi + 3)$. The authors remark that it might also be interesting to study solutions to this functional equation for $\xi$ on the real axis. The paper concludes with new formulas for the rational functions $f_m(z)$ studied in Berg and Meinardus (1994).

**(90-16)**   Daniel J. Bernstein (1994), *A Non-Iterative 2-adic Statement of the $3x+1$ Conjecture*, Proc. Amer. Math. Soc., **121** (1994), 405–408. (MR 94h:11108).

Let $\mathbb{Z}_2$ denote the 2-adic integers, and for $x \in \mathbb{Z}_2$ write $x = \sum_{i=0}^{\infty} 2^{d_i}$ with $0 \le d_0 < d_1 < d_2 < \cdots$. Set $\Phi(x) = -\sum_{j=0}^{\infty} \frac{1}{3^{j+1}} 2^{d_j}$. The map $\Phi$ is shown to be a homeomorphism of the 2-adic integers to itself, which is the inverse of the map $Q_\infty$ defined in Lagarias (1985). The author proves in Theorem 1 a result equivalent to $\Phi^{-1} \circ T \circ \Phi = S$, where $T$ is the $(3x+1)$-function on $\mathbb{Z}_2$, and $S$ is the shift map

$$S(x) = \begin{cases} \frac{x-1}{2} & \text{if } x \equiv 1 \pmod{2}, \\ \frac{x}{2} & \text{if } x \equiv 0 \pmod{2}. \end{cases}$$

He shows that the $3x+1$ Conjecture is equivalent to the conjecture that $\mathbb{Z}^+ \subseteq \Phi(\frac{1}{3}\mathbb{Z})$. He rederives the known results that $\mathbb{Q} \cap \mathbb{Z}_2 \subseteq Q_\infty(\mathbb{Q} \cap \mathbb{Z}_2)$, and that $Q_\infty$ is nowhere differentiable, cf. Müller (1991).

**(90-17)**   Daniel J. Bernstein and Jeffrey C. Lagarias (1996), *The $3x+1$ Conjugacy Map*, Canadian J. Math., **48** (1996), 1154-1169. (MR 98a:11027).

This paper studies the map $\Phi : \mathbb{Z}_2 \to \mathbb{Z}_2$ of Bernstein (1994) that conjugates the 2-adic shift map to the $3x+1$ function. This is the inverse of the map $Q_\infty$ in Lagarias (1985); see also Akin (2004). The map $\bar{\Phi}_n \equiv \Phi \pmod{2^n}$ is a permutation of $\mathbb{Z}/2^n\mathbb{Z}$. This permutation is shown to have order $2^{n-4}$ for $n \ge 6$. Let $\hat{\Phi}_n$ denote the restriction of this permutation to $(\mathbb{Z}/2^n\mathbb{Z})^* = \{x : x \equiv 1 \pmod{2}\}$. The function $\Phi$ has two odd fixed points $x = -1$ and $x = 1/3$ and the 2-cycle $\{1, -1/3\}$, hence each $\hat{\Phi}_n$ inherits two 1-cycles and a 2-cycle coming from these points. Empirical evidence indicates that $\hat{\Phi}_n$ has about $2n$ fixed points for $n \le 1000$. A heuristic argument based on this data suggests that $-1$ and $1/3$ are the only odd fixed points of $\Phi$. The analogous conjugacy map $\Phi_{25,-3}$ for the '$25x - 3$' problem is shown to have no nonzero fixed points.

**(90-18)**   Richard Blecksmith, Michael McCallum, and John L. Selfridge (1998), *3-Smooth Representations of Integers*, American Math. Monthly **105** (1998), 529–543. (MR 2000a:11019).

A *3-smooth representation* of an integer $n$ is a representation as a sum of distinct positive integers each of which has the form $2^a 3^b$, and no term divides any other term. This paper proves a conjecture of Erdos and Lewin that for each integer $t$ all sufficiently large integers have a 3-smooth representation with all individual terms larger than $t$. They note a connection of 3-smooth representations to the $3x+1$-problem, which is that a number $m$ iterates to 1 under the $3x+1$ function if and only if there are positive integers $e$ and $f$ such that $n = 2^e - 3^f m$ is a positive integer that has a 3-smooth representation with $f$ terms in which there is one term exactly divisible by each power of three from $0$ to $f - 1$. The choice of $e$ and $f$ is not unique, if it exists.

**(90-19)**   Stefano Brocco (1995), *A Note on Mignosi's Generalization of the $3x+1$ Problem*,  J. Number Theory, **52** (1995), 173–178. (MR 96d:11025).

F. Mignosi (1995) studied the function $T_\beta : \mathbb{N} \to \mathbb{N}$ defined by

$$T_\beta(n) = \begin{cases} \lceil \beta n \rceil & \text{if } n \equiv 1 \pmod{2} \\ \frac{n}{2} & \text{if } n \equiv 0 \pmod{2}, \end{cases}$$

where $\lceil x \rceil$ denotes the smallest integer $n \geq x$. He also formulated Conjecture $C_\beta$ asserting that $T_\beta$ has finitely many cycles and that every $n \in \mathbb{N}$ eventually enters a cycle under $T_\beta$. This paper shows that Conjecture $C_\beta$ is false whenever $\beta$ is a Pisot number or a Salem number. The result applies further to functions

$$T_{\beta,\alpha}(n) = \begin{cases} \lceil \beta n + \alpha \rceil & \text{if } n \equiv 1 \pmod{2} , \\ \frac{n}{2} & \text{if } n \equiv 0 \pmod{2} , \end{cases}$$

for certain ranges of values of $\beta$ and $\alpha$.

**(90-20)**   Serge Burckel (1994), *Functional equations associated with congruential functions,* Theoretical Computer Science **123** (1994), 397–406. (MR 94m.11147).

The author proves undecidability results for periodically linear functions generalizing those of Conway (1972). A periodically linear function $f : \mathbb{Z} \to \mathbb{Z}$ is one which is a linear function on each congruence class $(\bmod\, L)$ for some finite $L$. The author shows it is undecidable whether a given function has $f^{(k)}(1) = 0$ for some $k \geq 1$, and also whether a given function has the property: for each $n \geq 1$, some $f^{(k)}(n) = 0$. He also shows that the $3x + 1$ conjecture is equivalent to a certain functional equation having only the trivial solution over the set of all power-series $R(z) = \sum_{n=}^{\infty} a_n z^n$ with all $a_i = 0$ or 1. The functional equation is

$$3z^3 R(z^3) - 3z^9 R(z^6) - R(z^2) - R(\omega z^2) - R(\omega^2 z^2) = 0$$

where $\omega = \exp(\frac{2\pi i}{3})$.

**(90-21)**   Robert N. Buttsworth and Keith R. Matthews (1990), *On some Markov matrices arising from the generalized Collatz mapping,* Acta Arithmetica **55** (1990), 43–57. (MR 92a:11016).

This paper studies maps $T(x) = \frac{m_i x - r_i}{d}$ for $x \equiv i(\bmod\, d)$, where $r_i \equiv im_i(\bmod\, d)$. In the case where g.c.d. $(m_0, \ldots, m_{d-1}, d) = 1$ it gives information about the structure of $T$-ergodic sets $(\bmod\, m)$ as $m$ varies. A set $S \subseteq \mathbb{Z}$ is *T-ergodic* $(\bmod\, m)$ if it is a union of $k$ congruence classes $(\bmod\, m)$, $S = C_1 \cup \ldots \cup C_k$, such that $T(S) \subseteq S$ and there is an $n$ such that $C_j \cap T^{(n)}(C_i) \neq \phi$ holds for all $i$ and $j$. It characterizes them in many cases. As an example, for

$$T(x) = \begin{cases} \frac{x}{2} & \text{if } x \equiv 0 \pmod{2} , \\ \frac{5x - 3}{2} & \text{if } x \equiv 1 \pmod{3} , \end{cases}$$

the ergodic set $(\bmod\, m)$ is unique and is $\{n : n \in \mathbb{Z} \text{ and } (n, m, 15) = 1\}$, i.e. it is one of $\mathbb{Z}, \mathbb{Z} - 3, \mathbb{Z} - 5\mathbb{Z}$ or $\mathbb{Z} - 3\mathbb{Z} - 5\mathbb{Z}$ as $m$ varies. An example is given having infinitely many different ergodic sets $(\bmod\, m)$ as $m$ varies.

**(90-22)**   Charles C. Cadogan (1991), *Some observations on the $3x + 1$ problem,* Proc. Sixth Caribbean Conference on Combinatorics & Computing, University of the West Indies: St. Augustine Trinidad (C. C. Cadogan, Ed.) Jan. 1991, 84–91.

Cadogan (1984) reduced the $3x + 1$ problem to the study of its iterations on numbers $A_1 = \{n : n \equiv 1 \pmod{4}\}$. Here the author notes in particular the subclass $A = \{1, 5, 21, 85, \ldots\}$ where $x_{i+1} = 1 + 4x_i$ of $A_1$. He considers the successive odd numbers occurring in the iteration. He forms a two-dimensional table partitioning all odd integers in which $A_1$ is the first row and the first column of the table is called the anchor set (see Cadogan (1996) for more details) . He observes that Cadogan (1984) showed the iteration on higher rows successively moves down rows to the first row, but from row $A_1$ is flung back to higher rows, except for the subclass $A$, which remains on the first row. He comments that after each revisit to row $A_1$ "the path may become increasingly unpredictable." He concludes, concerning further work: "The target set $A_1$ is being vigorously investigated."

**(90-23)**   Charles C. Cadogan (1996), *Exploring the $3x + 1$ problem I.*, Caribbean J. Math. Comput. Sci. **6** (1996), 10–18. (MR 2001k:11032)

This paper studies itertion of the Collatz function $C(x)$, which is here denoted $f(x)$, and it includes most of the results of Cadogan (1991). The author gives various criteria under which trajectories will coalesce. He partitions the odd integers $2\mathbb{N} + 1 = \cup_{k=1}^{\infty} R_k$, in which $R_m = \{n \equiv 2^m - 1 \pmod{2^{m+1}}\}$. It enumerates their elements $R_{m,j} = (2^m - 1) + (j - 1)2^{m+1}$. and views these in an infinite two-dimensional array in which the $j$-th column $C_j$ consists of the numbers $\{R_{m,j} : m \geq 1\}$. Cadogan (1984) showed that if $x \in R_m$ for some $m \geq 2$, then $f^{(2)}(x) \in R_{m-1}$, thus after $2m$ iterations one reaches an element of $R_1$. Lemma 3.3 here observes that consecutive elements in columns are related by by $R_{m+1,j} = 1 + 2R_{m,j}$. For the first set $R_1 = \{n \equiv 1 \pmod 4\}$, Theorem 4.1 observes that if $y = 4x + 1$ then $f^{(3)}(y) = f(x)$. The author creates chains $\{x_n : n \geq 1\}$ related by $x_{n+1} = 4x_n + 1$ and calls these $S$-related elements. Theorem 4.2 then observes that the trajectories of $S$-related elements coalesce.

**(90-24)**   Marc Chamberland (1996), *A Continuous Extension of the $3x + 1$ Problem to the Real Line*, Dynamics of Continuous, Discrete and Impulsive Dynamical Systems, **2** (1996), 495–509. (MR 97f:39031).

This paper studies the iterates on $\mathbb{R}$ of the function

$$
\begin{aligned}
f(x) &= \frac{\pi x}{2} \left( \cos \frac{\pi x}{2} \right)^2 + \frac{3x + 1}{2} \left( \sin \frac{\pi x}{2} \right)^2 \\
&= x + \frac{1}{4} - \left( \frac{x}{2} + \frac{1}{4} \right) \cos \pi x .
\end{aligned}
$$

which interpolates the $3x + 1$ function $T(\cdot)$. A fact crucial to the analysis is that $f$ has negative Schwartzian derivative $Sf = \frac{f'''}{f'} - \frac{3}{2} \left( \frac{f''}{f'} \right)^2$ on $\mathbb{R}^+$. On the interval $[0, \mu_1)$, where $\mu_1 = 0.27773\ldots$ all iterates of $f$ contract to a fixed point 0. Here $\mu_n$ denotes the $n - th$ positive fixed point of $f$. The interval $[\mu_1, \mu_3]$ is invariant under $f$, where $\mu_3 = 2.44570\ldots$ and this interval includes the trivial cycle $A_1 = \{1, 2\}$. On this interval almost every point is attracted to one of two attracting cycles, which are $A_1$ and $A_2 = \{1.19253\ldots, 2.13865\ldots\}$. There is also an uncountable set of measure 0 on which the dynamics is "chaotic." On the interval $[\mu_3, \infty)$ the set of $x$ that do not eventually iterate to a point in $[\mu_1, \mu_3]$ is conjectured to be of measure zero. The point $\mu_3$ is proved to be a "homoclinic point," in the sense that for any $\epsilon > 0$ the iterates of $[\mu_3, \mu_3 + \epsilon)$ cover the whole interval $(\mu_1, \infty)$. It is shown that any nontrivial cycle of the

312          JEFFREY C. LAGARIAS

$3x + 1$ function on the positive integers would be an attracting periodic orbit of the function $f(x)$.

(90-25)  Busiso P. Chisala (1994), *Cycles in Collatz Sequences*, Publ. Math. Debrecen **45** (1994), 35–39. (MR 95h:11019).

The author shows that for any $m$-cycle of the Collatz map on positive *rationals*, the least element is at least as large as $(2^{[m\theta]/m} - 3)^{-1}$, where $\theta = \log_2 3$. Using this result, he derives a lower bound for $3x + 1$ cycle lengths based on the continued fraction of $\theta = [1, a_1, a_2, a_3, \ldots]$, in which the $n$-th convergent is $\frac{p_n}{q_n}$, and the intermediate convergent denominator $q_n^i$ is $iq_{n+1} + q_n$ for $0 \le i < a_{n+1}$. If the $3x + 1$ conjecture is true for $1 \le n \le N$, and $N \ge (2^{C(i,k)} - 3)^{-1}$, where $C(i,k) = \frac{[q_k^i \theta]}{q_k^i}$, then there are no nontrivial cycles of the $3x + 1$ function on $\mathbb{Z}^+$ containing less than $q_k^{i+1}$ *odd* terms. Using the known bound $N = 2^{40} \doteq 1.2 \times 10^{12}$, the author shows that there are at least $q_{15} = 10\,787\,915$ odd terms in any cycle of the $3x + 1$ function on $\mathbb{Z}^+$.

*Note.* Compare these results with those of Eliahou (1993).

(90-26)  Dean Clark (1995), *Second-Order Difference Equations Related to the Collatz $3n + 1$ Conjecture*, J. Difference Equations & Appl., **1** (1995), 73–85. (MR 96e:11031).

The paper studies the integer-valued recurrence $\frac{x_{n+1}+x_n}{2}$ if $x_{n+1} + x_n$ is even, and $x_n = \frac{b|x_{n+1}-x_n|+1}{2}$ if $x_{n+1} + x_n$ is odd, for $b \ge 1$ an odd integer. For $b = 1, 3, 5$ all recurrence sequences stabilize at some fixed point depending on $x_1$ and $x_2$, provided that $x_1 = x_2 \equiv \frac{b+1}{2} \pmod{b}$. For $b \ge 7$ there exist unbounded trajectories, and periodic trajectories of period $\ge 2$. In the "convergent" cases $b = 3$ or $5$ the iterates exhibit an interesting phenomenon, which the author calls *digital convergence*, where the low order digits in base $b$ of $x_n$ successively stabilize before the high order bits stabilize.

(90-27)  Dean Clark and James T. Lewis (1995), *A Collatz-Type Difference Equation*, Proc. Twenty-sixth Internationsal Conference on Combinatorics, Graph Theory and Computing (Boca Raton 1995), Congr. Numer. **111** (1995), 129-135. (MR 98b:11008).

This paper studies the difference equation

$$x_n = \begin{cases} \frac{x_{n-1}+x_{n-2}}{2} & \text{if } x_{n-1} + x_{n-2} \text{ is even}, \\ x_{n-1} - x_{n-2} & \text{if } x_{n-1} + x_{n-2} \text{ is odd}. \end{cases}$$

with integer initial conditions $(x_0, x_1)$. It suffices to treat the case that $gcd(x_0, x_1) = 1$. For such initial conditions the recurrence is shown to always converge to one of the 1-cycles 1 or $-1$ or to the 6-cycle $\{3, 2, -1, -3, -2, -1\}$.

(90-28)  Dean Clark and James T. Lewis (1998), *Symmetric solutions to a Collatz-like system of Difference Equations*, Proc. Twenty-ninth Internationsal Conference on Combinatorics, Graph Theory and Computing (Baton Rouge 1998), Congr. Numer. **131** (1998), 101-114. (see MR 99i:00021).

This paper studies the first order system of nonlinear difference equations

$$x_{n+1} = \lfloor \frac{x_n + y_n}{2} \rfloor$$
$$y_{n+1} = y_n - x_n,$$

where $\lfloor . \rfloor$ is the floor function (greatest integer function). Let $T(x,y) = (\lfloor \frac{x+y}{2} \rfloor, y-x)$ be a map of the plane, noting that $T(x_n, y_n) = (x_{n+1}, y_{n+1})$. The function $T$ is an invertible map of the plane, with inverse $S = T^{-1}$ given by $S(x,y) = (\lceil x - \frac{y}{2} \rceil, \lceil x + \frac{y}{2} \rceil)$, using the ceiling function $\lceil . \rceil$. One obtains an associated linear map of the plane, by not imposing the floor function above., i.e. $\tilde{T}(x,y) = (\frac{x+y}{2}, -x+y)$. The map $\tilde{T}$ is invertible, and for arbitrary real initial conditions $(x_0, y_0)$ the full orbit $\{\tilde{T}^{(k)}(x_0, y_0) : -\infty < k < \infty\}$ is bounded, with all points on it being confined to an invariant ellipse. The effect of the floor function is to perturb this linear dynamics. The authors focus on the question of whether all orbits having integer initial conditions $(x_0, y_0)$ remain bounded; however they don't completely resolve this question. Note that integer initial conditions imply the full orbit is integral; then invertibility implies that bounded orbits of this type must be periodic.

The difference equation given above for integer initial conditions $(x_0, y_0)$ can be transformed to a second order nonlinear recurrence by eliminating the variable $x_n$, obtaining.

$$y_{n+1} := \begin{cases} \frac{3y_n+1}{2} - y_{n-1} & \text{if } y_n \equiv 1 \pmod 2 \\ \frac{3y_n}{2} - y_{n-1} & \text{if } y_n \equiv 0 \pmod 2 \end{cases}.$$

with integer initial conditions $(y_0, y_1)$. They note a resemblance of this recurrence in form to the $3x+1$ problem, and view boundedness of orbits as a (vague) analogue of the $3x + 1$ Conjecture. Experimentally they observe that all integer orbits appear to be periodic, but the period of such orbits varies erratically with the initial conditions. For example the starting condition $(64, 0)$ for $T$ has period 87, but that of $(65, 0)$ has period 930. They give a criterion (Theorem 1) for an integer orbit to be unbounded, but conjecture this criterion is never satisfied.

The paper also studies properties of periodic orbits imposed by some symmetry operators. They introduce the operator $Q(x,y) := (\lfloor -x + \frac{y}{2}) \rfloor, y)$, observe it is an involution $Q^2 = I$ satisfying $(TQ)^2 = I$ and $S = QTQ^{-1}$. They also introduce a second symmetry operator $U(x,y) := (-x, 1 - y)$ which is is an involution $U^2 = I$ that commutes with $T$. These operators are used to imply some symmetry properties of periodic orbits, with respect to the line $x = y$. They also derive the result (Theorem 6): the sum of the terms $y_k$ that are even integers in a complete period of a periodic orbit is divisible by 4; the sum of all the $y_k$ over a cycle is strictly positive and equals the number of odd $y_k$ that appear in the cycle.

**(90-29)**   Philippe Devienne, Patrick Lebègue, Jean-Christophe Routier (1993), *Halting Problem of One Binary Horn Clause is Undecidable*, in: *Proceedings of STACS 1993*, Lecture Notes in Computer Science No. **665**, Springer–Verlag 1993, pp. 48–57. (MR 95e:03114).

The halting problem for derivations using a single binary Horn clause for reductions is shown to be undecidable, by encoding Conway's undecidability result on iterating periodically linear functions having no constant terms, cf. Conway

(1972). In contrast, the problem of whether or not ground can be reached using reductions by a single binary Horn clause is decidable. [M. Schmidt-Schauss, *Implication of Clauses is Undecidable*, Theor. Comp. Sci. **59** (1988), 287–296.]

**(90-30)**   Peter Eisele and Karl-Peter Hadeler (1990), *Game of Cards, Dynamical Systems, and a Characterization of the Floor and Ceiling Functions*, Amer. Math. Monthly **97** (1990), 466–477. (MR 91h:58086).

This paper studies iteration of the mappings $f(x) = a + \lceil \frac{x}{b} \rceil$ on $\mathbb{Z}$ where $a, b$ are positive integers. These are periodical linear functions (mod $b$). For $b \geq 2$, every trajectory becomes eventually constant or reaches a cycle of order 2.

**(90-31)**   Shalom Eliahou (1993), *The $3x + 1$ problem: New Lower Bounds on Nontrivial Cycle Lengths*, Discrete Math., **118**(1993), 45–56. (MR 94h:11017).

The author shows that any nontrivial cycle on $\mathbb{Z}+$ of the $3x+1$ function $T(x)$ has period $p = 301994A + 17087915B + 85137581C$ with $A, B, C$ nonnegative integers where $B \geq 1$, and at least one of $A$ or $C$ is zero. Hence the minimal possible period length is at least 17087915. The method uses the continued fraction expansion of $\log_2 3$, and the truth of the $3x + 1$ Conjecture for all $n < 2^{40}$. The paper includes a table of partial quotients and convergents to the continued fraction of $\log_2 3$.

**(90-32)**   Carolyn Farruggia, Michael Lawrence and Brian Waterhouse (1996), *The elimination of a family of periodic parity vectors in the $3x + 1$ problem*, Pi Mu Epsilon J. **10** (1996), 275–280.

This paper shows that the parity vector $10^k$ is not the parity vector of any integral periodic orbit of the $3x + 1$ mapping whenever $k \geq 2$. (For $k = 2$ the orbit with parity vector 10 is the integral orbit $\{1, 2\}$. )

**(90-33)**   Marc R. Feix, Amador Muriel, Danilo Merlini, and Remiglio Tartini (1995), *The $(3x + 1)/2$ Problem: A Statistical Approach*, in: *Stochastic Processes, Physics and Geometry II,* Locarno 1991. (Eds: S. Albeverio, U. Cattaneo, D. Merlini) World Scientific, 1995, pp. 289–300.

This paper formulates heuristic stochastic models imitating various behaviors of the $3x + 1$ function, and compares them to some data on the $3x + 1$ function. In Sect. 2 it describes a random walk model imitating "average" behavior of forward iterates of the $3x + 1$ function. In Sect. 3 it examines trees of inverse iterates of this function, and predicts that the number of leaves at level $k$ of the tree should grow approximately like $A(\frac{4}{3})^k$ as $k \to \infty$. In Sect. 4 it describes computer methods for rapid testing of the $3x + 1$ Conjecture. In Sect. 5 it briefly considers related functions. In particular, it considers the $3x - 1$ function and the function

$$\tilde{T}(x) = \begin{cases} \frac{x}{3} & \text{if } x \equiv 0 \pmod{3} \\ \frac{2x+1}{3} & \text{if } x \equiv 1 \pmod{3} \\ \frac{7x+1}{3} & \text{if } x \equiv 2 \pmod{3} \ . \end{cases}$$

Computer experiments show that the trajectories of $\tilde{T}(x)$ for $1 \leq n \leq 200,000$ all reach the fixed point $\{1\}$.

*Note.* Lagarias (1992) and Applegate and Lagarias (1995c) study more detailed stochastic models analogous to those given in §2, resp. §3 here.

**(90-34)**  Marc R. Feix, Amador Muriel and Jean-Louis Rouet (1994), *Statistical Properties of an Iterated Arithmetic Mapping,* J. Stat. Phys. **76** (1994), 725–741. (MR 96b:11021).

This paper interprets the iteration of the $3x + 1$ map as exhibiting a "forgetting" mechanism concerning the iterates (mod $2^k$), i.e. after $k$ iterations starting from elements it draws from a fixed residue class (mod $2^k$), the iterate $T^k(n)$ is uniformly distributed (mod $2^k$). It proves that certain associated $2^k \times 2^k$ matrices $M_k$ has $(M_k)^k = J_{2^k}$ where $J_{2^k}$ is the doubly-stochastic $2^k \times 2^k$ matrix having all entries equal to $2^{-k}$.

**(90-35)**  Piero Filipponi (1991), *On the $3n + 1$ problem: Something old, something new,* Rendiconti di Mathematica, Serie VII, Roma **11** (1991), 85–103. (MR 92i:11031).

This paper derives by elementary methods various facts about coalescences of trajectories and divergent trajectories. For example, the smallest counterexample $n_0$ to the $3x + 1$ Conjecture, if one exists, must have $n_0 \equiv 7, 15, 27, 31, 39, 43, 63, 75, 79, 91 \pmod{96}$. The final Theorem 16 has a gap in its proof, because formula (5.11) is not justified.

**(90-36)**  Leopold Flatto (1992) , *Z-numbers and $\beta$-transformations,* in: *Symbolic dynamics and its applications (New Haven, CT, 1991),* Contemp. Math. Vol. 135, American Math. Soc., Providence, RI 1992, 181–201.(MR94c:11065).

This paper concerns the $Z$-number problem of Mahler (1968). A real number $x$ is a *Z-number* if $0 \leq \{\{x(\frac{3}{2})^n\}\} < \frac{1}{2}$ holds for all $n \geq 0$, where $\{\{x\}\}$ denotes the fractional part of $x$. Mahler showed that there is at most one $Z$-number in each unit interval $[n, n + 1)$, for positive integer $n$, and bounded the number of such $1 \leq n \leq X$ that can have a $Z$-number by $X^{0.7}$. This paper applies the $\beta$-transformation of W. Parry [Acta Math. Acad. Sci. Hungar. **11** (1960), 401-416] to get an improved upper bound on the number of $n$ for which a $Z$-number exists. It studies symbolic dyanmics of this transformation for $\beta = \frac{3}{2}$, and deduces that the number of $1 \leq n \leq X$ such that there is a $Z$-number in $[n, n+1)$ is at most $X^\theta$ with $\theta = \log_2(3/2) \approx 0.59$. The paper also obtains related results for more general $Z$-numbers associated to fractions $\frac{p}{q}$ having $q < p < q^2$.

**(90-37)**  Zachary M. Franco (1990), *Diophantine Approximation and the $qx + 1$ Problem,* Ph.D. Thesis, Univ. of Calif. at Berkeley 1990. (H. Helson, Advisor).

This thesis considers iteration of the $qx + 1$ function defined by $C_q(x) = \frac{qx+1}{2^{\mathrm{ord}_2(qx+1)}}$, where $2^{\mathrm{ord}_2(y)} || y$, and both $q$ and $x$ are odd integers. The first part of the thesis studies a conjecture of Crandall (1978), and the results appear in Franco and Pomerance (1995). The second part of the thesis gives a method to determine for a fixed $q$ whether there are any orbits of period 2, i.e. solutions of $C_q^{(2)}(x) = x$, and it shows that for $|q| < 10^{11}$, only $q = \pm1, \pm3, 5, -11, -91,$

316     JEFFREY C. LAGARIAS

and 181 have such orbits. The method uses an inequality of F. Beukers [Acta Arith. **38** (1981) 389–410].

**(90-38)**   Zachary Franco and Carl Pomerance (1995), *On a Conjecture of Crandall Concerning the qx + 1 Problem,* Math. Comp. **64** (1995), 1333–1336. (MR9 5j:11019).

This paper considers iterates of the $qx+1$ function $C_q(x) = \frac{qx+1}{2^{\mathrm{ord}_2(qx+1)}}$, where $2^{\mathrm{ord}_2(y)} \| y$ and both $q$ and $x$ are odd integers. Crandall (1978) conjectured that for each odd $q \geq 5$ there is some $n > 1$ such that the orbit $\{C_q^{(k)}(n) : k \geq 0\}$ does not contain 1, and proved it for $q = 5, 181$ and $1093$. This paper shows that $\{q : \text{Crandall's conjecture is true for } q\}$ has asymptotic density 1, by showing the stronger result that the set $\{q : C^{(2)}(m) \neq 1 \text{ for all } m \in \mathbb{Z}\}$ has asymptotic density one.

**(90-39)**   David Gale (1991), *Mathematical Entertainments: More Mysteries,* Mathematical Intelligencer **13**, No. 3, (1991), 54–55.

This paper discusses the possible undecidability of the $3x+1$ Conjecture. It also discusses whether the orbit containing $n = 8$ under iteration of the original Collatz function is infinite.

**(90-40)**   Guo-Gang Gao (1993), *On consecutive numbers of the same height in the Collatz problem,* Discrete Math., **112** (1993), 261–267. (MR 94i:11018).

This paper proves that if there exists one $k$-tuple of consecutive integers all having the same height and same total stopping time, then there exists infinitely many such $k$-tuples. (He attributes this result to P. Penning.) There is a 35654-tuple starting from $2^{500} + 1$. He conjectures that the set $\{n : C^{(k)}(n) = C^{(k)}(n + 1)$ for some $k \leq \log_2 n\}$ has natural density one, and proves that it has a natural density which is at least 0.389.

**(90-41)**   Manuel V. P. Garcia and Fabio A. Tal (1999), *A note on the generalized 3n + 1 problem,* Acta Arith. **90**, No. 3 (1999), 245–250. (MR 2000i:11019).

This paper studies the generalized $3x + 1$ function, defined for $m > d \geq 2$ by

$$H(x) = \begin{cases} \frac{x}{d} & \text{if } x \equiv 0 \pmod{d}, \\ \frac{mx - \pi(ma)}{2} & \text{if } x \equiv a \pmod{d}, \ a \neq 0, \end{cases}$$

where $\pi(x)$ denotes projection (mod d) onto a fixed complete set of residues (mod d). The *Banach density* of a set $B \subset \mathbb{Z}^+$ is

$$\rho_b(B) = \limsup_{n \to \infty} (max_{a \in \mathbb{Z}^+} \frac{\sharp(B \cap \{a, a+1, \ldots, a+n-1\})}{n}).$$

The Banach density is always defined and is at least as large as the natural density of the set $B$, if it exists. Call two integers $m_1$ and $m_2$ *equivalent* if there is some positive integer $k$ such that $H^{(k)}(m_1) = H^{(k)}(m_2)$. The authors assume that $m < d^{d/d-1}$, a hypothesis which implies that almost all integers have some iterate which is smaller, and which includes the $3x+1$ function as a special case. They prove that if $\mathcal{P}$ is any complete set of representatives of equivalence classes of $\mathbb{Z}^+$ then the Banach density of $\mathcal{P}$ is zero. As a corollary they conclude that the Banach density of the orbit of any integer $n$ under such a map $H$ is zero.

In particular, the Banach density of any divergent trajectory for such a map is zero.

**(90-42)**   Gaston Gonnet (1991), *Computations on the $3n + 1$ Conjecture*, MAPLE Technical Newsletter **0**, No. 6, Fall 1991.

This paper describes how to write computer code to efficiently compute $3x + 1$ function iterates for very large $x$ using MAPLE. It displays a computer plot of the total stopping function for $n < 4000$, revealing an interesting structure of well-spaced clusters of points.

**(90-43)**   Lorenz Halbeisen and Norbert Hungerbühler (1997), *Optimal bounds for the length of rational Collatz cycles*, Acta Arithmetica **78** (1997), 227-239. (MR 98g:11025).

The paper presents "optimal" upper bounds for the size of a minimal element in a rational cycle of length $k$ for the $3x + 1$ function. These estimates improve on Eliahou (1993) but currently do not lead to a better linear bound for nontrivial cycle length than the value 17,087,915 obtained by Eliahou. They show that if the $3x + 1$ Conjecture is verified for $1 \leq n \leq 212, 366, 032, 807, 211$, which is about $2.1 \times 10^{14}$, then the lower bound on cycle length jumps to 102,225,496.

*Note.* The $3x + 1$ conjecture has now been verified well beyond the bound needed here, up to $2.7 \times 10^{16}$ by Oliveira e Silva (1999)., and he has since extended the verification to $5 \times 10^{18}$.

**(90-44)**   Gisbert Hasenjager (1990), *Hasse's Syracuse-Problem und die Rolle der Basen*, in: *Mathesis rationis. Festschrift für Heinrich Schepers*, (A. Heinekamp, W. Lenzen, M. Schneider, Eds.) Nodus Publications: Münster 1990 (ISBN 3-89323-229-X), 329–336.

This paper, written for a philosopher's anniversary, comments on the complexity of the $3x + 1$ problem compared to its simple appearance. It suggests looking for patterns in the iterates written in base 3 or base 27, rather than in base 2. It contains heuristic speculations and no theorems.

**(90-45)**   John A. Joseph (1998), *A chaotic extension of the Collatz function to $\mathbb{Z}_2[i]$*, Fibonacci Quarterly **36** (1998), 309–317. (MR 99f:11026).

This paper studies the function 1 $\mathcal{F} : \mathbb{Z}_2[i] \to \mathbb{Z}_2[i]$ given by

$$\mathcal{F}(\alpha) = \begin{cases} \frac{\alpha}{2} & \text{if } \alpha \in [0] \ , \\ \frac{3\alpha + 1}{2} & \text{if } \alpha \in [1] \ , \\ \frac{3\alpha + i}{2} & \text{if } \alpha \in [i] \ , \\ \frac{3\alpha + 1 + i}{2} & \text{if } \alpha \in [1 + i] \ , \end{cases}$$

where $[\alpha]$ denotes the equivalence class of $\alpha$ in $\mathbb{Z}_2[i]/2\mathbb{Z}_2[i]$. The author proves that the map $\tilde{T}$ is chaotic in the sense of R. L. Devaney [*A First Course in Dynamical Sysems: Theory and Experiment*, Addison-Wesley 1992]. He shows that $\tilde{T}$ is not conjugate to $T \times T$ via a $\mathbb{Z}_2$-module isomorphism, but is topologically conjugate to $T \times T$. This is shown using an analogue $\tilde{Q}_\infty$ of the $3x + 1$ conjugacy map $Q_\infty$ studied in Lagarias (1985), Theorem L and in Bernstein and Lagarias (1996).

**(90-46)**  Frantisek Kascak (1992), *Small universal one-state linear operator algorithms*, in: Proc. MFCS '92, Lecture Notes in Computer Science No. 629, Springer-Verlag: New York 1992, pp. 327–335. [MR1255147]

A *one-state linear operator algorithm* (OLOA) with modulus $m$, is specified by data $(a_r, b_r, c_r)$ a triple of integers, for each residue class $r$ $0 \bmod m$). Given an input integer $x$, the OLOA does the following in one step. It finds $x \equiv r \pmod m$ and based on the value of $r$, it does the following. If $c_r = 0$, the machine halts and outputs $x$, calling it a final number, If $c_r \neq 0$ it computes to $(a_r x + b_r \text{ DIV } c_r = \frac{a_r x + b_r}{c_r}$, and outputs this value, provided this output is a nonnegative integer. If $\frac{a_r x + b_r}{c_r}$ is a non-integer or is negative, the the number $x$ is called a terminal number, and the machine stops. An OLOA $L$ computes a partially defined function $m \to f_L(m)$ as follows. For initial input value $m$, the OLOA is iterated, and halts with $f_L(m) = x$ if $x$ is a final number. Otherwise $f_L(m) = \uparrow$ is viewed as undefined, and this occurs if the computation either reaches a terminal number or else runs for an infinite number of steps without stopping.

The author observes that the $3x + 1$ problem can be encoded as an OLOA $L$ with modulus $m = 30$. The increase of modulus of the $3x + 1$ function from $m = 2$ to $m = 30$ is made in order to encode the $3x + 1$ iteration arriving at value 1 as a halting state in the sense above. The $3x + 1$ function input $n$ is encoded as the integer $m = 5 \cdot 2^{n-1}$ to the OLOA. It is not known if the function $f_L$ computed by this OLOA has a recursive domain $D(f_L)$, but if the $3x + 1$ conjecture is true, then $D(F_L) = \mathbb{N}$ will be recursive.

The main result of the paper is that there exists an OLOA $L$ with modulus $m = 396$ which is a universal OLOA, i.e. there exists a unary recursive function $d$ such that $df_L$ is a universal unary recursive function. Here a unary partial recursive function $u$ is *universal* it there exists some binary recursive function $c$ such that $F_x(y) = uc(x, y)$ where $F_x(y)$ is an encoding of the $x$-th function in an encoding of all unary partial recursive function, evaluated at input $y$. In particular this machine $L$ has an unsolvable halting problem, i.e. the function $f_L$ has domain $D(f_L)$ which is not recursive. The encoding of the construction uses some ideas from Minsky machines, as in Conway (1972). The construction of Conway would give a universal OLOA with a much larger modulus.

**(90-47)**  Louis H. Kauffman (1995), *Arithmetic in the form,* Cybernetics and Systems **26** (1995), 1–57. [Zbl 0827.03033]

This interesting paper describes how to do arithmetic in certain symbolic logical systems developed by G. Spencer Brown, *Laws of Form*, George Unwin & Brown: London 1969. Spencer-Brown's formal system starts from a primitive notion of the additive void 0 and the multiplicative void 1, and uses division of space by boundaries to create numbers, which are defined using certain transformation rules defining equivalent expressions. Kauffman argues that these rules give operations more primitive than addition and multiplication. He develops a formal natural number arithmetic in this context, specifying replacement rules for rearrangements of symbolic expressions, and gives a proof of consistency of value (Theorem, p. 27).

In Appendix A he presents a second symbolic system, called *string arithmetic*. He discusses the $3x + 1$ problem in this context. This system has three kinds of symbols $*, \langle, \rangle$, where in representing integers the angles occur only in matching pairs. The integer $N$ is represented in unary as a string of $N$ asterisks. This integer has alternate expressions, in which the angles encode multiplication by 2. Addition is given by concatenation of expressions. The transformation rules between equivalent expressions are, if $W$ is any string of symbols

$$** \quad \longleftrightarrow \quad \langle * \rangle$$
$$\rangle\langle \quad \longleftrightarrow \quad \text{(blank)}$$
$$*W \quad \longleftrightarrow \quad W*$$

A natural number can be defined as any string $W$ equivalent under these rules to some string of asterisks. For examples, these rules give as equivalent representations of $N = 4$:

$$* * ** = \langle * \rangle \langle * \rangle = \langle ** \rangle = \langle \langle * \rangle \rangle.$$

He observes that the $3x + 1$ problem is particularly simple to formulate in this system. He gives transformation rules defining the two steps in the $3x + 1$ function. If $N = \langle W \rangle$, then it is even, and then $\frac{N}{2} = W$, while if $N = \langle W \rangle *$, then it is odd and $\frac{3N+1}{2} = \langle W * \rangle W$. These give the transformation rules

$$(T1): \qquad \langle W \rangle \quad \mapsto \quad W$$

$$(T2): \qquad \langle W \rangle * \quad \mapsto \quad \langle W * \rangle W$$

The $3x + 1$ Conjecture is equivalent to the assertion: starting from a string of asterisks of any length, and using the transformation rules plus the rules (T1) and (T2), one can reach the string of one asterisk $*$. Kauffman says: "A more mystical reason for writing the Collatz [problem] in string arithmetic is the hope that there is some subtle pattern right in the notation of string arithmetic that will show the secrets of the iteration."

(**90-48**)    Timothy P. Keller (1999), *Finite cycles of certain periodically linear functions*, Missouri J. Math. Sci. **11** (1999), no. 3, 152–157. (MR 1717767)

This paper is motivated by the original Collatz function $f(3n) = 2n$, $f(3n + 1) = 4n + 1$, $f(3n + 2) = 4n + 3$, which is a permutation of $\mathbb{Z}$, see Klamkin (1963). The class of periodically linear functions which are permutations. were characterized by Venturini (1997). The author defines a *permutation of type V* to be a periodically linear function $g_p$ defined for an odd integer $p > 1$ by $g_p(pn+r) = (p+(-1)^r)n+r$ for $0 \leq r \leq p-1$. The permutation $g_3$ is conjugate to the original Collatz function. This paper shows that for each $L \geq 1$ any permutation of type $V$ has only finitely many cycles of period $L$. The author conjectures that for each odd $p$ there is a constant $s_0(p)$ such that if $L \geq s_0(p)$ is the minimal period of a periodic orbit of $g_p$, then $L$ is a denominator of a convergent of the continued fraction expansion of $\gamma_p := \frac{\log(p-1)-\log p}{\log p - \log(p-1)}$. For $p = 7$ these denominators start $\Lambda(7) = \{1, 2, 13, 28, 265, 293, \dots\}$. He finds that $g_7$ has an orbit of period 265 with starting value $n_1 = 1621$, and an orbit of period 293 with starting value $n_2 = 293$. This question was raised for the original Collatz function by Shanks (1965).

**(90-49)**  David A. Klarner and Karel Post (1992), *Some fascinating integer sequences,* A collection of contributions in honour of Jack van Lint. Discrete Mathematics **106/107** (1992), 303–309. (MR 93i:11031).

This paper studies sets of integers $T$ that are closed under the iteration of certain multi-variable affine maps $\alpha(x_1, \cdots, x_r) = m_0 + m_1 x_1 + \cdots + m_r x_r$. It is related to Klarner (1988), which considered one-variable functions. Here the authors consider a single function of two variable. They fix an integer $m \geq 2$, and study sets of integers closed under action of the two-variable function

$$\gamma_m(x, y) := mx + my + 1.$$

Let $\langle \gamma_m : 0 \rangle$ denote the set generated starting from the element $\{0\}$. using iteration of this function, which may have a complicated structure. They set $G^{(1)} = \langle \gamma_m : 0 \rangle$ and consider the hierarchy of sumsets $G^{(k+1)} = G^{(1)} + G^{(k)}$. One finds that $G^{(2k)} = \{0\} \cup \left( \cup_{i=1}^{2k} (mG^{(2i)} + i) \right)$. They introduce the affine linear functions $\mu_i(x) = mx + i$, and prove that $M^{(2k)} := \langle \mu_{k+1}, \mu_{k+2}, \cdots, \mu_{2k} : \{0, 1, \ldots, 2k\} \rangle$ has $M^{(2k)} \subseteq G^{(2k)}$. They deduce that $M^{(2m-2)} = \mathbb{N}$ and hence in Theorem 1 that $G^{(2k)} = \mathbb{N}$ for all $k \geq m-1$. They deduce from this that $\langle \gamma_m : 0 \rangle$ is an $m$-recognizable set in the sense of Klarner (1988). They assert that for all $m$ that the sets $G^{(1)}$ and $G^{(2k)}$ all have positive limiting natural densities. They carry this out for the case $m = 5$, and find that $(G^{(1)}, G^{(2)}, G^{(4)}, G^{(6)}, G^{(8)})$ have natural densities $(\frac{1}{40}, \frac{1}{8}, \frac{1}{2}, \frac{7}{8}, 1)$, respectively. They then derive a finite automaton for generating the set $\langle \gamma_5 : 0 \rangle$.

**(90-50)**  Ivan Korec (1992), *The $3x+1$ problem, generalized Pascal triangles, and cellular automata,* Math. Slovaca, **42** (1992), 547–563. (MR 94g:11019).

This paper shows that the iterates of the Collatz function $C(x)$ can actually be encoded in a simple way by a one-dimensional nearest-neighbor cellular automaton with 7 states. The automaton encodes the iterates of the function $C(x)$ written in base 6. The encoding is possible because the map $x \to 3x+1$ in base 6 does not have carries propagate. (Compare the $C(x)$-iterates of $x = 26$ and $x = 27$ in base 6.) The $3x+1$ Conjecture is reformulated in terms of special structural properties of the languages output by such cellular automata.

**(90-51)**  Ivan Korec (1994), *A Density estimate for the $3x + 1$ problem,* Math. Slovaca **44** (1994), 85–89. (MR 95h:11022).

Let $S_\beta = \{n : \text{some } T^{(k)}(n) < n^\beta\}$. This paper shows that for any $\beta > \frac{\log 3}{\log 4} \doteq .7925$ the set $S_\beta$ has density one. The proof follows the method of E. Heppner (1978).

**(90-52)**  James R. Kuttler (1994), *On the $3x + 1$ problem,* Adv. Appl. Math. **15** (1994), 183–185. (Zbl. 803.11018.)

The author states the oft-discovered fact that $T^{(k)}(2^k n - 1) = 3^k n - 1$ and derives his main result that if $r$ runs over all odd integers $1 \leq r \leq 2^k - 1$ then $T^{(k)}(2^k n + r) = 3^p n + s$, in which $p \in \{1, 2, \ldots, k\}$ and $1 \leq s \leq 3^p$, and each value of $p$ occurs exactly $\binom{k-1}{p-1}$ times. Thus the density of integers with $T^{(k)}(n) > n$ is exactly $\frac{\alpha}{2^k}$, where $\alpha = \sum_{p > \theta k} \binom{k-1}{p-1}$, and $\theta = \log_2 3$. This counts the number of inflating vectors in Theorem C in Lagarias (1985).

**(90-53)**  Jeffrey C. Lagarias (1990), *The set of rational cycles for the $3x+1$ problem,* Acta Arithmetica **56** (1990), 33–53. (MR 91i:11024).

This paper studies the sets of those integer cycles of

$$T_k(x) = \begin{cases} \frac{3x+k}{2} & \text{if } x \equiv 1 \ (\text{mod } 2), \\ \frac{x}{2} & \text{if } x \equiv 0 \ (\text{mod } 2) , \end{cases}$$

for positive $k \equiv \pm 1$ (mod 6) which have $(x,k) = 1$. These correspond to rational cycles $\frac{x}{k}$ of the $3x+1$ function $T$. It conjectures that every $T_k$ has such an integer cycle. It shows that infinitely many $k$ have at least $k^{1-\epsilon}$ distinct such cycles of period at most log $k$, and infinitely many $k$ have no such cycles having period length less than $k^{1/3}$. Estimates are given for the counting function $C(k,y)$ counting the number of such cycles of $T_k$ of period $\leq y$, for all $k \leq x$ with $y = \beta \log x$. In particular $C(k, 1.01k) \leq 5k(\log k)^5$.

**(90-54)**  Jeffrey C. Lagarias (1999), *How random are 3x + 1 function iterates?,* in: *The Mathemagician and Pied Puzzler: A Collection in Tribute to Martin Gardner,* A. K. Peters, Ltd.: Natick, Mass. 1999, pp. 253–266.

This paper briefly summarizes results on extreme trajectories of 3x+1 iterates, including those in Applegate and Lagarias (1995a), (1995b),(1995c) and Lagarias and Weiss (1992). It also presents some large integers $n$ found by V. A. Vyssotsky whose trajectories take $c \log n$ steps to iterate to 1 for various $c > 35$. For example $n = 37\ 66497\ 18609\ 59140\ 59576\ 52867\ 40059$ has $\sigma_\infty(n) = 2565$ and $\gamma(n) = 35.2789$. It also mentions some unsolved problems.

The examples above were the largest values of $\gamma$ for $3x+1$ iterates known at the time, but are now superseded by examples of Roosendaal (2004+) achieving $\gamma = 36.716$.

**(90-55)**  Jeffrey C. Lagarias, Horacio A. Porta and Kenneth B. Stolarsky (1993), *Asymmetric tent map expansions I. Eventually periodic points,* J. London Math. Soc., **47** (1993), 542–556. (MR 94h:58139).

This paper studies the set of eventually periodic points $\text{Per}(T_\alpha)$ of the asymmetric tent map

$$T_\alpha(x) = \begin{cases} \alpha x & \text{if } 0 < x \leq \frac{1}{\alpha} \\ \frac{\alpha}{\alpha-1}(1-x) & \text{if } \frac{1}{\alpha} \leq x < 1 , \end{cases}$$

where $\alpha > 1$ is real. It shows that $\text{Per}(T_\alpha) = \mathbb{Q}(\alpha) \cap [0,1]$ for those $\alpha$ such that both $\alpha$ and $\frac{\alpha}{\alpha-1}$ are Pisot numbers. It finds 11 such numbers, of degree up to four, and proves that the set of all such numbers is finite.

It conjectures that the property $\text{Per}(T_\alpha) = \mathbb{Q}(\alpha) \cap [0,1]$ holds for certain other $\alpha$, including the real root of $x^5 - x^3 - 1 = 0$. The problem of proving that $\text{Per}(T_\alpha) = \mathbb{Q}(\alpha) \cap [0,1]$ in these cases appears analogous to the problem of proving that the $3x + 1$ function has no divergent trajectories.

*Note.* C. J. Smyth, *There are only eleven special Pisot numbers,* (Bull. London Math. Soc. **31** (1989) 1–5) proved that the set of 11 numbers found above is the complete set.

**(90-56)**  Jeffey C. Lagarias, Horacio A. Porta and Kenneth B. Stolarsky (1994), *Asymmetric tent map expansions II. Purely periodic points,* Illinois J. Math. **38** (1994), 574–588. (MR 96b:58093, Zbl 809:11042).

This paper continues Lagarias, Porta and Stolarsky (1993). It studies the set $\text{Fix}(T_\alpha)$ of purely periodic points of the asymmetric tent map $T_\alpha(\cdot)$ and the set $\text{Per}_0(T_\alpha)$ with terminating $T_\alpha$-expansion, in those cases wwhen $\alpha$ and $\frac{\alpha}{1-\alpha}$ are simultaneously Pisot numbers. It shows that $\text{Fix}(T_\alpha) \subseteq \{\gamma \in \mathbb{Q}(\alpha)$ and $\sigma(\alpha) \in A_\alpha^\sigma$ for all embeddings $\sigma : \mathbb{Q}(\alpha) \to \mathbb{C}$ with $\sigma(\alpha) \neq \alpha\}$, in which each set $A_\alpha^\sigma$ is a compact set in $\mathbb{C}$ that is the attractor of a certain hyperbolic iterated function system. It shows that equality holds in this inclusion in some cases, and not in others. Some related results for $\text{Per}_0(T_\alpha)$ are established.

**(90-57)**  Jeffrey C. Lagarias and Alan Weiss (1992), *The $3x+1$ problem: Two stochastic models,*  Annals of Applied Probability **2** (1992), 229–261. (MR 92k:60159).

This paper studies two stochastic models that mimic the "pseudorandom" behavior of the $3x+1$ function. The models are branching random walks, and the analysis uses the theory of large deviations.

For the models the average number of steps to get to 1 is $\alpha_0 \log n$, where $\alpha_0 = \left(\frac{1}{2} \log \frac{3}{4}\right)^{-1} \approx 6.9$ For both models it is shown that there is a constant $c_0 = 41.677...$ such that with probability one for any $\epsilon > 0$ only finitely many $m$ take $(c_0 + \epsilon) \log m$ iterations to take the value 1, while infinitely many $m$ take at least $(c_0 - \epsilon) \log m$ iterations to do so. This prediction is shown to be consistent with empirical data for the $3x+1$ function.

The paper also studies the maximum excursion

$$t(n) := \max_{k \geq 1} T^{(k)}(n)$$

and conjectures that $t(n) < n^{2+o(1)}$ as $n \to \infty$. An analog of this conjecture is proved for one stochastic model. The conjecture $t(n) \leq n^{2+o(1)}$ is consistent with empirical data of Leavens and Vermeulen (1992) for $n < 10^{12}$.

**(90-58)**  Gary T. Leavens and Mike Vermeulen (1992), $3x+1$ *search programs,* Computers & Mathematics, with Applications **24**, No. 11,(1992), 79–99. (MR 93k:68047).

This paper describes methods for computing $3x+1$ function iterates, and gives results of extensive computations done on a distributed network of workstations, taking an estimated 10 CPU-years in total. The $3x+1$ Conjecture is verified for all $n < 5.6 \times 10^{13}$. The paper presents statistics on various types of extremal trajectories of $3x+1$ iterates in this range. It gives a detailed discussion of techniques used for program optimization.

**(90-59)**  Simon Letherman, Dierk Schleicher and Reg Wood (1999), *On the $3X+1$ problem and holomorphic dynamics,* Experimental Math. **8**, No. 3 (1999), 241–251. (MR 2000g:37049.)

This paper studies the class of entire functions

$$f_h(z) := \frac{z}{2} + (z + \frac{1}{2})(\frac{1 + \cos \pi z}{2}) + \frac{1}{\pi}(\frac{1}{2} - \cos \pi z) \sin \pi z + h(z)(\sin \pi z)^2,$$

in which $h(z)$ is an arbitrary entire function. Each function in this class reproduces the $3x+1$-function on the integers, and the set of integers is contained in the set of critical points of the function. The simplest such function takes $h(z)$ to be identically zero, and is denoted $f_0(z)$. The authors study the iteration properties of this map in the complex plane $\mathbb{C}$. They show that $\mathbb{Z}$ is contained in the Fatou set of $f_h(z)$. There is a classification of the connected components of the Fatou set of an entire function into six categories: (1) (periodic) immediate basins of (super-) attracting periodic points, (2) (periodic) immediate basins of attraction of rationally indifferent periodic points, (3) (periodic) Siegel disks, (4) periodic domains at infinity (Baker domains), (5) preperiodic components of any of the above, and (6) wandering domains. The following results all apply to the function $f_0(z)$, and some of them are proved for more general $f_h(z)$. The Fatou component of any integer must be in the basin of attraction of a superattracting periodic point or be in a wandering domain; the authors conjecture the latter does not happen. The existence of a divergent trajectory is shown equivalent to $f_0(z)$ containing a wandering domain containing some integer. No two integers are in the same Fatou component, except possibly $-1$ and $-2$. The real axis contains points of the Julia set between any two integers, except possibly $-1$ and $-2$. It is known that holomorphic dynamics of an entire function $f$ is controlled by its critical points. The critical points of $f_0(z)$ on the real axis consist of the integers together with $-\frac{1}{2}$. The authors would like to choose $h$ to reduce the number of other critical points, to simplify the dynamics. However they show that any map $f_h(z)$ must contain at least one more critical point in addition to the critical points on the real axis (and presumably infinitely many.) The authors compare and contrast their results with the real dynamics studied in Chamberland (1996).

**(90-60)**    Jerzy Marcinkowski (1999), *Achilles, turtle, and undecidable boundedness problems for small DATALOG programs*, SIAM J. Comput. **29** (1999), 231–257. (MR 2002d:68035).

DATALOG is the language of logic programs without function symbols. A DATALOG program consists of a finite set of Horn clauses in the language of first order logic without equality and without functions.

The author introduces *Achilles-Turtle machines*, which model the iteration of *Conway functions*, as introduced in Conway (1972). These are functions $g : \mathbb{N} \to \mathbb{N}$ having the form

$$g(n) = \frac{a_j}{q_j} n \ \text{ if } n \equiv j (\text{mod } p), \ 0 \leq j \leq p - 1,$$

where for each $j$, $a_j \geq 0, q_j \geq 1$ are integers with $q_j | gcd(j, p)$ and with $\frac{a_j}{q_j} \leq p$. He cites Devienne, Lebégue and Routier (1993) for the idea of relating Conway functions to Horn clauses. In Section 2.3 he explicitly describes an Achilles-Turtle machine associated to computing the Collatz function.

The paper proves that several questions concerning the uniform boundedness of computations are undecidable. These include uniform boundedness for ternary linear programs; uniform boundedness for single recursive rule ternary programs; and uniform boundedness of single rule programs. These results give no information regarding the $3x + 1$ function itself.

**(90-61)**  Dănuţ Marcu (1991), *The powers of two and three,* Discrete Math. **89** (1991), 211-212. (MR 92h:11026).

This paper obtains a similar result to Narkiewicz (1980), with a similar proof, but obtains a slightly worse bound on the exceptional set $N(T) < 2.52T^\theta$ where $\theta = \frac{\log 2}{\log 3}$.

*Note.* This paper cites a paper of Gupta appearing in the same journal as Narkiewicz's paper [Univ. Beograd Elektrotech. Fak. Ser. Mat. Phys.] slightly earlier in the same year, but rather strangely fails to cite Narkiewicz (1980). For clarification, consult the Wikipedia article on D. Marcu.

**(90-62)**  Maurice Margenstern and Yuri Matiyasevich (1999), *A binomial representation of the $3X + 1$ problem,* Acta Arithmetica **91**, No. 4 (1999), 367–378. (MR 2001g:11015).

This paper encodes the $3x + 1$ problem as a logical problem using one universal quantifier and existential quantifiers, with an arithmetical formula using polynomials and binomial coefficients. The authors observe that the use of such expressions in a language with binomial coefficients often leads to shorter formulations than are possible in a language just allowing polynomial equations and quantifiers. They give three equivalent restatements of the $3x + 1$ Conjecture in terms of quantified binomial coefficient equations.

**(90-63)**  Keith R. Matthews (1992), *Some Borel measures associated with the generalized Collatz mapping,* Colloq. Math. **53** (1992), 191–202. (MR 93i:11090).

This paper studies maps of the form

$$T(x) = \frac{m_i x - r_i}{d} \text{ if } x \equiv i \pmod{d} .$$

These extend first to maps on the $d$-adic integers $\mathbb{Z}_d$, and then further to maps on the polyadic integers $\hat{\mathbb{Z}}$. Here $\hat{\mathbb{Z}}$ is the projective limit of the set of homomorphisms $\phi_{n,m} : \mathbb{Z} \to \mathbb{Z}/m\mathbb{Z}$ where $m|n$. The open sets $B(j,m) := \{x \in \hat{\mathbb{Z}} : x \equiv j(\text{mod } m)\}$ put a topology on $\hat{\mathbb{Z}}$, which has a Haar measure $\sigma(B(j,m)) = \frac{1}{m}$. The paper proves conjectures of Buttsworth and Matthews (1990) on the structure of all ergodic open sets (mod $m$). In particular the ergodic sets link together to give finitely many projective systems, each giving $T$-invariant measure on $\hat{\mathbb{Z}}$. The paper gives examples where there are infinitely many ergodic sets.

**(90-64)**  Danilo Merlini and Nicoletta Sala (1999), *On the Fibonacci's attractor and the long orbits in the $3n+1$ problem,* International Journal of Chaos Theory and Applications **4**, No. 2-3 (1999), 75–84.

This paper studies heuristic models for the Collatz tree of inverse iterates of the Collatz function $C(x)$, which the authors call the "chalice". They also consider the length of the longest trajectories for the Collatz map. The author's predictions are that the number of leaves in the Collatz tree at depth $k$ should grow like $Ac^k$ where $c = \frac{1}{2}(1 + \sqrt{\frac{7}{3}}) \approx 1.2637$ and $A = \frac{3c}{3c-2} \approx 0.6545$. This estimate is shown to agree closely with numerical data computed to depth $k = 32$. They predict that the longest orbits of the Collatz map should take no more than $67.1 \log n$ steps to reach 1. This is shown to agree with empirical data for the longest orbit found up to $10^{10}$ by Leavens and Vermeulen (1992).

*Note.* These models have some features like those given in Lagarias and Weiss (1992) for the $3x + 1$ map $T(n)$. The asymptotic constants they obtain differ from those in the models of Lagarias and Weiss (1992) because the Collatz function $C(x)$ takes one extra iteration each time an odd number occurs, compared to for the $3x + 1$ mapping $T(x)$. The Applegate and Lagarias estimate for the number of nodes in the $k$-th level of a $3x + 1$ tree is $(\frac{4}{3})^{k(1+o(1))}$, and their estimate for the maximal number of steps to reach 1 for the $3x + 1$ map is about $41.677 \log n$.

**(90-65)**   Karl Heinz Metzger (1995), *Untersuchungen zum $(3n + 1)$-Algorithmus. Teil I: Das Problem der Zyklen,* PM (Praxis der Mathematik in der Schule) **38** (1995), 255–257. (*Nachtrag zum Beweis de $(3n + 1)$-Problems,* PM **39** (1996), 217.)

The author studies cycles for the $3x + 1$ map, and more generally considers the $bx + 1$ map for odd $b$. He exhibits a cycle for the $5x + 1$ that does not contain 1, namely $n = 13$. He obtains a general formula (4.1) for a rational number $a$ to be in a cycle of the $3x + 1$ map, as in Lagarias (1990). At the end of the paper is a theorem asserting that the only cycle of the Collatz map acting on the positive integers is the trivial cycle $\{1, 2, 4\}$. The proof of this theorem has a gap, however. The author points this out in the addendum, and says he hopes to return to the question in future papers.

The author's approach to proving that the trivial cycle is the only cycle on the positive integers is as follows. He observes that the condition (4.1) for $c$ to be a rational cycle with $\nu$ odd elements can be rewritten as

$$c = \frac{A_\nu + x_0^*}{A_\nu + y_0^*}$$

for certain integers $(x_0^*, y_0^*)$, taking $A_\nu = 2^{2\nu} - 3^\nu$. He then observes that this can be viewed as a special case of a linear Diophantine equation in $(x, y)$,

$$c = \frac{A_\nu + x}{A_\nu + y}, \text{ with } A, c \text{ fixed},$$

which has the general solution $(x, y) = ((c^2 - 1)A + \lambda c, (c - 1)A + \lambda c)$, for some integer $\lambda$. His proof shows that the case where $(x_0^*, y_0^*)$ has associated value $\lambda = 0$ leads to a contradiction. The gap in the proof is that cases where $\lambda \neq 0$ are not ruled out.

**(90-66)**   Karl Heinz Metzger (1999), *Zyklenbestimmung beim $(bn+1)$-Algorithmus,* PM (Praxis der Mathematik in der Schule) **41** (1999), 25–27.

For an odd number $b$ the Collatz version of the $bx + 1$ function is

$$C_b(x) = \begin{cases} bx + 1 & \text{if } x \equiv 1 \pmod{2}, \\ \frac{x}{2} & \text{if } x \equiv 0 \pmod{2}. \end{cases}$$

The author studies cycles for the $bn + 1$ map. Let the *size* of a cycle count the number of distinct odd integers it contains. The author first observes that 1 is in a size one cycle for the $bn + 1$ problem if and only if $b = 2^\nu - 1$ for some $\nu \geq 1$. For the $5x + 1$ problem he shows that on the positive integers there is no cycle of size 1, a unique cycle of size 2, having smallest element $n = 1$, and exactly two cycles of size 3, having smallest elements $n = 13$ and $n = 17$, respectively.

At the end of the paper the author states a theorem asserting that the only cycle of the Collatz map acting on the positive integers is the trivial cycle $\{1, 2, 4\}$. This paper apparently is intended to repair the faulty proof in Metzger (1995). However the proof of this result still has a gap.

**(90-67)**  Pascal Michel (1993), *Busy beaver competition and Collatz-like problems*, Archive Math. Logic **32** (1993), 351–367. (MR 94f:03048).

The Busy Beaver problem is to find that Turing machine which, among all $k$-state Turing machines, when given the empty tape as input, eventually halts and produces the largest number $\Sigma(k)$ of ones on the output tape. The function $\Sigma(k)$ is well-known to be non-recursive. This paper shows that the current Busy Beaver record-holder for 5-state Turing machine computes a Collatz-like function. This machine $M_5$ of H. Marxen and G. Buntrock [Bulletin EATCS No. **40** (1990) 247–251] has $\Sigma(M_5) = 47,176,870$. Michel shows that $M_5$ halts on all inputs if and only if iterating the function $g(3m) = 5m + 6$, $g(3m + 1) = 5m + 9$, $g(3m + 2) = \uparrow$ eventually halts at $\uparrow$ for all inputs. Similar results are proved for several other 5-state Turing machines. The result of Mahler (1968) on $Z$-numbers is restated in this framework: If a $Z$-number exists then there is an input $m_0$ such that the iteration $g(2m) = 3m$, $g(4m + 1) = 6m + 2$, $g(4m + 3) = \uparrow$ never halts with $\uparrow$ when started from $m_0$.

**(90-68)**  Filippo Mignosi (1995), *On a generalization of the $3x + 1$ problem*, J Number Theory, **55** (1995), 28–45. (MR 96m:11016).

For real $\beta > 1$, define the function $T_\beta : \mathbb{N} \to \mathbb{N}$ by

$$T_\beta(n) = \begin{cases} \lceil \beta n \rceil & \text{if } n \equiv 1 \ (\text{mod } 2) \\ \frac{n}{2} & \text{if } n \equiv 0 \ (\text{mod } 2) . \end{cases}$$

Then $\beta = \frac{3}{2}$ gives the $(3x + 1)$-function. Conjecture $C_\beta$ asserts that $T_\beta$ has finitely many cycles and every $n \in \mathbb{N}$ eventually enters a cycle under iteration of $T_\beta$. The author shows that, for any fixed $0 < \epsilon < 1$, and for $\beta$ either transcendental or rational with an even denominator, if $1 < \beta < 2$, then the set $S(\epsilon, \beta) = \{n : \text{some } T_\beta^{(k)}(n) < \epsilon n\}$ has natural density one, while if $\beta > 2$ then $S^*(\epsilon, \beta) = \{n : \text{some } T_\beta^{(k)}(n) > \epsilon^{-1} n\}$ has natural density one. For certain algebraic $\beta$ different behavior may occur, and Conjecture $C_\beta$ can sometimes be settled. In particular Conjecture $C_\beta$ is true for $\beta = \sqrt{2}$ and false for $\beta = \frac{1+\sqrt{5}}{2}$.

**(90-69)**  Helmut A. Müller (1991), *Das '3n + 1' problem*, Mitteilungen der Math. Ges. Hamburg **12** (1991), 231–251. (MR 93c:11053).

This paper presents basic results on $3x + 1$ problem, with some overlap of Lagarias (1985), and it presents complete proofs of the results it states. it contains new observations on the $3x+1$ function $T$ viewed as acting on the 2-adic integers $\mathbb{Z}_2$. For $\alpha \in \mathbb{Z}_2$ the 2-adic valuation $|\alpha|_2 = 2^{-j}$ if $2^j || \alpha$. Müller observes that $T(\alpha) = \sum_{n=0}^\infty (-1)^{n-1}(n+1)2^{n-2}\binom{\alpha}{n}$ where $\binom{\alpha}{n} = \frac{\alpha(\alpha-1)\cdots(\alpha-n+1)}{n!}$. The function $T(\alpha)$ is locally constant but not analytic. Define the function $Q_\infty : \mathbb{Z}_2 \to \mathbb{Z}_2$ by $Q_\infty(\alpha) = \sum_{i=0}^\infty a_i 2^i$ where $a_i = 0$ if $|T^{(i)}(\alpha)|_2 < 1$ and $a_i = 1$ if

$|T^{(i)}(\alpha)|_2 = 1$. Lagarias (1985) showed that this function is a measure-preserving homeomorphism of $\mathbb{Z}_2$ to itself. Müller proves that $Q_\infty$ is nowhere differentiable.

(90-70)   Helmut A. Müller (1994), *Über eine Klasse 2-adischer Funktionen im Zusammenhang mit dem "$3x + 1$"-Problem,* Abh. Math. Sem. Univ. Hamburg **64** (1994), 293–302.
(MR 95e:11032).

Let $\alpha = \sum_{j=0}^{\infty} a_{0,j} 2^j$ be a 2-adic integer, and let $T^{(k)}(\alpha) = \sum_{j=0}^{\infty} a_{k,j} 2^j$ denote its $k$-th iterate. Müller studies the functions $Q_j : \mathbb{Z}_2 \to \mathbb{Z}_2$ defined by $Q_j(\alpha) = \sum_{k=0}^{\infty} a_{kj} 2^k$. He proves that each function $Q_j$ is continuous and nowhere differentiable. He proves that the function $f : \mathbb{Z}_2 \to \mathbb{Q}_2$ given by $f = \sum_{j=1}^{N} A_j Q_j(\alpha)$ with constants $A_j \in \mathbb{Q}_2$ is differentiable at a point $\alpha$ with $T^{(k)}(\alpha) = 0$ for some $k \geq 0$ if and only if $2A_0 + A_1 = 0$ and $A_2 = A_3 = \ldots = A_N = 0$.

(90-71)   Tomás Oliveira e Silva (1999), *Maximum excursion and stopping time record-holders for the $3x + 1$ problem: Computational results,* Math. Comp. **68** No. 1 (1999), 371-384, (MR 2000g:11015).

This paper reports on computations that verify the $3x + 1$ conjecture for $n < 3 \cdot 2^{53} = 2.702 \times 10^{16}$. It also reports the values of $n$ that are champions for the quantity $\frac{t(n)}{n}$, where

$$t(n) := \sup_{k \geq 1} T^{(k)}(n) \ .$$

In this range all $t(n) \leq 8n^2$, which is consistent with the conjecture $t(n) \leq n^{2+o(1)}$ of Lagarias and Weiss (1992).

*Note*: In 2004 he implemented an improved version of this algorithm. As of October 2008 his computation verified the 3x+1 conjecture up to $19 \cdot 2^{58} > 5.476 \times 10^{18}$. See his webpage at: `http://www.ieeta.pt/ẽtos/`; email: `tos@ieeta.pt`. This is the current record value for verifying the $3x + 1$ conjecture.

(90-72)   Elio Oliveri and Giuseppe Vella (1998), *Alcune Questioni Correlate al Problema Del "3D+1",* Atti della Accademia di scienze lettere e arti di Palermo, Ser. V, **28** (1997-98), 21–52. (Italian)

This paper studies the structure of forward and backward iterates of the $3x + 1$ problem, treating the iteration as proceeding from one odd integer to the next odd iterate. It obtains necessary conditions for existence of a nontrivial cycle in the $3x + 1$ problem. It observes that if $(D_1, \ldots, D_n)$ are the odd integers in such a cycle, then

$$\prod_{i=1}^{n} \frac{3D_i + 1}{D_i} = 2^{e_1 + \cdots + e_n} = 2^k.$$

If the $3x + 1$ conjecture is verified up to $D$ then $\frac{3D+1}{D} > \frac{3D_i+1}{D_i}$, for all $i$, whence

$$n \log_2(\frac{3D + 1}{2}) > k > n \log_2 3.$$

They conclude that there must be an integer in the interval $[n \log_2(\frac{3D+1}{2}), n \log_2 3]$, a condition which imposes a lower bound on $n$, the number of odd integers in the cycle. In fact one must have $k = 1 + \lfloor n \log_2 3 \rfloor$, and $Max_i\{D_i\} > (2^{\frac{k}{n}} - 3)^{-1}$, and $\frac{k}{n}$ is a good rational approximation to $\log_2 3$. Using the continued fraction expansion of $\log_2 3$, including the intermediate convergents, the authors conclude that if the $3x + 1$ conjecture is verified for all integers up to $D = 2^{40} + 1$, then one may conclude $n \geq 1078215$.

The paper also investigates the numbers $M(N, n)$ of parity-sequences of length $N$ which contain $n$ odd values, at positions $k_1, k_2, \ldots, k_n$ with $k_n = N$, and such that the partial sums $h_j = k_1 + k_2 + \cdots + k_j$ with all $h_j \leq L_j := \lfloor j \log_2 3 \rfloor$. Here $\log_2(3) \approx 1.585$. These are the possible candidate initial parity sequences for the smallest positive number in a nontrivial cycle of the $3x + 1$ iteration of period $N$ or longer.

*Note:* The method of the authors for getting a lower bound on the length of nontrivial cycles is similar in spirit to earlier methods, cf. Eliahou (1993) and Halbeisen and Hüngerbuhler (1997). Later computations of the authors show that if one can take $D = 2^{61} + 1 \approx 2.306 \times 10^{18}$, then one may conclude there are no nontrivial cycles containing less than $n = 6, 586, 818, 670$ odd integers. The calculations of T. Olivera e Silva, as of 2008, have verified the $3x + 1$ conjecture to $19 \times 2^{58} \approx 5.476 \times 10^{18}$, so this improved cycle bound result is now unconditional.

**(90-73)**  Margherita Pierantoni and Vladan Ćurčić (1996), *A transformation of iterated Collatz mappings,* Z. Angew Math. Mech. **76**, Suppl. 2 (1996), 641–642. (Zbl. 900.65373).

A generalized Collatz map has the form $T(x) = a_i x + b_i$, for $x \equiv i \pmod{n}$, in which $a_i = \frac{\alpha_i}{n}, b_i = \frac{\beta_i}{n}$, with $\alpha_i, \beta_i$ integers satisfying $i\alpha_i + \beta_i \equiv 0 \pmod{n}$. The authors note there is a continuous extension of this map to the real line given by

$$\hat{T}(x) = \sum_{m=0}^{n-1} (a_m x + b_m) \left( \sum_{h=0}^{n-1} e^{\frac{2\pi i h(m-x)}{n}} \right) = \sum_{h=0}^{n-1} (A_h x + B_h) e^{-\frac{2\pi i h x}{n}},$$

where $\{A_k\}, \{B_k\}$ are the discrete Fourier transforms of the $\{a_i\}$, res. $\{b_i\}$. This permits analysis of the iterations of the map using the discrete Fourier transform and its inverse.

They specialize to the case $n = 2$, where the data $(a_0, a_1, b_0, b_1)$ describing $T$ are half-integers, with $b_0$ and $a_1 + b_1$ integers. The $3x + 1$ function corresponds to $(a_0, a_1, b_0, b_1) = (\frac{1}{2}, \frac{3}{2}, 0, \frac{1}{2})$. One has $A_0 = \frac{1}{2}(a_0 + a_1)$, $A_1 = \frac{1}{2}(a_0 - a_1)$, $B_0 = \frac{1}{2}(b_0 + b_1)$, $B_1 = \frac{1}{2}(b_0 - b_1)$. The authors note that the recursion

$$x_{k+1} = \hat{T}(x_k) = (A_0 x_k + B_0) + (A_1 x_k + B_1)\cos(\pi x_k)$$

can be transformed into a two-variable system using the auxiliary variable $\xi_k = \cos(\pi x_k)$, as

$$
\begin{aligned}
x_{k+1} &= (A_0 x_k + B_0) + (A_1 x_k + B_1)\xi_k \\
\xi_{k+1} &= \cos\left(\pi(A_0 x_k + B_0) + \pi(A_1 x_k + B_1)\xi_k\right).
\end{aligned}
$$

They give a formula for $x_m$ in terms of the data $(x_0, \xi_0, \xi_1, \ldots, \xi_{m-1})$, namely

$$x_m = \left( \prod_{j=0}^{m-1} (A_0 + A_1 \xi_j) \right) x_0 + (B_0 + \xi_{m-1} B_1) + \sum_{k=0}^{m-2} (B_0 + \xi_k B_1) \prod_{j=k+1}^{m-1} (A_0 + \xi_k A_1).$$

Now they study the transformed system in terms of the auxiliary variables $\xi_j$ . When the starting values $(x_0, x_1)$ are integers, all subsequent values are integers, and then the auxiliary variables $\xi_k = \pm 1$. Then the recursion for $\xi_{k+1}$ above can be simplified by trigonometric sum of angles formulas. In particular the obtain recursions for the $\xi_j$ that are independent of the $x_k's$ whenever $a_0$ and $a_1$ are both integers, but not otherwise. Finally the authors observe that on integer orbits $\xi_m$ is a periodic function of $x_0$ (of period dividing $2^{m+1}$) which can be interpolated using a Fourier series in $\cos \frac{2\pi h}{2^{m+1}}, \sin \frac{2\pi h}{2^{m+1}}$, using the inverse discrete Fourier transform. They give explicit interpolations for $\xi_m$ for the $3x+1$ function for $m = 0, 1, 2$.

**(90-74)**   Nicholas Pippenger (1993), *An elementary approach to some analytic asymptotics*, SIAM J. Math. Anal. **24** (1993), No. 5, 1361–1377. (MR 95d:26004).

This paper studies asymptotics of recurrences of a type treated in Fredman and Knuth (1974). It treats those in §5 and §6 of their paper, in which $g(n) = 1$, and we set $h(x) := H(x) - 1$. These concern the recurrence $M(0) = 1$,

$$M(n + 1) = 1 + \min_{0 \le k \le n} \left( \alpha M(k) + \beta M(n - k) \right),$$

in the parameter range $\min(\alpha, \beta) > 1$. They reduced the problem to study of the function $h(x)$ satisfying $h(x) = 0$, $0 < x < 1$ satisfying, for $1 \le x < \infty$,

$$h(x) = 1 + h(\frac{x}{\alpha}) + h(\frac{x}{\beta}).$$

For $\alpha, \beta > 1$ let $\gamma$ be the unique positive solution to $\alpha^\gamma + \beta^{-\gamma} = 1$. Fredman and Knuth showed that $h(x) \sim Cx^\gamma$, when $\frac{\log \alpha}{\log \beta}$ is irrational, and $h(x) \sim D(x)x^\gamma$, where $D(x)$ is a periodic function of the variable $\log x$, if $\frac{\log \alpha}{\log \beta}$ is rational. Pippenger rederives these results using elementary arguments based on a geometric interpretation of $h(x)$ as a sum of binomial coefficients in a triangular subregion of the Pascal triangle. He obtains detailed information on the function $D(x)$ in the case that $\frac{\log \alpha}{\log \beta}$ is rational.

**(90-75)**   Qiu, Wei Xing (1997) , *Study on "3x+1" problem* (Chinese), J. Shanghai Univ. Nat. Sci. Ed. [Shanghai da xue xue bao. Zi ran ke xue ban] **3** (1997), No. 4, 462–464.

English Abstract: "The paper analyses the structure presented in the problem "$3x+1$" and points out there is no cycle in the problem except that $x = 1$."

**(90-76)**   Jacques Roubaud (1993), *N-ine, autrement dit quenine (encore), Réflexions historiques et combinatoires sur la n-ine, autrement dit quenine.* La Bibliothéque Oulipienne, numéro 66, Rotographie à Montreuil (Seine-Saint-Denis), November 1993.

This paper considers work on rhyming patterns in poems suggested by those of medieval troubadours. The $n$-ine is a generalization of the rhyme pattern of the sestina of Arnaut Daniel, a 12-th century troubadour, see Queneau (1963), Roubaud (1969). It considers a "spiral permutation" of the symmetric group

$S_N$, which he earlier gave mathematically as

$$\delta_n(x) := \begin{cases} \frac{x}{2} & \text{if } x \text{ is even} \\ \frac{2n+1-x}{2} & \text{if } x \text{ is odd} \end{cases}$$

restricted to the domain $\{1, 2, \ldots, n\}$.

Roubaud summarizes the work of Bringer (1969) giving restrictions on admissible $n$. She observed $p = 2n + 1$ must be prime and that 2 is a primitive root of $p$ is a sufficient condition for admissibility. He advances the conjectures that the complete set of admissible $n$ are those with $p = 2n + 1$ a prime, such that either $ord_p(2) = 2n$, so that 2 is a primitive root, or else $ord_p(2) = n$. (This conjecture later turned out to require a correction; namely, to allow those $n$ with $ord_p(2) = n$ only when $n \equiv 3 \pmod 4$.)

Roubaud suggests various generalizations of the problem. Bringer(1969) had studied the inverse permutation to $\delta_n$, given by

$$d_n(x) := \begin{cases} 2x & \text{if } 1 \le x \le \frac{n}{2} \\ 2n + 1 - 2x & \text{if } \frac{n}{2} < x \le n. \end{cases}$$

In Section 5 Roubaud suggests a generalization of this, to the permutations

$$d_n(x) := \begin{cases} 3x & \text{if } 1 \le x \le \frac{n}{3} \\ 2n + 1 - 3x & \text{if } \frac{n}{3} < x \le \frac{2n}{3} \\ 3x - (2n + 1) & \text{if } \frac{2n}{3} < x \le n. \end{cases}$$

Here $n = 8$ is a solution, and he composes a poem, "Novembre", a 3-octine, to this rhyming pattern, in honor of Raymond Queneau. Call these solutions 3-admissible, and those of the original problem 2-admissible. He gives the following list of 3-admissible solutions for $n \le 200$, which are not 2-admissible: $n = 8, 15, 21, 39, 44, 56, 63, 68, 111, 116, 128, 140, 165, 176, 200$.

He concludes with some other proposed rhyming schemes involving spirals, including the "spinine", also called "escargonine".

(90-77) Olivier Rozier (1990), *Demonstraton de l'absence de cycles d'une certain forme pour le probléme de Syracuse,* Singularité **1** no. 3 (1990), 9–12.

This paper proves that there are no cycles except the trivial cycle whose iterates (mod 2) repeat a pattern of the form $1^m 0^{m'}$. Such cycles are called *circuits* by R. P. Steiner, who obtained this result in 1978. Rozier's proof uses effective transcendence bounds similar to Steiner's, but is simpler since he can quote the recent bound

$$\left| \frac{\log 3}{\log 2} - \frac{p}{q} \right| > q^{-15},$$

when $(p, q) = 1$, which appears in M. Waldschmidt, Equationes diophantiennes et Nombres Transcendents, Revue Palais de la Découverte, **17**, No. 144 (1987) 10–24.

(90-78) Olivier Rozier (1991), *Probleme de Syracuse: "Majorations" elementaires des cycles,* Singularité **2**, no. 5 (1991), 8–11.

This paper proves that if the integers in a cycle of the $3x + 1$ function are grouped into blocks of integers all having the same parity, for any $k$ there are only a finite number of cycles having $\leq k$ blocks. The proof uses the Waldschmidt result $\left| \frac{\log 3}{\log 2} - \frac{p}{q} \right| > q^{-15}$.

**(90-79)**   Jürgen W. Sander (1990), *On the $(3N+1)$-conjecture*, Acta Arithmetica **55** (1990), 241–248. (MR 91m:11052).

The paper shows that the number of integers $\leq x$ for which the $3x + 1$ Conjecture is true is at least $x^{3/10}$, by extending the approach of Crandall (1978). (Krasikov (1989) obtains a better lower bound $x^{3/7}$ by another method.) This paper also shows that if the $3x + 1$ Conjecture is true for all $n \equiv \frac{1}{3}(2^{2k} - 1) \pmod{2^{2k-1}}$, for any fixed $k$, then it is true in general.

**(90-80)**   Jeffrey O. Shallit and David W. Wilson (1991), *The "$3x+1$" problem and finite automata,* Bulletin of the EATCS (European Association for Theoretical Computer Science), No. 46, 1991, pp. 182–185.

A set $S$ of positive integers is said to be *2-automatic* if the binary representations of the integers in $S$ form a regular language $L_S \subseteq \{0,1\}^*$. Let $S_i$ denote the set of integers $n$ which have some $3x + 1$ function iterate $T^{(j)}(n) = 1$, and whose $3x + 1$ function iterates include exactly $i$ odd integers $\geq 3$. The sets $S_i$ are proved to be 2-automatic for all $i \geq 1$.

**(90-81)**   Kenneth S. Stolarsky (1998), *A prelude to the $3x + 1$ problem,* J. Difference Equations and Applications **4** (1998), 451–461. (MR 99k:11037).

This paper studies a purportedly "simpler" analogue of the $3x + 1$ function. Let $\phi = \frac{1+\sqrt{5}}{2}$. The function $f : \mathbb{Z}^+ \to \mathbb{Z}^+$ is given by

$$\begin{cases} f(\lfloor n\phi \rfloor) = \lfloor n\phi^2 \rfloor + 2 \,, & \text{all } n \geq 1 \,, \\ f(\lfloor n\phi^2 \rfloor) = n \,. & \text{all } n \geq 1 \,. \end{cases}$$

This function is well-defined because the sets $A = \{\lfloor n\phi \rfloor : n \geq 1\}$ and $B = \{\lfloor n\phi^2 \rfloor : n \geq 1\}$ form a partition of the positive integers. (This is a special case of Beatty's theorem.) The function $f$ is analogous to the $3x + 1$ function in that it is increasing on the domain $A$ and decreasing on the domain $B$. The paper shows that almost all trajectories diverge to $+\infty$, the only exceptions being a certain set $\{3, 7, 18, 47, \ldots\}$ of density zero which converges to the two-cycle $\{3, 7\}$. Stolarsky determines the complete symbolic dynamics of $f$. The possible symbol sequences of orbits are $B^\ell (AB)^k A^\infty$ for some $\ell, k \geq 0$, for divergent trajectories, and $B^\ell (AB)^\infty$ for some $\ell \geq 0$, for trajectories that reach the two-cycle $\{3, 7\}$.

**(90-82)**   György Targónski (1991), *Open questions about KW-orbits and iterative roots,* Aequationes Math. **41** (1991), 277–278.

The author suggests the possibility of applying a result of H. Engl (An analytic representation for self-maps of a countably infinite set and its cycles, Aequationes Math. **25** (1982), 90-96, MR85d.04001) to bound the number of cycles of the $3x + 1$ problem. Engl's result expresses the number of cycles as the geometric multiplicity of 1 as an eigenvalue of a map on the sequence space $l^1$.

332                         JEFFREY C. LAGARIAS

**(90-83)**  Bryan Thwaites (1996), *Two conjectures, or how to win £ 1000*, Math. Gazette **80** (1996), 35–36.

One of the two conjectures is the $3x+1$ problem, for which the author offers the stated reward.

*Note.* See the author's earlier paper Thwaites (1985) for his assertion to have invented the $3x+1$ problem.

**(90-84)**  Toshio Urata and Katsufumi Suzuki (1996), *Collatz Problem*, (Japanese), Epsilon (The Bulletin of the Society of Mathematics Education of Aichi Univ. of Education) [Aichi Kyoiku Daigaku Sugaku Kyoiku Gakkai shi] **38** (1996), 123–131.

The authors consider iterating the Collatz function $C(x)$ (denoted $f(x)$) and the speeded-up Collatz function $\phi(x)$ mapping odd integers to odd integers by

$$\phi(x) = \frac{3x+1}{2^e}, \text{ with } 2^e \| 3x+1.$$

They let $i_n$ (resp. $j_n$) denote the number of iterations to get from $n$ to 1 of the Collatz function (resp. the function $\phi(x)$). They study the Cesaro means

$$h_n := \sum_{k=1}^{n} i_n, \ m_n := \sum_{k=1}^{n} j_k.$$

and observe empiricially that

$$h_n \approx 10.4137 \log n - 12.56$$

$$m_n \approx 2.406 \frac{\log n}{\log 2} - 2$$

fit the data up to $4 \times 10^8$. Then they introduce an entire function which interpolate the Collatz function at positive integer values

$$f(z) = \frac{1}{2} z (\cos \frac{\pi z}{2})^2 + (3z+1)(\sin \frac{\pi z}{2})^2.$$

They also introduce more complicated entire functions that interpolate the function $\phi(x)$ at odd integer values.

$$F(z) = \frac{4}{\pi^2} (\cos \frac{\pi z}{2})^2 \left( \sum_{n=0}^{\infty} [\alpha_n \frac{1}{(z-(2n+1))^2} - \frac{1}{(2n+1)^2}] \right)$$

in which $\alpha(n) := \phi(2n+1)$. They raise the problem of determining the Fatou and Julia sets of these functions. These functions are constructed so that the positive integer values (resp. positive odd integer values) fall in the Fatou set.

**(90-85)**  Toshio Urata, Katsufumi Suzuki and Hisao Kajita (1997), *Collatz Problem II*, (Japanese), Epsilon (The Bulletin of the Society of Mathematics Education of Aichi Univ. of Education) [Aichi Kyoiku Daigaku Sugaku Kyoiku Gakkai shi] **39** (1997), 121–129.

The authors study the speeded-up Collatz function $\phi(x) = \frac{3x+1}{2^e}$ which takes odd integers to odd integers. They observe that every orbit of $\phi(x)$ contains some integer $n \equiv 1 \pmod 4$. They then introduce entire functions that interpolate $\phi(x)$ at positive odd integers. They start with

$$F(z) = \frac{4}{\pi^2}(\cos\frac{\pi z}{2})^2 \left(\sum_{n=0}^{\infty}[\alpha_n \frac{1}{(z-(2n+1))^2} - \frac{1}{(2n+1)^2}]\right)$$

in which $\alpha(n) := \phi(2n+1)$, so that $F(2n+1) = \phi(2n+1)$ for $n \geq 0$. They observe $F(z)$ has an attracting fixed point at $z_0 = -0.0327$, a repelling fixed point at $z = 0$, and a superattracting fixed point at $z = 1$. They show that the positive odd integers are in the Fatou set of $F(z)$. They show that the immediate basins of the Fatou set around the positive odd integers of $F(z)$ are disjoint. Computer drawn pictures are included, which include small copies of sets resembling the Mandlebrot set.

The authors also introduce, for each integer $p \geq 2$, the entire functions

$$K_p(z) = \left(\frac{4}{\pi^2}(\cos\frac{\pi z}{2})^2\right)^p \left(\sum_{n=0}^{\infty}\frac{\alpha_n}{(z-(2n+1))^{2p}}\right),$$

and, for $p \geq 1$, the entire functions.

$$L_p(z) = \left(\frac{4}{\pi^2}\cos^2\frac{\pi z}{2}\right)^p \frac{\sin\pi z}{\pi} \left(\sum_{n=0}^{\infty}\frac{\alpha_n}{(z-(2n+1))^{2p+1}}\right)$$

These functions also interpolate $\phi(x)$ at odd integers, i.e.

$$\phi(2n+1) = K_p(2n+1) = L_p(2n+1) \text{ for } n \geq 0.$$

**(90-86)**  Toshio Urata and Hisao Kajita (1998), *Collatz Problem III*, (Japanese), Epsilon (The Bulletin of the Society of Mathematics Education of Aichi Univ. of Education) [Aichi Kyoiku Daigaku Sugaku Kyoiku Gakkai shi] **40** (1998), 57–65.

The authors study the speeded-up Collatz function $\phi(x) = \frac{3x+1}{2^e}$ which takes odd integers to odd integers. They describe the infinite number of preimages of a given $x$ as $\{\frac{4^n x - 1}{3} : n \geq 0\}$ if $x \equiv 1 \pmod 3$ and $\{\frac{4^n(2x)-1}{3} : n \geq 0\}$ if $x \equiv 2 \pmod 3$. They encode the trajectory of this function with a vector $(p_1, p_2, p_3, \dots)$ which keeps track of the powers of 2 divided out at each iteration. More generally they consider the speeded-up $(ax+d)$-function, with $a, d$ odd. For $y_0$ an initial value (with $y_0 \neq -\frac{d}{a}$), the iterates are $y_i = \frac{ay_{i-1}+d}{2^{p_i+1}}$. They encode this iteration for $n$ steps by an $n \times n$ matrix

$$\Theta := \begin{bmatrix} 2^{p_1} & 0 & 0 & \cdots & 0 & -a \\ -a & 2^{p_2} & 0 & \cdots & 0 & 0 \\ 0 & -a & 2^{p_3} & \cdots & 0 & 0 \\ & \cdots & & \cdots & & \\ 0 & 0 & 0 & \cdots & -a & 2^{p_n} \end{bmatrix},$$

which acts by

$$\Theta \begin{bmatrix} y_1 \\ y_2 \\ y_3 \\ \cdots \\ y_n \end{bmatrix} = \begin{bmatrix} d - a(y_n - y_0) \\ d \\ d \\ \cdots \\ d \end{bmatrix}.$$

The authors observe that $\det(\Theta) = 2^{p_1 + \cdots + p_n} - a^n$. A periodic orbit, one with $y_n = y_0$, corresponds to a vector of iterates mapping to the vector with constant entries $d$. They compute examples for periodic orbits of $5x + 1$ problem, and the $3x + 1$ problem on negative integers. Finally they study orbits with $y_0$ a rational number with denominator relatively prime to $a$, and give some examples of periodic orbits.

**(90-87)**   Toshio Urata (1999), *Collatz Problem IV*, (Japanese), Epsilon (The Bulletin of the Society of Mathematics Education of Aichi Univ. of Education) [Aichi Kyoiku Daigaku Sugaku Kyoiku Gakkai shi] **41** (1999), 111-116.

The author studies a 2-adic interpolation of the speeded-up Collatz function $\phi(n)$ defined on odd integers $n$ by dividing out all powers of 2, i.e. for an odd integer $n$, $\phi(n) = \frac{3n+1}{2^p(3n+1)}$, where $p(m) = ord_2(m)$. Let $\mathbb{Z}_2^* = \{x \in \mathbb{Z}_2 : x \equiv 1 \pmod{2}\}$ denote the 2-adic units. The author sets $OQ := \mathbb{Q} \cap \mathbb{Z}_2^*$, and one has $\mathbb{Z}_2^*$ is the closure $\overline{OQ}$ of $OQ \subset \mathbb{Z}_2$. The author shows that the map $\phi$ uniquely extends to a continuous function $\phi : \mathbb{Z}_2^* \smallsetminus \{-\frac{1}{3}\} \to \mathbb{Z}_2^*$. He shows that if $f(x) = 2x + \frac{1}{3}$ then $f(x)$ leaves $\phi$ invariant, in the sense that $\phi(f(x)) = \phi(x)$ for all $x \in \mathbb{Z}_2^* \smallsetminus \{-\frac{1}{3}\}$. It follows that $f(f(x)) = 4x + 1$ also leaves $\phi$ invariant.

To each $x \in \mathbb{Z}_2^* \smallsetminus \{-\frac{1}{3}\}$ he associates the sequence of 2-exponents $(p_1, p_2, \dots)$ produced by iterating $\phi$. He proves that an element $x \in \mathbb{Z}_2^* \smallsetminus \{\frac{-1}{3}\}$ uniquely determine $x$; and that every possible sequence corresponds to some value $x \in \mathbb{Z}_2^* \smallsetminus \{-\frac{1}{3}\}$ He shows that all periodic points of $\phi$ on $\mathbb{Z}_2^*$ are rational numbers $x = \frac{p}{q} \in OQ$, and that there is a unique such periodic point for any finite sequence $(p_1, p_2, \cdots, p_m)$ of positive integers, representing 2-exponents, having period $m$. If $C(p_1, p_2, \dots, p_m) = \sum_{j=0}^{m-1} 2^{p_1 + \cdots + p_j} 3^{m-1-j}$ then this periodic pont is

$$x = R(p_1, p_2 \dots, p_m) := \frac{C(p_1, \dots, p_m)}{2^{p_1 + \cdots + p_m} - 3^m}$$

He shows that an orbit is periodic if and only if its sequence of 2-exponents is periodic.

**(90-88)**   Giovanni Venturini (1992), *Iterates of Number Theoretic Functions with Periodic Rational Coefficients (Generalization of the 3x + 1 problem)*, Studies in Applied Math. **86** (1992), 185–218. (MR 93b:11102).

This paper studies iteration of maps $g : \mathbb{Z} \to \mathbb{Z}$ of the form $g(m) = \frac{1}{d}(a_r m + b_r)$ if $m \equiv r \pmod{d}$ for $0 \leq r \leq d-1$, where $d \geq 2$ is an arbitrary integer, and all $a_r$, $b_r \in \mathbb{Z}$. These maps generalize the $(3x+1)$-function, and include a wider class of such functions than in Venturini (1989). The author's methods, starting in section 3, are similar in spirit to the Markov chain methods introduced by Leigh (1985), which in turn were motivated by work of Matthews and Watts (1984, 1985).

The author is concerned with classifying $g$-ergodic sets $S$ of such $g$ which are finite unions of congruence classes. For example, the mapping $g(3m) = 2m$, $g(3m+1) = 4m+3$ and $g(3m+2) = 4m+1$ is a permutation and has $S = \{m : m \equiv 0 \text{ or } 5 (\mathrm{mod}\ 10)\}$ as a $g$-ergodic set. He then associates a (generally finite) Markov chain to such a $g$-ergodic set, whose stationary distribution is used to derive a conjecture for the distribution of iterates $\{g^{(k)}(n_0 : k \geq 1\}$ in these residue classes for a randomly chosen initial value $n_0$ in $S$. For the example above the stationary distribution is $p_0 = \frac{1}{3}$ and $p_5 = \frac{2}{3}$. One can obtain also conjectured growth rates $a(g|_S)$ for iterates of a randomly chosen initial value in a $g$-ergodic set $S$. For the example above one obtains $a(g|s) = (\frac{4}{3})^{2/3}(\frac{2}{3}) \cong 1.0583$.

The author classifies maps $g$ into classes $G_v(d)$, for $v = 0, 1, 2, \ldots$, with an additional class $G_\infty(d)$. The parameter $v$ measures the extent to which the numerators $a_r$ of iterates of $g(v)$ have common factors with $d$. The class $G_0(d)$ consists of those maps $g$ having $gcd(a_0 a_1 \ldots a_{d-1}, d) = 1$, which is exactly the *relatively prime case* treated in Matthews and Watts (1984) and the class $G_1(d)$ are exactly those maps having $gcd(a_r, d^2) = gcd(a_r, d)$ for all $r$, which were the class of maps treated in Matthews and Watts (1985). His Theorem 7 shows that for each finite $v$ both of Leigh's Markov chains for auxilary modulus $m = d$ are finite for all maps in $G_v(d)$ (strengthening Leigh's Theorem 7). The maps in the exceptional class $G_\infty(d)$ sometimes, but not always, lead to infinite Markov chains. The class $G_\infty(d)$ presumably contains functions constructed by J. H. Conway [Proc. 1972 Number Theory Conf., U. of Colorado, Boulder 1972, 39-42] which have all $b_r = 0$ and which can encode computationally undecidable problems.

The proof of the author's Corollary to Theorem 6 is incomplete and the result is not established: It remains an open problem whether a $\mathbb{Z}$-permutation having an ergodic set $S$ with $a(g|_S) > 1$ contains any orbit that is infinite. Sections 6 and 7 of the paper contain many interesting examples of Markov chains associated to such functions $g$; these examples are worth looking at for motivation before reading the rest of the paper.

**(90-89)**   Giovanni Venturini (1997), *On a generalization of the $3x + 1$ problem*, Adv. Appl. Math. **19** (1997), 295–305. (MR 98j:11013).

This paper considers mappings $T(x) = \frac{t_r x - u_r}{p}$, for $x \equiv r \pmod{p}$, having $t_r \in \mathbb{Z}^+$ and $u_r \equiv rt_r \pmod{p}$, with $p > 1$ being an integer. It shows that if gcd $(t_0 t_1 \ldots t_{p-1}, p) = 1$ with $t_0 t_1 \ldots t_{p-1} < p^p$ then for any fixed $\rho$ with $0 < \rho < 1$ almost all $m \in \mathbb{Z}$ have an iterate $k$ with $|T^{(k)}(m)| < \rho|m|$. The paper also considers the question: when are such mappings $T$ permutations of $\mathbb{Z}$? It proves they are if and only if $\sum_{r=0}^{p-1} \frac{1}{t_r} = 1$ and $T(r) \not\equiv T(s) \pmod{(t_r, t_s)}$ for $0 \leq r < s \leq p-1$. The geometric-harmonic mean inequality implies that $t_0 \ldots t_{p-1} > p^p$ for such permutations, except in the trivial case that all $t_i = p$. The paper states on page 303 that the conditions $t_0 \ldots t_{p-1} > p^p$ and $gcd(t_0 t_1 \cdots t_{p-1}, p) = 1$ imply that the density of every orbit is 0. However this is not established; that every orbit has density 0 remains an unproved conjecture.

**(90-90)**   Günther J. Wirsching (1993), *An improved estimate concerning $3N + 1$ predecessor sets*, Acta Arithmetica, **63** (1993), 205–210. (MR 94e:11018).

This paper shows that, for all $a \not\equiv 0 \pmod{3}$, the set $\theta_a(x) := \{n \leq x :$ some $T^{(k)}(x) = a\}$ has cardinality at least $x^{.48}$, for sufficiently large $x$. This is achieved by exploiting the inequalities of Krasikov (1989).

**(90-91)**   Günther J. Wirsching (1994), *A Markov chain underlying the backward Syracuse algorithm*, Rev. Roumaine Math. Pures Appl. **39** (1994), no. 9, 915–926. (MR 96d:11027).

The author constructs from the inverse iterates of the $3x + 1$ function ('backward Syracuse algorithm') a Markov chain defined on the state space $[0, 1] \times \mathbb{Z}_3^\times$, in which $\mathbb{Z}_3^\times$ is the set of invertible 3-adic integers. He lets $g_n(k, a)$

count the number of "small sequence" preimages of an element $a \in \mathbb{Z}_3^\times$ at depth $n + k$, which has $n$ odd iterates among its preimages, and with symbol sequence $0^{\alpha_0} 10^{\alpha_1} \cdots 10^{\alpha_{n-1}} 1$, with $\sum_{i=0}^{n} \alpha_i = k$ satisfying the "small sequence" condition $0 \le \alpha_j < 2 \cdot 3^{j-1}$ for $0 \le j \le n - 1$. These quantities satisfy a functional equation

$$g_n(k, a) = \frac{1}{2 \cdot 3^{n-1}} \sum_{j=0}^{2 \cdot 3^{n-1}} g_{n-1}\left(k - j, \; \frac{2^{j+1} a - 1}{3}\right) .$$

He considers the "renormalized" quantities

$$\hat{g}\left(\frac{k}{n}, a\right) := \frac{1}{\Gamma_n} g_n(k, a) ,$$

with

$$\Gamma_n = 2^{1-n} 3^{-\frac{1}{2}(n-1)(n-2)} (3^n - n) ,$$

He obtains, in a weak limiting sense, a Markov chain whose probability density of being at $(x, a)$ is a limiting average of $\hat{g}(x, a)$ in a neighborhood of $(x, a)$ as the neighborhood shrinks to $(x, \alpha)$. One step of the backward Syracuse algorithm induces (in some limiting average sense) a limiting Markovian transition measure, which has a density taking the form of a product measure $\frac{3}{2} \chi_{[\frac{x}{3}, \frac{x+2}{3}]} \otimes \phi$, in which $\chi_{[\frac{x}{3}, \frac{x+2}{3}]}$ is the characteristic function of the interval $\left[\frac{x}{3}, \frac{x+2}{3}\right]$ and $\phi$ is a nonnegative integrable function on $\mathbb{Z}_3^\times$ (which may take the value $+\infty$).

The results of this paper are included in Chapter IV of Wirsching (1996b).

**(90-92)** Günther J. Wirsching (1996), *On the combinatorial structure of* $3N + 1$ *predecessor sets,* Discrete Math. **148** (1996), 265–286. (MR 97b:11029).

This paper studies the set $P(a)$ of inverse images of an integer $a$ under the $3x + 1$ function. Encode iterates $T^{(k)}(n) = a$ by a set of nonnegative integers $(\alpha_0, \alpha_1, \ldots, \alpha_\mu)$ such that the 0-1 vector $\mathbf{v}$ encoding $\{T^{(j)}(n) \pmod{2}: 0 \le j \le k - 1\}$ has $\mathbf{v} = 0^{\alpha_0} 1\, 0^{\alpha_1} \cdots 1\, 0^{\alpha_\mu}$. Wirsching studies iterates corresponding to "small sequences," which are ones with $0 \le \alpha_i < 2 \cdot 3^{i-1}$. He lets $G_\mu(a)$ denote the set of "small sequence" preimages $n$ of $a$ having a fixed number $\mu$ of iterates $T^{(j)}(n) \equiv 1 \pmod 2$, and shows that $|G_\mu(a)| = 2^{\mu-1} 3^{1/2(\mu-1\cdots\mu-2)}$. He lets $g_a(k, \mu)$ denote the number of such sequences with $k = \alpha_0 + \alpha_1 + \cdots + \alpha_\mu$, and introduces related combinatorial quantities $\psi(k, \mu)$ satisfying $\sum_{0 \le l \le k} \psi(l, \mu - 1) \ge g_a(k, \mu)$. The quantities $\psi(k, \mu)$ can be asymptotically estimated, and have a (normalized) limiting distribution

$$\psi(x) = \lim_{\mu \to \infty} \frac{3^\mu - \mu}{2^\mu 3^{1/2\mu(\mu-1)}} \psi([3^\mu - \mu)x], \mu).$$

$\psi(x)$ is supported on $[0, 1]$ and is a $C^\infty$-function. He suggests a heuristic argument to estimate the number of "small sequence" preimages of $a$ smaller than $2^n a$, in terms of a double integral involving the function $\psi(x)$.     The results of this paper are included in Chapter IV of Wirsching (1998).

**(90-93)** Günther J. Wirsching (1997), $3n + 1$ *predecessor densities and uniform distribution in* $\mathbb{Z}_3^*$, in: (G. Nowak and H. Schoissengeier, Eds.) *Proc. Conference on Elementary and Analytic Number Theory (in honor of E. Hlawka)*, Vienna, July 18-20,1996, Wien: Universität Wien, Institut für Mathematik, 1996, pp. 230-240. (Zbl. 883.11010).

This paper formulates a kind of equidistribution hypothesis on 3-adic integers under backwards iteration by the $3x + 1$ mapping, which, if true, would imply that the set of integers less than $x$ which iterate under $T$ to a fixed integer $a \not\equiv 0 \pmod 3$ has size at least $x^{1-\epsilon}$ as $x \to \infty$, for any fixed $\epsilon > 0$.

(90-94) Günther J. Wirsching (1998a), *The Dynamical System Generated by the $3n + 1$ Function*, Lecture Notes in Math. No. 1681, Springer-Verlag: Berlin 1998. (MR 99g:11027).

This volume is a revised version of the author's Habilitationsscrift (Katholische Universität Eichstätt 1996). It studies the problem of showing that for each positive integer $a \not\equiv 0$ (mod 3) a positive proportion of integers less than $x$ iterate to $a$, as $x \to \infty$. It develops an interesting 3-adic approach to this problem.          Chapter I contains a history of work on the $3x + 1$ problem, and a summary of known results.          Chapter II studies the graph of iterates of $T$ on the positive integers, ("Collatz graph") and, in particular studies the graph $\Pi^a(\Gamma_T)$ connecting all the inverse iterates $P(a)$ of a given positive integer $a$. This graph is a tree for any noncyclic value $a$. Wirsching uses a special encoding of the symbolic dynamics of paths in such trees, which enumerates symbol sequences by keeping track of the successive blocks of 0's. He then characterizes which graphs $\Pi^a(\Gamma_T)$ contain a given symbol sequence reaching the root node $a$. He derives "counting functions" for such sequences, and uses them to obtain a formula giving a lower bound for the function

$$P_T^n(a) := \{m : 2^n a \le m < 2^{n+1}a \text{ and some iterate } T^{(j)}(m) = a\} \ ,$$

which states that

$$|P_T^n(a)| := \{m : 2^n(a) := \sum_{\ell=0}^{\infty} e_\ell(n + \lfloor \log_2 \left( \frac{3}{2} \right) \ell \rfloor \ , \ a) \ , \tag{1}$$

where $e_\ell(k, a)$ is a "counting function." (Theorem II.4.9). The author's hope is to use (1) to prove that

$$|P_T^n(a)| > c(a)2^n \ , \ n = 1, 2, \dots \ . \tag{2}$$

for some constant $c(a) > 0$.          Chapter III studies the counting functions appearing in the lower bound above, in the hope of proving (2). Wirsching observes that the "counting functions" $e_\ell(k, a)$, which are ostensibly defined for positive integer variables $k, \ell, a$, actually are well-defined when the variable $a$ is a 3-adic integer. He makes use of the fact that one can define a 'Collatz graph' $\Pi^a(\Gamma_T)$ in which $a$ is a 3-adic integer, by taking a suitable limiting process. Now the right side of (1) makes sense for all 3-adic integers, and he proves that actually $s_n(a) = +\infty$ on a dense set of 3-adic integers $a$. (This is impossible for any integer $a$ because $|P_T^n(a)| \le 2^n a$.) He then proves that $s_n(a)$ is a nonnegative integrable function of a 3-adic variable, and he proves that its expected value $\bar{s}_n$ is explicitly expressible using binomial coefficients. Standard methods of asymptotic analysis are used to estimate $\bar{s}_n$, and to show that

$$\liminf_{n\to\infty} \frac{\bar{s}_n}{2^n} > 0 \ , \tag{3}$$

which is Theorem III.5.2. In view of (1), this says that (2) ought to hold in some "average" sense. He then proposes that (2) holds due to the:

**Heuristic Principle.** *As $n \to \infty$, $s_n(a)$ becomes relatively close to $\bar{s}_n$, in the weak sense that there is an absolute constant $c_1 > 0$ such that*

$$s_n(a) > c_1 \bar{s}_n \ for \ all \ n > n_0(a) \ . \tag{4}$$

One expects this to be true for all positive integers $a$. A very interesting idea here is that it might conceivably be true for all 3-adic integers $a$. If so, there is some chance to rigorously prove it. Chapter IV studies this heuristic principle further by expressing the counting function $e_\ell(k, a)$ in terms of simpler counting functions $g_\ell(k, a)$ via a recursion

$$e_\ell(k, a) = \sum_{j=0}^{k} p_\ell(k - j) g_\ell(j, a) \ ,$$

in which $p_\ell(m)$ counts partitions of $m$ into parts of a special form. Wirsching proves that properly scaled versions of the functions $g_\ell$ "converge" (in a rather weak sense) to a limit function which is independent of $a$, namely

$$g_\ell(k, a) \approx 2^{\ell-1} 3^{\frac{1}{2}(\ell^2 - 5\ell + 2)} \psi\left(\frac{k}{3^\ell}\right) \ , \tag{5}$$

for "most" values of $k$ and $a$. (The notion of convergence involves integration against test functions.) The limit function $\psi : [0, 1] \to \mathbb{R}$ satisfies the functional-differential equation

$$\psi'(t) = \frac{9}{2}(\psi(3t) - \psi(3t - 2)) \ . \tag{6}$$

He observes that $\psi$ has the property of being $C^\infty$ and yet is piecewise polynomial, with infinitely many pieces, off a set of measure zero. There is a fairly well developed theory for related functional-differential equations, cf. V. A. Rvachev, Russian Math. Surveys **45** No. 1 (1990), 87–120. Finally, Wirsching observes that if the sense of convergence in (5) can be strengthened, then the heuristic principle can be deduced, and the desired bound (2) would follow.

**(90-95)** Günther J. Wirsching (1998b), *Balls in constrained urns and Cantor-like sets*, Z. für Analysis u. Andwendungen **17** (1998), 979-996. (MR 2000b:05007).

This paper studies solutions to an integral equation that arises in the author's analysis of the $3x + 1$ problem (Wirsching (1998a)). Berg and Krüppel (J. Anal. Appl. **17** (1998), 159–181) showed that the integral equation

$$\phi(x) = \frac{q}{q - 1} \int_{qx-q+1}^{qx} \phi(y) dy$$

subject to $\phi$ being supported in the interval [0, 1] and having $\int_0^1 \phi(y) dy = 1$ has a unique solution whenever $q > 1$. The function $\phi(y)$ is a $C^\infty$-function. In this paper Wirsching shows that a certain iterative procedure converges to this solution when $q > \frac{3}{2}$. He also shows that for $q > 2$ the function $\phi(y)$ is piecewise polynomial off a Cantor-like set of measure zero. The case of the $3x+1$ problem corresponds to the choice $q = 3$.

**(90-96)**   Wu, Jia Bang (1992), *A tendency of the proportion of consecutive num-*
*bers of the same height in the Collatz problem* (Chinese), J. Huazhong (Central
China) Univ. Sci. Technol. [Hua zhong gong xue yuan] **20**, No. 5, (1992),
171–174. (MR 94b:11024, Zbl. 766.11013).

English Abstract: "The density distribution and length of consecutive numbers of the same height in the Collatz problem are studied. The number of integers, $K$, which belong to $n$-tuples ($n \geq 2$) in interval $[1, 2^N)$ ($N = 1, 2, \ldots, 24$)
is accurately calculated. It is found that the density $d(2^N)$ ($= K/2^N - 1$) of
$K$ in $[1, 2^N)$ is increased with $N$. This is a correction of Garner's inferences
and prejudgments. The longest tuple in $[1, 2^{30})$, which is the 176-tuple with
initial number $722, 067, 240$ has been found. In addition, two conjectures are
proposed." (The author refers to Garner (1985).)

**(90-97)**   Wu, Jia Bang (1993), *On the consecutive positive integers of the same*
*height in the Collatz problem* (Chinese), Mathematica Applicata, suppl. **6** [Ying
yung shu hsüeh] (1993), 150–153. (MR 1 277 568)

English Abstract: "In this paper, we prove that, if a $n$-tuple ($n$ consecutive
positive integers of the same height) is found, an infinite number of $n$-tuples can
be found such that each of $n$ numbers in the same tuple has the same height.
With the help of a computer, the author has checked that all positive integers
up to $1.5 \times 10^8$, verified the $3x + 1$ conjecture and found a 120-tuple, which is the
longest tuple among all checked numbers. Thus there exists an infinite number
of 120-tuples."

*Note.* A table of record holders over intervals of length $10^7$ is included.

**(90-98)**   Wu, Jia Bang (1995), *The monotonicity of pairs of coalescence numbers*
*in the Collatz problem.* (Chinese), J. Huazong Univ. Sci. Tech. [Hua zhong
gong xue yuan] **23** (1995), suppl. II, 170–172.

English Abstract: "A 310-tuple with an initial number 6,622,073,000 has
been found. The concepts such as the pairs of coalescence numbers, conditional
pairs of coalescence numbers and unconditional pairs of coalescence numbers are
suggested. It is proved that, in interval $[1, 2^N]$, the density of conditional pairs
of coalescence numbers

$$\bar{d}(2^N) = \frac{1}{2^N} \#\{n < 2^N : n \text{ and } n+1 \text{ are } k(\leq N) \text{ times the pair of coalescence numbers}\}$$

is increased with $N$. A conjecture that, in interval $[1, 2^N]$, the density of

$$\bar{d}_0(2^N) = \frac{1}{2^N} \#\{n < 2^N : n \text{ and } n + 1 \text{ pair of coalescence number}\}$$

is also increased with $N$, is proposed."

**(90-99)**   Yang, Zhao Hua (1998), *An equivalent set for the $3x + 1$ conjecture* (Chinese), J. South China Normal Univ. Natur. Sci. Ed. [ Hua nan shi fan da xue
xue bao. Zi ran ke xue ban] 1998, no. 2, 66–68. (MR 2001f:11040).

The paper shows that the $3x + 1$ Conjecture is true if, for any fixed $k \geq 1$,
it is true for all positive integers in the arithmetic progression $n \equiv 3 + \frac{10}{3}(4^k -$
$1) \pmod{2^{2k+2}}$. Note that Korec and Znam (1987) gave analogous conditions

for the $3x + 1$ Conjecture being true, using a residue class (mod $p^k$) where $p$ is an odd prime for which 2 is a primitive root.

**(90-100)**  Zhang, Cheng Yu (1990), *A generalization of Kakutani's conjecture.* (Chinese) Nature Magazine [Zi ran za zhi (Shanghai)] , **13** No. 5, (1990), 267–269.

Let $p_1 = 2, p_2 = 3, \ldots$ list the primes in increasing order. This paper suggests the analysis of the mappings $T_{d,k}(x)$, indexed by $d \geq 1$ and $k \geq 0$, with

$$
T_{d,k}(x) = \begin{cases}
\frac{x}{2} & \text{if } 2 \mid x \\
\frac{x}{3} & \text{if } 3 \mid x \text{ and } 2 \nmid x \\
\ldots \\
\frac{x}{p_{d-1}} & \text{if } p_d \mid x \text{ and } p_j \nmid x, \ 1 \leq j \leq d-1, \\
p_{d+1}x + (p_{d-1})^k & \text{otherwise.}
\end{cases}
$$

This function for $(d, k) = (1, 0)$ is the Collatz function. The author actually does not give a tie-breaking rule for defining the function when several $p_j$ divide $x$. However, this is not important in studying the conjecture below.

The author formulates the *Kakutani $(d,k)$-Conjecture*, which states that the function $T_{d,k}(x)$ iterated on the positive integers always reaches the integer $(p_{d+1})^k$. Note that $(p_{d+1})^k$ always belongs to a cycle of the function $T_{d,k}(x)$. This conjecture for $(1, k)$ was presented earlier by the author, where it was shown equivalent to the $3x + 1$ Conjecture, and the name Kakutani conjecture comes from Yang and Zhang (1988).

The author shows this conjecture is false for $d = 4, 5, 6, 8$ because there is a nontrivial cycle. The cycles he finds, for given $p_{d+1}$ and starting value $n$ are: $p_5 = 11$ ($n = 17$), $p_6 = 13$ ($n = 19$), $p_7 = 17$ ($n = 243$), $p_9 = 29$ ($n = 179$).

The author then modifies the Kakutani $(d, k)$-Conjecture in these cases to say that every orbit on the positive integers enters one of the known cycles. For example he modifies Conjecture $(4, k)$ to say that all orbits reach the value $(11)^k$ or else enter the cycle containing $17 \cdot (11)^k$. He proves that the Conjecture $(d, k)$ is equivalent to the modified Conjecture $(d, 0)$ for $d = 1, 2, 3, 4$.

**(90-101)**  Zhang, Zhongfu and Yang, Shiming (1998), *Problems on mapping sequences* (Chinese), Mathmedia [Shu xue chuan bo ji kan] **22** (1998), No. 2, 76–88.

This paper presents basic results on a number of iteration problems on integers, represented as directed graphs, and lists twelve open problems. This includes a discussion of the $3x + 1$ problem. It also discusses iteration of the map $V(n)$ which equals the difference of numbers given by the decimal digits of $n$ rearranged in decreasing, resp. increasing order. On four digit numbers $V(n)$ iterates to a fixed point 6174, the Kaprekar constant, cf. D. R. Kaprekar, *An interesting property of the number 6174*, Scripta Math. **21** (1955) 244-245.

**(90-102)**  Zhou, Chuan Zhong (1995), *Some discussion on the $3x + 1$ problem* (Chinese), J. South China Normal Univ. Natur. Sci. Ed. [ Hua nan shi fan da xue xue bao. Zi ran ke xue ban] **1995**, No. 3, 103–105. (MR 97h:11021).

Let $H$ be the set of positive integers that eventually reach 1 under iteration by the Collatz function $C(n)$; the $3x + 1$ conjecture states that $H$ consists of all positive integers. This paper extends a result of B. Y. Hong (1986) by showing that if the $3x + 1$ Conjecture is not true the minimal positive $n \notin H$ that does not iterate to 1 satisfies $n \equiv 7, 15, 27, 31, 39, 63, 79$ or $91$ ( mod 96). It also proves that if $2^{2m+1} - 1 \in H$ then $2^{2m+2} - 1 \in H$. Finally the paper observes that the numerical computation of $C(n)$ could be simplified if performed in base 3.

**(90-103)**  Zhou, Chuan Zhong (1997), *Some recurrence relations connected with the $3x + 1$ problem* (Chinese), J. South China Normal Univ. Natur. Sci. Ed. [Hua nan shi fan da xue xue bao. Zi ran ke xue ban] **1997**, no. 4, 7–8.

English Abstract: "Let $\mathbb{N}$ denote the set of natural numbers, $J$ denote the $3x + 1$ operator, and $H = \{n \in \mathbb{N} : \text{there is } k \in \mathbb{N} \text{ so that } J^k(n) = 1\}$. The conjecture $H = \mathbb{N}$ is the so-called $3x+1$ problem. In this paper, some recurrence relations on this problems are given."

**Acknowledgements.** I am indebted to B. Bailey, A. M. Blas, M. Chamberland, G.-G. Gao, J. Goodwin, C. Hewish, S. Kohl, Wang Liang, D. Merlini, P. Michel, C. Reiter, T. Urata, G. Venturini, S. Volkov and G. J. Wirsching for references. I am indebted to the staff at Vinculum for supplying the Coxeter reference. For help with Chinese references I thank W.-C. Winnie Li, Xinyun Sun, Yang Wang and Chris Xiu. For help with Japanese references I thank T. Ohsawa.

# Index

Albeverio, Sergio 80-1
Allouche, Jean-Paul 70-1, 80-26
Amleh, Amal S. 90-1
Anderson, S. 80-2
Andrei, Stefan 90-2
Applegate, David 90-3, 90-4, 90-5
Arsac, Jacques 80-3
Ashbacher, Charles 90-6
Atkin, Arthur O. L. 60-1
Avidon , Michael R. 90-7

Baiocchi, Claudio 90-8
Baker, Alan 70-30, 90-12
Banerji, Ranan B. 90-9
Banks, Roger 70-2
Barone, Enzo 90-10
Beeler, Michel 70-2
Belaga, Edward 90-11, 90-12
Beltraminelli, Stefano 90-13
Berg, Lothar 90-14, 90-15, 90-95
Bergman, George 60-3
Bernstein, Daniel J. 90-16, 90-17
Beukers, Frits 90-37
Błażewitz, Jacek 80-4
Blecksmith, Richard 90-18
Böhm, Corrado 70-3
Boyd, David M. 80-5
Bringer, Monique 60-2
Brocco, Stefano 90-19
Buntrock, Gerhard 90-67
Burckel, Serge 90-20
Buttsworth, Robert N. 90-21

Cadogan, Charles 80-6, 90-22, 90-23
Chamberland, Marc 90-24
Chisala, Busiso P. 90-25
Chvatal, Vasik 70-4
Clark, Dean 90-26, 90-27, 90-28
Cloney, Thomas 80-7
Collatz, Lothar 70-34, 80-8
Conway, John Horton 70-5, 70-7, 80-9
Coppersmith, Don 70-6
Coxeter, Harold Scott MacDonald 70-7
Crampin, D. J. 80-25, 80-26
Crandall, Richard E. 70-8
Ćurčić, Vladan 90-73

Davison, J. Leslie 70-9
Devaney, Robert L. 90-45
Devienne, Philippe 90-29
Dodge, Clayton W. 70-33
Dolan, James M. 80-10
Dunn, Richard 70-10

Eisele, Peter 90-30
Eliahou, Shalom 90-31
Elliott, Peter D. T. A. 80-11

Engl, Heinz W. 90-82
Erdős, Paul 70-7, 70-11, 80-12, 80-35
Everett, C. J. 70-12

Farrar, J. Craig 70-25
Farruggia, Carolyn 90-32
Feix, Marc R. 90-33, 90-34
Filipponi, Piero 90-33
Flatto, Leopold 90-36
Franco, Zachary M. 90-37, 90-38
Fredman, Michael Lawrence 70-13, 70-14

Gale, David 90-39
Gao, Guo-Gang 90-40
Garcia, Manuel V. P. 90-41
Gardner, Martin 70-15
Garner, Lynn E. 80-13, 80-14
Gaschütz, Wolfgang 80-15
Gilman, Albert F. 80-10
Glaser, H. 80-16
Goles, Eric C. 80-7
Gonnet, Gaston 90-42
Gosper, William 70-2
Graham, Ronald L. 70-11, 80-12
Grove, Edward A. 90-1
Gupta, Hansraj 80-35
Guy, Richard K. 70-34, 80-17, 80-18, 80-19, 80-20

Hadeler, Karl-Peter 90-30
Halbeisen, Lorenz 90-43
Hampel, R. 80-41
Hasenjager, Gisbert 90-44
Hasse, Helmut 70-16, 80-8, 80-52
Hayes, Brian 80-21
Henneman, William 70-2
Heppner, Ernst 70-17
Hilton, Anthony J. W. 80-25, 80-26
Hoffman, Dean G. 70-18, 70-19
Hong, Bo Yang 80-22
Hungerbúhler, Norbert 90-43

Isard, Stephen D. 70-20

Jarvis, Frazer 80-23
Joseph, John A. 90-45

Kajita, Hisao 90-85, 90-86
Kaprekar, D. R. 80-52, 90-101
Kascak, Frantisek 90-46
Kauffman, Louis A. 90-47
Kay, David 70-21
Keller, Timothy P. 90-48
Kent, Candace M. 90-1
Klamkin, Murray S. 60-3
Klarner, David A. 70-4, 70-18, 70-19, 70-22, 70-23, 80-24, 80-25, 80-26, 90-49
Knuth, Donald E. 70-4, 70-13, 70-14

DEPARTMENT OF MATHEMATICS, UNIVERSITY OF MICHIGAN, ANN ARBOR, MI 48109–1109
*E-mail address*: lagarias@umich.edu